GEOGRAPHIC INFORMATION
SYSTEMS & SCIENCE

GEOGRAPHIC INFORMATION SYSTEMS & SCIENCE

THIRD EDITION

PAUL A. LONGLEY
UNIVERSITY COLLEGE LONDON, UK

MICHAEL F. GOODCHILD
UNIVERSITY OF CALIFORNIA, SANTA BARBARA, USA

DAVID J. MAGUIRE
BIRMINGHAM CITY UNIVERSITY, UK

DAVID W. RHIND
CITY UNIVERSITY, LONDON, UK

WILEY

John Wiley & Sons, Inc

EXECUTIVE EDITOR	Ryan Flahive
ASSOCIATE EDITOR	Veronica Armour
EDITORIAL ASSISTANT	Meredith Leo
MARKETING MANAGER	Margaret Barret
SENIOR DESIGNER	James O'Shea
PRODUCTION MANAGER	Micheline Frederick
SENIOR PHOTO EDITOR	Sheena Goldstein
SENIOR ILLUSTRATION EDITOR	Sandra Rigby
MEDIA EDITOR	Bridgett O'Lavin
PRODUCTION EDITOR	Richard DeLorenzo
OUTSIDE PRODUCTION SERVICES	Furino Production
COVER PHOTO	Background photo: Ali Ender Birer/NASA Visible Earth/iStockphoto
	Top photo: Courtesy of Paul Longley
	Center photo: Alex Segre/Alamy
	Bottom photo: Eliza Snow/iStockphoto

This book was typeset by Aptara. The text was printed and bound by RR Donnelley/Jefferson City who also printed the cover.

This book is printed on acid free paper. ∞

ISBN: 978-0-470-72144-5

Printed in the United States of America

10 9 8 7 6 5 4 3

CONTENTS

18 Operating Safely with GIS   451

19 GIS Partnerships   477

20 Epilogue: GIS&S in the Service of Humanity   501

FOREWORD

Joe Lobley here, again.

Things aren't always what they seem.

You may have heard about that "Best Job in the World" campaign to find a lucky person to be in charge of an island on the Great Barrier Reef. Well, I won the parallel competition run by the Pacific Islands Geography Society to create a GIS for the island. So here I am sitting on a God-forsaken plot of sand, desperate to get back to civilization and recover my life. How did all this come about, I hear you ask?

It all started about two years ago. I had been making boatloads of money as a GIS consultant and owner of a small GIS software firm with an exclusive contract to survey (at speed) the street centerlines of a damaged Middle East trouble spot. Working for the U.S. armed forces and security services was fun not least because these guys are doing all the leading-edge GIS stuff. I've kept a copy of the street files to sell on to a satnav company when normal life returns (can you believe what these guys pay?), and I've downloaded a half-decent line generalization routine to erase the craters that I had to drive around. Anyway, life was good, but then I entered the stupid competition and won, in recognition of my sustained contribution to making GIS work in places like Senegal and the Middle East.

Did I say island? More like a half-mile square bucket of sand, and barely a mosaic tile on the Google Datum (whatever that is . . .). Before I left, I thought I'd find out about it and gather some base-map data to get the GIS off to a good start. Like me you're thinking spatial data infrastructures, aren't you? When I started to search all the geoportals, at first I thought I was dumb, but then I realized that these systems don't match the hype. There's no useful data, the services are unreliable because it's all grassroots stuff, and there are no real quality standards. At least the GYM (Google, Yahoo!, Microsoft) Club services work, and you know where you are with them, even if you've got to be careful with the small print in their contracts.

Just before I left, the "gang of four" sent me a draft of their manuscript for the third edition of their GIS&S book. I've interacted a bit with them over the years and got to like them. To cut to the chase: the one who is clever with money and fixing things has told me that they can get me a transfer to the High Oceans Geodetic Survey on a project mapping ice sheet breakup in Antarctica. The price: they want me to pen a reality check on GIS to include in the book.

So here goes. GIS technology is great but only when it works: I've got all the tools, three e-mail systems and 5,000 e-books, but no reliable electricity and a pathetic dial-up modem. Using technology without understanding some science is like trying to build a house without foundations and the best way to ensure garbage-in, garbage-out (I finally get the Systems and Science thing in the title, by the way). Governments are vitally important, in GIS and everything else.

Surprised at all this coming from a Good Ol' Boy? You don't have to like governments to realize that they've rescued the world from financial meltdown. They have also generated the need for my software by having lots of wars. And someone has done a great selling job: every department in every government now wants GIS, and that's just how it should be with the best decisions, data and purchasing directives coming straight from the top. Of course there's still lots of c*** talked by government employees who say they are leading this initiative or that: the private sector just humors them and takes their money. Talking of the private sector, they're going to get the shock of their lives when they wake up and find that open-source software and data providers have eaten their lunch. Meanwhile academics are just philanderers who use their precious papers (that no one reads) to critique anything and everything that tries to breathe: even those who can add seem to while away their time using techniques that don't work to tackle problems that don't matter. Try charging some of my clients for telling them that "ontology prescribes epistemology": if you ask me, more has been written about ontologies than is known. My real world, by contrast, is all about getting hold (by fair means or foul) of the right software and the right data to solve people's real problems, and deploying appropriate "shock and awe" graphics to deliver the message that the client wants to see.

So that's it. These guys actually seem quite good at writing about GIS, and you get the impression that all the time they spent working in or with industry and governments wasn't completely wasted. Even I learned a thing or two from some of their real-world case studies. If they get me off this mound of sand I may learn a bit more.

Fellahs? . . .

Joe Lobley

DEDICATION

We dedicate this third edition to three colleagues who each had a remarkable and lasting impact on the field of geographic information science and systems, and on our own work.

Peter Burrough (1944–2009) was professor of physical geography at the Geographical Institute of Utrecht University in the Netherlands. Trained as a soil scientist in Oxford, he is perhaps best known for the two editions of his textbook Principles of Geographical Information Systems (1986 and, in 1998, with Rachael McDonnell, both published by Oxford University Press). He made seminal contributions to research on uncertainty, the representation of indeterminate objects, and the modeling of environmental processes. He was featured in Box 15.2 of our first edition.

Shupeng Chen (1920–2008) pioneered the study of GIS and remote sensing in China. He founded China's State Key Laboratory of Resources and Environmental Information Systems, and was elected to the Chinese Academy of Sciences in 1980. Later he was elected to membership of the World Academy for the Developing World and the International Eurasian Academy of Sciences. He became a prominent and well-known member of the international geography community, playing an active role in the International Geographical Union's Commission on Geographical Data Sensing and Processing. He was a major influence on the creation of the International Society for Digital Earth (see Section 19.3.2.2).

Reginald Golledge (1937–2009) was professor of geography at the University of California, Santa Barbara. He almost single-handedly established the field of behavioral geography, with its scientific study of how people perceive and find their way in the world. After losing much of his sight in the early 1980s, he dedicated his career to the study of disability, developing many novel ideas to aid visually impaired people to navigate and understand their surroundings. The personal navigation system that he and his colleagues developed is featured in Section 11.3.1, and he was featured in Box 1.8 of our second edition.

PREFACE

The field of geographic information systems, GIS, is concerned with the description, explanation, and prediction of patterns and processes at geographic scales. GIS is a science, a system technology, a discipline, and an applied problem-solving methodology. In many parts of the world we interact with GIS, directly or indirectly, many times during every day of our lives—using any of a plethora of new location-aware hardware devices, using rapidly developing Web services, or simply using goods and services that are provided at specific locations. Inevitably, it is sometimes difficult to penetrate the hyperbole that sometimes surrounds the diffusion and use of new ideas, technologies, media, and software, making it difficult for even the experienced user to take a balanced perspective of what can be achieved, subject to what safeguards. Guiding the user through this maze in an authoritative and independent way is the preserve of textbooks such as this, and there must now be well over 50 books on GIS on the world market. We believe that this one has become perhaps the fastest selling and most used because we see GIS as providing a gateway to science and problem solving (geographic information systems "and science" in general) and because we relate available technologies for handling geographic information to the scientific principles that should govern its use (geographic information: "systems and science"). In the decade since this book first appeared, many aspects of the technology have changed almost beyond recognition, yet the underlying changes to the science of problem solving have been much more subtle.

Our own (collective and individual) abilities to create, access, use, and evaluate geographic information have been transformed over the same time period. Yet GIS remains of enduring importance because of its central coordinating principles, the specialist tools that have been developed to handle spatial data, the unique analysis methods that are key to understanding spatial data, and the distinctive management techniques required for geographical information (GI) handling. Each section of this book investigates the unique, complex, and difficult problems that are posed by geographic information, and together they build into a holistic understanding of all that is important about GIS.

Our Approach

GISystems are a proven technology, and the basic operations of GIS today provide secure and established foundations for measurement, management, mapping, and analysis of the real world. GIScience provides us with the ability to devise GIS-based analysis that is robust and defensible. GI technology facilitates analysis and continues to evolve rapidly, especially in relation to mobile computing and the Internet, and their likely successors and spin-offs. Better technology, better systems, and better science make for better management and effective exploitation of GI.

Fundamentally, GIS is an applications-led technology, yet successful applications need appropriate scientific foundations. Effective use of GISystems is impossible if they are simply seen as black boxes producing magic. GIS is rarely applied in controlled, laboratory-like conditions. Our messy, inconvenient, and apparently haphazard real world is the laboratory for GIS, and the science of real-world application is the difficult kind—it can rarely control for, or assume away, things that we would prefer were not there and that get in the way of our applications. Scientific understanding of the inherent uncertainties and imperfections in representing the world makes us able to judge whether the conclusions of our analysis are sustainable, and is essential for everything except the most trivial use of GIS. GIScience is also founded on a search for understanding and predictive power in a world where human factors interact with those relating to the physical environment. Good science is also ethical and clearly communicated science, and thus the ways in which we analyze and depict geography also play an important role.

Digital geographic information is central to the practicality of GIS. If it does not exist, it is expensive to collect, edit, or update. If it does exist, it cuts costs and time. But it is the subtleties of GI that often present the greatest challenges: without an understanding of aspects such as projection, accuracy, previous transformations, and licensing we can never be confident (and surely never certain) about the results we can obtain from GIS projects. GI underpins the rapid growth of trading in geographic information. It provides possibilities not only for local business but also for entering new markets or for forging new relationships with other organizations. It is a foolish individual who sees it only as a commodity like baked beans or shaving foam. Its value relies on its coverage, on the strengths of its representation of diversity, on its truth within a constrained definition of that word, and on its availability.

Few of us are hermits. The way in which geographic information is created and exploited through

GIS affects us as citizens, as owners of enterprises, and as employees. It has increasingly been argued that GIS is only a part—albeit a part growing in importance and size—of the Information, Communications, and Technology (ICT) industry. This is a limited perception, typical of the ICT supply-side industry that tends to see itself as the sole progenitor of change in the world (wrongly). Actually, it is much more sensible to take a balanced demand- and supply-side perspective: GIS and geographic information can and do underpin many operations of many organizations, but how GIS works in detail differs between different cultures and can often also partly depend on whether an organization is in the private or public sector. Seen from this perspective, management of GIS facilities is crucial to the success of organizations, and GIS is integral to enterprise-wide management of information. "Doing GIS" is ultimately about management of the organization with appropriate tools, information, knowledge, skills, and commitment, whether locally or globally.

For this reason we devote an entire section of this book to management issues. We go far beyond how to choose, install, and run a GIS; that is only one part of the enterprise. We try to show how to use GIS and geographic information to contribute to the success of any organization, and have it recognized as doing just that. Defining "success" requires an understanding of what drives organizations and how they operate in the reality of their operating environments. "Reality" entails thinking about geographic information as an asset, but one that must be handled within an appropriate ethical framework. The new discussion of the business case for GIS presented in this third edition provides a detailed treatment of the trade-offs in decision making in which GIS can play a major role.

Our Audience

We have variously, both alone and in collaboration, spent much of our careers researching and writing about geographic information. As the field has grown, so more and more people have become concerned to find out about the enduring GIS principles, techniques, and management practices that underlie the effective use of geographic information technologies. Most obviously, this is the first edition of our book to take on board the innovation of Google Maps and Earth (and their equivalents from Microsoft and Yahoo!), and the consequential ballooning of

interest in accessible mapping and new applications driven by the map interfaces. If mapping has become seamless, so too has computing, following the ever wider use of low-cost, hand-held location-aware devices. These developments are the mainstays of what is currently described as "neogeography" (a term that has also been used with different connotations at various times in the past), with its focus on the use of Web 2.0 and the GeoWeb to adapt GIS to personal and community activities.

In other respects, however, the pace of change in GIS is more sedate, and we hope this book remains essential to anyone desiring a rich understanding of how GIS is used in the real world. Wider developments in technology and in society make GIS today a strategically important interdisciplinary meeting place. It is taught as a component of a huge range of undergraduate and graduate courses throughout the world, to students who already have different skills, who seek different disciplinary perspectives on the world, and who assign different priorities to practical problem solving and the intellectual curiosities of science.

GIS is popular and widespread, but is it professional? As GIS matures, so its constituent activities become increasingly codified. The U.S. University Consortium for GIScience (UCGIS: www.ucgis.org) has formalized a curriculum for GIS, and some of us have also been involved in institution-specific initiatives (Spatial Literacy in Teaching: www.spatial-literacy.org; and the Santa Barbara Center for Spatial Studies: www.spatial.ucsb.edu). We believe that the essence of these developments is assembled in this book. In revising our book, we have also been cognizant of moves toward professional GIS certification schemes and the accreditation of qualifications, much in the way that architects or engineers are required to attain licensed practitioner status prior to working on particular projects. The methods and techniques assembled here would form a core and an essential part of any formal requirement for licensing practitioner expertise in geographic information systems and technologies (GIST).

In responding to all of these developments—newly accessible neogeography, the evolving GIScience curriculum, and professional accreditation—what follows can be thought of as a textbook, though perhaps not in a conventionally linear sense. We have not structured the book according to any GIS curriculum for students or professionals beyond the core organizing principles, techniques, analysis methods, and management practices that we believe to be important. The material in each section of the book progresses in a cumulative way, yet we envisage that very few students will start at Chapter 1 and systematically work through to Chapter 20. Learning is not like that anymore (if ever it was), and most instructors will navigate a course between sections and chapters of the book that serves their particular disciplinary, curricular, and practical priorities.

That said, we have prepared a number of pointers to the ways in which the course might be used in education and training. Some of the ways in which three of us have used the book in our own undergraduate and postgraduate settings are posted on the book's Web site (www.wiley.com/college/longley), and we hope that other instructors will share their best practices with us as time goes on (please see the Web site for instructions on how to upload instructor lists and offer feedback on those that are already there!). Our Instructor Manual (see www.wiley.com/college/longley) provides suggestions as to the use of this book in a range of disciplines and educational settings.

Inevitably, it is impossible to provide a comprehensive treatment of all that is distinctive about GIScience in a book of this length. We have provided a primer on the most important geographical methods and techniques, and in seeking to go further, users may find themselves faced with the many options presented by an online search engine. A very positive characteristic of the evolution of GIS in recent years has been the move to open-source software and creative-commons data, much of which is principally documented and disseminated online. Yet a search engine is not the same as the index of a book, and search-engine returns may be dictated more by the market share of leading vendors than by relevance to the problem at hand. Accordingly, two of us have been involved in a related project that aims to provide an authoritative and independent guide to the principles and techniques of geospatial analysis, and the best software tools that are available to put them into practice. The result is a free-to-access Web site, www.spatialanalysisonline.com, with a comprehensive search facility, and related printed and PDF versions of the Web materials (see Further Reading). In many cases, this also provides the reader with the opportunity to further pursue the technical discussion of material discussed in this book, and we reference it at numerous points for this reason.

The format of this book is intended to make learning about GIS fun. GIS is an important transferable skill because people successfully use it to solve real-world problems. We thus convey this success through use of real (not contrived, conventional textbook-like) applications, in clearly identifiable boxes throughout the text. But even this cannot convey the excitement of learning about GIS, which only comes from doing. With this in mind, an online series of laboratory classes have been created to accompany the book. These are available, free of charge, to any individual working in an institution that has an ESRI Inc. (Redlands, CA) site license (see training.esri.com/campus/catalog/licenses/courselist.cfm?id=43). They are cross-linked in detail to individual chapters and sections in the book, and provide learners with the opportunity to refresh the concepts and techniques they have acquired through classes and reading, as well as the opportunity to work through extended examples using ESRI ArcGIS. As de Smith et al. (2009) make clear, this is by no means the only available software for learning GIS: we have chosen it for our own lab exercises because it is widely used, because one of us used to work for ESRI in California, and because ESRI's cooperation enabled us to tailor the lab exercises to our own material. The course catalog for the six-module 'Turning Data into Information' course can be viewed at training.esri.com/acb2000/showdetl.cfm?DID=6&Product_ID=821. There are, of course, many other options for lab teaching and distance learning from private and publicly funded bodies such as the UNIGIS consortium (www.unigis.net), the Worldwide Universities Network (www.wun.ac.uk/ggisa), and Pennsylvania State University World Campus (www.worldcampus.psu.edu).

GIS is not just about machines, but also about people. It is very easy to lose touch with what is new in GIS, such are the scale and pace of development. Many of these developments have been, and continue to be, the outcome of work by motivated and committed individuals—many an idea or implementation of GIS would not have taken place without an individual to champion it. In the first edition of this book, we used boxes highlighting the contributions of a number of its champions to convey the idea that GIS is a living, breathing subject. The second edition set the precedent of removing all of the living champions of GIS and replaced them with a completely new set. We have followed this precedent in this third edition—not as any intended slight of the remarkable contributions that these individuals have made, but as a necessary way of freeing up space to present vignettes of an entirely new set of committed, motivated individuals whose contributions have also made a difference to GIS.

Students today are seemingly required to digest ever-increasing volumes of material. We have tried to summarize some of the most important points in this book using short "factoids," such as the following one, which we think assist students in recalling core points.

Short, pithy, statements can be memorable.

We hope that instructors will be happy to use this book as a core teaching resource. We have tried to provide a number of ways in which they can encourage their students to learn more about GIS through a range of assessments. At the end of each chapter we provide four questions that typically entail:

▶ Student-centered learning by doing.

▶ A review of material contained in the chapter.

▶ A review and research task—involving integration of issues discussed in the chapter with those discussed in additional external sources.

▶ A compare and research task—similar to the review and research task above, but additionally entailing linkage with material from one or more other chapters in the book.

The online lab classes at the ESRI Virtual Campus have also been designed to allow learning in a self-paced way, and there are self-test exercises at the end of each section for use by learners working alone or by course evaluators at the conclusion of each lab class.

As the title implies, this is a book about geographic information systems, the practice of science in general, and the principles of geographic information science (GIScience) in particular. We remain convinced of the need for high-level understanding, and our book deals with ideas and concepts—as well as with actions. Just as scientists need to be aware of the complexities of interactions between people and the environment, so managers must be well informed by a wide range of knowledge about issues that might impact their actions. Success in GIS often comes from dealing as much with people as with machines.

The New Learning Paradigm

This is not a traditional textbook because:

▶ It recognizes that GISystems and GIScience do not lend themselves to traditional classroom teaching alone. Only by a combination of approaches can such crucial matters as principles, technical issues, practice, management, ethics, and accountability be learned. Thus the book is complemented by a Web site (www.wiley.com/college/longley) and by exercises that can be undertaken in laboratory or self-paced settings.

▶ It presents principles and techniques of GIScience alongside the latest trends in data, software, societal uses of technologies, and industry trends to those learning about GIS for the first time—bringing the reader into close contact with the continuing evolution of GIS.

▶ It does not see the textbook as the beginning and the end of learning about GIS. Rather, it provides a series of hooks with the evolving online resource, www.spatialanalysisonline.com. This provides an informed and independent view on how, in practice, it is possible to take GIS further, in advanced teaching and in research.

▶ Together, these resources make it possible to tailor learning to individual or small-group needs. These needs are also addressed in the Instructor Manual to the book (www.wiley.com/college/longley).

▶ We have recognized that GIS is driven by real-world applications and real people that respond to real-world needs. Hence, information on a range of applications and GIS champions is threaded throughout the text.

▶ We have linked our book to online learning resources throughout, notably the ESRI Virtual Campus (training.esri.com/gateway/index.cfm).

▶ The book that you have in your hands has been heavily restructured and revised, in the light of the many changes of the past five years, while retaining the best features of the (highly successful) first two editions.

Summary

This is a book that recognizes the growing commonality between the concerns of science and technology, and government and business. The examples of GIS people and problems that are scattered throughout this book have been chosen deliberately to illuminate this commonality, as well as the interplay between organizations and people from different sectors. To differing extents, the five sections of the book develop common concerns with effectiveness and efficiency by bringing together information from disparate sources, acting within regulatory and ethical frameworks, adhering to scientific principles, and preserving good reputations. This, then, is a book that combines the basics of GIS with the solving of problems that often have no single, ideal solution—the world of interdisciplinary, mission-oriented science applied to business, government, and community action. In short, we have tried to create a book that remains attuned to the way the world works now, that understands the ways in which most of us increasingly operate as knowledge workers, and that grasps the need to face complicated issues that do not have ideal solutions.

Conventions and Organization

We use the acronym GIS in many ways in the book, partly to emphasize one of our goals—the interplay between geographic information systems and geographic information science—and at times we use two other possible interpretations of the three-letter acronym: geographic information studies and geographic information services. We distinguish between the various meanings where appropriate, or where the context fails to make the meaning clear, especially in Section 1.7 and in the Epilogue. We also use the acronym in both singular and plural senses, following what is now standard practice in the field, to refer as appropriate to a single geographic information system or to geographic information systems in general. To complicate matters still further, we have noted the increasing use of "geospatial" rather than "geographic". We use "geospatial" where other people use it as a proper noun/title, but elsewhere use the more elegant and widely understood "geographic."

We have organized the book in five major but interlocking sections: after two chapters that establish the foundations of GI Systems and Science and the real world of applications, the sections appear as Principles (Chapters 3 through 6), Techniques (Chapters 7 through 11), Analysis (12 through 16), and Management and Policy (Chapters 17 through 19). We cap the book off with an Epilogue that summarizes the main topics and looks to the future. The boundaries between these sections are in practice permeable but remain in large part predicated on separating persistent principles—ideas that will be around long after today's technology has been relegated to the museum—from knowledge that is necessary to an understanding of today's technology, and likely near-term developments. In a similar way, many of the analytic methods have had reincarnations through different manual and computer technologies in the past and will doubtless metamorphose further in the future. We hope you find the book stimulating and helpful. Please tell us—either way!

Acknowledgments

We take complete responsibility for all the material contained herein. But much of it draws on contributions made by friends and colleagues from across the world, many of them outside the academic GIS community. We thank them all for those contributions and the discussions we have had over the years. We cannot mention all of them but would particularly like to mention the following. We thanked the following for their direct and indirect inputs to the first edition of this book: Mike Batty, Clint Brown, Nick Chrisman, Keith Clarke, Andy Coote, Martin Dodge, Danny Dorling, Jason Dykes, Max Egenhofer, Pip Forer, Andrew Frank, Rob Garber, Gayle Gaynor, Peter Haggett, Jim Harper, Rich Harris, Les Hepple, Sophie Hobbs, Andy Hudson-Smith, Karen Kemp, Chuck Killpack, Robert Laurini, Vanessa Lawrence, John Leonard, Bob Maher, David Mark, David Martin, Elanor McBay, Ian McHarg, Scott Morehouse, Nick Mann, Lou Page, Peter Paisley, Cath D'Alton (née Pyke), Jonathan Raper, Helen Ridgway, Jan Rigby, Christopher Roper, Garry Scanlan, Sarah Sheppard, Karen Siderelis, David Simonett, Roger Tomlinson, Carol Tullo, Dave Unwin, Sally Wilkinson, David Willey, Jo Wood, and Mike Worboys. We each noted our indebtedness in different ways to Stan Openshaw, for his insight, his energy, his commitment to GIS, and his compassion for geography. The second time around we also acknowledged the support of David Ashby, Elena Besussi, Nancy Chin, Sonja Curtis, Martin Dodge, Mike de Smith, Keily Larkins, Daryl Lloyd, Lyn Roberts, Kevin Schürer, Paul Torrens, Tom Veldkamp, and Peter Verburg. This time around we also thank Mohammad Adnan, James Cheshire, Helen Couclelis, Tom Cova, Joel Deardon, Sarah Elwood, Ryan Flahive, Maurizio Gibin, Yi Gong, Muki Haklay, Miles Irving, Don Janelle, Kate Jones, Daniel Lewis, Pablo Mateos, Oliver O'Brien, Martin Raubal, Duncan Smith, Dan Sui, and May Yuan. Special thanks go to Alex Singleton for inspiration on new topics for the book and perspiration in assembling the artwork for the final product. Finally, special thanks go to Tao, Fiona, Heather, and Christine.

Paul Longley, University College London
Michael Goodchild, University of California,
Santa Barbara
David Maguire, Birmingham City University
David Rhind, At Large

July 2009

Further Reading

De Smith, M.J., Goodchild, M.F., and Longley, P.A. 2009. Geospatial Analysis: A Comprehensive Guide to Principles, Techniques and Software Tools (3rd ed.). Leicester, UK: Troubador. See also www.spatialanalysisonline.com.

List of Acronyms and Abbreviations

ABM   agent-based model
AEGIS   advanced engineering GIS
AGI   Association for Geographic Information
AGILE   Association of Geographic Information Laboratories in Europe
AHP   analytical hierarchy process
AJAX   asynchronous Javascript and XML
ALSM   airborne laser swath mapping
AM   automated mapping
AML   Arc Macro Language
ANZLIC   Australia New Zealand Land Information Council
API   application programming interface
AR   augmented reality
ARPANET   Advanced Research Projects Agency Network
ASCII   American Standard Code for Information Interchange
AVHRR   advanced very high resolution radiometer
AVRIS   Airborne Visible InfraRed Imaging Spectrometer
BLOB   binary large object
BRIC   Large countries—Brazil, Russia, India, and China—with rapidly growing economies
CA   cellular automata
CAD   computer-aided design and drafting
CAMA   Computer Assisted Mass Appraisal
CAMS   capacity area management system
CARS   computer-aided routing system
CASA   Centre for Advanced Spatial Analysis
CASE   computer-aided software engineering
CBERS   China-Brazil Earth Resources Satellites
CCTV   closed-circuit television
CD   compact disc
CEN   Comité Européen de Normalisation
CERCO   group of European NMOs (Comité Européen de Responsibles de la Cartographie Officielle, or European Committee of Persons in Charge of Official Cartography)
CERN   Conseil Européen pour la Recherche Nucléaire
CGIS   Canada Geographic Information System
CGS   Czech Geological Survey
CIA   Central Intelligence Agency
COGO   coordinate geometry
COM   component object model
COTS   commercial off-the-shelf
CPU   central processing unit
CSDGM   Content Standards for Digital Geospatial Metadata
CSE   computer science and engineering
DBA   database administrator
DBMS   database management system
DCL   data control language

DCM   digital cartographic model
DCW   Digital Chart of the World
DDL   data definition language
DEM   digital elevation model
DGPS   Differential Global Positioning System
DIME   Dual Independent Map Encoding
DLM   digital landscape model
DML   data manipulation language
DPI   dots per inch
DRG   digital raster graphic
DWD   Germany's National Meteorological Service (Deutscher Wetterdienst)
EC   European Commission
ECU   Experimental Cartography Unit
EDA   exploratory data analysis
EOSDIS   Earth Observing System Data and Information System
EPA   U.S. Environmental Protection Agency
ERDAS   Earth Resource Data Analysis System
ESDA   Exploratory Spatial Data Analysis
ESRI   Environmental Systems Research Institute
ESRI BIS   ESRI Business Information Solutions
EU   European Union
FAO   Food and Agriculture Organization
FEMA   Federal Emergency Management Agency
FGDC   Federal Geographic Data Committee
FM   facility management
FOIA   Freedom of Information Act
FSA   Forward Sortation Area
GA   genetic algorithm
GDP   gross domestic product
GIS   geographic information system
GIS&S   geographic information systems and science
GIScience   geographic(al) information science
GIS-T   GIS in transporation
GLONASS   the Russian global positioning system (Global Orbiting Navigation Satellite System)
GML   Geography Markup Language
GOS   geospatial one-stop
GPS   Global Positioning System
GRASS   Geographic Resources Analysis Support System
GSDI   global spatial data infrastructure
GUI   graphical user interface
GWR   geographically weighted regression
HLS   hue, lightness, and saturation
HTML   hypertext markup language
HTTP   hypertext transmission protocol
IBRU   International Boundaries Research Unit
ICC   Cartographic Institute of Catalonia (Institut Cartogràfic de Catalunya)

ID   identifier
IDE   integrated development environment
IDEC   Catalan SDI (Infraestructura de Dades Espaciales de Catalunya)
IDW   inverse-distance weighting
IGN   Institut Géographique National
IM   instant messenger
INPE   Brazil's National Institute for Space Research (Instituto Nacional de Pesquisas Espaciais)
IPCC   Intergovernmental Panel on Climate Change
INSPIRE   Infrastructure for Spatial Information in Europe
IP   Internet protocol
IPR   intellectual property rights
IRBD   International River Boundaries Database
IS   information system
ISCGM   International Steering Committee for Global Mapping
ISO   International Standards Organization
IT   information technology
ITT   invitation to tender
KML   Keyhole Markup Language
LAN   local area network
LBS   location-based services
LDO   local delivery office
LiDAR   light detection and ranging
LIESMARS   State Key Laboratory for Information Engineering in Surveying, Mapping and Remote Sensing
LISA   local indicators of spatial association
LIST   Land Information System of Tasmania
LMIS   Land Management Information System
MAT   point of minimum aggregate travel
MAUP   Modifiable Areal Unit Problem
MBR   minimum bounding rectangle
MCDM   multicriteria decision making
MOCT   Ministry of Construction and Transportation
MODIS   Moderate Resolution Imaging Spectroradiometer
MrSID   Multiresolution Seamless Image Database
MSC   Mapping Science Committee
NAD27   North American Datum of 1927
NAD83   North American Datum of 1983
NASA   National Aeronautics and Space Administration
NATO   North Atlantic Treaty Organization
NAVTEQ   Navigation Technolgies
NCGIA   National Center for Geographic Information and Analysis
NGA   National Geospatial-Intelligence Agency
NGIA   National Geospatial-Intelligence Agency
NGO   nongovernmental organization
NICT   new information and communication technologies

NIMA   National Imagery and Mapping Agency
NIMBY   not in my back yard
NLS   National Land Survey
NMO   national mapping organization
NMP   National Mapping Program
NOAA   National Oceanic and Atmospheric Administration
NSDI   National Spatial Data Infrastructure
NSF   National Science Foundation
OAC   UK Office for National Statistics Output Area Classification
OCR   optical character recognition
ODBMS   object database management system
OGC   Open Geospatial Consortium
OLM   object-level metadata
OLS   ordinary least squares
OMB   Office of Management and Budget
OS   Ordnance Survey (Great Britain, GB, or Northern Ireland, NI)
OSM   OpenStreetMap
PAF   postcode address file
PC   personal computer
PCC   percent correctly classified
PCGIAP   Permanent Committee on GIS Infrastructure for Asia and Pacific
PCMCIA   Personal Computer Memory Card International Association
PDA   personal digital assistant
PDF   portable document format
PERT   Program, Evaluation, and Review Techniques
PIP   Victoria Property Information Project
PLI   Queensland Property Location Index
PLSS   Public Land Survey System
PPGIS   public participation in GIS
PSI   public sector information
RDBMS   relational database management system
R&D   research and development
RFI   Request for Information
RFID   radio frequency identification
RFP   Request for Proposals
RGB   red-green-blue
RGS   Royal Geographical Society (UK)
RMSE   root mean square error
ROI   return on investment
RS   remote sensing
RSS   really simple syndication
SARS   severe acute respiratory syndrome
SDE   Spatial Database Engine
SDI   spatial data infrastructure
SDSS   spatial decision support systems
SDTS   spatial data transfer standard
SETI   Search for Extraterrestrial Intelligence
SG SPACE   Singapore Geospatial Collaborative Environment

SOA   service-oriented architecture
SOAP   simple object access protocol
SOHO   small office/home office
SPC   State Plane Coordinates
SPOT   Systéme Probatoire d'Observation
       de la Terre
SQL   Structured/Standard Query Language
SWMM   Storm Water Management Model
SWOT   strengths, weaknesses, opportunities, threats
TB   terabyte
TIGER   Topologically Integrated Geographic
       Encoding and Referencing
TIN   triangulated irregular network
TML   Trans Manche Link
TOID   Topographic Identifier
TSP   traveling-salesman problem
UCSB   University of California, Santa Barbara
UCGIS   University Consortium for Geographic
       Information Science
UML   Unified Modeling Language
UN   United Nations
UNIGIS   University GIS Consortium
UNRCC-AP   United Nations Regional Cartographic
       Conference for Asia and the Pacific
UPS   Universal Polar Stereographic
URI   uniform resource identifier

URL   uniform resource locator
URISA   Urban and Regional Information Systems
       Association
USGS   United States Geological Survey
USLE   Universal Soil Loss Equation
UTM   Universal Transverse Mercator
VBA   Visual Baisic for Applications
VfM   value for money
VGA   video graphics array
VGI   volunteered geographic information
ViSC   visualization in scientific computing
VR   virtual reality
WAN   wide area network
WCS   Web coverage service
WFS   Web feature service
WGS84   World Geodetic System of 1984
WHO   World Health Organization
WiFi   wireless networking technology
WIMP   windows, icons, menus, and pointers
WMO   World Meteorological Organization
WMS   Web mapping service
WTO   World Trade Organization
WWF   World Wide Fund for Nature
WWW   World Wide Web
WYSIWYG   what you see is what you get
XML   extensible markup language

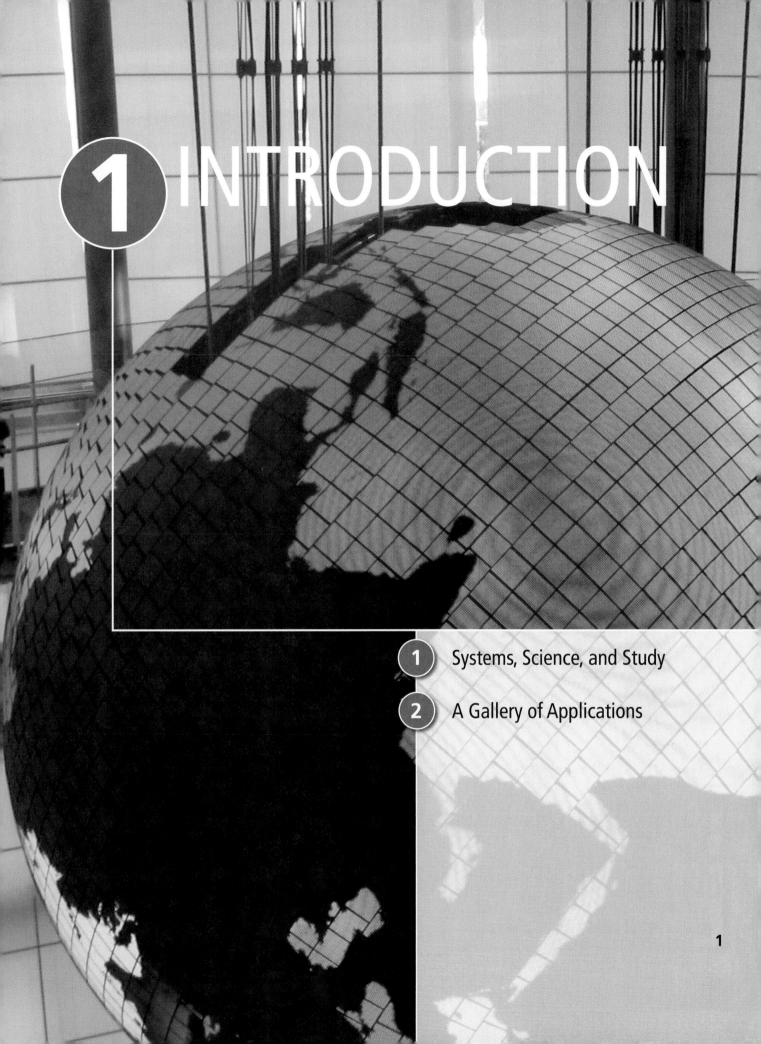

1 INTRODUCTION

1

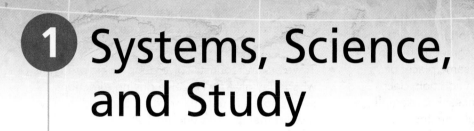

1 Systems, Science, and Study

This chapter introduces the conceptual framework for the book by addressing several major questions:

- What exactly is geographic information, and why is it important? What is special about it?

- What is information generally, and how does it relate to data, knowledge, evidence, wisdom, and understanding?

- What kinds of decisions make use of geographic information?

- What is a geographic information system (GIS), and how would I know one if I saw one?

- What is geographic information science, and how does it relate to the use of GIS for scientific purposes?

- How do scientists use GIS, and why do they find it helpful?

- How do companies make money from GIS?

LEARNING OBJECTIVES

After studying this chapter you will:

- Know definitions of the terms used throughout the book, including GIS itself.

- Be familiar with a brief history of GIS.

- Recognize the sometimes invisible roles of GIS in everyday life and the roles of GIS in business.

- Understand the significance of geographic information science and how it relates to geographic information systems.

- Understand the many impacts GIS is having on society and the need to study those impacts.

1.1 Introduction: Why Does GIS Matter?

Almost everything that happens, happens somewhere. We humans confine our activities largely to the surface and near surface of the Earth. We travel over it and in the lower levels of the atmosphere, and we go through tunnels dug just below the surface. We dig ditches and bury pipelines and cables, construct mines to get at mineral deposits, and drill wells to access oil and gas. Keeping track of all of this activity is important, and knowing where it occurs can be the most convenient basis for tracking. Knowing where something happens is of critical importance if we want to go there ourselves or send someone there, to find other information about the same place, or to inform people who live nearby. In addition, decisions have geographical consequences. For example, adopting a particular funding formula creates geographical winners and losers, most obviously when the outcome is a "zero-sum game." Therefore geographic location is an important attribute of activities, policies, strategies, and plans. Geographic information systems (GIS) are a special class of information systems that keep track not only of events, activities, and things, but also of where these events, activities, and things happen or exist.

> **Almost everything that happens, happens somewhere. Knowing where something happens can be critically important.**

Because location is so important, it is an issue in many of the problems society must solve. Some of these problems are so routine that we almost fail to notice them—the daily question of which route to take to and from work, for example. Others are quite extraordinary occurrences and require rapid, concerted, and coordinated responses by a wide range of individuals and organizations—such as the events of August 29, 2005 in New Orleans (Box 1.1). Problems that involve an aspect of location, either in the information used to solve them or in the solutions themselves, are termed geographic problems. Here are some more examples:

- Health care managers solve geographic problems (and may create others) when they decide where to locate new clinics and hospitals.

- Delivery companies solve geographic problems when they decide the routes and schedules of their vehicles, often on a daily basis.

- Transportation authorities solve geographic problems when they select routes for new highways.

- Geodemographics consultants solve geographic problems when they assess the performance of retail outlets and recommend where to expand or rationalize store networks.

- Forestry companies solve geographic problems when they determine how best to manage forests, where to cut, where to locate roads, and where to plant new trees.

- National park authorities solve geographic problems when they schedule recreational path maintenance and improvement (Figure 1.1).

- Governments solve geographic problems when they decide how to allocate funds for building sea defenses.

- Travelers and tourists solve geographic problems when they give and receive driving directions, select hotels in unfamiliar cities, and find their way around theme parks (Figure 1.2).

- Farmers solve geographic problems when they employ new information technology to make better decisions about the amounts of fertilizer and pesticide to apply to different parts of their fields.

Figure 1.1 Maintaining and improving footpaths in national parks is a geographic problem.

Figure 1.2 Navigating tourist destinations is a geographic problem.

If so many problems are geographic, what distinguishes them from each other? Here are three bases for classifying geographic problems. First, there is the question of **scale**, or level of geographic detail. The architectural design of a building can present geographic

problems, as in the case of disaster management (Box 1.1), but only at a very detailed or local scale. The information needed to service the building is also local—the size and shape of the parcel, the vertical and subterranean extent of the buildings, the slope of the land, and its accessibility using normal and emergency infrastructure. At the other end of the scale range, the global diffusion of the 2003 severe acute respiratory syndrome (SARS) epidemic and of bird flu in 2004 were problems at a much broader and coarser scale, involving information about entire national populations and global transport patterns.

Scale or level of geographic detail is an essential property of any GIS project.

Second, geographic problems can be distinguished on the basis of intent, or **purpose**. Some problems are strictly practical in nature—they must often be solved as quickly as possible and/or at minimum cost, in order to achieve such practical objectives as saving money, avoiding fines by regulators, or coping with an emergency. Others are better characterized

Applications Box 1.1

Hurricane Katrina, August 29, 2005

Hurricane hazards come in many forms: storm surges, high winds, tornadoes, and flooding. This means it is important for households and communities to have plans of safety actions that anticipate these hazards. Hurricane Katrina (Figure 1.3) hit the City of New Orleans, Louisiana, with the full force of a Category 5 storm on August 29, 2005, having

Figure 1.3 (A) Hurricane Katrina as at August 28, 2005. (Courtesy NOAA/NESDIS: www.nnvl.noaa.gov)

Figure 1.3 (B) Its aftermath in New Orleans on August 29, 2005, showing the flooding of the I-10 Interstate Highway, directly caused by the breaching of the levees of the 17th Street Canal. (Image from Wikipedia: http://en.wikipedia. org/wiki/Image:KatrinaNewOrleansFlooded_edit2.jpg)

already cut a swath across the Deep South of the United States from Florida through Mississippi. Meteorologists were very successful in predicting its trajectory and strength, and much of the affected area had been evacuated prior to the storm's arrival. New Orleans was declared a federal emergency area on August 24. The physical damage caused by its passage and aftermath breached flood protection levees in more than 50 places, leading to flooding of 80% of the City of New Orleans (Figure 1.4). Hurricane Katrina caused an estimated $81 billion (2005 US dollars) of damage and 1836 lives were lost.

Dealing with the aftermath of this emergency posed a number of geographic problems. Many of the GIS maps that were used to deal with the situation were produced by volunteers, as well as official agencies. The initial demand for GIS maps was from first responders and emergency staff on the ground, who needed customized street maps for search and rescue. These included street maps showing the density of resident population, or the key urban features that were cited in emergency calls, or the latitude and longitude coordinates needed for helicopter rescues, or the last known locations of

missing persons. Other "situational awareness" maps were required for use by incident commanders and other decision makers working at scales ranging from the county to the federal level. These identified sites that were key to power outage or restoration, the changing availability of cell phone bandwidth as towers came back online, the areas that had likely experienced (or would experience) flooding, road closures and access restrictions, availability of shelters and kitchens, locations of water and ice distribution points, and the locations of environmentally hazardous sites.

In these operational and tactical applications, GIS deployment ensured that the days and hours before and immediately after the impact of the storm were used productively. Yet, the government's relief efforts and the delayed response to the flooding of New Orleans came in for considerable subsequent criticism. In a more strategic sense, it has also been pointed out that Hurricane Katrina was an avoidable disaster, caused by funding cutbacks and failure to understand the design strength of the levees. Strategic intervention failed, but GIS was nevertheless of utmost importance in short-term disaster management and medium-term cleanup operations.

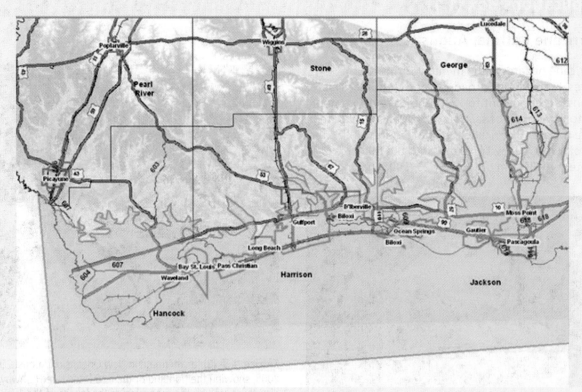

Figure 1.4 A prediction of the effects of a 32-foot storm surge with 20-foot wave action, as modeled using a GIS. The city boundary is shown in green, and the limit of the storm surge in red. (Source: ArcNews)

as driven by human curiosity. When geographic data are used to verify the theory of continental drift, or to map distributions of glacial deposits, or to analyze the historic movements of people in anthropological or archaeological research (Box 1.2 and Figure 1.5), there is no sense of an immediate problem that needs to be solved. Rather, the intent is the advancement of human understanding of the world, which we often recognize as the intent of science.

Although science and practical problem solving can be thought of as distinct human activities, it is often argued that there is no longer any effective distinction between their methods. The tools and methods used by a scientist in a government agency to ensure the protection of an endangered species are essentially the same as the tools used by an academic ecologist to advance our scientific knowledge of biological systems. Both use the most accurate measurement devices, employ terms whose meanings have been widely shared and agreed, insist that their results be replicable by others, and in general follow all of the principles of science that have evolved over the past centuries.

The use of GIS for both forms of activity reinforces the idea that science and practical problem solving are no longer distinct in their methods, as does the fact that GIS is used widely in all kinds of organizations, from academic institutions to government agencies and corporations. The use of similar

Figure 1.5 Store location principles are very important in developing markets across the world, as with Tesco's investment in Beijing, China. (© Lou-Foto/Alamy Limited)

tools and methods right across science and problem solving is part of a shift from the pursuit of curiosity within traditional academic disciplines to solution-centered, interdisciplinary team work.

In this book we distinguish between uses of GIS that focus on design, or so-called normative uses, and uses that advance science, or so-called positive uses (a rather confusing meaning of that term, unfortunately, but the one commonly used by philosophers of science—its use implies that science confirms theories by finding positive evidence in support of them, and rejects theories when negative evidence is found). Finding new locations for

Where Did Your Ancestors Come From?

As individuals, many of us are interested in *where* we came from—not just geographically, but possibly also in terms of social standing, or our inherited genes. Some of the best clues to our ancestry come from our (family) surnames, and Western surnames have different types of origins, many of which are explicitly or implicitly geographic in origin. (Such clues are less important in some Eastern societies where family histories are generally much better documented.) Research at University College London is using GIS and historic censuses and records to investigate the changing local and regional geographies of surnames within the UK since the late Nineteenth Century (Figure 1.6). Aggregating names into cultural, ethnic, or linguistic groups can tell us quite a lot about migration, about changes in local and regional economies, and even about measures of local economic health and vitality. Similar GIS-based analysis can be used to generalize about the intergenerational characteristics of international emigrants (for example, to North America, Australia, and New Zealand—see Figure 1.7), or the regional naming patterns of immigrants to the United States from the Indian subcontinent or China. This helps us understand our place in the world. Fundamentally, this is curiosity-driven research: it is interesting to us as individuals to understand more about our origins, and it is interesting to everyone with planning or policy concerns with any particular place to understand the social and cultural mix of people who live there. But it is not central to resolving any specific problem within a specific timescale.

▶

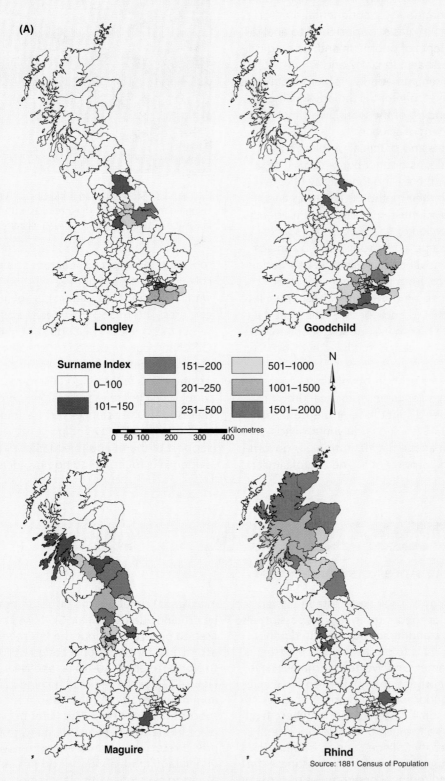

Figure 1.6 The UK geography of the Longleys, the Goodchilds, the Maguires, and the Rhinds in (A) 1881 and (B) 1998. (Reproduced with permission of Daryl Lloyd)

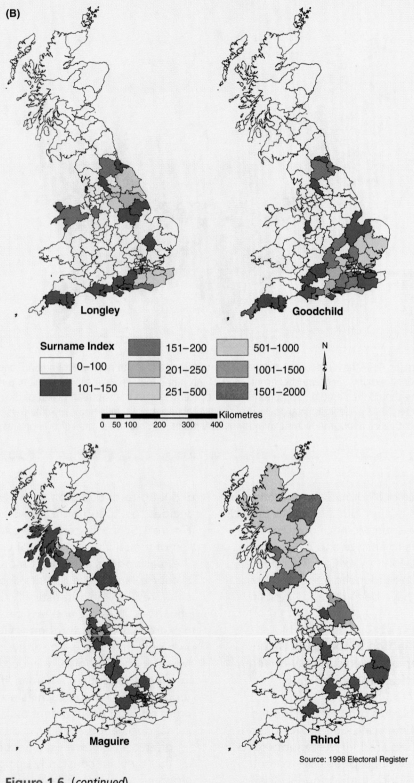

(B)

Longley

Goodchild

Surname Index

0–100
101–150
151–200
201–250
251–500
501–1000
1001–1500
1501–2000

Kilometres
0 50 100 200 300 400

Maguire

Rhind

Source: 1998 Electoral Register

Figure 1.6 (*continued*)

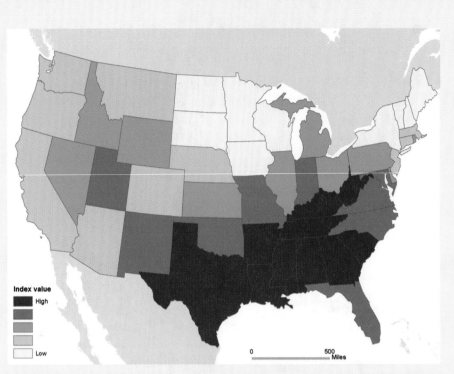

Figure 1.7 The Singleton family name derives from a place in northwest England, and understandably the greatest concentration of this name today remains in this region. But why should the name be disproportionately concentrated in the south and west of the United States? Geographical analysis of the global pattern of family names can help us to hypothesize about the historic migrations of families, communities, and cultural groups.

retailers is an example of a normative application of GIS, with its focus on design. But in order to predict how consumers will respond to new locations, it is necessary for retailers to analyze and model the actual patterns of behavior they exhibit. Therefore, the models they use will be grounded in observations of messy reality that have been tested in a positive manner.

With a single collection of tools, GIS is able to bridge the gap between curiosity-driven science and practical problem solving

Third, geographic problems can be distinguished on the basis of their **time scale**. Some decisions are operational and are required for the smooth functioning of an organization, such as how to control electricity inputs into grids that experience daily surges and troughs in usage (see Section 10.9). Others are tactical and concerned with medium-term decisions, such as where to cut trees in next year's forest harvesting plan. Still other decisions are strategic and are required to give an organiza-

tion long-term direction, as when a retailer decides to expand or rationalize its store net-work (Figure 1.5). These terms are explored in the context of logistic applications of GIS in Section 2.3.4. The real world is somewhat more complex than this, of course, and these distinctions may blur—what is theoretically and statistically the 1000-year flood influences strategic and tactical considerations but may arrive a year after the previous one! Other problems that interest geophysicists, geologists, or evolutionary biologists may occur on timescales that are much longer than a human lifetime, but are still geographic in nature, such as predictions about the future physical environment of Japan or about the animal populations of Africa. Geographic databases are often *transactional* (see Section 10.9.1), meaning that they are constantly being updated as new information arrives, unlike paper maps, which stay the same once printed. Chapter 2 contains a more detailed discussion of the range and remit of GIS applications, and provides a view of how GIS pervades many aspects of our daily lives. Other applications are discussed to illustrate particular principles, techniques,

Some Technical Reasons Why Geographic Information Is Special

- Geographic information is multidimensional, because *two* coordinates must be specified to define a location, whether they be *x* and *y* or latitude and longitude.

- It is voluminous, since a geographic database can easily reach a terabyte in size (see Table 1.1).

- It may be represented at different levels of spatial resolution, for example, using a representation equivalent to a 1:1 million scale map and a 1:24,000 scale map (Section 3.7).

- It may be represented in different ways inside a computer (Chapter 3), and how this is done can

strongly influence the ease of analysis and the end results.

- It must often be projected onto a flat surface, for reasons identified in Section 5.8.

- It requires many special methods for its analysis (see Chapters 14 and 15).

- It can be time-consuming to analyze.

- Although much geographic information is static, the process of updating is complex and expensive.

- Display of geographic information in the form of a map requires the retrieval of large amounts of data.

---

analytic methods, and management practices as these arise throughout the book.

1.1.1 Spatial Is Special

The adjective *geographic* refers to the Earth's surface and near surface, and defines the subject matter of this book, but other terms have similar meaning. *Spatial* refers to any space, not only the space of the Earth's surface; this term is used frequently in the book, almost always with the same meaning as *geographic*. But many of the methods used in GIS are also applicable to other nongeographic spaces, including the surfaces of other planets, the space of the cosmos, and the space of the human body that is captured by medical images. GIS techniques have even been applied to the analysis of genome sequences on DNA. So the discussion of analysis in this book is of *spatial* analysis (Chapters 14 and 15), not geographic analysis, to emphasize this versatility.

Another term that has been growing in usage in recent years is *geospatial*—implying a subset of spatial applied specifically to the Earth's surface and near surface. The former National Intelligence and Mapping Agency was renamed the National Geospatial-Intelligence Agency in late 2003 by President George W. Bush, and the Web portal for federal government data is called Geospatial One-Stop. In this book we have tended to avoid geospatial, preferring geographic, and we use spatial where we need to emphasize generality.

People who encounter GIS for the first time are sometimes driven to ask why geography is so important;

why, they ask, is spatial special? After all, there is plenty of information around about geriatrics, for example, and in principle one could create a geriatric information system. So why has geographic information spawned an entire industry, if geriatric information has not done so to anything like the same extent? Why are there no courses in universities specifically in geriatric information systems? Part of the answer should be clear already: almost all human activities and decisions involve a geographic component, and the geographic component is important. Another reason will become apparent in Chapter 3, where we will see that working with geographic information involves complex and difficult choices that are also largely unique. Other, more technical reasons will become clear in later chapters and are briefly summarized in Box 1.3.

1.2 Data, Information, Knowledge, Evidence, Wisdom

Information systems help us to manage *what we know* by making it easy to organize and store, access and retrieve, manipulate and synthesize, and apply to the solution of problems. We use a variety of terms to describe what we know, including the five that head this section and that are shown in Table 1.2. There are no universally agreed-upon definitions of these terms, the first two of which are used frequently in the GIS arena. Nevertheless, it is worth trying to come to grips

Table 1.1 Potential GIS database volumes for some typical applications (volumes estimated to the nearest order of magnitude). Strictly, bytes are counted in powers of 2 – 1 kilobyte is 1024 bytes, not 1000.

1 megabyte	1 000 000	Single dataset in a small project database
1 gigabyte	1 000 000 000	Entire street network of a large city or small country
1 terabyte	1 000 000 000 000	Elevation of entire Earth surface recorded at 30 m intervals
1 petabyte	1 000 000 000 000 000	Satellite image of entire Earth surface at 1 m resolution
1 exabyte	1 000 000 000 000 000 000	A future 3-D representation of entire Earth at 10 m resolution?

with their various meanings because the differences between them can often be significant. What follows draws on many sources and thus provides the basis for using these terms throughout the book. Data clearly refer to the most mundane kinds of information, and wisdom to the most substantive. *Data* consist of numbers, text, or symbols that are in some sense neutral and almost context-free. Raw geographic facts (see Section 18.5.1.2), such as the temperature at a specific time and location, are examples of data. When data are transmitted, they are treated as a stream of bits; a crucial requirement is to preserve the integrity of the dataset. The internal meaning of the data is irrelevant in such considerations. Data (the plural of datum) are assembled together in a database (see Chapter 10), and the volumes of data that are required for some typical applications are shown in Table 1.1.

The term *information* can be used either narrowly or broadly. In a narrow sense, information can be treated as devoid of meaning, and therefore as essentially synonymous with data, as defined in the previous paragraph. Others define information as *anything* that can be digitized, that is, represented in digital form (see Chapter 3), but also argue that information is differentiated from data by implying some degree of selection, organization, and preparation for particular purposes—information is data serving some

purpose, or data that have been given some degree of interpretation. Information is often costly to produce, but once digitized it is cheap to reproduce and distribute. Geographic datasets, for example, may be very expensive to collect and assemble, but very cheap to copy and disseminate. One other characteristic of information is that it is easy to add value to it through processing and through merger with other information. GIS provides an excellent example of the latter because of the tools it provides for combining information from different sources (see Section 18.5).

GIS does a better job of sharing data and information than knowledge, which is more difficult to detach from the knower.

Knowledge does not arise simply from having access to large amounts of information. It can be considered as information to which value has been added by interpretation based on a particular context, experience, and purpose. Put simply, the information available in a book or on the Internet or on a map becomes knowledge only when it has been read and understood. How the information is interpreted and used will be different for different readers depending on their previous experience, expertise, and needs. It is important to distinguish two types of knowledge: *codified*

Table 1.2 A ranking of the support infrastructure for decision making.

Decision-making support infrastructure	Ease of sharing with everyone	GIS example
Wisdom ↑	*Impossible*	Policies developed and accepted by stakeholders
Knowledge ↑	*Difficult, especially tacit knowledge*	Personal knowledge about places and issues
Evidence ↑	*Often not easy* datasets or scenarios	Results of GIS analysis of many
Information ↑	*Easy*	Contents of a database assembled from raw facts
Data	*Easy*	Raw geographic facts

and *tacit*. Knowledge is codifiable if it can be written down and transferred with relative ease to others. Tacit knowledge is often slow to acquire and much more difficult to transfer. Examples include the knowledge built up during an apprenticeship, understanding of how a particular market works, or familiarity with using a particular technology or language. This difference in transferability means that codified and tacit knowledge need to be managed and rewarded quite differently. Because of its nature, tacit knowledge is often a source of competitive advantage.

Some have argued that knowledge and information are fundamentally different in at least three important respects:

1. Knowledge entails a knower. Information exists independently, but knowledge is intimately related to people.

2. Knowledge is harder to detach from the knower than information; shipping, receiving, transferring it between people, or quantifying it are all much more difficult than for information.

3. Knowledge requires much more assimilation— we digest it rather than hold it. While we may hold conflicting information, we rarely hold conflicting knowledge.

Evidence is considered a halfway house between information and knowledge. It seems best to regard it as a multiplicity of information from different sources, related to specific problems, and with a consistency that has been validated. Major attempts have been made in medicine to extract evidence from a welter of sometimes contradictory sets of information, drawn from worldwide sources, in what is known as meta-analysis, or the comparative analysis of the results of many previous studies.

The definition of *wisdom* is even more elusive than that of the other terms. Normally, it is used in the context of decisions made or advice given which is disinterested, based on all the evidence and knowledge available, but given with some understanding of the likely consequences. Almost invariably, it is highly individualized rather than being easy to create and share within a group. Wisdom is in a sense the top level of a hierarchy of decision-making infrastructure.

1.3 Systems and Science

Geographic information systems are computer-based systems for storing and processing geographic information. They are tools that improve the efficiency and effectiveness of handling information about geographic objects and events. They can be used to carry out many useful tasks, including storing vast amounts of geographic information in databases, conducting analytical operations in a fraction of the time it would take to do it by hand, and automating the process of making useful maps. Geographic information systems also process information, but there are limits to the kinds of procedures and practices that can be automated when turning data into information. Moreover, it is more the realm of evidence, knowledge, and wisdom to assess whether this selectivity and preparation for purpose actually adds any value, or whether the results add insight to interpretation in geographic applications. Such issues are the realm of geographic information science. This rapidly developing field is concerned with the scientific context and underpinnings of geographic information systems. It provides a framework within which new evidence, knowledge, and ultimately wisdom about the Earth can be created, in ways that are efficient, effective, and safe to use.

1.3.1 The Science of Problem Solving

Like all sciences, an essential requirement of geographic information science is a method for discovering new knowledge. The GI scientific method must support:

- Transparency of assumptions and methods so that other GI scientists can determine how previous knowledge has been discovered and how they might add to the existing body of knowledge.

- Objectivity through a detached and independent perspective that avoids or accommodates bias (unintended or otherwise).

- The ability of any other qualified scientist to reproduce the results of an analysis.

- Methods of validation using the results of the analysis (internal validation) or other information sources (external validation).

How then are problems solved using a scientific method, and are geographic problems solved in ways different from other kinds of problems? We humans have accumulated a vast storehouse about the world, including information both on how it *looks*, or its *forms*, and how it *works*, or its dynamic *processes*. Some of those processes are natural and built into the design of the planet, such as the processes of tectonic movement that lead to earthquakes and the processes of atmospheric circulation that lead to hurricanes. Others are human in origin, reflecting the increasing

Figure 1.8 Social processes, such as carbon dioxide emissions, modify the Earth's environment. (Digital Vision)

influence that we have on our natural environment, through the burning of fossil fuels, the felling of forests, and the cultivation of crops (Figure 1.8). Others are imposed by us, in the form of laws, regulations, and practices. For example, zoning regulations affect the ways in which specific parcels of land can be used.

Knowledge about how the world works is more valuable than knowledge about how it looks because it can be used to predict.

These two types of information differ markedly in their degree of generality. Form varies geographically, and the Earth's surface looks dramatically different in different places; compare the settled landscape of northern England with the deserts of the U.S. Southwest (Figure 1.9). But processes can be very general. The ways in which the burning of fossil fuels affects the atmosphere are essentially the same in China as in Europe, although the two landscapes look very different. Science has always valued such general knowledge over knowledge of the specific, and hence has valued process knowledge over knowledge of form. Geographers in particular have witnessed a long-standing debate, lasting centuries, between the competing needs of *idiographic* geography, which focuses on the description of form and emphasizes the unique characteristics of places, and *nomothetic* geography, which seeks to discover general processes. Both are essential, of course, since knowledge of general process is useful in solving specific problems only if it can be combined effectively with knowledge of form. For example, we can only assess the risk of roadside landslip in New South Wales if we know both how slope stability is generally impacted by such factors as shallow subsurface characteristics and porosity, and where slopes at risk are located (Figure 1.10).

One of the most important merits of GIS as a problem-solving tool lies in its ability to combine the general with the specific, as in this example from New South Wales. A GIS designed to solve this problem would contain knowledge of New South Wales' slopes, in the form of computerized maps, and the programs executed by the GIS would reflect general knowledge of how slopes affect the probability of mass movement under extreme weather conditions. The *software* of a GIS captures and implements general knowledge, while the *database* of a GIS represents specific information. In that sense, a GIS resolves the old debate between nomothetic and idiographic camps by accommodating both.

GIS solves the ancient problem of combining general scientific knowledge with specific information and gives practical value to both.

General knowledge comes in many forms. Classification is perhaps the simplest and most rudimentary,

Figure 1.9 The form of the Earth's surface shows enormous variability, for example, between the deserts of the southwest United States and the settled landscape of northern England. (A: Courtesy ImageState)

Figure 1.10 Predicting landslides requires general knowledge of processes and specific knowledge of the area—both are available in a GIS. (© Chris Selby/AlamyLimited)

perfectly predicted. From Newton's laws we are able to predict the motions of the planets almost perfectly, although Einstein later showed that certain observed deviations from the predictions of the laws could be explained with his Theory of Relativity. Laws of this level of predictive quality are few and far between in the geographic world of the Earth's surface. The real world is the only geographic-scale laboratory that is available for most GIS applications, and considerable uncertainty is generated when we are unable to control for all conditions. These problems are compounded in the socioeconomic realm, where the role of human agency makes it almost inevitable that any attempt to develop rigid laws will be frustrated by isolated exceptions. Thus, while market researchers use spatial interaction models, in conjunction with GIS, to predict how many people will shop at each shopping center in a city, substantial errors will occur in the predictions. Nevertheless, the results are of great value in developing location strategies for retailing. The Universal Soil Loss Equation, used by soil scientists in conjunction with GIS to predict soil erosion, is similar in its rather low predictive power, but again the results are sufficiently accurate to be very useful in the right circumstances.

Solving problems involves several distinct components and stages. First, there must be an *objective*, or a goal that the problem solver wishes to achieve. Often this is a desire to maximize or minimize—find the solution of least cost, or shortest distance, or least time, or greatest profit; or make the most accurate prediction possible. These objectives are all expressed

and is widely used in geographic problem solving. In many parts of the United States and other countries, efforts have been made to limit the development of wetlands in the interest of preserving them as natural habitats and avoiding excessive impact on water resources. To support these efforts, resources have been invested in mapping wetlands, largely from aerial photography and satellite imagery. These maps simply classify land, using established rules that define what is and what is not a wetland (Figure 1.11).

More sophisticated forms of knowledge include *rule sets*—for example, rules that determine what use can be made of wetlands, or what areas in a forest can be legally logged. The U.S. Forest Service has rules to define wilderness and to impose associated regulations regarding the use of wilderness, including prohibition on logging and road construction.

Much of the knowledge gathered by the activities of scientists suggests the term *law*. The work of Sir Isaac Newton established the Laws of Motion, according to which all matter behaves in ways that can be

Figure 1.11 A wetland map of part of the Amazon region of Brazil. The map has been made by classifying Landsat imagery at 30 meter resolution. (Courtesy National Institute for Space Research (INPE), Brazil)

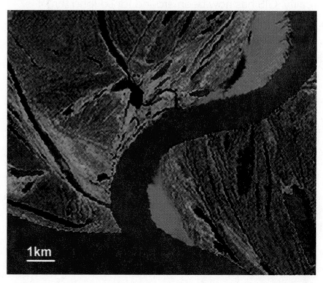

1km

intangible form; that is, they can be measured on some well-defined scale. Others are said to be *intangible* and involve objectives that are much harder, if not impossible, to measure. They include maximizing *quality of life* and *satisfaction* and minimizing *environmental impact*. Sometimes the only way to work with such intangible objectives is to involve human subjects, through surveys or focus groups, by asking them to express a preference among alternatives. A large body of knowledge has been acquired about such human-subjects research, and much of it has been employed in connection with GIS. For discussion of the use of such mixed objectives see Section 16.4. This topic is taken up again in Chapter 17 in the context of estimating the return on investment of GIS.

Often a problem will have *multiple objectives*. For example, a company providing a mobile snack service to construction sites will want to maximize the number of sites that can be visited during a daily operating schedule, and will also want to maximize the expected returns by visiting the most lucrative sites. An agency charged with locating a corridor for a new power transmission line may decide to minimize cost, while at the same time seeking to minimize environmental impact. Such problems employ methods known as *multicriteria decision making* (MCDM).

Many geographic problems involve multiple goals and objectives, which often cannot be expressed in commensurate terms.

1.3.2 The Technology of Problem Solving

The previous sections have presented GIS as a technology to support both science and problem solving, using both specific and general knowledge about geographic reality. GIS has now been around for so long that it is, in many senses, a background technology, like word processing. Yet in other important respects, GIS is more than a technology, and continues to attract attention as a focus for scientific journals and conferences.

Many definitions of GIS have been suggested over the years, and none of them is entirely satisfactory. Today, the label GIS is attached to many things, including a collection of software tools (sometimes bought from a vendor) to carry out certain well-defined functions (*GIS software*); digital representations of various aspects of the geographic world, in the form of datasets (*GIS data*); a community of people who use and perhaps advocate the use of these tools for various purposes (the *GIS community*); and the activity of using a GIS to solve problems or advance science (*doing GIS*). The basic label works in all of these

Table 1.3 Definitions of a GIS, and the groups who find them useful.

A container of maps in digital form	The general public
A computerized tool for solving geographic problems	Decision makers, community groups, planners
A spatial decision support system	Management scientists, operations researchers
A mechanized inventory of geographically distributed features and facilities	Utility managers, transportation officials, resource managers
A tool for revealing what is otherwise invisible in geographic information	Scientists, investigators
A tool for performing operations on geographic data that are too tedious or expensive or inaccurate if performed by hand	Resource managers, planners

ways, and its meaning surely depends on the context in which it is used.

Certain definitions of GIS (and the audiences for which they may be helpful) are summarized in Table 1.3. As we describe in Chapter 3, GIS is much more than *a container of maps in digital form*. This description can be misleading, but it may nonetheless be helpful to give to someone looking for a simple explanation—a guest at a cocktail party, a relative, or a seat neighbor on an airline flight. We all know and appreciate the value of maps, and the notion that maps could be processed by a computer is directly analogous to the use of word processing or spreadsheets to handle other types of information. A GIS is also *a computerized tool for solving geographic problems*. This definition speaks to the purposes of GIS rather than to its functions or physical form—an idea that is expressed in another definition, *a spatial decision support system*. A GIS is *a mechanized inventory of geographically distributed features and facilities*. This definition explains the value of GIS to the utility industry, where it is used to keep track of such entities as underground pipes, transformers, transmission lines, poles, and customer accounts. A GIS is *a tool for revealing what is otherwise invisible in geographic information*. This interesting definition emphasizes the power of a GIS, as an analysis engine, to examine data and reveal its patterns, relationships, and anomalies—things that might not be apparent to someone looking at a map. A GIS is *a tool for performing operations on geographic data that are too tedious or expensive or inaccurate if performed by hand*. This definition speaks to the problems associated with manual analysis of maps, particularly the extraction of simple measures, of area for example.

Everyone has a favorite definition of a GIS, and there are many to choose from.

1.4 A Brief History of GIS

As might be expected, some controversy surrounds the history of GIS since parallel developments occurred in North America, Europe, and Australia (at least). Much of the published history focuses on the U.S. contributions. We therefore do not yet have a well-rounded history of our subject. What is clear, however, is that the extraction of simple geographic measures largely drove the development of the first real GIS, the Canada Geographic Information System, or CGIS, in the mid-1960s (see Section 17.3). The Canada Land Inventory was a massive effort by the federal and provincial governments to identify the nation's land resources and their existing and potential uses. The most useful results of such an inventory are measures of area, yet area is notoriously difficult to measure accurately from a map (see Section 15.1.1). CGIS was planned and developed as a measuring tool, a producer of tabular information, rather than as a mapping tool.

The first GIS was the Canada Geographic Information System, designed in the mid-1960s as a computerized map-measuring system.

A second burst of innovation occurred in the late 1960s in the U.S. Bureau of the Census, in planning the tools needed to conduct the 1970 Census of Population. The DIME program (Dual Independent Map Encoding) created digital records of all U.S. streets in order to support automatic referencing and aggregation of census records. The similarity of this technology to that of CGIS was recognized immediately and led to a major program at Harvard University's Laboratory for Computer Graphics and Spatial Analysis to develop a general-purpose GIS that could handle the needs of both applications. This project eventually led to the ODYSSEY GIS software of the late 1970s.

Early GIS developers recognized that the same basic needs were present in many different application areas, from resource management to the census.

In a largely separate development during the latter half of the 1960s, cartographers and mapping agencies had begun to ask whether computers might be adapted to their needs, possibly reducing costs and shortening the time of map creation. The UK Experimental Cartography Unit (ECU) pioneered high-quality computer mapping in 1968; it published the world's first computer-made map in a regular series in 1973 with the British Geological Survey (Figure 1.12). The ECU also pioneered GIS work in education, post- and ZIP codes as geographic references, visual perception of maps, and much else. National mapping agencies, such as Britain's Ordnance Survey, France's Institut Géographique National, and the U.S. Geological Survey and the Defense Mapping Agency (now the National Geospatial-Intelligence Agency) began to investigate the use of computers to support the editing of maps that would avoid the expensive and slow process of hand correction and redrafting. The first automated cartography developments occurred in the 1960s, and by the late 1970s most major cartographic agencies were already computerized to some degree. But the magnitude of the task ensured that it would not be until 1995 that the first country (Britain) achieved complete digital map coverage in a database.

Remote sensing also played a part in the development of GIS, as a source of technology as well as a source of data. The first military satellites of the 1950s were developed and deployed in great secrecy to gather intelligence, but the declassification of much of this material in recent years has provided interesting insights into the role played by the military and intelligence communities in the development of GIS. Although the early spy satellites used conventional film cameras to record images, digital remote sensing began to replace them in the 1960s, and by the early 1970s civilian remote-sensing systems such as Landsat were beginning to provide vast new data resources on the appearance of the planet's surface from space, as well as to exploit the technologies of image classification and pattern recognition that had been developed earlier for military applications. The military was also responsible for the development in the 1950s of the world's first uniform system of measuring location, driven by the need for accurate targeting of intercontinental ballistic missiles. This development led directly to the methods of positional control in use today. Military needs were also responsible for the initial development of the Global Positioning System (GPS; see Section 5.9).

Many technical developments in GIS originated during the Cold War.

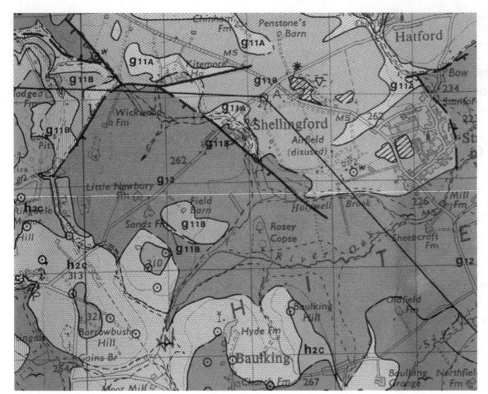

GIS really began to take off in the early 1980s, when the price of computing hardware had fallen to a level that could sustain a significant software industry and cost-effective applications. Among the first customers were forestry companies and natural resource agencies, driven by the need to keep track of vast timber resources and to regulate their use effectively. At the time a modest computing system—far less powerful than today's personal computer—could be obtained for about $250,000, and the associated software for about $100,000. Even at these prices the benefits of consistent management using GIS, and the decisions that could be made with these new tools, substantially exceeded the costs. The market for GIS software continued to grow, computers continued to fall in price and increase in power, and the GIS software industry has been growing ever since.

> **The modern history of GIS dates from the early 1980s, when the price of sufficiently powerful computers fell below a critical threshold.**

As indicated earlier, the history of GIS is a complex story, much more complex than can be described in this brief history, but Table 1.4 summarizes some of the major events.

1.5 Views of GIS

As the previous discussion shows, GIS is a complex beast, with many distinct appearances. To some it is a way to automate the production of maps, whereas to others this application seems far too mundane compared to the complexities associated with solving geographic problems and supporting spatial decisions, and with the power of a GIS as an engine for analyzing data and revealing new insights. Others see a GIS as a tool for maintaining complex inventories, one that adds geographic perspectives to existing information systems and allows the geographically distributed resources of a forestry or utility company to be tracked and managed. The sum of all of these perspectives is clearly too much for any one software package to handle. As a result, GIS has grown from its initial commercial beginnings as a simple off-the-shelf package to a complex of software, hardware, people, institutions, networks, and activities, all of which can be very confusing to the novice. Thus, for example, the major software vendor ESRI, Inc. (Redlands, California) today sells a family of products under the ArcGIS brand name in order to service the disparate needs of its diverse user-base: ArcInfo is the high-end full-feature desktop system; ArcView is a

Table 1.4 Major events that shaped GIS.

Date	Type	Event	Notes
The Era of Innovation			
1957	Application	First known automated mapping produced	Swedish meteorologists and British biologists
1963	Technology	CGIS development initiated	Canada Geographic Information System is developed by Roger Tomlinson and colleagues for Canadian Land Inventory. This project pioneers much technology and introduces the term GIS.
1963	General	URISA established	The Urban and Regional Information Systems Association founded in the US. Soon becomes point of interchange for GIS innovators.
1964	Academic	Harvard Lab established	The Harvard Laboratory for Computer Graphics and Spatial Analysis is established under the direction of Howard Fisher at Harvard University. In 1966 SYMAP, the first raster GIS, is created by Harvard researchers.
1967	Technology	DIME developed	The US Bureau of Census develops DIME-GBF (Dual Independent Map Encoding—Geographic Database Files), a data structure and street-address database for 1970 census.
1967	Academic and general	UK Experimental Cartography Unit (ECU) formed	Pioneer organization in a range of computer cartography and GIS areas.
1969	Commerical	ESRI Inc. formed	Jack Dangermond, a student from the Harvard Lab, and his wife Laura form ESRI to undertake projects in GIS.
1969	Commercial	Intergraph Corp. formed	Jim Meadlock and four others that worked on guidance systems for Saturn rockets form M&S Computing, later renamed Intergraph.
1969	Academic	'Design With Nature' published	Ian McHarg's book is the first to describe many of the concepts in modern GIS analysis, including the map overlay process (see Section 14.2.4).
1969	Academic	First technical GIS textbook	Nordbeck and Rystedt's book details algorithms and software developed for spatial analysis.
1972	Technology	Landsat 1 launched	Originally named ERTS (Earth Resources Technology Satellite), this is the first of many major Earth remote sensing satellites.
1973	General	First digitizing production line	Set up by Ordnance Survey, Britain's national mapping agency.
1974	Academic	AutoCarto 1 Conference	Held in Reston, Virginia, this is the first in an important series of conferences that set the GIS research agenda.
1976	Academic	GIMMS now in worldwide use	Written by Tom Waugh (a Scottish academic), this vector-based mapping and analysis system is run at 300 sites worldwide.
1977	Academic	Topological Data Structures	Harvard Lab organizes a major conference and develops the ODYSSEY GIS.
The Era of Commercialization			
1981	Commercial	ArcInfo launched	ArcInfo is the first major commerical GIS software system. Designed for minicomputers and based on the vector and relational database data model, it sets a new standard for the industry.

(continued overleaf)

Table 1.4 (*continued*)

Date	Type	Event	Notes
1984	Academic	'Basic Readings in Geographic Information Systems' published	This collection of papers published in book form by Duane Marble, Hugh Calkins, and Donna Peuquet is the first accessible source of information about GIS.
1985	Technology	GPS operational	The Global Postitioning System gradually becomes a major source of data for navigation, surveying, and mapping.
1986	Academic	'Principles of Geographical Infomation Systems for Land Resources Assessment' published	Peter Burrough's book is the first specifically on GIS principles. It quickly becomes a worldwide reference text for GIS students.
1986	Commercial	MapInfo Corp. formed	MapInfo software develops into first major desktop GIS product. It defines a new standard for GIS products, complementing earlier software systems.
1987	Academic	*International Journal of Geographical Information Systems,* now *IJGI Science,* launched	Terry Coppock and others publish the first journal on GIS. The first issue contains papers from the USA, Canada, Germany, and UK.
1987	General	Chorley Report	'Handling Geographical Information' is an influential report from the UK government that highlights the value of GIS.
1988	General	*GISWorld* begins	*GISWorld,* now *GeoWorld,* the first worldwide magazine devoted to GIS, is published in the USA.
1988	Technology	TIGER announced	TIGER (Topologically Integrated Geographic Encoding and Referencing), a follow-on from DIME, is described by the US Census Bureau. Low-cost TIGER data stimulate rapid growth in US business GIS.
1988	Academic	US and UK Research Centers announced	Two separate initiatives, the US NCGIA (National Center for Geographic Information and Analysis) and the UK RRL (Regional Research Laboratory) Initiative show the rapidly growing interest in GIS in academia.
1991	Academic	*Big Book 1* published	Substantial two-volume compendium *Geographical Information Systems: principles and applications,* edited by David Maguire, Mike Goodchild, and David Rhind documents progress to date.
1992	Technical	DCW released	The 1.7 GB Digital Chart of the World, sponsored by the US Defense Mapping Agency, (now NGA), is the first integrated 1:1 million scale database offering global coverage.
1994	General	Executive Order signed by President Clinton	Executive Order 12906 leads to creation of US National Spatial Data Infrastructure (NSDI), clearinghouses, and the Federal Geographic Data Committee (FGDC).
1994	General	OpenGIS Consortium® born (now Open Geospatial Consortium®)	The OpenGIS® Consortium of GIS vendors, government agencies, and users is formed to improve interoperability.
1995	General	First complete national mapping coverage	Great Britain's Ordnance Survey completes creation of its initial database—all 230,000 maps covering country at largest scale (1:1,250, 1:2,500 and 1:10,000) encoded.

Table 1.4 (*continued*)

Date	Type	Event	Notes
1996	Technology	Internet GIS products introduced	Several companies, notably Autodesk, ESRI, Intergraph, and MapInfo, release new generation of Internet-based products at about the same time.
1996	Commercial	MapQuest	Internet mapping service launched, producing over 130 million maps in 1999. Is subsequently purchased by AOL for $1.1 billion.
1999	General	GIS Day	First GIS Day attracts over 1.2 million global participants who share an interest in GIS.
The Era of Exploitation			
1999	Commercial	IKONOS	A new generation of very high resolution satellite sensors: IKONOS claims 90 cm ground resolution; Quickbird (launched 2001) claims 62 cm resolution.
2000	Commercial	GIS passes $7 bn	Industry analyst Daratech reports GIS hardware, software, and services industry at $6.9 bn, growing at more than 10% per annum.
2000	General	GIS has 1 million users	GIS has more than 1 million core users, and there are perhaps 5 million casual users of GI.
2002	General	Launch of online National Atlas of the United States	Online summary of US national-scale geographic information with facilities for map making (www.nationalatlas.gov)
2003	General	Launch of online national statistics for the UK	Exemplar of new government Web sites describing economy, population, and society at local and regional scales (www.statistics.gov.uk)
2003	General	Launch of Geospatial One-Stop	A US Federal E-government initiative providing access to geospatial data and information (www.geodata.gov/gos)
2004	General	National Geospatial-Intelligence Agency (NGA) formed	Biggest GIS user in the world, National Imagery and Mapping Agency (NIMA), renamed NGA to signify emphasis on geo-intelligence
2006	Technology	Launch of Google Earth	First major virtual globe—a Web-based 3D GIS application. 150 million downloads in first 12 months.
2007	Commercial	Pitney Bowes, Inc. purchases MapInfo	Maker of mail-handling machines buys MapInfo Corp. for $408 million.
2007	Commercial	Navtech purchased by Nokia	Mobile phone company purchases street data provider for $8.1 billion.
2008	Commercial	TeleAtlas purchased by TomTom	Relatively new consumer GIS company purchases street data provider for $2.9 billion.

simpler system designed for viewing, analyzing, and mapping data; ArcGIS Engine is a set of software components that developers can embed in their applications; ArcGIS Mobile is a lightweight software system that can be deployed on hand-held, mobile devices; ArcGIS Server can support sophisticated data management and GIS-oriented Web sites; ArcExplorer is a free-access Web browser; and ArcGIS Online is an ESRI-hosted data and applications resource that can be accessed over the Web. Other vendors specialize in certain niche markets, such as the utility industry, or military and intelligence applications. GIS is a dynamic and evolving field, and its future is certain to be exciting, but speculations on where it might be headed are reserved for the final chapter.

Today a single GIS vendor offers many different products for distinct applications.

1.5.1 Anatomy of a GIS

1.5.1.1 The *Network*

Despite the complexity noted in the previous section, a GIS does have its well-defined component parts. Today, the most fundamental of these is probably the *network*, without which no rapid communication or sharing of digital information could occur, except between a small group of people crowded around a computer monitor. GIS today relies heavily on the Internet, and on its limited-access cousins, the *intranets* of corporations, agencies, and the military. The Internet was originally designed as a network for connecting computers, but it has grown to become society's mechanism of information exchange, handling everything from personal messages to massive shipments of data, and increasing numbers of business transactions.

It is no secret that the Internet in its many forms has had a profound effect on technology, science, and society in the last 20 or so years. Who could have foreseen in 1990 the impact that the Web, e-commerce, digital government, mobile systems, and information and communication technologies would have on our everyday lives (see Chapter 18)? These technologies have radically changed forever the way we conduct business, how we communicate with our colleagues and friends, the nature of education, and the value and transitory nature of information.

The Internet began life as a U.S. Department of Defense communications project called ARPANET (Advanced Research Projects Agency Network) in 1972. In 1980 Tim Berners-Lee, a researcher at CERN, the European organization for nuclear research, developed the hypertext capability that underlies today's World Wide Web—a key application that has brought the Internet into the realm of everyday use. Uptake and use of Web technology have been remarkably quick, diffusion being considerably faster than almost all comparable innovations (for example, the radio, the telephone, and the television). By 2008, 1.4 billion people worldwide used the Internet, and the fastest growth rates were to be found in the Middle East, Africa, and Latin America (www.internetworldstats.com). However, the global penetration of the medium remained very uneven; for example, as of 2009 74% of North Americans used the medium, but only 5% of Africans (Figure 1.13).

The use of the WWW to give access to maps dates from 1993.

In the early years of the Internet, GIS turned out to be a compelling application that prompted many people to take advantage of the Web. At the same time, GIS benefited greatly from adopting the Internet paradigm and the momentum that the Web has generated. Many of the early Internet applications of GIS on the Internet remain in use, in updated form, today. They range from using GIS on the Internet to disseminate information, a type of electronic yellow pages (e.g., www.yell.com), to selling goods and services (e.g., www.nestoria.com; Figure 1.14), to direct revenue generation through subscription services (e.g., www.mapquest.com/solutions/), to helping members of the public to participate in important local, regional, and national debates. The Internet became very popular as a vehicle for delivering GIS applications for several reasons. It provides an established, widely used platform and accepted standard for interacting with information of many types. It also offers a cost-effective

Figure 1.13 The world geography of Internet usage as of April 2008. (Image courtesy Kwintessential Ltd.)

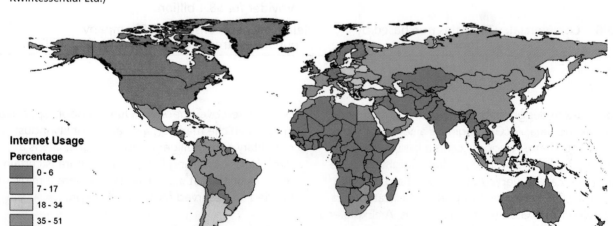

Internet Usage
Percentage

- 0 - 6
- 7 - 17
- 18 - 34
- 35 - 51
- 52 - 78

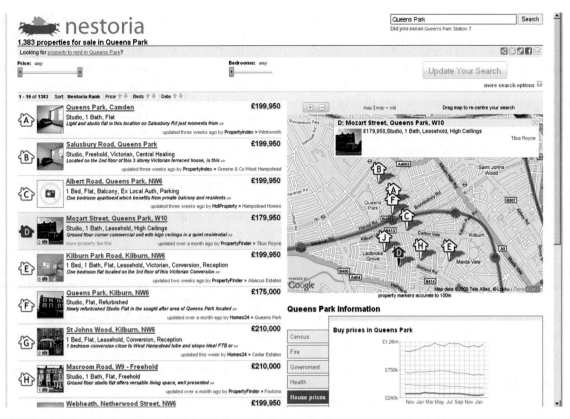

Figure 1.14 Marketing of residential real estate in the UK. (Source: www.nestoria.co.uk)

way of linking together distributed users (for example, telecommuters and office workers, customers and suppliers, students and teachers). From the early days onward, the interactive and exploratory nature of navigating linked information became a great hit with users.

In the early 2000s, Internet technology became increasingly portable (see Section 11.3.3). This meant not only that port-able GIS-enabled devices could be used in conjunction with the wireless networks available in public places such as airports and railway stations, but also that such devices could be connected through broadband in order to deliver GIS-based representations on the move. This technology was exploited in the burgeoning GIService (yet another use of the three-letter acronym GIS) sector, which offered distributed users access to centralized GIS capabilities. Today, many GIServices are made available for personal use through mobile and handheld applications as *location-based services* (see Chapter 11). Personal devices, from pagers to mobile phones to personal digital assistants (PDAs), fill the briefcases and adorn the clothing of people in many walks of life. These devices are able to provide real-time geographic services such as mapping, routing, and geographic yellow pages. These services are often funded through advertisers, or can be purchased

on a pay-as-you go or subscription basis; they have changed the business GIS model for many types of applications.

The past 10 years have also seen the development of themed geographic networks, such as the U.S. Geospatial One-Stop (www.geo-one-stop.gov; see Section 11.2.2), which is one of 24 U.S. federal e-government initiatives to improve the coordination of government at local, state, and federal levels. Its geoportal (www.geodata.gov/gos) identifies an integrated collection of geographic information providers and users that interact via the medium of the Internet. Online content can be located using the interactive search capability of the portal, and then content can be directly used over the Internet. This form of Internet application is explored further in Chapter 11.

The Internet is core to most aspects of GIS use, and the days of standalone GISystems are mostly over.

In all of these applications, the Internet provides a unidirectional flow of data and information to a large number of users from the growing number of sites that make up the World Wide Web. Over time, this system has evolved into what has been termed Web

2.0, which today facilitates bidirectional collaboration between users and sites, the outcome of which is that information is collated and made available to others. The geographical embodiment of these ideas, the GeoWeb, has even been termed GeoWeb 2.0 (see Chapter 11). The two main technologies that have stimulated the Web 2.0 paradigm are Asynchronous Javascript And XML (AJAX) and Application Programming Interfaces (API). AJAX enables the development of Web sites that have a look and feel more akin to desktop applications. They have improved the usability of Web mapping significantly by enabling direct manipulation of map data where user interactions (such as "click and drag") are visualized instantaneously.

Web2.0 enables two-way collaborations between users and Web sites.

AJAX-enabled Web sites compare favorably to first-generation Web GIS applications in which users typically click a pan or zoom control and then wait for the page to reload before visualizing the result. APIs are available from a variety of Web sites, both spatial (e.g., Google Maps, Yahoo! Maps, Microsoft Live Maps) and nonspatial (e.g., Flickr, Facebook) and provide a series of functions to third-party applications. For example, the API of Google Maps provides basic GIS operations, such as the ability to draw shapes, place points, geocode locations, and display these on top of

high-resolution base-map or satellite data. Figure 1.15 illustrates the www.londonprofiler Web site, which uses the Google Maps API to display a variety of high-resolution sociodemographic data about London on top of Google's base-map and satellite data. The construction of such sites using APIs still requires quite a high level of technical proficiency, but is far simpler than learning to install, manage, and configure more traditional Web GIS platforms.

1.5.1.2 The Other Five Components of a GIS

The second piece of the GIS anatomy (Figure 1.16) is the user's hardware, the device that the user interacts with directly in carrying out GIS operations, by typing, pointing, clicking, or speaking, and that returns information by displaying it on the device's screen or generating meaningful sounds. Traditionally, this device sat on an office desktop, but today's user has much more freedom because GIS functions can be delivered through laptops, PDAs, in-vehicle devices, and even cellular telephones. Section 11.3 discusses the currently available technologies in greater detail. In the language of the network, the user's device is the *client*, connected through the network to a *server* that is probably handling many other user clients simultaneously. The client may be *thick*, if it performs a large part of the work locally, or *thin* if it does little more than link the user to the server. A high-specification PC or Macintosh is an instance of a thick client, with

Figure 1.15 A Web2.0 GIS implementation: the Londonprofiler site, showing a mashup of social deprivation data against a background of satellite data. (Image courtesy of Maurizio Gibin)

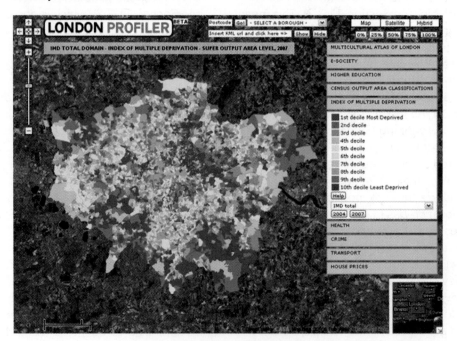

Six parts of a GIS

People · Software · Network · Data · Hardware · Procedures

Figure 1.16 The six component parts of a GIS.

powerful local capabilities, while devices attached to TVs that offer little more than Web-browser capabilities are instances of thin clients.

The third piece of the GIS anatomy is the software that runs locally in the user's machine. This can be as simple as a standard Web browser (e.g., Microsoft Explorer or Safari) if all work is done remotely using assorted digital services offered on large servers. More likely, it is a package bought from one of the GIS vendors, such as Autodesk Inc. (San Rafael, California; www.autodesk.com), Environmental Systems Research Institute, Inc. (ESRI; Redlands, California; www.esri.com), Intergraph Corp. (Huntsville, Alabama; www.ingr.com), or MapInfo Corp. (Troy, New York; www.mapinfo.com). Each vendor offers a range of products, designed for different levels of sophistication, different volumes of data, and different application niches. Idrisi (Clark University, Worcester, Massachusetts, www.clarklabs.org) is an example of a GIS produced and marketed by an academic institution rather than by a commercial vendor. GIS software packages have become quite sophisticated in recent years and can handle all the requirements of standard GIS projects. Equally, however, Internet searches provide a means of identifying reusable software routines or non-GIS sources that facilitate spatial operations, although the provenance and applicability of less known utilities may not be immediately obvious.

Commercial software is very rarely "open source"—meaning that users can modify the code that was used to create it—for a range of reasons, including protecting copyright and maintaining user support. Even public domain software (such as the GRASS GIS suite, or more specialized software such as SWARM or Repast) is often not open source, for similar reasons. This sometimes makes it difficult for users to ascertain how fit GIS software may be for particular GIS applications. Thus, although the Internet makes it possible to assemble information about software, Internet searches do not necessarily enable objective comparison of software functions. Michael de Smith, along with two of the authors of this book, have produced an online guide (www.spatialanalysisonline.com) and book that is intended to raise awareness of the range of software options that are available and the quality of the results that may be produced.

The fourth piece of the anatomy is the database, which consists of a digital representation of selected aspects of some specific area of the Earth's surface or near surface, built to serve some problem-solving or scientific purpose. A database might be built for one major project, such as the location of a new high-voltage power transmission corridor, or it might be continuously maintained, fed by the daily transactions that occur in a major utility company (installation of new underground pipes, creation of new customer accounts, daily service crew activities). It might be as small as a few megabytes (a few million bytes, easily stored on a DVD) or as large as several terabytes (a terabyte is roughly a trillion bytes, stored on a large hard disk or many DVDs). Table 1.1 gives some sense of potential GIS database volumes.

GIS databases can range in size from a megabyte to a petabyte or more.

In addition to these four components—network, hardware, software, and database—a GIS also requires management. An organization must establish procedures, lines of reporting, control points, and other mechanisms for ensuring that its GIS activities meet its needs, stay within budgets, maintain high quality, and generally meet the needs of the organization. These issues are explored in Chapters 17, 18, and 19.

Finally, a GIS is useless without the people who design, program, and maintain it, supply it with data, and interpret its results. The people of GIS will have various skills, depending on the roles they perform. Almost all will have the basic knowledge needed to work with geographic data—knowledge of such topics as data sources, scale and accuracy, and software products—and will also have a network of acquaintances in the GIS community. Most important of all, they will have a capacity for critical spatial thinking, allowing them to filter the message of spatial data through the medium of GIS. The next section outlines some of the roles played by the people of GIS and the industries in which they work.

1.6 The Business of GIS

Very many people play many roles in GIS, from software development to software sales and from teaching about GIS to using its power in everyday activities. Many of these people are able to provide "open-source" software or data that can be used or adapted free of charge because the costs of production are met through volunteer activity or are underwritten by research institutions such as universities. Such activity has mushroomed with the ever wider use of the Internet to disseminate software and data, as well as the innovation and use of search engines, blogs, and social networking sites to spread news about what is available. This makes a huge, yet largely unquantifiable, contribution to economic activity. Yet even in narrower and more clearly bounded terms, the activity of GIS remains very big business. This section looks at the diverse roles that people play in the business of GIS, and is organized by the major areas of human activity associated with it.

1.6.1 The Software Industry

Perhaps the most conspicuous GIS business sector, though by no means the largest in either economic or human terms, is the GIS software industry. There is a diverse collection of software developers with a wide range of backgrounds: Autodesk and Intergraph from CAD, ESRI from GIS, GE Smallworld from electricity, and Pitney Bowes MapInfo from business. This has led to a range of interesting software products. Measured in economic terms, the GIS software industry currently accounts for well over $1 billion in annual sales, although estimates vary, in part because of the difficulty of defining GIS precisely. The software industry employs several thousand programmers, software designers, systems analysts, application specialists, and sales staff, with backgrounds that include computer science, geography, and many other disciplines.

The GIS software industry accounts for over $1 billion in annual sales.

1.6.2 The Data Industry

The acquisition, creation, maintenance, dissemination, and sale of GIS data also account for a huge volume of economic activity. Traditionally, a large proportion of GIS data have been produced centrally by national mapping agencies, such as Great Britain's Ordnance Survey. In most countries the funds needed to support national mapping come from sales of products to customers. Until changes introduced in 2010 (see Section 18.5), sales accounted for almost all of the Ordnance Survey's annual turnover of approximately $200 million. However, federal government policy in the United States requires that prices be set at no more than the cost of reproduction of data; sales are therefore only a small part of the income of the U.S. Geological Survey, the nation's premier civilian mapping agency. The innovation of free-to-view mapping services such as Google Maps (maps.google.com) and Microsoft Bing Maps and Virtual Earth, with its business model based on advertising revenues, is having profound implications for the provision of map data (see Chapter 19). Volunteer-driven, open-source approaches to online cartography such as OpenStreetMap (www.openstreetmap.org) are also revolutionizing online cartography with their novel approach to map production.

In value of annual sales, the GIS data industry is much more significant than the software industry.

In recent years improvements in GIS and related technologies, and reductions in prices, along with various kinds of government stimuli, have led to the rapid growth of a private GIS data industry, serving businesses and service providers (see Section 2.3.3). Data may also be integrated with software in order to offer packaged solutions, as with ESRI's Business Analyst products and services. Private companies are now also licensed to collect high-resolution data using satellites and to sell them to customers—as, for example, with GeoEye (www.geoeye.com) and IKONOS satellite data (see Table 1.4). Other companies collect similar data from aircraft. Still other companies specialize in the production of high-quality data on street networks, a basic requirement of many delivery companies. TeleAtlas (www.teleatlas.com) is an example of this industry, employing over 1800 staff in producing, maintaining, and marketing high-quality street network data worldwide.

As developments in the information economy gather still further momentum, many organizations are becoming focused on delivering integrated business solutions rather than raw or value-added data. The Internet makes it possible for GIS users to access routinely collected data from sites that may be remote from locations where more specialized analysis and interpretation functions are performed. In these circumstances, it is no longer incumbent on an organization to manage either its own data or those that it buys from value-added resellers. For example, ESRI offers a data management service, in which client data are managed and maintained for a range of clients that are at liberty to analyze them in quite separate locations. This may in time lead to greater vertical integration of the software and data industry—for example, ESRI Inc. has developed an e-bis division and acquired its own geodemographic system (called Tapestry) to service a range of business needs. As GIS-based data handling becomes increasingly commonplace, so GIS is finding increasing application in new areas of public-sector service provision, particularly where large amounts of public money are disbursed at the local level—as in policing, education provision, and public health. Many data warehouses and start-up organizations are beginning to develop public-sector data infrastructures, particularly where greater investment in public services is taking place. The GeoWeb (see Chapter 11) is creating fertile environments in which a very wide range of public- and private-sector data sources can be combined, analyzed, and displayed.

1.6.3 The GIServices Industry

Some GISystems are described as having an enterprise-wide scope; that is, they serve many users, with many applications, in multiple geographic locations. Planning, procuring, implementing, administering, and learning how to use such systems require specialist knowledge, and many specialist services providers have come to offer such knowledge and expertise. The types of services offered include consulting

(of both a strategic and an operational nature: see Section 2.3.3), data collection, system customization, specialist hardware and software sales and support (e.g., supply of scanners, plotters, and GPS receivers), training, and value-added applications. GIService providers range from small businesses to the giants such as Accenture, IBM, and Oracle. It is difficult to be precise about the exact size of the GIServices market, but when stripped of multifunction organizations, it is likely to be even larger than the GIS software market.

1.6.4 The GeoWeb Services Industry

The Internet also allows GIS users to access specific functions that are provided by remote sites. For example, MapQuest in the United States (www.mapquest.com), Yellow Pages in the UK (www.yell.com), and the international Google Maps site (maps.google.com) all provide mapping, geocoding, and routing services that are used every day by millions of people to find the best route between two points. By typing a pair of street addresses, the user instructs a routing analysis (see Section 15.4.2) and receives the results in the form of a map and a set of written driving or walking directions (see Figure 1.17).

Figure 1.17 Routing instructions. (Source: Google Maps service)

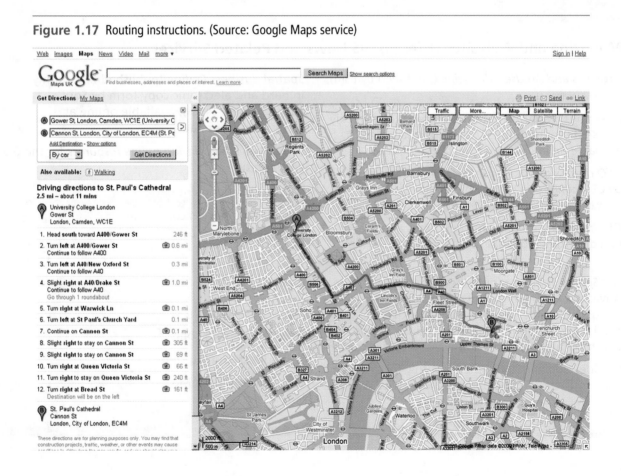

This has several advantages over performing the same analysis on one's own PC: there is no need to buy software to perform the analysis or to buy the necessary data, the data are routinely updated by the GIService provider, and the calculations may even be adjusted in the light of day-to-day variability in traffic conditions. There are clear synergies of interest between GIService providers and organizations providing location-based services (see Chapter 11), and many sites that provide access to raw GIS data also provide GIServices.

GIServices are a rapidly growing form of electronic commerce.

GIServices continue to develop rapidly. In today's world one of the most important commodities is attention: the fraction of a second of attention given to a billboard or sponsored Web page link, or the audience attention that a TV station sells to its advertisers. The value of attention also depends on the degree of fit between the message and the recipient—an advertiser will pay more for the attention of a small number of people if it knows that they include a large proportion of its typical customers. Advertising directed at the individual, based on an individual pro-file, is even more attractive to the advertiser. Direct mail companies have exploited the power of geographic location to target specific audiences for many years, basing their strategies on neighborhood profiles constructed from census records (see Section 2.3.3). But new technologies offer to take this much further. The technology already exists to identify the buying habits of a customer who stops at a gas pump and uses a credit card, and to direct targeted advertising through a TV screen at the pump. As online advertising comes to take an increasing share of most organizations' advertising budgets, so the extent to which niche geographical markets map onto the virtual communities of social networking sites (such as Facebook and Bebo) is set to become an increasingly important focus of marketing initiatives.

1.6.5 The Publishing Industry

Much smaller, but nevertheless highly influential in the world of GIS, is the publishing industry, with its magazines, books, and journals. Several online portals and print magazines are directed at the GIS community (see Box 1.4). Several scholarly journals serve the GIS community, by publishing new advances in

---

Technical Box (1.4)

Magazine and Web sites Offering GIS News and Related Services

ArcNews and ArcUser Magazine (published by ESRI); see www.esri.com

Directions Magazine (Internet-centered and weekly newsletter publication by directionsmag.com), available online at www.directionsmag.com

GEO:connexion UK Magazine published quarterly by GEO:connexion Ltd., with Web site at www.geoconnexion.com

GEOInformatics published eight times a year by Cmedia Productions BV, with Web site at www.geoinformatics.com

GeoSpatial Solutions (published monthly by Advanstar Communications), with Web site www.geospatial-online.com. The company (GEOTEC media) also publishes *GPSWorld* and *GEOWorld,* which are available online at www.geoplace.com

GIS@development (published monthly for an Asian readership by GIS Development, India), with Web site at www.GISDevelopment.net

Spatial Business Online (published bi-monthly in hard and electronic copy form by South Pacific Science Press), available online at www.gisuser.com.au

Some Web sites offering online resources for the GIS community:

www.gisdevelopment.net

www.geoconnexion.com

www.gis.com

www.giscafe.com

gis.about.com

www.geocomm.com

www.spatialnews.com

www.directionsmag.com

www.opengis.org/press

Some Scholarly Journals Emphasizing GIS Research

Annals of the Association of American Geographers

Applied Spatial Analysis and Planning

Cartography and Geographic Information Science

Cartography—The Journal

Computers and Geosciences

Computers, Environment and Urban Systems

Geographical Analysis

GeoInformatica

International Journal of Geographical Information Science (formerly International Journal of Geographical Information Systems)

ISPRS Journal of Photogrammetry and Remote Sensing

Journal of Geographical Systems

Photogrammetric Engineering and Remote Sensing (PE&RS)

The Photogrammetric Record

Terra Forum

Transactions in GIS

URISA Journal

GIS research. The oldest journal specifically targeted at the community is the *International Journal of Geographical Information Science*, established in 1987. Other still older journals in areas such as cartography and geographic analysis regularly accept GIS articles, and several have changed their names and shifted focus significantly. Box 1.5 presents a list of the journals that emphasize GIS research.

1.6.6 GIS Education

The first courses in GIS were offered in universities in the early 1970s, often as an outgrowth of courses in cartography or remote sensing. Today, thousands of such courses can be found in universities and colleges all over the world. Training courses are offered by the vendors of GIS software, and increasing use is made of the Web in various forms of remote GIS education and training (see Box 1.6).

Often, a distinction is made between education and training in GIS: training in the use of a particular software product is contrasted with education in the fundamental principles of GIS. In many university courses, lectures are used to emphasize fundamental principles, while computer-based laboratory exercises emphasize training. In our view, an education should be for life, and the material learned during an

Sites Offering Web-based Education and Training Programs in GIS

Birkbeck College (University of London) GIScOnline M.Sc. in Geographic Information Science at www.bbk.ac.uk

Curtin University's distance learning programs in geographic information science at www.cage.curtin.edu.au

ESRI's Virtual Campus at training.esri.com

Kingston Centre for GIS, Distance Learning Programme at www.kingston.ac.uk

Pennsylvania State University Certificate Program in Geographic Information Systems at www.worldcampus.psu.edu

UNIGIS International, Postgraduate Courses in GIS at www.unigis.org University of Southern California GIS distance learning certificate program at www.usc.edu

education should be applicable as far into the future as possible. Fundamental principles tend to persist long after software has been replaced with new versions, and the skills learned in running one software package may be of very little value when a new technology arrives. On the other hand, much of the fun and excitement of GIS comes from actually working with it, and fundamental principles can be very dry and dull without hands-on experience.

1.7 GISystems, GIScience, and GIStudies

Geographic information systems are useful tools, helping everyone from scientists to citizens to solve geographic problems. But like many other kinds of tools, such as computers themselves, their use raises questions that are sometimes frustrating and sometimes profound. For example, how does a GIS user know that the results obtained are accurate? What principles might help a GIS user to design better maps? How can location-based services be used to help users to navigate and understand human and natural environments? Some of these questions concern GIS design, and others are about GIS data and methods. Taken together, we can think of them as questions that arise from the use of GIS—that are stimulated by exposure to GIS or to its products. Many of them are addressed in detail at many points in this book, and the book's title emphasizes the importance of both systems and science.

The term *geographic information science* was coined in a paper by Michael Goodchild published in 1992. In it, the author argued that these questions and others like them were important and that their systematic study constituted a science in its own right. Information science studies the fundamental issues arising from the creation, handling, storage, and use of information. Similarly, GIScience should study the fundamental issues arising from geographic information, as a well-defined class of information in general. Other terms have much the same meaning: *geomatics* and *geoinformatics*, *spatial information science*, *geoinformation engineering*. All suggest a scientific approach to the fundamental issues raised by the use of GIS and related technologies, though they all have different roots and emphasize different ways of thinking about problems (specifically geographic or more generally spatial, emphasizing engineering or science, etc.).

GIScience has evolved significantly in recent years. It is now the principle focus of several renamed research journals (see Box 1.5), as well as the focus of the

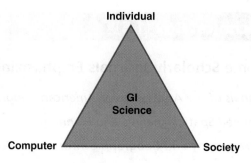

Figure 1.18 The remit of GIScience, according to Project Varenius (www.ncgia.org).

U.S. University Consortium for Geographic Information Science (www.ucgis.org), an organization of roughly 70 research universities that engages in research agenda setting, lobbying for research funding, and related activities. An international conference series on GIScience has been held biannually since 2000 (see www.giscience.org). The Varenius project (www.ncgia.org) provides one disarmingly simple way to view developments in GIScience (Figure 1.18). Here, GIScience is viewed as anchored by three concepts: the individual, the computer, and society. These form the vertices of a triangle, and GISc lies at its core. The various terms that are used to describe GISc activity can be used to populate this triangle. Thus research about the individual is dominated by cognitive science, with its concern for understanding of spatial concepts, learning and reasoning about geographic data, and interaction with the computer. Central to such research is the desire to engineer more readily intelligible user interfaces, in the interest of improved human—computer interaction. Allied to this is the desire to improve geovizualization and render intelligible the results of spatiotemporal analysis. Research about the computer is dominated by issues of representation, the adaptation of new technologies, computation, and visualization. And finally, research about society addresses issues of impact and societal context. Others have developed visions of how higher education should prepare students for success in the variety of professions that rely on geospatial technologies, such as the U.S. University Consortium for Geographic Information Science (Box 1.7). It is possible to imagine how the themes presented in Box 1.7 could be used to populate Figure 1.18 in relation to the three vertices of this triangle.

In some important respects, GIScience is about using the software environment of GIS as an environment in which old problems can be redefined, reshaped, and resolved. Many of the research topics in GIScience are actually much older than GIS. The

The 2006 Geographic Information Science and Technology Higher Education Agenda of the U.S. University Consortium for Geographic Information Science (www.ucgis.org), and Related Chapters in this Book

I. Analytical Methods

1 Academic and analytical origins (Chapter 1)

2 Query operations and query languages (Chapters 14, 8, 7)

3 Geometric measures (Chapter 14)

4 Basic analytical operations (Chapter 14)

5 Basic analytical methods (Chapter 15)

6 Analysis of surfaces (Chapter 15)

7 Spatial statistics (Chapter 15)

8 Geostatistics (Chapter 16)

9 Spatial regression and econometrics (Chapters 15, 16)

10 Data mining (Chapters 15, 16)

11 Network analysis (Chapters 14, 15)

12 Optimization and location-allocation modeling (Chapters 14, 15)

II. Conceptual Foundations

1 Philosophical foundations (Chapters 1, 3, 4)

2 Cognitive and social foundations (Chapter 1)

3 Domains of geographic information (Chapter 3)

4 Elements of geographic information (Chapter 3)

5 Relationships (Chapters 4, 6)

6 Imperfections in geographic information (Chapter 6)

III. Cartography and Visualization

1 History and trends (Chapter 12)

2 Data considerations (Chapters 5, 9, 12)

3 Principles of map design (Chapter 12)

4 Graphic representation techniques (Chapters 11, 12, 13, 14)

5 Map production (Chapter 12)

6 Map use and evaluation (Chapters 12, 6, 3)

IV. Design Aspects

1 The scope of GIS&T system design (Chapters 7, 17, 1)

2 Project definition (Chapters 2, 17, 18)

3 Resource planning (Chapters 18, 19)

4 Database design (Chapters 8, 9, 7)

5 Analysis design (Chapter 8)

6 Application design (Chapters 2, 17, 18)

7 System implementation (Chapters 17, 18, 19)

V. Data Modeling

1 Basic storage and retrieval structures (Chapter 8)

2 Database management systems (Chapter 8)

3 Tessellation data models (Chapters 8, 3)

4 Vector and object data models (Chapters 3, 8)

5 Modeling 3D, uncertain, and temporal phenomena (Chapters 13, 6, 16)

VI. Data Manipulation

1 Representation transformation (Chapters 3, 4, 8)

2 Generalization and aggregation (Chapters 4, 3)

3 Transaction management (Chapters 8, 7)

VII. Geocomputation

1 Emergence of geocomputation (Chapter 16)

2 Computational aspects and neurocomputing (Chapters 15, 16)

3 Cellular Automata (CA) (Chapter 16)

4 Heuristics (Chapters 15, 16)

5 Genetic algorithms (GA) (Chapter 16)

6 Agent-based models (Chapter 16)

7 Simulation modeling (Chapter 16)

8 Uncertainty (Chapter 6)

9 Fuzzy sets (Chapter 16)

▶

More detail on all of these topics, and additional topics presented at more recent UCGIS assemblies, can be found at www.ucgis.org/priorities/research/2002researchagenda.htm. Note that we give some of the advanced topics listed above are discussed in more detail at www.spatialanalysisonline.com.

need for methods of spatial analysis, for example, dates from the first maps, and many methods were developed long before the first GIS appeared on the scene in the mid-1960s. Another way to look at GIScience is to see it as the body of knowledge that GISystems implement and exploit. Map projections (Chapter 5), for example, are part of GIScience and are used and transformed in GISystems. A further area of great importance to GIS is cognitive science, particularly the scientific understanding of how people think about their geographic surroundings. If GISystems are to be easy to use, they must fit with human ideas about such topics as driving directions or how to construct useful and understandable maps. Box 1.8 introduces Peter Gould, a quantitative and behavioral geographer whose various research activities did much to prepare geography for the GIS revolution and the inception of GIScience.

Many roots to GIS can be traced to the spatial analysis tradition in the discipline of geography.

In the 1970s it was easy to define or delimit a geographic information system: it was a single piece of software residing on a single computer. With time, and particularly with the development of the Internet, Web 2.0 and new approaches to software engineering, the old monolithic nature of GIS has been replaced by something much more fluid. GIS is no longer an activity confined to the desktop (Chapter 11). The emphasis throughout this book is on this new vision of GIS, as the set of coordinated parts discussed earlier in Section 1.5. Perhaps the *system* part of GIS is no longer necessary. Certainly the phrase *GIS data* suggests some redundancy, and various people have suggested that we could drop the "S" altogether in favor of GI, for geographic information. GI*Systems* are only one part of the GI whole, which also includes the fundamental issues of GI*Science*. Much of this book is also about GI*Studies*, which can be defined as the systematic study of society's use of geographic information, including its institutions, standards, and procedures. Many of these topics are addressed in the later chapters. Several of the

Peter Gould, Geographer

Peter R. Gould (1932–2000) (Figure 1.19A) taught in the Department of Geography at Pennsylvania State University for 35 years following the award of a Ph.D. from Northwestern University, until his retirement in 1998. He was the author of many influential books, including two that focused on his passion for the subject: *The Geographer at Work* (Routledge, 1985) and *Becoming a Geographer* (Syracuse University Press, 1999). Among many other topics, he undertook pathbreaking work on the use of statistical methods with spatial data; on the nature of the cognitive maps people construct in their minds; on the impacts of the Chernobyl disaster; and on the mechanisms that spread the AIDS epidemic (he published an early column on this subject in *Playboy* magazine).

In the late 1980s Gould and Professor Waldo Tobler of the University of California, Santa Barbara, conducted a small experiment designed to demonstrate the power of alternative modes of georeferencing (see Chapter 5). Using nothing more than the latitude and longitude of Tobler's house, Gould and several co-conspirators mailed a series of letters and postcards from various parts of the world. A postcard mailed at Cape Hatteras in North Carolina, one of several that eventually found their way to Tobler through the mail system, is reproduced in Figure 1.19B. In an era in which many social scientists have concerns about the "surveillance society" and the use of geographic information technologies to enforce it, it is perhaps salutary to reflect that the different georeferencing systems that can so readily be adapted to surveillance have their roots in previous eras

Figure 1.19 (A) Peter Gould, Geographer. (James Collins, Photographer, Penn State University)

Figure 1.19 (B) A postcard from Peter to geographer Waldo Tobler, identifying Tobler's address using precise latitude and longitude coordinates.

UCGIS research topics (Box 1.7) suggest this kind of focus, including *GIS&T and Society* and *Organizational and Institutional Aspects*. In recent years the role of GIS in society—its impacts and its deeper significance—has become the focus of extensive writing in the academic literature, particularly in the discipline of geography, and much of it has been critical of GIS. We explore these critiques in some detail in the next section.

The importance of social context is nicely expressed by Nick Chrisman's definition of GIS, which might also serve as an appropriate final comment on the earlier discussion of definitions:

"The organized activity by which people:

1) measure aspects of geographic phenomena and processes; 2) represent these measure- ments, usually in the form of a computer database, to emphasize spatial themes, entities, and relationships; 3) operate upon these representations to produce more measurements and to discover new relationships by integrating disparate sources; and 4) transform these representations to conform to other frameworks of entities and relationships. These activities reflect the larger context (institutions and cultures) in which these people carry out their work. In turn, the GIS may influence these structures."** (Chrisman 2003, p. 13)

Chrisman's social structures are clearly part of the GIS whole, and as students of GIS we should be aware of the ethical issues raised by the technology we study. Social structures are core to *GIStudies*.

1.8 GIS and the Study of Geography

GIS has always had a special relationship to the academic discipline of geography, as it has to other disciplines that deal with the Earth's surface, including geodesy, landscape architecture, planning, and surveying. This section explores that special relationship and its sometimes tense characteristics, as well as some of the responses of the GIS field to criticism.

In the 1980s GIS technology began to offer a solution to the problems of inadequate computation and limited data handling. However, the quite sensible priorities of vendors at the time might be described as solving the problems of 80% of their customers 80% of the time, and the integration of techniques based on higher-order concepts was a low priority. Today's GIS vendors can probably be credited with solving the problems of at least 90% of their customers 90% of the time, while the opportunity for users to assemble software solutions using the Web relieves dependence on business solutions for both specialist and general-purpose applications. The remit of GIScience remains to diffuse improved, curiosity-driven scientific understanding into the knowledge base of existing successful applications. But the development of improved applications has also been driven to a significant extent by the advent of GPS and other digital data infrastructure initiatives by the late 1990s. New data-handling technologies and new rich sources of digital data open up prospects for refocusing and reinvigorating academic interest in applied scientific problem solving. Although repeat purchases of GIS technology leave the field with a buoyant future in the IT mainstream, there is enduring unease in some academic quarters about GIS applications and their social implications. Much of this unease has been expressed in the form of critiques, notably from geographers, and John Pickles's 1993 edited volume *Ground Truth: The Social Implications of Geographic Information Systems* remains an enduring consolidation of these concerns. Several types of arguments have surfaced:

- The ways in which GIS represents the Earth's surface, and particularly human society, favor certain people, phenomena and perspectives, at the expense of others. For example, GIS databases tend to emphasize homogeneity, partly because of the limited space available on the computer screen and partly because of the costs of more accurate data collection (see Chapters 3, 4, and 8).

Minority views, and the views of individuals, can be submerged in this process, as can information that differs from the official or consensus view. For example, a soil map represents the geographic variation in soils by depicting areas of constant class, separated by sharp boundaries. This is clearly an approximation, and in Chapter 6 we explore the role of uncertainty in GIS. GIS often forces knowledge into forms that are more likely to reflect the view of the majority, or the official view of government, and as a result *marginalizes* the opinions of minorities or the less powerful. Seen from this perspective, even the widely available mapping and analysis functions of Google Maps (maps.google.com) or virtual Earths (www.microsoft.com/virtualearth; earth.google.com) offer a privileged view of the world.

- Although in principle it is possible to use GIS for any purpose, in practice it is often used for purposes that may be ethically questionable or may invade individual privacy, such as surveillance and the gathering of military and industrial intelligence. The technology may appear neutral, but it is always used in a social context. As with the debates over the atomic bomb in the 1940s and 1950s, the scientists who develop and promote the use of GIS surely bear some responsibility for how it is eventually used. The idea that a tool can be inherently neutral, and its developers therefore immune from any ethical debates, is strongly questioned in this literature.

- The very success of GIS is a cause of concern. There are qualms about a field that appears to be led by technology and the marketplace, rather than by human need. There are fears that GIS has become *too* successful in modeling socioeconomic distributions and that as a consequence GIS has become a tool of the "surveillance society." Countering this outlook is the view that greater access to information, especially over the Web, has enlivened debate and has gone some considerable way toward leveling the playing field in terms of data access.

- There are concerns that GIS remains a tool in the hands of the already powerful—notwithstanding the diffusion of technology that has accompanied the plummeting cost of computing and wide adoption of the Internet. As such, it is seen as maintaining the status quo in terms of power structures. By implication, any vision of GIS for all of society is viewed as unattainable. The falling cost of GIS and the greater level of awareness are allowing a much greater proportion of the population to access GIS directly or indirectly via other media channels. Today, the GeoWeb is empowering more and more individuals and groups to present their perspectives to a wide audience.

- There appears to be an underrepresentation of applications of GIS in *critical* research. This academic perspective is centrally concerned with the connections between human agency and particular social structures and contexts. Some of its protagonists maintain that such connections are not amenable to digital representation in whole or in part. A few studies of this type are beginning to be reported that investigate, for example, the compatibility between GIS and feminist epistemologies and politics.

- Some view the association of GIS with the scientific and technical project as fundamentally flawed. More narrowly, there is a view that GIS applications (like spatial analysis before it) are inextricably bound to the philosophy and assumptions of the approach to science known as *logical positivism* (see also the reference to "positive" in Section 1.1). As such, the argument goes, GIS can never be more than a positivist tool and a normative instrument, and cannot enrich other more critical perspectives in geography. This is a criticism not just of GIS, but also of the application of the scientific method to the analysis of social systems.

Many geographers remain suspicious of the use of GIS in geography.

Recent years have seen the popularization of the term *neogeography* to describe developments in Web mapping technology and spatial data infrastructures that have greatly enhanced our abilities to assemble, share, and interact with geographic information online. Allied to this is the increased crowd sourcing

Figure 1.20 A crowd-sourced street map of part of São Paulo, Brazil. (Source: OpenStreetMap.org)

by online communities of *volunteered geographic information* (VGI). Neogeography is founded on the two-way, many-to-many interactions between users and Web sites that have emerged under Web 2.0, as embodied in projects such as Wikimapia (www.wikimapia.org) and OpenStreetMap (www.openstreetmap.org). Today, Wikimapia contains user-generated entries for more places than are available in any official list of place-names, while OpenStreetMap is well on the way to creating a free-to-use global map database through assimilation of digitized satellite photographs with GPS tracks supplied by volunteers (see Figure 1.20). This has converted many new users to the benefits of creating, sharing, and using geographic information, often through ad hoc collectives and interest groups. As such, Web 2.0 simultaneously facilitates crowd sourcing of VGI while making basic GIS functions increasingly accessible to an ever broader community of users. The creation, maintenance, and distribution of databases is no less than a "wikification of GIS." Neogeography provides both a partial response to the earlier social critiques of GIS, in that it brings GIS and some use of spatial data infrastructures to the masses, and a reinforcement of them, in that many Web 2.0 applications present new challenges to citizen privacy and confidentiality. The empowerment of many nonexpert GIS users also brings with it the new challenges of ensuring that tools are used efficiently, effectively, and safely, and reemphasizes that Web 2.0 can never be more than a partial technological solution to the effective deployment of GIS.

We wonder where all this discussion will lead. We have chosen a title for this book that includes both systems and science, and certainly much more of it is about the broader concept of geographic information than about isolated, monolithic software systems per se. We believe strongly that effective users of GIS require some awareness of *all* aspects of geographic information, from the basic principles and techniques to concepts of management and familiarity with applications. We hope this book provides that kind of awareness. On the other hand, we have chosen not to include GIStudies in the title. Although the later chapters of the book address many aspects of the social context of GIS, including issues of privacy, the context to GIStudies is rooted in social theory. GIStudies need the kind of focused attention that we cannot give, and we recommend that students interested in more depth in this area explore the specialized texts listed in the references.

Questions for Further Study

1. Examine the geographic data available for the area within 50 mi (80 km) of either where you live or where you study. Use it to produce a short (2500-word) illustrated profile of either the socioeconomic or the physical environment. (See, for example, www.geodata.gov/gos; www.geographynetwork.com; eu-geoportal.jrc.it; or www.magic.gov.uk.)

2. What are the distinguishing characteristics of the scientific method? Discuss the relevance of each to GIS.

3. We argued in Section 1.4.3.1 that the Internet has dramatically changed GIS. What are the arguments for and against this view?

4. Locate each of the issues identified in Box 1.7 in two triangular "GIScience" diagrams like that shown in Figure 1.18—one for themes that predominantly relate to the developments in science and one for themes that predominantly relate to developments in technology. Give short written reasons for your assignments. Compare the distribution of issues within each of your triangles in order to assess the relative importance of the individual, the computer, and society in the development of GIScience and of GI technologies.

Chrisman, N.R. 2003. *Exploring Geographical Information Systems* (2nd ed.). Hoboken, NJ: Wiley.

Curry, M.R. 1998. *Digital Places: Living with Geographic Information Technologies*. London: Routledge.

De Smith, M., Goodchild, M.F., and Longley, P.A. 2009. *Geospatial Analysis: A Comprehensive Guide to Principles, Techniques and Software Tools* (3rd ed.). Leicester: Troubador and www.spatialanalysisonline.com.

DiBiase, D., deMers, M., Johnson, A., Kemp, K., Taylor Luck, A., Plewe, B., and Wentz, E. (eds.). 2006. *Geographic Information Science & Technology Body of Knowledge*. Washington, DC: Association of American Geographers.

Foresman, T.W. (ed.). 1998. *The History of Geographic Information Systems: Perspectives from the Pioneers*. Upper Saddle River, NJ: Prentice Hall.

Goodchild, M.F. 1992. Geographical information science. *International Journal of Geographical Information Systems* 6: 31–45.

McMaster, E.B., and Usery, E.L. 2004. *A Research Agenda for Geographic Information Science*. Boca Raton, FL: CRC Press.

Pickles, J. 1993. *Ground Truth: The Social Implications of Geographic Information Systems*. New York: Guilford Press.

2 A Gallery of Applications

Fundamentally, GIS is about workable applications. This chapter gives a flavor of the breadth and depth of real-world GIS implementations. It considers:

- How GIS affects our everyday lives.

- How GIS applications have developed and how they embody scientific principles.

- The goals of applied problem solving.

- How GIS can be used to study and solve problems in government, business, transportation and the environment.

LEARNING OBJECTIVES

After studying this chapter you will:

- Grasp the many ways in which we interact with GIS in everyday life.

- Appreciate the range and diversity of GIS applications in environmental and social science.

- Be able to identify many of the scientific assumptions that underpin real-world applications.

- Understand how GIS is applied in the representative application areas of government, business, transportation and the environment.

2.1 Introduction

2.1.1 One Day of Life with GIS

7:00 My alarm goes off . . . The energy to power the alarm comes from the local energy company, which uses a GIS to manage all its assets (e.g., electrical conductors, devices, and structures) so that it can deliver electricity continuously to domestic and commercial customers (Figure 2.1). It also uses GIS to locate renewable energy sources such as wind turbines.

7:05 I jump in the shower . . . The water for the shower is provided by the local water company, which uses a hydraulic model linked to its GIS to predict water usage and ensure that water is always available to its valuable customers (Figure 2.2).

7:35 I open the mail . . . The property tax bill is based on the valuation of our house. The local government department that sent it uses an enterprise-wide GIS for valuation, public service delivery, and management. Outside I can hear that the garbage is being collected by a different part of the enterprise, which uses GIS to route its vehicles and minimize transit distance to local recycling plants and the landfill site. The local government enterprise has pegged increases in property taxes to levels below retail price inflation and has hit targets for recycling waste.

There are also a small number of circulars addressed to me, sometimes called "junk mail" (my partner doesn't get any because he always opts out of mailing lists). A surprising number of our neighbors have vacationed in Richmond, and we made the trip for the first time last year; now the travel company is using its GIS to market similar suggestions. A second circular is a special offer for property insurance, from a firm that uses its GIS to target neighborhoods with low past-claims histories. We get less junk mail than we used to because geodemographic and lifestyles GIS is used to target mailings more precisely, thus reducing waste.

8:00 The other half leaves for work . . . He teaches GIS at one of the city community colleges. As a lecturer in one of the college's most popular classes, he has a full workload and likes to get to work early.

Figure 2.1 An electrical utility application of GIS.

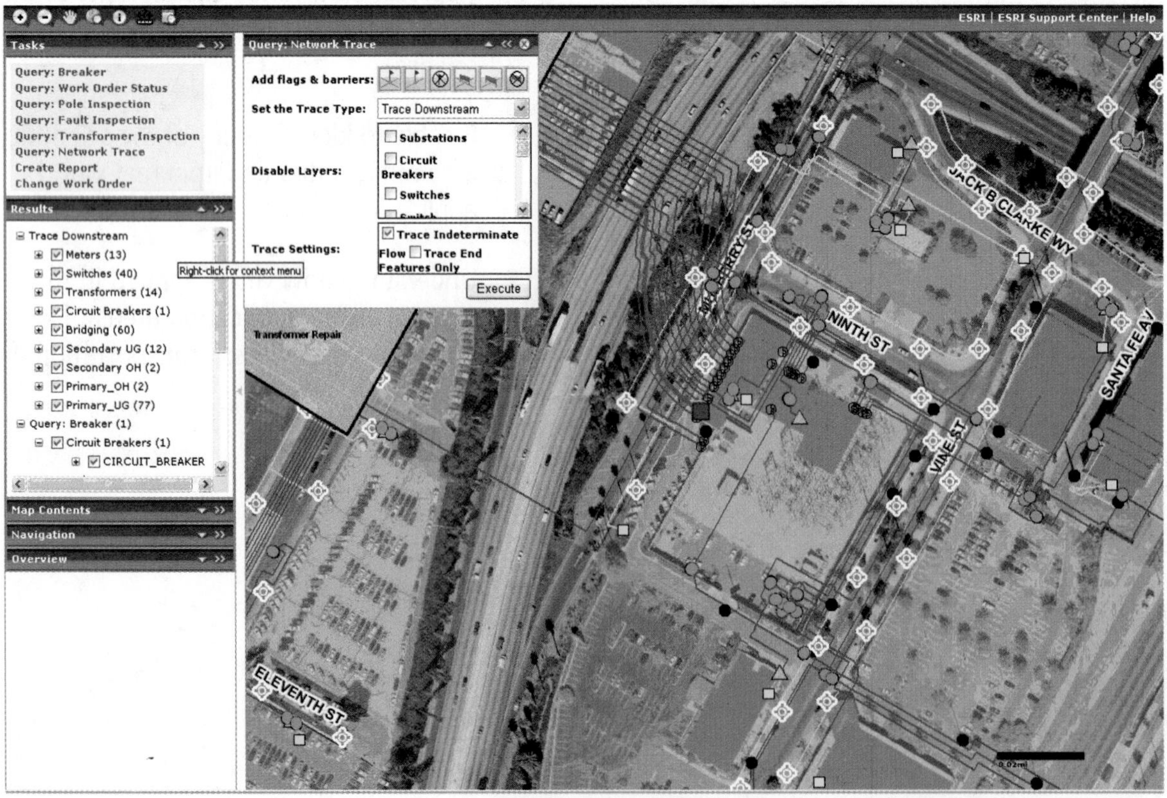

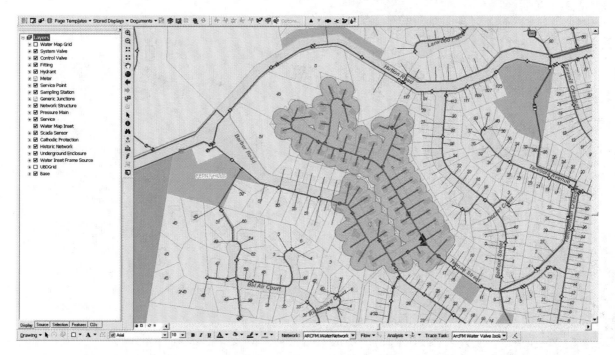

Figure 2.2 Application of a GIS for managing the assets of a water utility.

8:15 I walk the kids to the bus stop . . . Our local authority seems able to predict how many parents have kids of school age, so thankfully we have been spared the horror story of the "school place lottery" that has plagued a neighboring jurisdiction. Our children attend the local middle school that is 3 miles away. The school district administrators use a GIS to optimize the routing of school buses. Introduction of this service enabled the district to cut their annual school busing costs by 16%, and the fuel consumption and time it takes the kids to get to school have also been reduced.

8:45 I catch a train to work . . . A text alert warns me that there are severe delays on the City line, so I walk instead to the District line station that is a 15-minute walk from the house. I have been meaning to get an extra set of keys cut for the house, and so I take the opportunity of calling a specialist locksmith on the way. I find the location of the nearest locksmith using my iPhone, and I use the same device to navigate my way to the store. Arriving at the station, the current locations of trains are displayed on electronic maps on the platforms using a real-time feed from Global Positioning System (GPS) receivers mounted on the trains.

9:15 I read the newspaper on the train . . . The paper for the newspaper comes from sustainable forests managed by a GIS. The forestry information system used by the forest products company indicates which areas are available for logging, the best access routes, and the likely yield (Figure 2.3).

9:30 I arrive at work . . . I am GIS manager for the local City GIS. Today I have meetings to review annual budgets, plan for the next round of hardware and software acquisition, and deal with a nasty copyright infringement claim.

12:00 I grab a sandwich for lunch . . . The price of bread fell in real terms for much of the last decade. In some small part this is because of the increasing use of GIS in precision agriculture. This has allowed real-time mapping of soil nutrients and yield, and means that farmers can apply just the right amount of fertilizer in the right location and at the right time.

6:30 Shop till you drop . . . After work we drive to the downtown stores to try out some of the discount offers that were in the morning mail. The traffic is still heavy, but we use the real-time feeds of traffic conditions of our in-vehicle GPS device to avoid the traffic bottlenecks, and we arrive at a parking lot that is flagged as having plenty of free spaces. We actually bump into a few of our neighbors at the renovated Tesbury Center—I suspect the promotion was targeted by linking a marketing GIS to Tesbury's own store loyalty card data. We get home in time for the scheduled evening delivery of our new bathroom suite, ordered online and delivered by a company that uses GIS to minimize its delivery costs while maximizing convenience to customers like us who cannot wait around all day for deliveries.

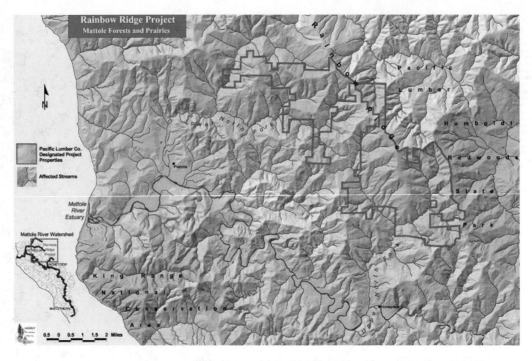

Figure 2.3 Forestry management GIS.

10:30 The kids are in bed . . . I'm on the Internet to try and find a new house . . . We live in a good neighborhood with many similarly articulate, well-educated folk, but it has become noisier since the new distributor road was routed close by. Our resident association mounted a vociferous campaign of protest, and its members filed numerous complaints to the Web site where the draft proposals were posted. But the benefit-cost analysis carried out using the local authority's GIS clearly demonstrated that it was either a bit more noise for us or the physical dissection of a vast swathe of housing elsewhere, and that we would have to grin and bear it. Post-GIS, I guess that narrow-interest NIMBY (Not In My Back Yard) protests don't get such a free run as they once did, and online planning makes proposals available to people who, unlike me, are not able to pop out during their lunch hours to peruse the proposals in the planning office.

So here I am using an online GIS-powered Web site that presents consolidated area-wide maps of what is available on the market. Not only does this show many more properties than any single realtor or estate agent can access, but it can also be easily cross-referenced with official statistics that record sales prices achieved over the past 10 years. I can pinpoint properties that precisely match our criteria, and I can also make sure that we don't pay too much! The site also tells us about crime rates, the quality of schools, and the walking time to the nearest rail lines.

I also use a streetview service to get a feel for the different neighborhoods that are within our price range (see Box 2.1).

GIS is used to improve many of our day-to-day working and living arrangements.

This diary is fictitious of course, but most of the activities described in it are everyday occurrences repeated hundreds and thousands of times around the world. It highlights a number of key things about GIS. GIS

- affects each of us, every day.
- can be used to foster effective short- and long-term decision making.
- has great practical importance.
- can be applied to many socioeconomic and environmental problems.
- encourages public participation in decision making.
- supports mapping, measurement, management, monitoring, and modeling operations.
- generates measurable economic benefits.
- requires key management skills for effective implementation.
- provides a challenging and stimulating educational experience for students.

- can be used as a source of direct income.

- can be combined with other technologies.

- is a dynamic and stimulating area in which to work.

At the same time, some of these examples point to some of the critiques that have been leveled at GIS in recent years (see Section 1.8). Even today, only a small fraction of the world's population has access to information technologies of any kind, let alone high-speed access to the Internet. At the global scale, information technology can exacerbate the differences between developed and less developed nations, across what have been termed *digital divides*, and there is also *digital differentiation* between rich and poor communities within nations. Uses of GIS for marketing often involve practices that are construed as invasion of privacy, since they allow massive databases to be constructed from what many would regard as confidential personal information that may have been unwittingly divulged. It is important that we understand, reflect on, and sometimes act on issues like these while exploring GIS.

2.1.2 Why GIS?

Our day of life with GIS illustrates the unprecedented frequency with which, directly or indirectly, we interact with digital machines. Today, more and more individuals and organizations find themselves using GIS to answer the fundamental question, *where*? This is because of:

- Wider availability of GIS through the Internet, as well as through organization-wide local area networks.

- Increasingly wide availability of low-cost, locationally aware, hand-held devices.

- Reductions in the price of GIS hardware and software because economies of scale are realized by a fast-growing market.

- Greater awareness that decision making has a geographic dimension—in part stimulated by the availability of seamless "virtual Earth" data sources, such as the products of Google and Microsoft.

- Wider realization that geography provides the most effective way of organizing enterprise-wide information systems.

- Greater ease of user interaction, using standard Windows environments, and greater ease of uploading data to contribute to geographic databases.

- Better technology to support applications, specifically in terms of visualization, data management and analysis, and linkage to other software.

- Proliferation of geographically referenced digital data, such as those generated using Global Positioning System (GPS) technology or supplied by value-added resellers (VARs) of data.

- Availability of open-source, free to use software across the Web, in addition to sophisticated packaged GIS solutions that are ready to run out of the box.

- The accumulated experience of applications that *work*.

Geography is the best way of managing enterprise-wide information systems.

2.2 Science, Geography, and Applications

2.2.1 Scientific Questions and GIS Operations

As we saw in Section 1.3, one objective of science is to solve problems that are of real-world concern. Clearly, the range and complexity of scientific principles and techniques that are brought to bear upon problem solving will vary between applications. Within the spatial domain, the goals of applied problem solving include, but are not restricted to:

- Managing spatial operations and inventories as part of an enterprise-wide strategy for information management.

- Rational, effective, and efficient allocation of resources, in accordance with clearly stated criteria—whether, for example, it be physical construction of infrastructure in utilities applications, or scattering fertilizer in precision agriculture.

- Monitoring and understanding observed spatial distributions of attributes—such as variation in soil nutrient concentrations, or the geography of environmental health.

- Understanding the difference that *place* makes—identifying which characteristics are inherently similar between places, and what is distinctive and possibly unique about them. For example, there are regional and local differences in people's surnames (see Box 1.2), and regional variations in voting patterns are the norm in most democracies.

- Understanding of processes in the natural and social environments, such as processes of coastal erosion or river delta deposition in the natural environment, and understanding of changes in residential preferences or store patronage in the social.

- Prescription of strategies for environmental maintenance and conservation, as in national park management.

Understanding and resolving these diverse problems entails a number of general data-handling operations—such as inventory compilation and analysis, mapping, and spatial database management—that may be successfully undertaken using GIS. It also requires appropriate data, and recent years have seen a revolution in the supply and availability of "virtual Earth" data by Google and Microsoft, among others.

GIS is fundamentally about solving real-world problems.

GIS has always been fundamentally an applications-led area of activity. The accumulated experience of applications has led to the borrowing and creation of particular conventions for representing, visualizing, and to some extent analyzing data for particular classes of applications. Over time, some of these conventions have become useful in application areas quite different from those for which they were originally intended, and software suppliers have developed general-purpose routines that may be customized in application-specific ways, as in the way that spatial data are visualized. The way that accumulated experience and borrowed practice become formalized into standard conventions makes GIS essentially an inductive field.

In terms of the definition and remit of GIScience (see Sections 1.3 and 1.7), the conventions used in applications are based on very straightforward concepts. Many data-handling operations are routine and are available as adjuncts to Web mapping products (e.g., **maps.google.com** and **www.microsoft.com/maps/**) or popular word-processing packages (e.g., Microsoft MapPoint: **www.microsoft.com/mappoint**). They work and are very widely used (e.g., see Figure 2.4), yet they may not always be readily adaptable to scientific problem solving in the sense developed in Section 1.3.

2.2.2 GIScience Applications

Early GIS was successful in depicting how the world looks, but shied away from most of the bigger

---

Figure 2.4 Microsoft MapPoint Europe mapping of spreadsheet data of burglary rates in Exeter, England using an adjunct to a standard office software package. (Courtesy D. Ashby. © 1988–2001 Microsoft Corp. and/or its suppliers. All rights reserved. © 2000 Navigation Technologies B.V. and its suppliers. All rights reserved. Selected Road Maps 2000 by AND International Publishers N.V. All rights reserved. © Crown Copyright 2000. All rights reserved. License number 100025500. Additional demographic data courtesy of Experian Limited. © 2004 Experian Limited. All rights reserved.)

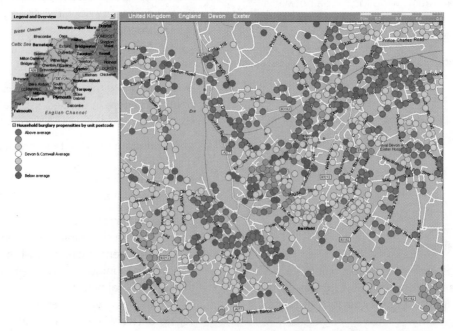

questions concerning how the world works. Today GIScience is developing this extensive experience of applications into a bigger agenda and is embracing fuller conceptual underpinning to successful problem solving. GIS nevertheless remains fundamentally an applications-led technology, and many applications remain modest in both the technology that they utilize and the scientific tasks that they set out to accomplish. There is nothing fundamentally wrong with this, of course, as the most important test of geographic science and technology is whether or not it is useful for exploring and understanding the world around us. Indeed, the broader relevance of geography as a discipline can only be sustained in relation to this simple goal, and no amount of scientific and technological ingenuity can salvage geographic representations of the world that are too inaccurate, expensive, cumbersome, or opaque to reveal anything new about it. In practice, this means that GIS applications must be grounded in sound concepts and theory if they are to resolve any but the most trivial of questions.

GIS applications need to be grounded in sound concepts and theory.

2.3 Representative Application Areas and Their Foundations

2.3.1 Introduction and Overview

There is, quite simply, a huge range of applications of GIS, and indeed several pages of this book could be filled with a list of application areas. They include topographic base mapping, socioeconomic and environmental modeling, global (and interplanetary!) modeling, and education. Applications generally set out to fulfill the five *Ms* of GIS: mapping, measurement, monitoring, modeling, and management.

The five Ms of GIS application are mapping, measurement, monitoring, modeling, and management.

In very general terms, GIS applications may be classified as traditional, developing, and new. The longest-established GIS application fields include military, government, education, and utilities. The past 20 years have seen the wide development of business uses, such as banking and financial services, transportation logistics, real estate, and market analysis. The past

10 years have seen the emergence of applications in small office/home office (SOHO) settings, personal and consumer applications using location-based services, and new applications concerned with planning for disaster, emergency management, security, intelligence, and counterterrorism. This is a somewhat rough-and-ready classification, however, because the applications of some agencies (such as utilities) fall into more than one class. Geography is also increasingly seen as key to enterprise-wide management of information.

A further way to view trends in GIS applications is to examine the diffusion of GIS use. Figure 2.5 shows the classic model of diffusion originally developed by Everett Rogers. Rogers's model divides the adopters of an innovation into five categories:

- Venturesome Innovators—willing to accept risks and sometimes regarded as oddballs.

- Respectable Early Adopters—regarded as opinion formers or role models.

- Deliberate Early Majority—willing to consider adoption only after peers have adopted.

- Skeptical Late Majority—overwhelming pressure from peers needed before adoption occurs.

- Traditional Laggards—people oriented to the past.

In an era of ubiquitous GIS on the Internet and increasingly wide availability of mobile GIS, only laggard organizations have yet to adopt GIS as integral to enterprise-wide management of information, although some areas of application are more comprehensively developed than others. The Innovators who dominated the field in the 1970s were typically based in universities and research organizations. The Early Adopters were the users of the 1980s, many of whom were in government and military establishments. The Early Majority,

---

Figure 2.5 The classic Rogers model of innovation diffusion applied to GIS.

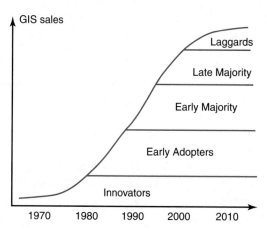

typically in private businesses, came to the fore in the mid-1990s. Further technological developments in the 2000s, linked to changed business models for the creation and dissemination of geographic information, saw Late Majority adoption during that decade.

A wide range of motivations underpins the use of GIS, although it is possible to identify a number of common themes. Applications dealing with day-to-day issues typically focus on very practical concerns such as cost-effectiveness, service provision, system performance, competitive advantage, and database creation, access, and use. Other, more strategic applications are concerned with creating and evaluating scenarios under a range of circumstances. Many applications involve use of GIS by large numbers of people. It is not uncommon for a large government agency, university, or utility to have more than 100 GIS seats, and a significant number have more than 1000. Once GIS applications become established within an organization, usage often spreads widely. Integration of GIS with corporate information system (IS) policy, planning, and systems is an essential prerequisite for success in many organizations.

The scope of these applications is best illustrated with respect to representative application areas, and in the remainder of this chapter we consider:

1. Government and public service (Section 2.3.2)

2. Business and service planning (Section 2.3.3)

3. Logistics and transportation (Section 2.3.4)

4. Environment (Section 2.3.5)

We begin by identifying the range of applications within each of the four domains. Next, we focus on one application within each domain. Each application is chosen, first, for simplicity of exposition but also, second, for the scientific questions that it raises. In this book, we try to relate science and application in two ways. First, we flag the sections elsewhere in the book where the scientific issues raised by the applications are discussed. Second, the applications discussed here, and others like them, provide the illustrative material for our discussion of principles, techniques, analysis, and practices in the other chapters of the book. A recurrent theme in each of the application classes is the importance of geographic location, and hence what is *special* about the handling of georeferenced data (Section 1.1.1). The gallery of applications that we set out here is intended to show how geographic data can provide crucial context to decision making.

2.3.2 Government and Public Service

2.3.2.1 Applications Overview

Government users were among the first to discover the value of GIS. Indeed, the first recognized GIS—the Canada Geographic Information System (CGIS)—was developed for natural resource inventory and management by the Canadian government (see Section 1.4). CGIS was a national system, and, unlike today, in the early days of GIS only national or federal organizations could afford the technology. Today GIS is used at all levels of government from the national to the neighborhood, though government users still comprise the biggest single group of GIS professionals. It is helping to supplement traditional top-down government decision making with bottom-up representation of real communities in government decision making at all levels (Figure 2.6). We will see in later chapters how this deployment of GIS applications is consistent with greater supplementation of top-down deductivism with bottom-up inductivism in science. The diagram also illustrates the importance of increasing online participation by the public in decision making and the supplementation of established official and statutory sources of information with volunteered geographic information (Section 11.2). The importance of spatial variation to government and public service should not be underestimated—70 to 80% of local government work should involve GIS in some way.

> **As GIS has become cheaper, so it has come to be used in government decision making at all levels from the national to the neighborhood.**

Today, local government organizations are acutely aware of the need to improve the quality of their products, processes, and services through ever-increasing efficiency of resource usage (see Section 17.2). Thus GIS is used to inventory resources and infrastructure, plan transportation routing, improve public service delivery, manage land development, and generate revenue by increasing economic activity. Local governments also use GIS in unique ways. Because governments are responsible for the long-term health, safety, and welfare of citizens, wider issues need to be considered, including incorporating public values in decision making, delivering services in a fair and equitable manner, and representing the views of citizens by working with elected officials. Typical GIS applications include monitoring public health risk, managing public housing stock, allocating welfare assistance funds, and tracking crime. Allied to analysis using geodemographics (see Section 2.3.3), they are also used for operational, tactical, and strategic decision making in law enforcement, health care planning, and management of education systems.

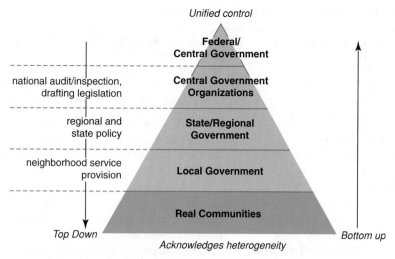

Figure 2.6 The changing use of GIS at different levels of government decision making. "Top-down" edicts from the central government are increasingly supplemented by "bottom-up" initiatives arising from the contributions of local communities.

It is convenient to group local government GIS applications on the basis of their contribution to asset inventory, policy analysis, and strategic modeling/planning. Table 2.1 summarizes GIS applications in this way. These applications can be implemented as centralized GIS or distributed desktop applications. Some will be designed for use by highly trained GIS professionals, while citizens will access others as "front counter," "store-front," or Internet systems. Chapter 8 discusses the different implementation models for GIS.

2.3.2.2 Case Study Application: GIS in Tax Assessment

Tax mapping and assessment is a classic example of the value of GIS in local government. In many countries local government agencies have a mandate to raise revenue from property taxes. The amount of tax payable is partly or wholly determined by the value of taxable land and property. A key part of this process is evaluating the value of land and property fairly to ensure equitable distribution of a community's tax burden. In the United States the task of determining the taxable value of land and property is performed by the Tax Assessor's Office, which is usually a separate local government department. The Valuation Office Agency fulfills a similar role in the UK. The tax department can quickly get overwhelmed with requests for valuation of new properties and protests about existing valuations.

> **The Tax Assessor's Office is often the first home of GIS in local government.**

Essentially, a tax assessor's role is to assign a value to properties using three basic methods, alone or in combination: cost, income, and market. The cost method is based on the replacement cost of the property and the value of the land. The tax assessor must examine data on construction costs and vacant land values. The income method takes into consideration how much income a property would generate if it were rented. This requires details on current market rents, vacancy rates, operating expenses, taxes, insurance, maintenance, and other costs. The market method is the most popular. It compares the property to other recent sales that have a similar location, size, condition, and quality. Until quite recently, access to whole market data was the preserve of the few (almost invariably drawn from government administration), but online government is making housing market data increasingly widely available in many countries (see Box 2.1). Collecting, storing, managing, analyzing, and displaying all this information is a very time-consuming activity, and not surprisingly GIS has had a major impact on the way tax assessors go about their business.

2.3.2.3 Method

Tax assessors, working in a Tax Assessor's Office, are responsible for accurately, uniformly, and fairly judging the value of all taxable properties in their jurisdictions. Details about properties are maintained on a tax assessment roll that includes information such as ownership, address, land and building value, and tax exemptions. The Assessor's Office is also responsible for processing applications for tax abatement, in cases of overvaluation, and exemptions for surviving spouses, veterans, and the elderly. Figure 2.7 shows some aspects of a tax assessment GIS in the State of Washington, USA.

A GIS is used to collect and manage the geographic boundaries and associated information about

Table 2.1 GIS applications in local government. (*GIS and Decision Making in Local Government.*)

	Inventory Applications (locating property information such as ownership and tax assessments by clicking on a map)	Policy Analysis Applications (e.g., number of features per area, proximity to a feature or land use, correlation of demographic features with geological features)	Management/Policy-making Applications (e.g., more efficient routing, modeling alternatives, forecasting future needs, work scheduling)
Economic Development	Location of major businesses and their primary resource demands	Analysis of resource demand by potential local supplier	Informing businesses of availability of local suppliers
Transportation and Services Routing	Identification of sanitation truck routes, capacities and staffing by area; identification of landfill and recycling sites	Analysis of potential capacity strain given development in certain areas; analysis of accident patterns by type of site	Identification of ideal high-density development areas based on criteria such as established transportation capacity
Housing	Inventory of housing stock age, condition, status (public, private, rental, etc.), durability, and demographics	Analysis of public support for housing by geographic area, drive time from low-income areas to needed service facilities,.etc.	Analysis of funding for housing rehabilitation, location of related public facilities; planning for capital investment in housing based on population growth projections
Infrastructure	Inventory of roads, sidewalks, bridges, utilities (locations, names, conditions, foundations, and most recent maintenance)	Analysis of infrastructure conditions by demographic variables such as income and population change	Analysis to schedule maintenance and expansion
Health	Locations of persons with particular health problems	Spatial, time-series analysis of the spread of disease; effects of environmental conditions on disease	Analysis to pinpoint possible sources of disease
Tax Maps	Identification of ownership data by land plot	Analysis of tax revenues by land use within various distances from the city center	Projecting tax revenue change arising from land-use changes
Human Services	Inventory of neighborhoods with multiple social risk indicators; location of existing facilities and services designated to address these risks	Analysis of match between service facilities and human services needs and capacities of nearby residents	Facility siting, public transportation routing, program planning, and place-based social intervention
Law Enforcement	Inventory of location of police stations, crimes, arrests, convicted perpetrators, and victims; plotting police beats and patrol car routing; alarm and security system locations	Analysis of police visibility and presence; officers in relation to density of criminal activity; victim profiles in relation to residential populations; police experience and beat duties	Reallocation of police resources and facilities to areas where they are likely to be most efficient and effective; creation of random routing maps to decrease predictability of police beats

Table 2.1 (*continued*)

	Inventory Applications	Policy Analysis Applications	Management/Policy-making Applications
Land-use Planning	Parcel inventory of zoning areas, floodplains, industrial parks, land uses, trees, green space, etc.	Analysis of percentage of land used in each category, density levels by neighborhoods, threats to residential amenities, proximity to locally unwanted land uses	Evaluation of land-use plan based on demographic characteristics of nearby population (e.g., will a smokestack industry be sited upwind of a respiratory disease hospital?)
Parks and Recreation	Inventory of park holdings/playscapes, trails by type, etc.	Analysis of neighborhood access to parks and recreation opportunities, age-related proximity to relevant playscapes	Modeling population growth projections and potential future recreational needs/playscape uses
Environmental Monitoring	Inventory of environmental hazards in relation to vital resources such as groundwater; layering of nonpoint pollution sources	Analysis of spread rates and cumulative pollution levels; analysis of potential years of life lost in a particular area due to environmental hazards	Modeling potential environmental harm to specific local areas; analysis of place-specific multilayered pollution abatement plans
Emergency Management	Location of key emergency exit routes, their traffic flow capacity and critical danger points (e.g., bridges likely to be destroyed by an earthquake)	Analysis of potential effects of emergencies of various magnitudes on exit routes, traffic flow, etc.	Modeling effect of placing emergency facilities and response capacities in particular locations
Citizen Information/ Geodemographics	Location of persons with specific demographic characteristics such as voting patterns, service usage and preferences, commuting routes, occupations	Analysis of voting characteristics of particular areas	Modeling effect of placing information kiosks at particular locations

(Source: Adapted from J. O'Looney, 200, *Beyond Maps: GIS and Decision Making in Local Government*. Redlands, CA: ESRI Press).

properties. Typically, data associated with properties are held in a Computer Assisted Mass Appraisal (CAMA) system that is used for sale analysis, evaluation, data management, and administration, and for generation of notices to owners. CAMA systems are usually implemented on top of a database management system (DBMS) and can be linked to the parcel database using a common key (see Section 10.3 for further discussion of how this works). The basic tax assessment task involves a geographic database query to locate all sales of similar properties within a predetermined distance of a given property. The property to be valued is first identified in the property database. Next, a geographic query is used to ascertain the values of all comparable properties within a predetermined search radius (typically one mile) of the property. These properties are then displayed on the assessor's screen. The assessor can then compare the characteristics of these properties (lot size, sales price and date of sale, neighborhood status, property improvements, etc.) and value the property.

2.3.2.4 Scientific Foundations: Principles, Techniques, and Analysis

Scientific Foundations Critical to the success of the tax assessment process is a high-quality, up-to-date geographic database that can be linked to a CAMA system. Considerable effort must be expended to design, implement, and maintain the geographic database. Even for a small community of 50,000

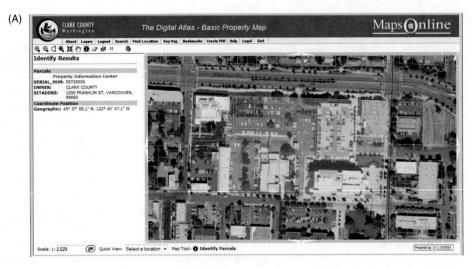

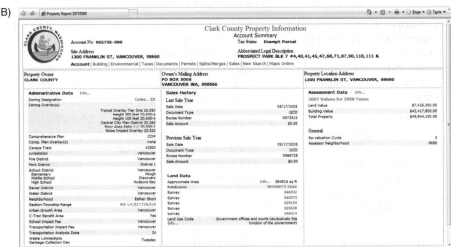

Figure 2.7 Clark County, Washington, U.S. tax assessment GIS: (A) tax map; (B) property attributes, sales history, and assessed market value.

properties, it can take several months to assemble the geographic descriptions of property parcels with their associated attributes. Chapters 9 and 10 explain the processes involved in managing geographic databases such as this. Linking GIS and CAMA systems can be quite straightforward, providing that both systems are based on DBMS technology and use a common identifier to effect linkage between a map feature and a property record. Typically, a unique parcel number (in the United States) or unique property reference number (in the UK) is used.

A high-quality geographic database is essential to tax assessment.

Clearly, the system is dependent on an unambiguous definition of parcels, and common standards about how different characteristics (such as size, age,

and value of improvements) are represented. The GIS can help enforce coding standards and can be used to derive some characteristics automatically in a transparent and an objective fashion. For example, GIS makes it very straightforward to calculate the area of properties using boundary coordinates. Fundamentally, this application, like many others in GIS, depends on an unambiguous and accurate inventory of geographic extent. To be effective it must link this with clear, intelligible, and stable attribute descriptors. These are all core characteristics of scientific investigation, and although the application is driven by results rather than scientific curiosity, it nevertheless follows scientific procedures of controlled comparison.

Principles Tax assessment makes the assumption that, other things being equal, properties close together in space will have similar values. This is an application of Tobler's First Law of Geography (see also Section 3.1). However, it is left to the assessor to identify comparator properties and to weight their relative importance. This seems rather straightforward, but in practice can prove very difficult—particularly where the exact extent of the effects of good and bad neighborhood attributes cannot be precisely delineated. In practice, the value of location in a given neighborhood is often assumed to be uniform (see Section 6.2 for a discussion of uniform zones), and properties of a given construction type are also assumed to be identical. This assumption may be valid in areas where houses were constructed at the same time according to common standards. However, in older areas where infill has been common, properties of a given type vary radically in quality over short distances.

Techniques Tax assessment requires a good database, a plan for system management and administration, and a workflow design. These procedures are set out in Chapters 9, 10, and 17. The alternative of manually sorting paper records, or even tabular data in a CAMA system, is very laborious and time-consuming, and thus the automated approach of GIS is very cost-effective.

Analysis Tax assessment actually uses standard GIS techniques such as proximity analysis, geographic and attribute query, mapping, and reporting. These must be robust and defensible when challenged by individuals seeking reductions in assessments. The basis to such challenge has been strengthened by wide public access to online house sales data. Chapter 14 sets out appropriate procedures, while Chapter 12 describes appropriate conventions for representing properties and neighborhoods cartographically.

2.3.2.5 Generic Scientific Questions Arising from the Application

This is not perhaps the most glamorous application of GIS, but its operational value in tax assessment cannot be overestimated. It requires an up-to-date inventory of properties and information from several sources about sales and sale prices, improvements, and building programs. To help tax assessors understand geographic variations in property characteristics, it is also possible to use GIS for more strategic modeling activities. The many tools in GIS for charting, reporting, mapping, and exploratory data analysis help assessors to understand the variability of property value within their jurisdictions. Some assessors have also built models of property valuations and have clustered properties based on multiple criteria (see Section 16.4). Thus, with the availability of more and more data on the Web, evermore increasingly sophisticated housing market models can be created by anyone in government organizations, property services, business, and even individuals (see Box 2.1). Housing market models help assessors to gain knowledge of the structure of communities and highlight unusually high or low valuations. Once a property database has been created, it becomes a very valuable asset, not just for the tax assessor's department, but also for many other departments in a local government agency. Public works departments may seek to use it to label access points for repairs and meter reading, housing departments may use it to maintain data on property condition, and many other departments may like shared access to a common address list for record keeping and mailings.

A property database is useful for many purposes besides tax assessment.

2.3.2.6 Management and Policy

Tax assessment is a key local government application because it is a direct revenue generator. It is easy to develop a cost-benefit case for this application (Chapter 17), and it can pay for a complete department or corporate GIS implementation quickly (Chapter 18). Tax assessment is a service offered directly to members of the public. As such, the service must be reliable and achieve a quick turnaround (usually within one week). It is quite common for citizens to question the assessed value for their property, since this is the principal determinant of the amount of tax they will have to pay. A tax assessor must, therefore, be able to justify the method and data used to determine property values. A GIS is a great help and often convinces people of the objectivity involved (sometimes overimpressing people that it is totally scientific). As such, GIS is an important tool for efficiency and equitable local government.

2.3.3 Business and Service Planning

2.3.3.1 Applications Overview

Business and service planning (sometimes called retailing) applications focus on the use of geographic data to provide operational, tactical, and strategic context to decisions that involve the fundamental question, *where*? *Geodemographics* is a shorthand term for composite indicators of consumer behavior that are available at the small-area level (e.g., census output area, or postal zone). The most important data sources for creating geodemographic indicators are censuses of population or similar official sources that

Housing Market Search: Google Streetview, Nestoria, Zillow, and Zoopla

Housing market search (finding out where to buy or rent a home) is an inherently geographic problem. Until quite recently, the prospective purchaser or tenant would approach one or more realtors (estate agents) in order to ascertain which properties were available, would typically view a small number of properties, and would then select from the available alternatives. Choice was made based on direct experience (property viewings) of a small number of options, but largely in ignorance of recent transactions and of what might be available elsewhere in the market.

Consolidator sites such as Zillow in the United States (www.zillow.com), Nestoria in the UK, Germany, Italy, and Spain (www.nestoria.com), and Zoopla in the UK (www.zoopla.com) have changed all of this (Figure 2.8). In the early 2000s, realtors devoted increasing amounts of energy to online advertising, aware that the medium allowed them to market more properties to more clients, at lower cost than is possible through office footfall and accompanied viewings alone. At the same time, savvy house purchasers have become aware that sales data are now available online, and so they are much more aware of local housing market conditions than was previously the case.

Sites such as Nestoria work by stitching together the digital listings of all properties that are for sale or

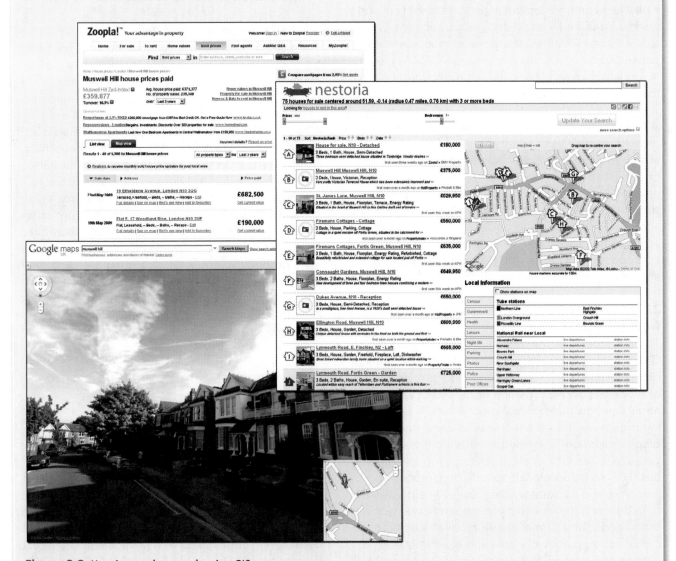

Figure 2.8 Housing market search using GIS.

rent, and use zip (post-) codes or address gazetteers in order to precisely georeference all properties on the market. The site uses feeds of Google Maps data in order to display all properties available within a search space that is defined by user-specified neighborhood names or zip codes. The user can adjust the search space using the standard Google Maps pan and zoom functions, as well as the price range and property attributes that are deemed most relevant.

Sites such as Zoopla and Zillow use actual sales data in the same way as local tax assessors in order to model the values of all properties, and not just those that are currently for sale. These models are usually commercial secrets or "black-box" predictors, but it is likely that modeled prices are derived by inflating or deflating data on prices paid for properties in the (sometimes distant) past, or using area-wide government or bank statistics on house prices paid. These are rather simplistic because in practice house prices depend on a host of other property attributes (such as state of repair) and neighborhood effects (termed "externalities" in the academic literature). The Zoopla site does, however, allow users to refine price estimates by entering data on property condition as well as other attributes that may be unavailable through government statistics.

Such Web sites also include data on local amenities such as schools, various leisure facilities and public transport options, as well as crime rates and

other environmental services. A recent development is that the www.londonprofiler.com site in turn takes real-time RSS (Really Simple Sundication—a Web feed format) feeds from Nestoria, making it possible for the user to assess house prices in relation to other geographic variables, such as the paths of rail lines, local health indicators, and patterns of educational attainment. A research challenge for the future is to formally incorporate as many salient (individual property and neighborhood) factors into these models as possible, without making the housing market model overly complicated—in other words, the quest is for a parsimonious model of house prices.

Take a look at www.zillow.com to check out the discrepancies between the actual recent house sales prices and the Zillow model estimates. Do you discern any systematic discrepancies between them at the state or local level?

In carrying out online housing market searches, users are not restricted to sales or even just numeric data. Figure 2.8 illustrates how housing market search returns and neighborhood information from Nestoria may be viewed alongside house price predictions from Zoopla, and the detail of the street environment may be viewed through Google Streetview. Together, these sites illustrate how direct experience of housing market search is increasingly supplemented by online sources that provide indirect experience of the market. We discuss the issues that this raises in Chapter 3.

are available at the small-area level. Data-reduction techniques such as cluster analysis are used to identify similarities in patterning of population and neighborhood characteristics (such as population age, the sizes of properties, or employment structure) and to assign a shorthand label that summarizes these population characteristics and neighborhood attributes. These aggregations (sometimes called subgroups) are themselves then aggregated into smaller numbers of categories in order to help managers generalize and understand the typology. Table 2.2 presents the 7 Supergroups and 21 Groups that make up the UK Office for National Statistics Output Area Classification (OAC), and Figure 2.9 shows the geodemographic composition of the City of Nottingham at the Supergroup level of disaggregation. Even at this coarse classification level, students of geography will recognize

many aspects of the patterning of urban social areas (such as concentrations of blue-collar and ethnic minority areas in the inner city and prospering suburbs around the periphery).

The OAC geodemographic system is based on census data alone and is freely available for download (from www.statistics.gov.uk/about/methodology_by_theme/area_classification/default.asp). Other systems are available in the UK and in most developed markets worldwide. Procurement of commercial systems is usually very costly, in part because of the addition of other commercial datasets to the census mix. Some of these relate to issues such as household income and finance, or online behavior, or purchasing behavior with respect to big-ticket items such as overseas holidays. Public services may also use their own data in order to create geodemographic classifications for

Table 2.2 The typology of a geodemographic system: The UK Output Area Classification.

Supergroup	Group
1 Blue Collar Communities	1a Terraced Blue Collar 1b Younger Blue Collar 1c Older Blue Collar
2 City Living	2a Transient Communities 2b Settled in the City
3 Countryside	3a Village Life 3b Agricultural 0.0 3c Accessible Countryside
4 Prospering Suburbs	4a Prospering Younger Families 4b Prospering Older Families 4c Prospering Semis 4d Thriving Suburbs
5 Constrained by Circumstances	5a Senior Communities 5b Older Workers 5c Public Housing
6 Typical Traits	6a Settled Households 6b Least Divergent 6c Young Families in Terraced Homes 6d Aspiring Households
7 Multicultural	7a Asian Communities 7b Afro-Caribbean Communities

use in particular applications such as policing, health care, or managing the education sector, or may examine the match between geodemographic indicators and their own indicators of hardship or education attainment (Figure 2.10)

Geodemographic classifications are frequently used in business applications to begin to understand geographic variations in the purchasing behavior of different customer types. The term *market area analysis* describes the activity of assessing the distribution of retail outlets relative to the greatest concentrations of potential customers. The approach is increasingly being adapted to improving public service planning, in areas such as health, education, and law enforcement.

Geodemographic data are the basis for much market area analysis.

The tools of business applications typically range from simple desktop mapping to sophisticated decision support systems. Tools are used to analyze and inform the range of an organization's *operational*, *tactical*, and *strategic* functions. These tools may be part of standard GIS software, or they may be developed in-house by the organization, or they may be purchased (with or without accompanying data) as a business solution product. We noted in Section 1.1 that operational functions concern the day-to-day processing of routine transactions and inventory analysis in an organization, such as stock management. Tactical functions require the allocation of resources to address specific (usually short-term) problems, such as store sales promotions. Strategic functions contribute to the organization's longer term goals and mission, and entail problems such as opening new stores or rationalizing existing store networks.

Early business applications were simply concerned with mapping spatially referenced data, as a general descriptive indicator of the retail environment. This remains the first stage in most business applications and in itself adds an important dimension to analysis of organizational function. More recently, decision support tools based on geodemographics have acquired mainstream research and development roles for business GIS applications.

Some of the operational roles of GIS in business are discussed under the heading of logistics applications in Section 2.3.4. These include stock flow management systems and distribution network management, the specifics of which vary from industry sector to sector. Geodemographic analysis is an important operational tool in market area analysis, where it is used to plan marketing campaigns. Each of these applications can be described as assessing the circumstances of an organization. The most obvious strategic application concerns the spatial *expansion* of a new entrant across a retail market. Expansion in a market poses fundamental spatial problems—such as whether to expand through *contagious diffusion* across space or *hierarchical diffusion* down a settlement structure, or whether to pursue some combination of the two (Figure 2.11). Many organizations periodically experience spatial *consolidation* and branch *rationalization*. Consolidation and rationalization may occur: (a) when two organizations with overlapping networks merge; (b) in response to competitive threat; or (c) in response to changes in the retail environment. Changes in the retail environment may be short term and cyclic, as in the response to the recession phase of business cycles, or structural, as with the rationalization of clearing bank branches following the wide adoption of Internet or telephone banking (see Section 18.5.5). Still other organizations undergo spatial *restructuring*, as in the market repositioning of bank branches to supply a wider range of more profitable financial services. Spatial restructuring is often the consequence of technological change. For example, recent years have seen the development of "clicks and mortar" strategies by retailers, whereby the customer

Figure 2.9 The geodemographic structure of Nottingham, according to the OAC.

uses an Internet gateway to identify stock availability at stores convenient to the customer for pickup—a service that may be used in association with location-based services (Section 11.3.2). A final type of strategic operation involves *distribution* of goods and services, as in the case of so-called e-tailers, who use the Internet for merchandizing, but must create or buy into viable distribution networks. These various strategic operations require a range of spatial analytic tools and data types, and entail a move from what-is visualization to what-if forecasts and predictions.

2.3.3.2 Case Study Application: Hierarchical Diffusion and Convenience Shopping

Tesco is, by some margin, the most successful grocery (food) retailer in the UK and has used its knowledge of the home market to launch successful initiatives in Asia and the developing markets of Eastern Europe (see Figure 1.5). Achieving real sales growth in its core business of groceries is difficult, particularly in view of the strict planning regime in its home market, which prevents widespread development of new stores, and legislation to prevent the emergence, through acquisitions, of local spatial monopolies of supply.

One way in which Tesco has adapted to these constraints on growth in the domestic market is through strategic diversification into consumer durables, clothing, financial services, and mobile telephony (cell phones), particularly in its largest (high-order, colored dark blue in Figure 2.11) stores and online. A second driver to growth has been a successful store loyalty card program, which rewards members with money-off coupons or leisure experiences according to their in-store and online purchases. This program generates *lifestyles* data as a very useful by-product, which enables the retailer to identify consumption profiles of its customers, as well as the neighborhoods from which they are drawn. This enables the company to identify, for example, whether customers are value driven and stores should be focused on budget food offerings, or whether customers are principally motivated by quality and might be encouraged to purchase goods from the company's Finest range. This is a very powerful marketing tool, though unlike geodemographic discriminators these data tell the company little about those households that are not their customers, or the products that their own customers buy elsewhere.

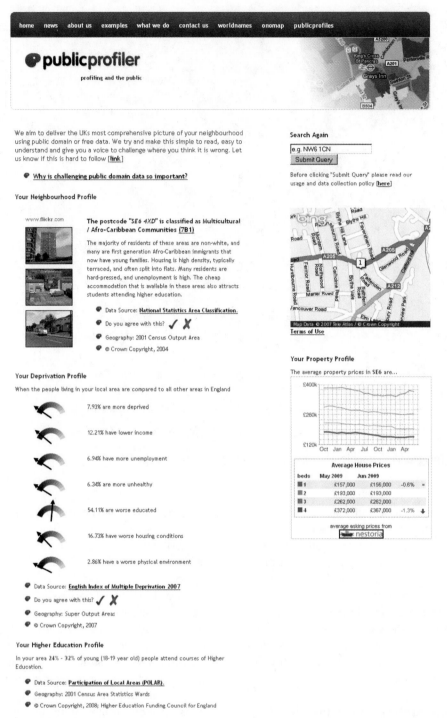

Figure 2.10 Neighborhood profiling using ancillary data sources to the OAC classification (www.publicprofiler.org).

A third driver to growth entails the building or acquisition of much smaller neighborhood stores (low-order, colored light blue in Figure 2.11). These provide a local community service and are readily assimilated into the existing retail landscape; thus they are much easier to create within the constraints of the planning system. Figure 2.12A shows one such Express format store. Such stores are planned by Tesco's in-house store location team using GE Smallworld GIS. Figure 2.12B shows an Express store location (labeled T3) in Bournemouth, UK, in relation to the edge of the town and the locations of five competitor chains.

GIS can be used to predict the success of a retailer in penetrating a local market area.

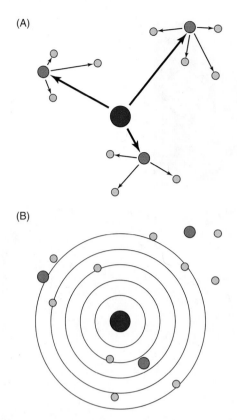

Figure 2.11 (A) Hierarchical and (B) contagious spatial diffusion.

2.3.3.3 Method

The location of the new store is in a suburban residential neighborhood, and as such it was anticipated that its customer base would be mainly local—that is, resident within a 1-km radius of the store. A budget was allocated for promoting the new establishment, in order to encourage repeat patronage. An established means of promoting new stores is through leaflet drops or enclosures with free local newspapers. However, such tactical interventions are limited by the coarseness of distribution networks—most organizations that deliver circulars will only undertake deliveries for complete postal sectors (typically 20,000 population size), and so this is likely to be a crude and wasteful targeting medium.

A second strategy for promoting new stores is to use the GIS to identify all of the households resident within a 1-km radius of the store. Each UK unit postcode (roughly equivalent to a US Zip+4 code, and typically comprising 18 to 22 addresses) is assigned the grid reference of the first mail delivery point on the mail deliverer's walk. Thus one of the quickest ways of identifying the relevant addresses entails plotting the unit postcode addresses and selecting those that lie within the search radius. Matching the unit postcodes with the full postcode address file (PAF) then suggests that there are approximately 3,236 households resident

in the search area. Each of these addresses might then be mailed a circular, thus eliminating the largely wasteful activity of contacting the other 16,800 households in the postal sector that were unlikely to use the store.

Yet even this tactic can be refined. Sending the same packet of money-off coupons to all 3,236 households assumes that each has identical disposable incomes and consumption habits. There may be little point, for example, in incentivizing a domestic-beer drinker to buy premium champagne, or vice versa. Thus it makes sense to overlay the pattern of geodemographic profiles onto the target area, in order to tailor the coupon offerings to the differing consumption patterns of "blue-collar communities" versus those classified as belonging to "prospering suburbs," for example. A final stage of refinement can be developed for this analysis. Using its lifestyles (storecard) data, Tesco can identify those households that already prefer to use the chain, despite the previous nonavailability of a local convenience store. Some of these customers will use Tesco for their main weekly shop, but may top up with convenience or perishable goods (such as bread, milk, or cut flowers) from a competitor. Such households might be offered stronger incentives to purchase particular ranges of goods from the new store, without the wasteful cannibalizing activity of offering coupons toward purchases that are already made from Tesco in the weekly shop.

2.3.3.4 Scientific Foundations: Geographic Principles, Techniques, and Analysis

The following assumptions and organizing principles are inherent to this case study.

Scientific Foundations Fundamental to the application is the assumption that the closer a customer lives to the store, the more likely he or she is to patronize it. This is formalized as Tobler's First Law of Geography in Section 3.1 and is accommodated into our *representations* as a distance decay effect. The *nature* (Chapter 4) of such distance decay effects does not have to be linear; in Section 4.5 we will introduce a range of *nonlinear* effects. The science of geodemographic profiling can be stated succinctly as "birds of a feather flock together"—that is, the differences in the observed social and economic characteristics of residents *between* neighborhoods are greater than the differences observed *within* them.

The use of small-area geodemographic profiles to mix the coupon incentives that might be sent to prospective customers assumes that each potential customer is equally and utterly typical of the postcode in which he or she resides. The individual resident *in* an area is thus assigned the characteristics *of* the area.

(A)

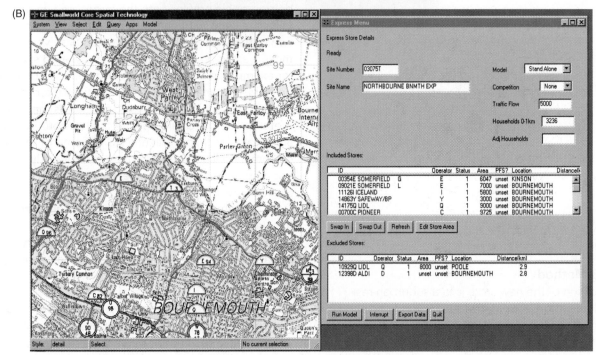

(B)

Figure 2.12 (A) The format and (B) the location of a new Tesco Express store. (Courtesy Alex Singleton)

In practice, of course, individuals within households have different characteristics, as do households within streets, zones, and any other aggregation. The practice of confounding characteristics of areas with individuals resident in them is known as committing the *ecological fallacy*. The term does not refer to any branch of biology, but shares with ecology a primary concern with describing the *linkage* of living organisms (individuals) to their geographic surroundings. This is inevitable in most socioeconomic GIS applications because data that relate to the potentially sensitive characteristics of individuals must be kept confidential (see the point about GIS and the surveillance society in Section 1.8). While few could take offense at any error in mistargeting money-off coupons as in this example, ecological analysis has the potential to cause distinctly unethical outcomes if individuals are penalized because of where they live— for example, should individuals find it difficult to gain credit because of the credit histories of their neighborhoods. Such discriminatory activity is usually prevented by industry codes of conduct or even legislation.

The use of lifestyles data culled from store loyalty card records enables individuals to be targeted precisely, but such individuals might well be geographically or socially *unrepresentative* of the population at large or the market as a whole (see Section 4.4). More generally, geography is a science that has very few natural units of analysis. What, for example, is the natural unit for measuring a soil profile? In socioeconomic applications, even if we have disaggregate data we might remain uncertain as to whether we should consider the individual or the household as the basic unit of analysis. In some households one individual always makes the important decisions, while in others this is a shared responsibility. (We return to this issue in our discussion of *uncertainty* in Chapter 6.)

The use of lifestyle data from store loyalty programs allows the retailer to enrich geodemographic profiling with information about his or her own customers. This is a cutting-edge marketing activity, but one in which there is plenty of scope for relevant research that can take what-is information about existing customer characteristics and use it to conduct what-if analysis of behavior, given a different constellation of retail outlets. (We return to the issue of defining appropriate *predictor variables* in Section 16.1 and *measurement error* in our discussion of uncertainty in Chapter 6). More fundamentally still, is it acceptable (predictively and ethically) to represent the behavior of consumers using any measurable socioeconomic variables? And can there ever be a "best," but subjectively defined cocktail of predictor variables? And if there is, is it ethically acceptable to use any commercial cocktail of variables, the weighting of which is a closely guarded secret? Can the use of black-box commercial solutions ever be described as scientific?

Principles The definition of the primary market area that is to receive incentives assumes that a *linear* radial distance measure is intrinsically meaningful. In practice, there are a number of severe shortcomings in using this measure. Perhaps the most obvious shortcoming is that *spatial structure* will distort the radial measure of market area—the market is likely to extend further along the more important travel arteries, for example, and will be restricted by physical obstacles such as blocked-off streets and rivers, and by traffic-management devices such as stop lights. Some of the impediments may principally be perceived rather than real; it may be that residents of West Parley (Figure 2.12B) would never think of going into north Bournemouth to shop, for example, and that the store's customers will remain overwhelmingly drawn from the area south of the store. Such perceptions of *psychological distance* can be very important, yet difficult to accommodate in representations. We return to the issue of appropriate distance metrics in Section 14.3.1 and to issues of *cognitive engineering* in Section 13.2.1.

Techniques The assignment of unit postcode coordinates to the catchment zone is performed through a procedure known as *point in polygon* analysis, which is considered in Section 14.2.3. The analysis as described here assumes that the principles that underpin consumer behavior in Bournemouth, UK, are essentially the same as those operating anywhere else on Planet Earth. There is no attempt to accommodate regional and local factors such as adjusting the attenuating effect of distance (see above) to accommodate the different distances people are prepared

to travel to find a convenience store (e.g., as between an urban and a rural area); and adjusting the likely attractiveness of the outlet to take account of ease of access, availability of parking at different times of day, or a range of qualitative factors such as layout or branding. Many spatial techniques are now available for making the general properties of spatial analysis more *sensitive to context*.

Analysis Our stripped-down account of the store location problem has not considered the *competition* from the other stores shown in Figure 2.12B—despite that fact that all residents almost certainly already purchase convenience goods somewhere! Our description does, however, address the phenomenon of *cannibalizing*, whereby new outlets of a chain poach customers from its existing sites. In practice, both of these issues may be addressed through analysis of the *spatial interactions* between stores. Although this issue is beyond the scope of this book, the tradition of *spatial interaction modeling* is ideally suited to the problems of defining realistic catchment areas and estimating store revenues. A range of analytic solutions can be devised in order to accommodate the fact that store catchments often overlap.

2.3.3.5 Generic Scientific Questions Arising from the Application

A dynamic retail sector is fundamental to the functioning of all advanced economies, and many investments in location are so huge that they cannot possibly be left to chance. Doing nothing is simply not an option. Intuition tells us that the effects of distance to outlet, and the organization of existing outlets in the retail hierarchy, *must* have some kind of impact on patterns of store patronage. But in intensely competitive consumer-led markets, the important question is *how much* impact?

> **Human decision making is complex, but predicting even a small part of it can be very important to a retailer.**

Consumers are sophisticated beings, and their shopping behavior is often complex. Understanding local patterns of convenience shopping is perhaps quite straightforward, when compared with other retail decisions that involve stores that have a wider range of attributes, in terms of floor space, range and quality of goods and services, price, and customer services offered. Different consumer groups find different retailer attributes attractive. Hence it is the mix of *individuals* with particular characteristics that largely determines the likely store turnover of a particular location. Our example illustrates the kinds of simplifying assumptions that we may choose to make using the best available data in order to represent consumer characteristics and store attributes. However, it is important to remember that even blunt-edged tools can increase the effectiveness of operational and strategic R&D (research and development) activities manyfold. An untargeted leafleting campaign might typically achieve a 1% hit rate, whereas a campaign informed by even quite rudimentary market area analysis might conceivably achieve a rate that is five times higher. The pessimist might dwell on the 95% failure rate that a supposedly scientific approach entails, yet the optimist should be more than happy with the fivefold increase in the efficiency of use of the marketing budget!

2.3.3.6 Management and Policy

The geographic development of retail and business organizations has sometimes taken place in a haphazard way. However, the competitive pressures of today's markets require an understanding of branch location networks, as well as their abilities to anticipate and respond to threats from new entrants. The role of Internet technologies in the development of e-tailing (electronic retailing) is important too, and these introduce further spatial problems to retailing—for example, in developing an understanding of the geographies of engagement with new information and communications technologies, or in devising the logistics of delivering goods and services ordered online (see Section 2.3.4).

For the time being, the clustered world of geodemographics remains preoccupied with mapping in physical space, yet more and more of the public spend more and more of their time interacting, transacting, and collaborating online. It remains to be seen whether, and if so to what extent, online social networking has lasting implications for the development of geodemographic indicators and their use in business and service planning.

> **Some similarities between individuals are more apparent in online social networks than in geographic neighborhoods.**

Simple mapping packages alone provide insufficient scientific grounding to resolve retail location problems. Thus a range of *GIServices* have been developed—some in-house by large retail corporations (such as Tesco), some by software vendors that provide analytical and data services to retailers, and some by specialist consultancy services. There is an ongoing debate as to which of these solutions is most appropriate to retail applications. The resolu-

tion of this debate lies in understanding the nature of particular organizations, their range of goods and services, and the priority that organizations assign to operational, tactical, and strategic concerns.

Finally, the case study in this section has focused on the supply of conventional goods by a private-sector organization, but many of the same principles and issues characterize the effective provision of public services by local and central government. Yet the use of geodemographics in public service delivery is different in three important respects:

- It is naïve to suggest that the particular cocktails of variables that experience shows make good geodemographic indicators for the private sector also provide a magic bullet for the public. By definition, we consume public goods (such as policing, education, or the fire service) collectively, and no one has yet suggested clear reasons why the variables that predict, for example, propensity to take a vacation to an exotic location should prove invaluable in understanding fear of crime.

- Geodemographic thematic maps such as Figure 2.9 are based on the clustering of social similarities, usually with little attempt to incorporate explicit neighborhood size and scale measures: these are likely to be important when considering local preferences for school services or attitudes to local policing, for example.

- Many commercial systems are sold to private-sector organizations as black boxes that have a track record of usefulness in selling private goods. The organization will never know precisely what ingredients were used to create the summary geodemographic indicator, or how the raw data were weighted. This information may be trivial and unimportant in marketing, and may actually get in the way of management decision making, but a much clearer audit trail is required in public service delivery. Citizens and taxpayers are all stakeholders in public services, and public accountability requires that resource allocation decisions are made in a transparent manner, using methods and techniques that are scientifically reproducible by any interested party.

2.3.4 Logistics and Transportation

2.3.4.1 Applications Overview

Knowing where things are can be of enormous importance for the fields of logistics and transportation, which deal with the movement of goods from one place to another, and the infrastructure (highways,

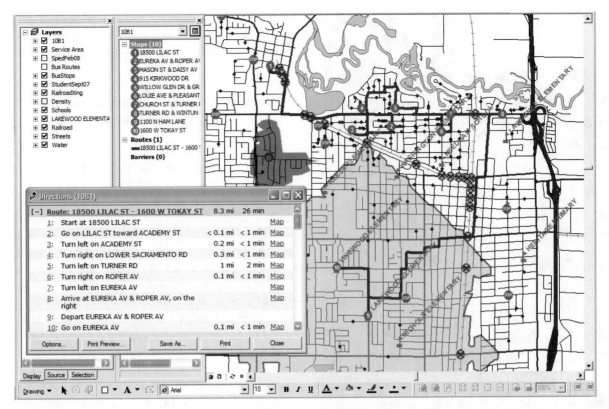

Figure 2.13 Use of GIS for school bus routing, showing school pupil home locations and elementary schools to be serviced.

railroads, canals) that moves them. Logistics are similarly important when moving people between locations, as with school bus routing (Figure 2.13). Highway authorities need to decide what new routes are needed and where to build them, and later need to keep track of highway conditions. Logistics companies (e.g., parcel delivery companies, shipping companies) need to organize their operations, deciding where to place their central sorting warehouses and the facilities that transfer goods from one mode to another (e.g., from truck to ship), how to route parcels from origins to destinations, and how to route delivery trucks. Transit authorities need to plan routes and schedules, to keep track of vehicles and to deal with incidents that delay them, and to provide information on the system to the traveling public. All of these fields employ GIS, in a mixture of operational, tactical, and strategic applications.

The field of logistics addresses the shipping and transportation of goods.

Each of these applications has two parts: the static part that deals with the fixed infrastructure and the dynamic part that deals with the vehicles, goods, and people that move on the static part. Of course, not even a highway network is truly static, since highways are often rebuilt, new highways are added, and some highways are even moved. But the minute-to-minute timescale of vehicle movement is sharply different from the year-to-year changes in the infrastructure that ultimately make movement possible. Historically, GIS has been easier to apply to the static part, but recent developments in the technology are making it much more powerful as a tool to address the dynamic part as well. Today, it is possible to use cheap radio frequency identification (RFID) tags or GPS receivers (Section 5.9) to track goods and vehicles as they move around; transit authorities increasingly use such systems to inform their users of the locations of buses and trains (Section 11.3.3).

GPS is also finding applications in dealing with emergency incidents that occur on the transportation network (Figure 2.14). The OnStar system (www.onstar.com) is one of several products that make use of the ability of GPS to accurately determine location virtually anywhere. When installed in a vehicle, the system is programmed to transmit location automatically

Figure 2.14 Systems such as OnStar (www.onstar.com) allow information on the location of an accident, determined by a GPS unit in the vehicle, to be sent to a central office and compared to a GIS database of highways and streets in order to determine the incident location, so that emergency teams can respond. (Large photo © Slobo Mitic/iStockphoto; inset © Vladimir Kondrachov/iStockphoto)

to a central office whenever the vehicle is involved in an accident and its airbags deploy. This can be lifesaving if the occupants of the vehicle do not know where they are or are otherwise unable to call for help.

Many applications in transportation and logistics involve optimization, or the design of solutions to meet specified objectives. Section 15.4 discusses this type of analysis in detail and includes several examples dealing with transportation and logistics. For example, a delivery company may need to deliver parcels to 200 locations in a given shift, dividing the work between 10 trucks. Different ways of dividing the work and routing the vehicles can result in substantial differences in time and cost, so it is important for the company to use the most efficient solution (see Box 15.4 for an example of the daily workload of an appliance

repair company). Logistics and related applications of GIS have been known to save substantially over traditional, manual ways of determining routes.

GIS has helped many service and delivery companies to substantially reduce their operating costs in the field.

2.3.4.2 Case Study Application: Planning for Emergency Evacuation

Modern society is at risk from numerous types of disasters, including terrorist attacks, extreme weather events such as hurricanes, accidental spills of toxic chemicals resulting from truck collisions or train derailments, and earthquakes. In recent years several major events have required massive evacuation of civilian populations; the preparations for and aftermath of Hurricane Katrina (Box 1.1) provide some of the most salutary lessons of recent years.

In response to the threat of such events, most communities attempt to plan. But planning is made particularly difficult because the magnitude and location of the event can rarely be anticipated. Suppose, for example, that we attempt to develop a plan for dealing with a spill of a volatile toxic chemical resulting from a train derailment. It might make sense to plan for the worst case, for example, the spillage of the entire contents of several cars loaded with chlorine gas. But the derailment might occur anywhere on the rail network, and the impact will depend on the strength and direction of the wind. Possible scenarios might involve people living within tens of kilometers of any point on the track network (see Section 14.3.2 for details of the buffer operation, which would be used in such cases to determine areas lying within a specified distance). Locations can be anticipated for some disasters, such as those resulting from fire in buildings known to be storing toxic chemicals, but hurricanes and earthquakes can impact almost anywhere within large areas.

The magnitude and location of a disaster can rarely be anticipated.

To illustrate the value of GIS in evacuation planning, we have chosen some of the classic work of Tom Cova, an academic expert on GIS in emergency management. Tom's early work was strongly motivated by the problems that occurred in wildfires in California. Figure 2.15 shows the extent of the Zaca fire of 2007, some five weeks after it started: we discuss the issues of involving the public in emergencies such as this in Section 13.4.1. Such fires can have very severe human

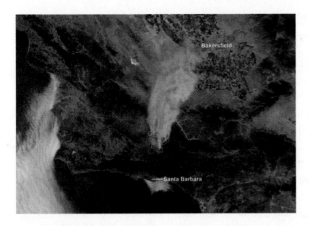

Figure 2.15 Satellite imagery of the 2007 Zaca fire, taken some five weeks after it started on a private ranch. Thick smoke blows nearly due north, and the actively burning areas are outlined in red. The settlements of Goleta and Santa Barbara were threatened by the fire, and plans were put in place to evacuate large parts of the area. (Courtesy NASA)

and economic consequences: the 1991 Oakland Hills fire, for example, destroyed approximately 1,580 acres and over 2,700 structures in the East Bay Hills and became the most expensive fire disaster in California's history, taking 25 lives and causing over $1.68 billion in damages.

Over the years, Cova has developed a number of tools that allow neighborhoods to assess the potential for problems associated with evacuation and to develop plans accordingly. One such tool uses a GIS database containing information on the distribution of population in a neighborhood and on the street pattern. The result is an evacuation vulnerability map, such as that shown in Figure 2.16. Because the magnitude of a disaster cannot be known in advance, the method works by identifying the worst-case scenario that could affect a given location.

Suppose a specific household is threatened by an event that requires evacuation, such as a wildfire, and assume for the moment that one vehicle is needed to evacuate each household. If the house is in a cul-de-sac, the number of vehicles needing to exit the cul-de-sac will be equal to the number of households on the street. If the entire neighborhood of streets has only one exit, all vehicles carrying people from the neighborhood will need to use that one exit. Cova's method works by looking further and further from the household location, to find the most important bottleneck—the one that has to handle the largest amount of traffic. In an area with a dense network of streets, traffic will disperse among several exits, reducing the bottleneck effect. But a densely packed neighborhood with only a single exit can be the source of massive evacuation problems, if a disaster requires the rapid evacuation of the entire neighborhood. In the Oakland Hills fire there were several critical bottlenecks—one-lane streets that normally carry traffic in both directions but became hopelessly clogged in the emergency.

Figure 2.16 Evacuation vulnerability map of the area of Santa Barbara, California. Colors denote the difficulty of evacuating an area based on the area's worst-case scenario. (Courtesy Tom Cova)

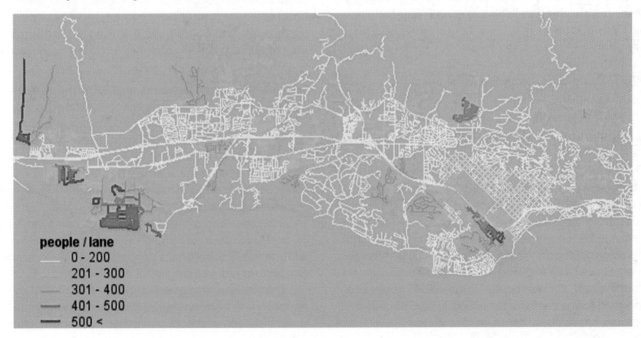

people / lane
— 0 - 200
201 - 300
— 301 - 400
— 401 - 500
— 500 <

Figure 2.16 shows a map of Santa Barbara, California, with streets colored according to Cova's measure of evacuation vulnerability. The color assigned to any location indicates the number of vehicles that would have to pass through the critical bottleneck in the worst-case evacuation, with red indicating that over 500 vehicles per lane would have to pass through the bottleneck. The red area near the shore in the lower left is a densely packed area of student housing, with very few routes out of the neighborhood. An evacuation of the entire neighborhood would produce a very heavy flow of vehicles on these exit routes. The red area in the upper left has a much lower population density, but has only one narrow exit.

2.3.4.3 Method

Two types of data are required for the analysis. Census data are used to determine population and household counts, and to estimate the number of vehicles involved in an evacuation. Census data are available as aggregate counts for areas of a few city blocks, but not for individual houses. Thus there will be some uncertainty regarding the exact numbers of vehicles needing to leave a specific street, though estimates for entire neighborhoods will be much more accurate. The locations of streets are obtained from so-called *street centerline* files, which give the geographic locations, names, and other details of individual streets (see Section 10.5 for an overview of geographic data types). The TIGER (Topologically Integrated Geographic Encoding and Referencing) files, produced by the U.S. Bureau of the Census and the U.S. Geological Survey and readily available from many sites on the Internet, are one free source of such data for the United States. Many private companies also offer such data, adding new information such as traffic flow volumes or directions (for U.S. sources, see TeleAtlas, s'Hertogenbosch, the Netherlands, navigation.teleatlas.com; and NAVTEQ, Chicago, Illinois, www.navteq.com).

Street centerline files are essential for many applications in transportation and logistics.

The analysis proceeds by beginning at every street intersection and working outward following the street connections to reach new intersections. Every connection is tested to see if it presents a bottleneck, by dividing the total number of vehicles that would have to move out of the neighborhood by the number of exit lanes. After all streets have been searched out to a specified distance from the start, the worst-case value (vehicles per lane) is assigned to the starting intersection. Finally, the entire network is colored by the worst-case value.

2.3.4.4 Scientific Foundations: Geographic Principles, Techniques, and Analysis

Scientific Foundations Cova's example is one of many applications that have been found for GIS in the general areas of logistics and transportation. As a planning tool, it provides a way of rating areas against a highly uncertain form of risk, a major evacuation. Although the worst-case scenario that might affect an area may never occur, the tool nevertheless provides very useful information to planners who design neighborhoods, giving them graphic evidence of the problems that lack of foresight in street layout can cause. The approach points to a major problem with the modern style of street layout in subdivisions that, somewhat ironically, limits the number of entrances to subdivisions from major streets in the interests of creating a sense of community and of limiting high-speed through traffic. Cova's analysis shows that such limited entrances can also be bottlenecks in major evacuations.

The analysis demonstrates the value of readily available sources of geographic data, since both major inputs—demographics and street layout—are available in digital form. At the same time, we should note the limitations of using such sources. Census data are aggregated to areas that, while small, nevertheless provide only aggregated counts of population. The street layouts of TIGER and other sources can be out of date and inaccurate, particularly in new developments, although users willing to pay higher prices can often obtain current data from the private sector. And the essentially geometric approach cannot deal with many social issues: evacuation of the disabled and elderly, and issues of culture and language that may impede evacuation. In Chapter 16 we look at this problem using the tools of dynamic simulation modeling, which are much more powerful and provide ways of addressing such issues.

Principles Central to Cova's analysis is the concept of *connectivity*. Very little would change in the analysis if the input maps were stretched or distorted because what matters is how the network of streets is connected to the rest of the world. Connectivity is an instance of a *topological* property, a property that remains constant when the spatial framework is stretched or distorted. Other examples of topological properties are *adjacency* and *intersection*, both of which cannot be destroyed by stretching a map. We discuss the importance of topological properties and their representation in GIS in Section 9.2.3.

The analysis also relies on being able to find the *shortest path* from one point to another through a street network, and it assumes that people will follow

Figure 2.17 Many in-vehicle navigation services now offer dynamic updating of routing information, in the light of real-time traffic updates or subscriber information about road blockages.

such paths when they evacuate. Many forms of GIS analysis rely on being able to find shortest paths, and we discuss some of them in Section 15.4. Many Web sites allow users to find shortest paths between two street addresses (Figure 1.17). In practice, people will often not use the shortest path, preferring routes that may be quicker but longer, or routes that are more scenic. Moreover, in the general case, routing analysis needs to accommodate variability in journey times according to the time of day or time of week, or temporary breaks in connectivity. Many of the traffic routing services hosted by in-vehicle devices are able to accommodate temporal variations by taking real-time feeds of traffic conditions from sensors, and some use uploaded citizen information about blocked passages in order to reroute other users (Figure 2.17).

Techniques   The techniques used in this example are widely available in GIS. They include *spatial interpolation* techniques, which are needed to assign worst-case values to the streets, since the analysis only produces values for the intersections. Spatial interpolation is widely used in GIS to use information obtained at a limited number of sample points to guess values at other points; this technique is discussed in general in Section 4.5 and in detail in Section 14.3.6. The shortest-path methods used to route traffic are also widely available in GIS, along with other functions needed to create, manage, and visualize information about networks.

Analysis   Cova's technique is an excellent example of the use of GIS analysis to *make visible what is otherwise invisible* (see Table 1.3). By processing the data and mapping the results in ways that would be impossible by hand, he succeeds in exposing areas that are difficult to evacuate and drawing attention to potential problems.

This idea is so central to GIS that it has sometimes been claimed as the primary purpose of the technology, though that seems a little strong and ignores many of the other applications discussed in this chapter.

2.3.4.5 Generic Scientific Questions Arising from the Application

The logistic and transportation applications of GIS rely heavily on representations of networks and often must ignore off-network movement. Drivers who cut through parking lots, children who cross fields on their way to school, houses in developments that are not aligned along linear streets, and pedestrians in underground shopping malls all confound the network-based analysis that GIS makes possible. Humans are endlessly adaptable, and their behavior will often confound the simplifying assumptions that are inherent to a GIS model. For example, suppose a system is developed to warn drivers of congestion on freeways and to recommend alternative routes on neighborhood streets. While many drivers might follow such recommendations, others will reason that the result could be severe congestion on neighborhood streets and reduced congestion on the freeway, and so they will ignore the recommendation. Residents of the neighborhood streets might also be tempted to try to block the use of such systems, arguing that they result in unwanted and inappropriate traffic, as well as risk to themselves. Arguments such as these are based on the notion that the transportation system can only be addressed as a whole and that local modifications based on limited perspectives, such as the addition of a new freeway or bypass, may create more problems than they solve.

2.3.4.6 Management and Policy

GIS is used in all three modes—operational, tactical, and strategic—in logistics and transportation. This section concludes with some examples in all three categories.

  In *operational* systems, GIS is used:

- To monitor the movement of mass-transit vehicles in order to improve performance and to provide improved information to system users.

- To route and schedule delivery and service vehicles on a daily basis to improve efficiency and reduce costs.

In *tactical* systems:

- To design and evaluate routes and schedules for public bus systems, school bus systems, garbage collection, and mail collection and delivery.

- To monitor and inventory the condition of highway pavement, railroad track, and highway signage, and to analyze traffic accidents.

In *strategic* systems:

- To plan locations for new highways and pipelines, and associated facilities.
- To select locations for warehouses, intermodal transfer points, and airline hubs.

2.3.5 Environment

2.3.5.1 Applications Overview

Although it is the last area to be discussed here, the environment drove some of the earliest applications of GIS and was a strong motivating force in developing the very first GIS in the mid-1960s (Section 1.4). Environmental applications are the subject of several GIS texts, so only a brief overview will be given here for the purposes of illustration.

The development of the Canada Geographic Information System in the 1960s was driven by the need for policies over the use of land. Every country's land base is strictly limited (although the Dutch have managed to expand theirs very substantially by damming and draining), and alternative uses must compete for space. Measures of area are critical to effective strategy—for example, to identify how much land is being lost to agriculture through urban development and sprawl, and how this will impact the ability of future generations to feed themselves. Today, we have very effective ways of monitoring land-use change through remote sensing from space, and we are able to get frequent updates on the loss of tropical forest in such critical areas as the Amazon Basin.

Thus far, the Twenty-First Century has seen increasing proportions of the world's population gravitating toward cities and towns, and so understanding of the environmental impacts of urban settlements is an increasingly important focus of attention in science and policy. Researchers have used GIS to investigate how urban sprawl occurs in order to understand the environmental consequences of sprawl and to predict its future consequences. Such predictions can be based on historic patterns of growth, together with information on the locations of roads, steeply sloping land unsuitable for development, land that is otherwise protected from urban use, and other factors that encourage or restrict urban development. Each of these factors may be represented in map form, as a layer in the GIS, while specialist software can be designed to simulate the processes that drive growth.

Simulated land-use patterns are the outputs of *dynamic simulation models*, or computer programs

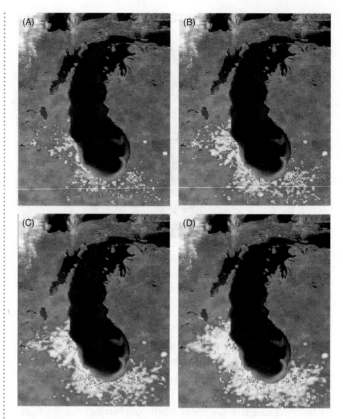

Figure 2.18 Growth in the American Midwest under four different urban growth scenarios. Horizontal extent of image is 400 km. (Source: Paul Torrens, 2005 "Simulating sprawl with geographic automata models," reproduced courtesy of the author)

designed to emulate the operation of some part of the human or environmental system. Figure 2.18, taken from the work of geographer Paul Torrens, presents a simple simulation of urban growth in the American Midwest under four rather different growth scenarios: (A) uncontrolled suburban sprawl; (B) growth restricted to existing travel arteries; (C) leapfrog development, occurring because of local zoning controls; and (D) development that is constrained to some extent. Other applications are concerned with the simulation of processes principally in the natural environment. Many models have been coupled with GIS in the past decade in order to simulate such processes as soil erosion, forest growth, groundwater movement, and runoff. Dynamic simulation modeling is discussed in detail in Chapter 16.

2.3.5.2 Case Study Application: Deforestation on Sibuyan Island, the Philippines

If the increasing extent of urban areas, described earlier, is one side of the development coin, then the reduction in the extent of natural land cover is frequently the other. Deforestation is one important manifestation of land-use change and poses a threat to the habitat of many species in tropical and temperate

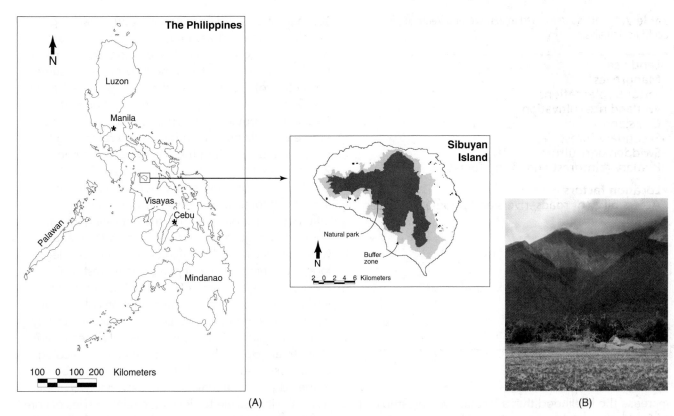

Figure 2.19 (A) Location of Sibuyan Island in the Philippines, showing location of the National Park and buffer zone, and (B) typical forested mountain landscape of Sibuyan Island. (Image B: Courtesy Peter Verberg)

forest areas alike. Ecologists, environmentalists, and urban geographers are therefore using GIS in interdisciplinary investigations to understand the local conditions that lead to deforestation and to understand its consequences. Important evidence of the rate and patterning of deforestation has been provided through analysis of remote-sensing images. These analyses of *pattern* need to be complemented by analysis at detailed levels of the causes and underlying driving factors of the *processes* that lead to deforestation.

The negative environmental impacts of deforestation can be ameliorated by adequate spatial planning of natural parks and land development schemes. But the more strategic objective of sustainable development can only be achieved if a holistic approach is taken to ecological, social, and economic needs. GIS provides the medium of choice for integrating knowledge of natural and social processes in the interests of integrated environmental planning. Working at the University of Wageningen in the Netherlands, Peter Verburg and Tom Veldkamp coordinate a research program that is using GIS to understand the sometimes complex interactions that exist between socio-economic and environmental systems, and to gauge their impact on land-use change in a range of different regions of the world (see www.cluemodel.nl).

GIS allows us to compare the environmental conditions prevailing in different parts of the world.

One of Verburg and Veldkamp's case study areas is Sibuyan Island in the Philippines (Figure 2.19A), where deforestation poses a major threat to biodiversity. Sibuyan is a small island (area 456 km²) of steep forested mountain slopes (Figure 2.19B) and gently sloping coastal land that is used mainly for agriculture, mining, and human settlement. The island has remarkable biodiversity: an estimated 700 plant species, of which 54 occur only on Sibuyan Island, and a unique local fauna. The objective of this case study was to identify a range of different development scenarios that make it possible to anticipate future land-use and habitat change, and hence also anticipate changes in biodiversity.

2.3.5.3 Method

The initial stage of Verburg and Veldkamp's research was a *qualitative* investigation involving interviews with different stakeholders on the island to identify a list of factors likely to influence land-use patterns. Table 2.3 lists the data that provided direct or indirect indicators of pressure for land-use change. For example, the suitability of the soil for agriculture or the accessibility of a location to local markets can

Table 2.3 Data sources used in Verburg and Veldkamp's ecological analysis.

Land use
Mangroves
Coconut plantations
Wetland rice cultivation
Grassland
Secondary forest
Swidden agriculture
Primary rainforest and mossy forest

Location factors
Accessibility of roads, rivers and populated places
Altitude
Slope
Aspect
Geology
Geomorphology
Population density
Population pressure
Land tenure
Spatial policies

increase the likelihood that a location will be stripped of forest and used for agriculture. They then used these data in a *quantitative* GIS-based analysis to calculate the probabilities of land-use transition under three different scenarios of land-use change; each scenario was based on different spatial planning policies. Scenario 1 assumes no effective protection of the forests on the island (and a consequent piecemeal pattern of illegal logging); Scenario 2 assumes protection of the designated natural park area alone; and Scenario 3 assumes protection not only of the natural park but also of a GIS-defined buffer zone. Figure 2.20 illustrates the forecasted remaining forest area under each of the scenarios at the end of a 20-year simulation period ending in 2019.

The three different scenarios not only resulted in different forest *area extents* by 2019 but also different *spatial patterning* of the remaining forest. For example, gaps in the forest area under Scenario 1 were mainly caused by shifting cultivation and illegal logging within the area of primary rainforest, while most deforestation under Scenario 2 occurred in the lowland areas. Qualitative interpretations of the outcomes and aggregate statistics are supplemented by numerical spatial indices such as *fractal dimensions* in order to anticipate the effects of changes on ecological processes—particularly the effects of disturbance at the edges of the remaining forest area. Such statistics make it possible to define the relative sizes of core and fragmented forest areas (for example, Scenario 1 in Figure 2.20 leads to the greatest fragmentation of the forest area). This in turn makes it possible to measure the effects of development on biodiversity.

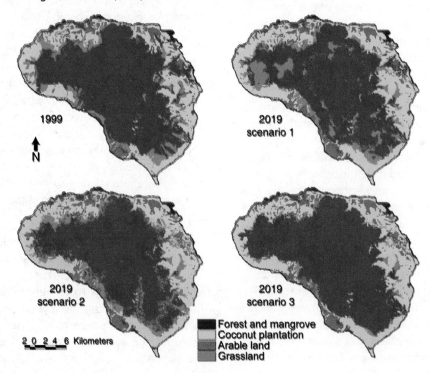

Figure 2.20 Forest area (dark green) in 1999 and at the end of the land-use change simulations (2019) for three different scenarios.

2.3.5.4 Scientific Foundations: Geographic Principles, Techniques, and Analysis

Scientific Foundations   The goal of the research was to predict changes in the biodiversity of the island. Existing knowledge of a set of ecological *processes* led the researchers to the view that biodiversity would be compromised by changes in overall size of natural forest and by changes in the *patterning* of the forest that remained. It was hypothesized that changes in patterning would be *caused* by the combined effects of a further set of physical, biological, and human *processes*. Thus the researchers took the observed existing land-use pattern and used their understanding of the physical, biological, and human processes to predict future land-use changes. The different forecasts of land use (based on different scenarios) could then be used to identify how the ecology of the island might change in the future. This theme of inferring *process* from *pattern*, or *function* from *form*, is a common characteristic of GIScience applications.

It is not possible to identify a uniform set of physical, biological, and human processes that is valid in all regions of the world (the pure *nomothetic* approach of Section 1.3.1). Conversely, it does not make sense to treat every location as unique (*idiographic*, in terms of Section 1.3.1) in terms of the processes extant upon it. The art and science of ecological modeling requires us to make a good call not just on the range of relevant determining factors, but also on their importance in the specific case study, with due consideration to the appropriate scale range at which each is relevant.

Geographic Principles   GIS makes it possible to incorporate diverse physical, biological, and human elements, and to forecast the likely size, shape, scale, and dimension of land-use parcels. The emphasis in this study is on prediction of what the overall pattern of habitat change (fragmentation) is likely to look like and on its overall effects on biodiversity, rather than on predicting the land uses at specific unique locations on the island. As such, the motivation for the scenarios is focused more on identifying the overall outcome of different processes than on making successful per-land-parcel predictions as measured by a confusion matrix (Section 6.3.2.1).

Fundamental to this outcome is the assumption that the ecological consequences of future deforestation (e.g., for biodiversity) can be reliably inferred from a map of predicted land uses. The forecasting procedure also assumes that the various indicators of land development pressure are *robust, accurate,* and *reliable*; yet in practice any measure of land

development pressure is likely to be ambiguous, controversial, and *uncertain*. Further uncertainties are generated by the scale of analysis that is carried out, and, as in the retailing example of Section 2.3.3.2, the qualitative importance of local context may play an important role.

Land-use change is deemed to be a measurable response to a wide range of locationally variable factors. These factors have traditionally been the remit of different disciplines that have different intellectual traditions of measurement and analysis. Peter Verburg's view is that his research assumes that GIS can provide a sort of 'Geographic Esperanto'—that is, a common language to integrate diverse, geographically variable factors. As such, it makes use of the core GIS idea that the world can be understood as a series of layers of different types of information, that can be added together meaningfully through overlay analysis to arrive at conclusions.

Techniques   The *multicriteria* techniques used to harmonize the different location factors into a composite spatial indicator of development pressure are widely available in GIS and are discussed in detail in Chapter 16. The individual component indicators are acquired using techniques such as *on-screen digitizing* and *classification of imagery* to obtain a land-use map. All data are converted to a *raster* data structure with common resolution and extent. Relations between the location factors and land use are quantified using *correlation* and *regression* analysis based on the spatial dataset.

Analysis   The GIS is used to simulate scenarios of future land-use change based on different spatial policies. The application is predicated on the premise that changes in ecological *process* can be reliably inferred from predicted changes in land-use *pattern*. Process is inferred not just through size measures, but also through spatial measures of connectivity and fragmentation—since these latter aspects affect the ability of a species to mix and breed without disturbance. The analysis of the extent and ways in which different land uses fill space is performed using specialized software. Although such spatial indices are useful tools, they may be more relevant to some aspects of biodiversity than others, since different species are vulnerable to different aspects of habitat change.

2.3.5.5 Generic Scientific Questions Arising from the Application

GIS applications need to be based on sound science. In environmental applications, this knowledge base is unlikely to be the preserve of any single academic discipline. Many environmental applications require recourse to use of GIS in the field, and field researchers often require multidisciplinary understanding of the full range of processes leading to land-use change.

Regardless of the quality of the measurements that are made, uncertainty will always creep into any prediction, for a number of reasons. Data are never perfect, being subject to measurement error and uncertainty (see Chapter 6) arising out of the need selectively to generalize, abstract, and approximate (see Section 3.4). Furthermore, simulations of land-use change are subject to changes in exogenous forces such as the world economy. Any forecast can only be a selectively simplified representation of the real world and the processes operating within it. GIS users need to be aware of this because the forecasts produced by a GIS will always appear to be precise in numerical terms, and spatial representations will usually be displayed using crisp lines and clear mapped colors.

GIS users should not think of systems as *black boxes* and should be aware that explicit spatial forecasts may have been generated by invoking assumptions about process and data that are not as explicit. User awareness of these important issues can be improved through appropriate *metadata* and documentation of research procedures—particularly when interdisciplinary teams may be unaware of the disciplinary conventions that govern data creation and analysis in parts of the research.

Interdisciplinary science and the cumulative development of algorithms and statistical procedures can lead GIS applications to conflict with an older principle of scientific reporting, that the results of analysis *should always be reported in sufficient detail to allow someone else to replicate them*. Today's science is complex, and all of us from time to time may find ourselves using tools developed by others that we do not fully understand. It is up to all of us to demand to know as many of the details of GIS analysis as is reasonably possible. This is recognized by GIScientists the world over (see, for example, Box 2.2).

Users of GIS should always know exactly what the system is doing to their data.

2.3.5.6 Management and Policy

GIS is now widely used in all areas of environmental science, from ecology to geology and from oceanography to alpine geomorphology. GIS is also helping to reinvent environmental science as a discipline grounded in field observation, as data can be captured using widely available personal data assistants (PDAs) and notebooks, before being analyzed on a battery-powered laptop in a field tent, and then uploaded via

---

Deren Li, an Academic Leader in GIScience

Professor Deren Li (Figure 2.21) is chair of the Academic Committee of LIESMARS, the State Key Laboratory for Information Engineering in Surveying, Mapping, and Remote Sensing, at Wuhan University, China. He is a member of both the Chinese Academy of Sciences and the Chinese Academy of Engineering. He obtained his Ph.D. in 1985, having entered the university following the end of the Cultural Revolution, and is now a leader in the rapidly developing field of GIScience in China. Best known for his contributions to photogrammetry and remote sensing, he has led LIESMARS to a prominent role in Chinese research, has served as president of Wuhan Technical University of Surveying and Mapping prior to its merger into Wuhan University, and has been heavily engaged in the development of national policy regarding remote sensing and the use of its products. The students who have emerged from Wuhan University's programs have in many instances gone on to play leading roles themselves in the international GIScience community. In 2008 LIESMARS played a leading role in assembling and organizing the glut of

Figure 2.21 Deren Li, GIScientist. (Courtesy Deren Li)

remotely sensed imagery that helped China to assess and to begin to recover from the devastating Wenchuan earthquake, which killed almost 70,000 people. Much honored around the world, Deren Li is a beacon to China's rapidly growing international role in GI systems and science.

---

a satellite link to a home institution. The art of scientific forecasting (by no means a contradiction in terms) is developing in a cumulative way, as interdisciplinary teams collaborate in the development and sharing of applications that range in sophistication from simple composite mapping projects to intensive numerical and statistical simulation experiments.

 ## 2.4 Concluding Comments

This chapter has presented a selection of GIS application areas and specific instances within each of them. Throughout, the emphasis has been on the range of contexts, from day-to-day problem solving to curiosity-driven science. The principles of the scientific method have been stressed throughout—the need to maintain an inquiring mind, constantly asking questions about what is going on, and what it means; the need to use terms that are well defined and understood by others, so that knowledge can be communicated; the need to describe procedures in sufficient detail so that they can be replicated by others; and the need for accuracy in observations, measurements, and predictions. These principles are valid whether the context is a simple inventory of the assets of a utility company, or the simulation of complex biological systems.

Questions for Further Study

1. Devise a diary for your own activity patterns for a typical (or a special) day, like that described in Section 2.1.1, and speculate how GIS might affect your own daily activities. What activities are not influenced by GIS, and how might its use in some of these contexts improve your daily quality of life?

2. Compare and contrast the operational, tactical, and strategic priorities of the GIS specialists responsible for the specific applications described in Sections 2.3.2, 2.3.3, 2.3.4 and 2.3.5.

3. Examine one of the applications chapters in the Longley et al. (2005) volume in the guide to Further Reading (below). To what extent do you believe that the author of your chapter has demonstrated that GIS has been successful in application? Suggest some of the implicit and explicit assumptions that are made in order to achieve a successful outcome.

4. Look at one of the applications areas in the Longley et al. (2005) volume in the guide to Further Reading (below). Then reexamine the list of critiques of GIS at the end of Section 1.8. To what extent do you think that the critiques are relevant to the applications you have studied?

Further Reading

Geertman, S. and Stillwell, J.C.H. (eds.). 2009. *Planning Support Systems: Best Practice and New Methods*. Dordrecht: Springer.

Kok, K., Verburg, P.H., and Veldkamp, T.A. 2007. Integrated assessment of the land system: the future of land use. Special issue of *Land Use Policy* 24(3): 517–610.

Longley, P.A. and Goodchild, M.F. 2008. The use of geodemographics to improve public service delivery. In Hartley, J., Donaldson, C., Skelcher, C., and Wallace, M. (eds.). *Managing to Improve Public Services*, Chapter 9, pp. 176–194. Cambridge: Cambridge University Press.

Longley, P.A., Goodchild, M.F., Maguire, D.J., and Rhind, D.W. (eds.). 2005. *Geographical Information Systems: Principles; Techniques; Management and Applications (abridged edition)*. Hoboken; NJ: Wiley.

Pultar, E., Raubal, M., Cova, T.J., and Goodchild, M.F. Dynamic GIS case studies: wildfire evacuation and volunteered geographic information, *Transactions in GIS* 13: 85–104.

Rogers, E.M. 2003. *Diffusion of Innovations* (5th ed.). New York: Simon and Schuster.

2 PRINCIPLES

73

3 Representing Geography

This chapter introduces the concept of representation, or the construction of a digital model of some aspect of the Earth's surface. Representations have many uses, allowing us to learn, think, and reason about places and times that are outside our immediate experience. This is the basis of scientific research, planning, and many forms of day-to-day problem solving.

The geographic world is extremely complex, revealing more detail the closer one looks, seemingly ad infinitum. In order to build a representation of any part of it, it is necessary to make choices about what to represent, at what level of detail, and over what time period. The large number of possible choices creates many opportunities for designers of GIS software.

Generalization methods are used to remove detail that is unnecessary for an application, in order to reduce data volume and speed up operations.

○ LEARNING OBJECTIVES

After studying this chapter you will understand:

- The importance of understanding representation in GIS.

- The concepts of fields and objects and their fundamental significance.

- What raster and vector representation entails and how these data structures affect many GIS principles, techniques, and applications.

- Why the paper map is a particular instance of a GIS product and data source.

- Why map generalization methods are important and how they are based on the concept of representational scale.

- The art and science of representing real-world phenomena in GIS.

3.1 Introduction

We live on the surface of the Earth and spend most of our lives in a relatively small fraction of that space. Of the approximately 500 million sq km of surface, only one-third is land, and only a fraction of that is occupied by the cities and towns in which most of us live. The rest of the Earth, including the parts we never visit, the atmosphere, and the solid ground under our feet, remains unknown to us except through the information that is communicated to us through books, newspapers, television, the Web, or the spoken word. We live lives that are almost infinitesimal in comparison with the 4.5 billion years of Earth history, or the over 10 billion years since the universe began, and we know about the Earth before we were born only through the evidence compiled by geologists, archaeologists, historians, and other specialists. Similarly, we know nothing about the world that is to come, where we have only predictions to guide us.

Because we can observe so little of the Earth directly, we rely on a host of methods for learning about its other parts, deciding where to go as tourists or shoppers, choosing where to live, running the operations of corporations, agencies, and governments, and many other activities. Almost all human activities at some time require knowledge (see Section 1.2) about parts of the Earth that are outside our direct experience because they occur either elsewhere in space or elsewhere in time.

Sometimes this knowledge is used as a *substitute* for directly sensed information, creating a *virtual* reality (see Section 13.4.4). Increasingly, it is used to *augment* what we can see, touch, hear, feel, and smell, through the use of mobile information systems that can be carried around. Our knowledge of the Earth is not created entirely freely, but must fit with the mental concepts we began to develop as young children—concepts such as containment (Paris is *in* France) or proximity (Dallas and Fort Worth are *close*). In digital representations, we formalize these concepts through *data models* (Chapter 8), the structures and rules that are programmed into a GIS to accommodate data. These concepts and data models together constitute our *ontologies*, the frameworks that we use for acquiring knowledge of the world.

> **Almost all human activities require knowledge about the Earth—past, present, or future.**

One such ontology, a way to structure knowledge of movement through time, is a three-dimensional diagram, in which the two horizontal axes denote location on the Earth's surface, and the vertical axis denotes time. Figure 3.1 presents the *time space aquarium* that largely contains the activity space of three children living in Cheshunt, UK. The icons on the three trajectories identify the travel modes (walking, cycling, or automobile) that they use as

Figure 3.1 Schematic representation of weekend activities of three children in Cheshunt, UK. The horizontal dimensions represent geographic space (rendered using OpenStreetMap) and the vertical dimension represents time of day. Each person's track plots as a three-dimensional line, beginning at the base in the morning and ending at the top in the evening. (Reproduced with permission of Yi Gong: base image Courtesy www.openstreetmap.org)

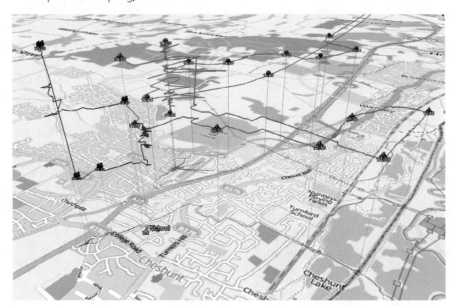

they move through space (the base of the aquarium, rendered using OpenStreetMap) and time (the vertical dimension) on a single weekend day. The spatial and temporal granularity of the diagram is set in such a way that even quite small events are recorded, but the granularity could be increased, for example, by taking only a single GPS signal on the hour or by recording only those changes in location that exceed a predetermined distance from the preceding point. When we view this diagram on the page of a book, much of the fine detail of the three activity patterns is lost: it is not possible to discern, for example, the precise trajectory of a child playing football on a field, or whether the cyclist dismounts for short sections of a journey, or how long a car waits at traffic lights before continuing its journey. Indeed, some of these details cannot be recovered if the GPS is set to record changes in location every 5 minutes. Closer perspectives could display more information, but such detail rarely adds much that is useful, and a vast storehouse would be required to capture the precise trajectories of many children throughout even a single day.

The real trajectories of the individuals shown in Figure 3.1 are complex, yet the figure is only a representation of them—a model on a piece of paper, generated by a computer from a database. We use the terms *representation* and *model* because they imply a simplified relationship between the contents of the figure and the database, and the real-world trajectories of the individuals.

Such representations or models serve many useful purposes and occur in many different forms. For example, representations occur:

- in the human mind, when our senses capture information about our surroundings, such as the images captured by the eye, or the sounds captured by the ear, and memory preserves such representations for future use;

- in photographs, which are two-dimensional models of the light emitted or reflected by objects in the world into the lens of a camera;

- in spoken descriptions and written text, in which people describe some aspect of the world in language, in the form of travel accounts or diaries; or

- in the numbers that result when aspects of the world are measured, using such devices as thermometers, rulers, or speedometers.

By building representations, we humans can assemble far more knowledge about our planet than we ever could as individuals. We can build representations that serve such purposes as planning, resource management and conservation, travel, or the day-to-day operations of a parcel delivery service.

Representations help us assemble far more knowledge about the Earth than is possible on our own.

Representations are reinforced by the rules and laws that we humans have learned to apply to the unobserved world around us. When we encounter a fallen log in a forest, we are willing to assert that it once stood upright, and once grew from a small shoot, even though no one may actually have observed or reported either of these stages. We predict the future occurrence of eclipses based on the laws we have discovered about the motions of the Solar System. In GIS applications, we often rely on methods of spatial interpolation to guess the soil characteristics, rainfall, or other conditions that exist in places where no observations were made, based on the rule (often elevated to the status of a First Law of Geography and attributed to Waldo Tobler) that all places are similar, but nearby places are more similar than distant places.

Tobler's First Law of Geography: Everything is related to everything else, but near things are more related than distant things.

3.2 Digital Representation

This book is about one particular form of representation that has become very important in our society—representation in digital form. Today, almost all communication between people through such media as the telephone, Web pages, music, television, newspapers and magazines, or e-mail is at some time in its life in digital form. Information technology based on digital representation is moving into all aspects of our lives, from science to commerce to daily existence. The personal computer is the most obvious digital information processing device, and there are now well over 75 computers per 100 population in the dozen or so most industrialized societies; computers are the mainstay of most office work; and digital technology has pervaded many devices that we use every day, from the microwave oven to the automobile.

One important characteristic of digital technology is that the representation itself is rarely if ever seen

The Binary Counting System

The binary counting system uses only two symbols, 0 and 1, to represent numerical information. A group of eight binary digits is known as a *byte*, and volume of storage is normally measured in bytes rather than bits (see Table 1.1). There are only two options for a single digit, but there are four possible combinations for two digits (00, 01, 10, and 11), eight possible combinations for three digits (000, 001, 010, 011, 100, 101, 110, 111), and 256 combinations for a full byte. Digits in the binary system (known as binary digits, or *bits*) behave like digits in the decimal system but use powers of two. The rightmost digit denotes units, the next digit to the left denotes twos, the next to the left denotes fours, and so on. For example, the binary number 11001 denotes one unit, no twos, no fours, one eight, and one sixteen, and is equivalent to 25 in the normal (decimal) counting system. We call this the *integer* digital representation of 25 because it represents 25 as a whole number and is readily amenable to arithmetic operations. Whole numbers are commonly stored in GIS using either *short* (2-byte or 16-bit) or *long* (4-byte or 32-bit) options. Short integers can range from −65535 to +65535, and long integers from −4294967925 to +4294967295.

The 8-bit ASCII (American Standard Code for Information Interchange) system assigns codes to each symbol of text, including letters, numbers, and common symbols. The number 2 is assigned ASCII code 48 (00110000 in binary), and the number 5 is 53 (00110101), so if 25 were coded as two characters using 8-bit ASCII its digital representation would be 16 bits long (0011000000110101). The characters 2 = 2 would be coded as 48, 61, 48 (001100000011 110100110000). ASCII is used for coding text, which consists of mixtures of letters, numbers, and punctuation symbols.

Numbers with decimal places are coded using *real* or *floating-point* representations. A number such as 123.456 (three decimal places and six significant digits) is first transformed by powers of 10 so that the decimal point is in a standard position, such as the beginning (e.g., 0.123456×10^3). The fractional part (0.123456) and the power of 10 (3) are then stored in separate sections of a block of either 4 bytes (32 bits, *single* precision) or 8 bytes (64 bits, *double* precision). This gives enough precision to store roughly 7 significant digits in single precision, or 14 in double precision.

Integer, ASCII, and real conventions are adequate for most data, but in some cases it is desirable to associate images or sounds with places in GIS, rather than text or numbers. To allow for this, GIS designers have included a BLOB option (standing for binary large object), which simply allocates a sufficient number of bits to store the image or sound, without specifying what those bits might mean.

by the user because only a few technical experts ever see the individual elements of a digital representation. What we see instead are *views* designed to present the contents of the representation in a form that is meaningful to us. The term *digital* derives from *digits*, or the fingers, and our system of counting based on the 10 digits of the human hand. But while the counting system has 10 symbols (0 through 9), the representation system in digital computers uses only two (0 and 1). In a sense, then, the term *digital* is a misnomer for a system that represents all information using some combination of the two symbols 0 and 1, and the more exact term *binary* is more appropriate. In this book we follow the convention of using *digital* to refer to electronic technology based on binary representations.

Computers represent phenomena as binary digits. Every item of useful information about the Earth's surface is ultimately reduced by a GIS to some combination of 0s and 1s.

Over the years many standards have been developed for converting information into digital form. Box 3.1 shows the standards that are commonly used in GIS to store data, whether they consist of whole or decimal numbers or text. There are many competing coding standards for images and photographs (GIF, JPEG, TIFF, etc.) and for movies (e.g., MPEG) and sound (e.g., MIDI, MP3). Much of this book is about the coding systems used to represent geographic data, especially Chapter 8, and as you might guess this turns out to be comparatively complicated.

Digital technology is successful for many reasons, not the least of which is that all kinds of information share a common basic format (0s and 1s) and can be handled in ways that are largely independent of their actual meaning (see Box 3.1). The Internet, for example, operates on the basis of packets of information, consisting of strings of 0s and 1s, which are sent through the network based on the information contained in the packet's header. The network needs to know only what the header means and how to read the instructions it contains regarding the packet's destination. The rest of the contents are no more than a collection of bits, representing anything from an e-mail message to a short burst of music or highly secret information on its way from one military installation to another, and are almost never examined or interpreted during transmission. This allows one digital communications network to serve every need, from electronic commerce to social networking sites, and it allows manufacturers to build processing and storage technology for vast numbers of users who have very different applications in mind. Compare this to earlier ways of communicating, which required printing presses and delivery trucks for one application (newspapers) and networks of copper wires for another (telephone).

Digital representations of geography hold enormous advantages over previous types—paper maps, written reports from explorers, or spoken accounts. We can use the same cheap digital devices—the components of PCs, the Internet, or mass storage devices—to handle every type of information, inde-pendent of its meaning. Digital data are easy to copy, they can be transmitted at close to the speed of light, they can be stored at high density in very small spaces, and they are less subject to the physical deterioration that affects paper and other physical media.

Perhaps more importantly, data in digital form are easy to transform, process, and analyze. Geographic information systems allow us to do things with digital representations that we were never able to do with paper maps: to measure accurately and quickly, to overlay and combine, and to change scale, zoom, and pan without respect to map sheet boundaries. The vast array of possibilities for processing that digital representation opens up is reviewed in Chapters 14 through 16, and is also covered in the applications that are distributed throughout the book.

Digital representation has many uses because of its simplicity and low cost.

3.3 Representation of What and for Whom?

Thus far we have seen how humans are able to build representations of the world around them, but we have not yet discussed why representations are useful, and why humans have become so ingenious at creating and sharing them. The emphasis here and throughout the book is on one type of representation, termed *geographic*, and defined as a representation of some part of the Earth's surface or near-surface, at scales ranging from the architectural to the global.

Geographic representation is concerned with the Earth's surface or near surface at scales from the architectural to the global.

Geographic representations are among the most ancient, having their roots in the needs of very early societies. The tasks of hunting and gathering can be much more efficient if hunters are able to communicate the details of their successes to other members of their group—the locations of edible roots or game, for example. Maps must have originated in the sketches early people made in the dirt of campgrounds or on cave walls, long before language became sufficiently sophisticated to convey equivalent information through speech. We know that the peoples of the Pacific built representations of the locations of islands, winds, and currents out of simple materials to guide each other, and that social insects such as bees use very simple forms of representation to communicate the locations of food resources.

Hand-drawn maps and speech are effective media for communication between members of a small group, but much wider communication became possible with the invention of the printing press in the Fifteenth Century. Now large numbers of copies of a representation could be made and distributed, and for the first time it became possible to imagine that something could be known by every human being— that knowledge could be the common property of humanity. Only one major restriction affected what could be distributed using this new mechanism: the representation had to be flat. If one were willing to accept that constraint, however, paper proved to be enormously effective; it was cheap, light and thus easily transported, and durable. Only fire and water proved to be disastrous for paper, and human history is replete with instances of the loss of vital

Prince Henry the Navigator and Admiral Zheng He

Prince Henry of Portugal (Figure 3.2), who died in 1460, was known as Henry the Navigator because of his keen interest in exploration. In 1433 Prince Henry sent a ship from Portugal to explore the west coast of Africa in an attempt to find a sea route to the Spice Islands. This ship was the first to travel south of Cape Bojador (latitude 26 degrees 20 minutes North). To make this and other voyages Prince Henry assembled a team of mapmakers, sea captains, geographers, ship builders, and many other skilled craftsmen. Prince Henry showed the way for Vasco de Gama and other famous Fifteenth Century explorers. His management skills could be applied in much the same way in today's GIS projects.

Admiral Zheng He was born in 1371 in what is now China's Yunnan Province. In a series of seven expeditions between 1405 and 1433 he explored the coasts of Thailand, Indonesia, India, Arabia, and East Africa, using massive fleets of up to 200 ships (Figure 3.3), mapping, trading, and settling Chinese along the way. His last two voyages were the most extensive, and speculation persists about the areas that he might have discovered (see the book *1421: The Year the Chinese Discovered America* by Gavin Menzies). Unfortunately, the Ming Emperor destroyed the records of these expeditions.

Figure 3.2 Prince Henry the Navigator, originator of the Age of Discovery in the Fifteenth Century, and promoter of a systematic approach to the acquisition, compilation, and dissemination of geographic knowledge.

Figure 3.3 A human representation, from the opening ceremony of the 2008 Beijing Olympics, of the oars of Admiral Zheng's ships. (© Jiao Weiping/Xinhua Press/© Corbis)

information through fire or flood, from the burning of the Alexandria Library in the Seventh Century that destroyed much of the accumulated knowledge of classical times to the major conflagrations of London in 1666, San Francisco in 1906, or Tokyo in 1945, and the flooding of the Arno that devastated Florence in 1966.

One of the most important periods for geographic representation began in the early Fifteenth Century in Portugal. Henry the Navigator (Box 3.2) is often credited with originating the Age of Discovery, the period of European history that led to the accumulation of large amounts of information about other parts of the world through sea voyages and land explorations. Maps became the medium for sharing information about new discoveries and for administering vast

colonial empires, and their value was quickly recognized. Although detailed representations now exist of all parts of the world, including Antarctica, in a sense the spirit of the Age of Discovery continues in the explorations of the oceans, caves, and outer space, and in the constant process of re-mapping that is needed to keep up with constant changes in the human and natural worlds.

It was the creation, dissemination, and sharing of accurate representations that distinguished the Age of Discovery from all previous periods in human history (and it would be unfair to ignore its distinctive negative consequences, notably the spread of European diseases and the growth of the slave trade). Information about other parts of the world was assembled in the form of maps and journals, reproduced in large numbers using the recently invented printing press, and distributed on paper. Even the modest costs associated with buying copies were eventually addressed through the development of free public lending libraries in the Nineteenth Century, which gave access to virtually everyone. Today, we benefit from what is now a long-standing tradition of free and open access to much of humanity's accumulated store of knowledge about the geographic world, in the form of paper-based representations, through the institution of libraries and the copyright doctrine that gives people rights to material for personal use (see Chapter 18 for a discussion of laws affecting ownership and access). The Internet is the present-day delivery channel that provides distributed access to geographic information through online virtual Earths (specifically those of Google and Microsoft) and other GIServices (Section 1.6.3).

In the Age of Discovery maps became extremely valuable representations of the state of geographic knowledge.

Not surprisingly, representation also lies at the heart of our ability to solve problems using the digital tools that are available in GIS. Any application of GIS requires clear attention to questions of *what* should be represented and *how*. There is a multitude of possible ways of representing the geographic world in digital form, none of which is perfect and none of which is ideal for all applications.

The key GIS representation issues are what to represent and how to represent it.

One of the most important criteria for the usefulness of a representation is its *accuracy*. Because the geographic world is seemingly of infinite complexity, choices are always to be made in building any representation—what to include and what to leave out. When U.S. President Thomas Jefferson dispatched Meriwether Lewis to explore and report on the nature of the lands from the upper Missouri to the Pacific, he said Lewis possessed "a fidelity to the truth so scrupulous that whatever he should report would be as certain as if seen by ourselves." But Jefferson clearly did not expect Lewis to report everything he saw in complete detail: Lewis exercised a large amount of judgment about what to report and what to omit. (The related question of the accuracy of what is reported is taken up at length in Chapter 6.)

One more vital interest drives our need for representations of the geographic world, as well as the need for representations in many other human activities.

When a pilot must train to fly a new type of aircraft, it is much cheaper and less risky for him or her to work with a flight simulator than with the real aircraft. Flight simulators can represent a much wider range of conditions than a pilot will normally experience in flying. Similarly, when decisions have to be made about the geographic world, it is effective to experiment first on models or representations, exploring different scenarios. Of course, this works only if the representation behaves as the real aircraft or world does, and a great deal of knowledge must be acquired about the world before an accurate representation can be built that permits such simulations. But the use of representations for training, exploring future scenarios, and re-creating the past is now common in many fields, including surgery, chemistry, and engineering, and with technologies like GIS it is becoming increasingly common in dealing with the geographic world.

Many plans for the real world can be tried out first on models or representations.

3.4 The Fundamental Problem

Geographic data are built up from atomic elements or from facts about the geographic world. At its most primitive, an atom of geographic data (strictly, a datum) links a place, often a time, and some descriptive property. The first of these, place, is specified in one of several ways that are discussed at length in Chapter 5, and there are also many ways of specifying the second, time. We often use the term *attribute* to refer to the last of these three descriptive properties. For example, consider the statement "The temperature at local noon on December 2, 2010 at latitude 34 degrees 45 minutes north, longitude 120 degrees 0 minutes west, was 19 degrees Celsius." It ties location and time to the property or attribute of atmospheric temperature.

Geographic data link place, time, and attributes.

Other facts can be broken down into their primitive atoms. For example, the statement "Mount Everest is 8848 m high" can be derived from two atomic geographic facts, one giving the location of Mount Everest in latitude and longitude, and the other giving the elevation at that latitude and longitude. Note, however, that the statement would not be a geographic fact to a community that had no way of knowing where Mount Everest is located.

Many aspects of the Earth's surface are comparatively static and slow to change. Height above

Types of Attributes

The simplest type of attribute, termed *nominal*, is one that serves only to identify or distinguish one entity from another. Place-names are a good example, as are names of houses, or the numbers on a driver's license—each serves only to identify the particular instance of a class of entities and to distinguish it from other members of the same class. Nominal attributes include numbers, letters, and even colors. Even though a nominal attribute can be numeric, it makes no sense to apply arithmetic operations to it: adding two nominal attributes, such as two drivers' license numbers, creates nonsense.

Attributes are *ordinal* if their values have a natural order. For example, Canada rates its agricultural land by classes of soil quality, with Class 1 being the best, Class 2 not so good, and so on. Adding or taking the ratios of such numbers makes little sense, since 2 is not twice as much of anything as 1, but at least ordinal attributes have inherent order. Averaging makes no sense either, but the *median*, or the value such that half of the attributes are higher-ranked and half are lower-ranked, is an effective substitute for the average for ordinal data as it gives a useful central value.

Attributes are *interval* if the differences between values make sense. The scale of Celsius temperature is interval because it makes sense to say that 30 and 20 are as different as 20 and 10.

Attributes are *ratio* if the ratios between values make sense. Weight is ratio because it makes sense to say that a person of 100 kg is twice as heavy as a person of 50 kg; but Celsius temperature is only interval because 20 is not twice as hot as 10 (and this argument applies to all scales that are based on similarly arbitrary zero points, including longitude).

In GIS it is sometimes necessary to deal with data that fall into categories beyond these four. For example, data can be directional or *cyclic*, including flow direction on a map, or compass direction, or longitude, or month of the year. The special problem here is that the number following 359 degrees is 0. Averaging two directions such as 359 and 1 yields 180, so the average of two directions close to North can appear to be South. Because cyclic data sometimes occur in GIS, and few designers of GIS software have made special arrangements for them, it is important to be alert to the problems that may arise.

---

sea level changes slowly because of erosion and the movements of the Earth's crust, but these processes operate on scales of hundreds or thousands of years, and for most applications except geophysics we can safely omit time from the representation of elevation. In contrast, atmospheric temperature changes daily, and dramatic changes sometimes occur in minutes with the passage of a cold front or thunderstorm. Thus time is distinctly important, though such climatic variables as mean annual temperature can be represented as static.

There is a vast range of attributes in geographic information. We have already seen that some attributes vary slowly and some rapidly. Some attributes are physical or environmental in nature, while others are social or economic. Some simply *identify* a place or an entity, distinguishing it from all other places or entities; examples include street addresses, social security numbers, or the parcel numbers used for recording land ownership. Other attributes measure something at a location and perhaps at a time (e.g., atmospheric temperature or elevation), while others classify into categories (e.g., the class of land use,

differentiating between agriculture, industry, or residential land). Because attributes are important outside the domain of GIS, there are standard terms for the different types (see Box 3.3).

Geographic attributes are classified as nominal, ordinal, interval, ratio, and cyclic.

But this idea of recording atoms of geographic information, combining location, time, and attribute, misses a fundamental problem, which is that the world is in effect infinitely complex, and the number of atoms required for a complete representation is similarly infinite. The closer we look at the world, the more detail it reveals—and it seems that this process extends ad infinitum. The shoreline of Maine appears complex on a map, but even more complex when examined in greater detail, and as more detail is revealed the shoreline appears to get longer and longer, and more and more convoluted (see Figure 4.15). To characterize the world completely, we would have to specify the location of every person, every blade of grass, and every grain of sand—in fact, every sub-atomic particle, which is clearly an impossible task, since the Heisenberg Uncertainty Principle places limits on the ability to measure precise positions of subatomic particles. Thus in practice any representation must be partial—it must limit the level of detail provided, or ignore change through time, or ignore certain attributes, or simplify in some other way.

The world is infinitely complex, but computer systems are finite. Representations must somehow limit the amount of detail captured.

One very common way of limiting detail is to throw away or ignore information that applies only to small areas—in other words not look too closely. The image you see on a computer screen is composed of a million or so picture elements or *pixels*, and if the whole Earth were displayed at once each pixel would cover an area roughly 10 km on a side, or about 100 sq km. At this level of detail the island of Manhattan occupies roughly 10 pixels, and virtually everything on it is a blur. We would say that such an image has a *spatial resolution* of about 10 km, and know that anything much less than 10 km across is virtually invisible. Figure 3.4 shows Manhattan at a spatial resolution of 250 m, detailed enough to pick out the shape of the island and Central Park.

It is easy to see how this helps with the problem of too much information. The Earth's surface covers about 500 million sq km, so if this level of detail (a spatial resolution of 10 km) is sufficient for an application, a property of the surface such as elevation can be described with only 5 million pieces of information, instead of the 500 million it would take to describe elevation with a resolution of 1 km, and the 500 trillion (500,000,000,000,000) it would take to describe elevation with 1-m resolution.

Another strategy for limiting detail is to observe that many properties remain constant over large areas. For example, in describing the elevation of the Earth's surface we could take advantage of the fact that roughly two-thirds of the surface is covered by water, with its surface at sea level. Of the 5 million pieces of information needed to describe elevation at 10-km resolution, approximately 3.4 million will be recorded as zero, a colossal waste. If we could find an efficient way of identifying the area covered by water, then we would need only 1.6 million real pieces of information.

Humans have found many ingenious ways of describing the Earth's surface efficiently because the problem we are addressing is as old as representation itself and as important for paper-based representa-

Figure 3.4 An image of Manhattan taken by the MODIS instrument on board the TERRA satellite the day after the attack on the World Trade Center on September 11, 2001. MODIS has a spatial resolution of about 250 m, detailed enough to reveal the coarse shape of Manhattan and to identify the Hudson and East rivers, the burning World Trade Center (white spot), and Central Park (the gray blur with the Jacqueline Kennedy Onassis Reservoir visible as a black dot). (Courtesy NOAA)

tions as it is for binary representations in computers. But this ingenuity is itself the source of a substantial problem for GIS: there are many ways of representing the Earth's surface, and users of GIS thus face difficult and at times confusing choices. This chapter discusses some of those choices, and the issues are pursued further in subsequent chapters on uncertainty (Chapter 6) and data modeling (Chapter 8). Representation remains a major concern of GIScience, and researchers are constantly looking for ways to extend GIS representations to accommodate new types of information.

3.5 Discrete Objects and Continuous Fields

3.5.1 Discrete Objects

The level of detail as a fundamental choice in representation has already been mentioned. Another, perhaps even more fundamental, choice is between two conceptual schemes. There is good evidence that we as humans like to simplify the world around us by naming things and by seeing individual things as instances of broader categories. We prefer a world of black and white, of good guys and bad guys, to the real world of shades of gray.

The two fundamental ways of representing geography are discrete objects and continuous fields.

This preference is reflected in one way of viewing the geographic world, known as the *discrete object* view. In this view, the world is empty, except where it is occupied by objects with well-defined boundaries that are instances of generally recognized categories. Just as the desktop is littered with books, pencils, or computers, the geographic world is littered with cars, houses, lampposts, and other discrete objects. In a similar vein the landscape of Minnesota is littered with lakes, and that of Scotland with mountains. One characteristic of the discrete object view is that objects can be counted, so license plates issued by the State of Minnesota carry the legend "10,000 lakes," and climbers know that there are exactly 284 mountains in Scotland over 3,000 ft (the so-called Munros, from Sir Hugh Munro, who originally listed 277 of them in 1891—the count was expanded to 284 in 1997).

The discrete object view represents the geographic world as objects with well-defined boundaries in otherwise empty space.

Biological organisms fit this model well, allowing us to count the number of residents in an area of a city or to describe the behavior of individual bears. Manufactured objects also fit the model, and we have little difficulty counting the number of cars produced in a year or the number of airplanes owned by an airline. But

other phenomena are messier. It is not at all clear what constitutes a mountain, for example, or exactly how a mountain differs from a hill, or when a mountain with two peaks should be counted as two mountains.

Geographic objects are identified by their dimensionality. Objects that occupy area are termed two-dimensional and are generally referred to as areas. The term *polygon* is also common, for the technical reasons explained later. Other objects are more like one-dimensional lines, including roads, railways, or rivers; they are often represented as one-dimensional objects and generally referred to as lines. Other objects are more like zero-dimensional points, such as individual animals or buildings, and are referred to as points.

Of course, in reality, all objects that are perceptible to humans are three dimensional, and their representation in fewer dimensions can be at best an approximation. But the ability of GIS to handle truly three-dimensional objects as volumes with associated surfaces remains limited. GIS increasingly allows for a third (vertical) coordinate to be specified for all point locations. Buildings are sometimes represented by assigning height as an attribute, though if this option is used it is impossible to distinguish flat roofs from any other kind. Various strategies have been used for representing overpasses and underpasses in transportation networks because this information is vital for navigation but not normally represented in strictly two-dimensional network representations. One common strategy is to represent turning options at every intersection, so that an overpass appears in the database as an intersection with no turns (Figure 3.5).

The discrete object view leads to a powerful way of representing geographic information about objects. Think of a class of objects of the same dimensionality—for example, all of the Brown bears (Figure 3.6) in the Kenai Peninsula of Alaska. We would naturally think of these objects as points. We might want to know the sex of each bear and its date

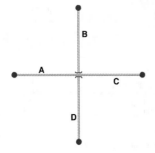

Figure 3.5 The problems of representing a three-dimensional world using a two-dimensional technology. The intersection of links A, B, C, and D is an overpass, so no turns are possible between such pairs as A and B. (Hal Gage/Alaskastock/Photolibrary Group Limited)

From link	To link	Turn?
A	B	No
A	C	Yes
A	D	No
B	C	No
B	D	Yes
B	A	No
C	D	No
C	A	Yes
C	B	No
D	A	No
D	B	Yes
D	C	No

Figure 3.6 Bears are easily conceived as discrete objects, maintaining their identity as objects through time and surrounded by empty space. (Hal Gage/Alaskastock/Photolibrary Group Limited)

of birth, if our interests were in monitoring the bear population. We might also have a collar on each bear that transmitted the bear's location at regular intervals. All of this information could be expressed in a table, such as the one shown in Table 3.1, with each row corresponding to a different discrete object, and each column to an attribute of the object. To reinforce a point made earlier, this is a very efficient way of capturing raw geographic information on brown bears.

But it is not perfect as a representation for all geographic phenomena. Imagine visiting the Earth from another planet and asking the humans what they chose as a representation for the infinitely complex and beautiful environment around them. The visitor would hardly be impressed to learn that they chose tables, especially when the phenomena represented were natural phenomena such as rivers, landscapes,

or oceans. Nothing on the natural Earth looks remotely like a table. It is not at all clear how the properties of a river or the properties of an ocean should be represented as a table. So while the discrete object view works well for some kinds of phenomena, it misses the mark badly for others.

3.5.2 Continuous Fields

Although we might think of terrain as composed of discrete mountain peaks, valleys, ridges, slopes, and the like, and might list them in tables and count them, there are unresolvable problems of definition for all of these objects. Instead, it is much more useful to think of terrain as a continuous surface in which elevation can be defined rigorously at every point (see Box 3.4). Such continuous surfaces form the basis of the other common view of geographic phenomena, known as the *continuous field* view (not to be confused with other meanings of the word field). In this view the geographic world can be described by a number of *variables*, each measurable at any point on the Earth's surface and changing in value across the surface.

> **The continuous field view represents the real world as a finite number of variables, each one defined at every possible position.**

Objects are distinguished by their dimensions and naturally fall into categories of points, lines, or areas. Continuous fields, on the other hand, can be distinguished by what varies and how smoothly. A continuous field of elevation, for example, varies much more smoothly in a landscape that has been worn down by glaciation or flattened by blowing sand than one recently created by cooling lava. Cliffs are places in continuous fields where elevation changes suddenly rather than smoothly. Population density is a kind of continuous field, defined everywhere as the number of people per unit area, though the definition breaks down if the field is examined so closely that the individual people become visible.

Table 3.1 Example of representation of geographic information as a table: the locations and attributes of each of four brown bears in the Kenai Peninsula of Alaska. Locations have been obtained from radio collars. Only one location is shown for each bear, at noon on July 31, 2009 (imaginary data).

Bear ID	Sex	Estimated year of birth	Date of collar installation	Location, noon on 31 July 2009
001	M	2005	02242009	−150.6432, 60.0567
002	F	2003	03312009	−149.9979, 59.9665
003	F	2000	04212009	−150.4639, 60.1245
004	F	2001	04212009	−150.4692, 60.1152

Dimensions

Areas are two-dimensional objects, and volumes are three dimensional, but GIS users sometimes talk about 2.5-D. Almost without exception the elevation of the Earth's surface has a single value at any location (exceptions include overhanging cliffs). So elevation is conveniently thought of as a continuous field, a variable with a value everywhere in two dimensions, and a full 3-D representation is only necessary in areas with an abundance of overhanging cliffs or caves, if these are important features. The idea of dealing with a three-dimensional phenomenon by treating it as a single-valued function of two horizontal variables gives rise to the term 2.5-D. Figure 3.7B shows an example, in this case an elevation surface.

Continuous fields can also be created from classifications of land into categories of land use or soil type. Such fields change suddenly at the boundaries between different classes. Other types of fields can be defined by continuous variation along lines rather than across space. Traffic density, for example, can be defined everywhere on a road network, and flow volume can be defined everywhere on a river. Figure 3.7 shows some examples of field-like phenomena.

Continuous fields can be distinguished by *what* is being measured at each point. As described in Box 3.3, the variable may be nominal, ordinal, interval, ratio, or cyclic. A *vector* field assigns two variables, magnitude and direction, at every point in space and is used to represent flow phenomena such as winds or currents; fields of only one variable are termed *scalar* fields.

Here is a simple example illustrating the difference between the discrete object and field conceptualizations. Suppose you were hired for the summer to count the number of lakes in Minnesota, and were promised that your answer would appear on every license plate issued by the state. The task sounds simple, and you were happy to get the job. But on the first day you started to run into difficulty (Figure 3.8). What about small ponds—do they count as lakes? What about wide stretches of rivers? What about swamps that dry up in the summer? And is a lake with a narrow section connecting two wider parts one lake or two? Your biggest dilemma concerns the scale of mapping, since the number of lakes shown on a map clearly depends on the map's level of detail; a more detailed map almost certainly will show more lakes.

Your task clearly reflects a discrete object view of the phenomenon. The action of counting implies that lakes are discrete, two-dimensional objects littering an otherwise empty geographic landscape. In a continuous field view, however, all points are either lake or nonlake. Moreover, we could refine the scale a little to take account of marginal cases; for example, we

Figure 3.7 Examples of field-like phenomena. (A) Image of part of the Dead Sea in the Middle East. The lightness of the image at any point measures the amount of radiation captured by the satellite's imaging system. (B) A simulated image derived from the Shuttle Radar Topography Mission. The image shows the Carrizo Plain area of Southern California, with a simulated sky and with land cover obtained from other satellite sources. (Courtesy NASA/ JPL–Caltech)

(A)

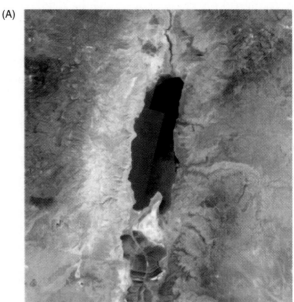

(B)

Figure 3.8 Lakes are difficult to conceptualize as discrete objects because it is often difficult to tell where a lake begins and ends, or to distinguish a wide river from a lake. (Oliviero Olivieri/Getty Images, Inc.)

might define the scale shown in Table 3.2, which has five degrees of lakeness. The complexity of the view would depend on how closely we looked, of course, and so the scale of mapping would still be important. But all of the problems of defining a lake as a discrete object would disappear (though there would still be problems in defining the levels of the scale). Instead of counting, our strategy would be to lay a grid over the map, and assign each grid cell a score on the lakeness scale. The size of the grid cell would determine how accurately the result approximated the value we could

Table 3.2 A scale of lakeness suitable for defining lakes as a continuous field.

Lakeness	Definition
1	Location is always dry under all circumstances.
2	Location is sometimes flooded in spring.
3	Location supports marshy vegetation.
4	Water is always present to a depth of less than 1 m.
5	Water is always present to a depth of more than 1 m.

theoretically obtain by visiting every one of the infinite number of points in the state. At the end, we would tabulate the resulting scores, counting the number of cells having each value of lakeness, or averaging the lakeness score. We could even design a new and scientifically more reasonable license plate—"Minnesota, 12% lake" or "Minnesota, average lakeness 2.02."

The difference between objects and fields is also well illustrated by photographs (e.g., Figure 3.7A). Paper images produced from old-fashioned photographic film are created by variation in the chemical state of the material in the photographic film. In early photography, minute particles of silver were released from molecules of silver nitrate when the unstable molecules were exposed to light, thus darkening the image in proportion to the amount of incident light. We think of the image as a field of continuous variation in color or darkness. But when we look at the image, the eye and brain begin to infer the presence of discrete objects, such as people, rivers, fields, cars, or houses, as they interpret the content of the image.

3.6 Rasters and Vectors

Continuous fields and discrete objects define two conceptual views of geographic phenomena, but they do not solve the problem of digital representation. A continuous field view still potentially contains an infinite amount of information if it defines the value of the variable at every point, since there are an infinite number of points in any defined geographic area. Discrete objects can also require an infinite amount of information for full description. For example, a coastline contains an infinite amount of information if it is mapped in infinite detail. Thus continuous fields and discrete objects are no more than conceptualizations, or ways in which we think about geographic phenomena; they are not designed to deal with the limitations of computers.

Two methods are used to reduce geographic phenomena to forms that can be coded in computer databases, and we call these methods raster and vector. In principle, both can be used to code both fields and discrete objects, but in practice a strong association exists between raster and fields, and between vector and discrete objects.

Raster and vector are two methods of representing geographic data in digital computers.

3.6.1 Raster Data

In a raster representation space is divided into an array of rectangular (usually square) cells (Figure 3.9). All geographic variation is then expressed by assigning

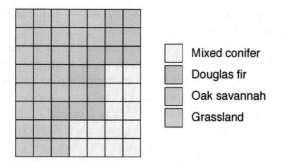

Legend:
- Mixed conifer
- Douglas fir
- Oak savannah
- Grassland

Figure 3.9 Raster representation. Each color represents a different value of a nominal-scale variable denoting land cover class.

properties or attributes to these cells. The cells are sometimes called pixels (short for *picture elements*).

Raster representations divide the world into arrays of cells and assign attributes to the cells.

One of the most common forms of raster data comes from remote-sensing satellites, which capture information in this form and send it to the ground to be distributed and analyzed. Data from the Landsat Thematic Mapper, for example, which are commonly used in GIS applications, come in cells that are 30 m a side on the ground, or approximately 0.1 ha (hectare) in area. Similar data can be obtained from sensors mounted on aircraft. Imagery varies according to the spatial resolution (expressed as the length of a cell side as measured on the ground) and also according to the timetable of image capture by the sensor. Some satellites are in *geostationary* orbit over a fixed point on the Earth and capture images constantly. Others pass over a fixed point at regular intervals (e.g., every 12 days). Finally, sensors vary according to the part or parts of the spectrum that they sense. The visible parts of the spectrum are most important for remote sensing, but some invisible parts of the spectrum are particularly useful in detecting heat and the phenomena that produce heat, such as volcanic activities. Many sensors capture images in several areas of the spectrum, or *bands*, simultaneously, because the relative amounts of radiation in different parts of the spectrum are often useful indicators of certain phenomena, such as green leaves, or water, on the Earth's surface. The AVIRIS (Airborne Visible InfraRed Imaging Spectrometer) captures no fewer than 224 different parts of the spectrum and is being used to detect particular minerals in the soil, among other applications. Remote sensing is a complex topic, and further details are available in Chapter 9.

Square cells fit together nicely on a flat table or a sheet of paper, but they will not fit together neatly on the curved surface of the Earth. So just as representations on

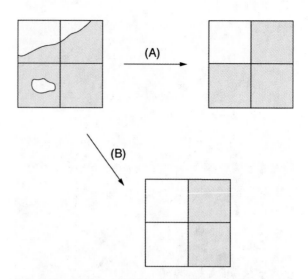

Figure 3.10 Effect of a raster representation using (A) the largest share rule and (B) the central point rule.

paper require that the Earth be flattened, or projected, so too do rasters. (Because of the distortions associated with flattening, the cells in a raster can never be perfectly equal in shape or area on the Earth's surface.) Projections, or ways of flattening the Earth, are described in Section 5.8. Many of the terms that describe rasters suggest the laying of a tile floor on a flat surface—we talk of raster cells *tiling* an area, and a raster is said to be an instance of a *tesselation*, derived from the word for a mosaic. The mirrored ball hanging above a dance floor recalls the impossibility of covering a spherical object like the Earth perfectly with flat, square pieces.

When information is represented in raster form, all detail about variation within cells is lost, and instead the cell is given a single value. Suppose we wanted to represent the map of the counties of Texas as a raster. Each cell would be given a single value to identify a county, and we would have to decide on the rule to apply when a cell falls in more than one county. Often the rule is that the county with the *largest share* of the cell's area gets the cell. Sometimes the rule is based on the *central point* of the cell, and the county at that point is assigned to the whole cell. Figure 3.10 shows these two rules in operation. The largest share rule is almost always preferred, but the central point rule is sometimes used in the interests of faster computing and is often used in creating raster datasets of elevation.

3.6.2 Vector Data

In a vector representation, all lines are captured as points connected by precisely straight lines. (Some GIS software allows points to be connected by curves rather than straight lines, but in most cases curves have to be approximated by increasing the density of points.) An area is captured as a series of points

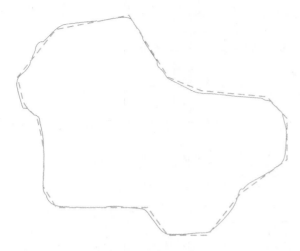

Figure 3.11 An area (red line) and its approximation by a polygon (dashed blue line).

or *vertices* connected by straight lines as shown in Figure 3.11. The straight edges between vertices explain why areas in vector representation are often called *polygons*, and in GIS-speak the terms *polygon* and *area* are often used interchangeably. Lines are captured in the same way, and the term *polyline* has been coined to describe a curved line represented by a series of straight segments connecting vertices.

To capture an area object in vector form, we need only specify the locations of the points that form the vertices of a polygon. This seems simple and is also much more efficient than a raster representation, which would require us to list all of the cells that form the area. These ideas are captured succinctly in the comment "Raster is vaster, but vector is correcter." To create a precise approximation to an area in raster, it would be necessary to resort to using very small cells, and the number of cells would rise proportionately (with every halving of the width and height of each cell resulting in a quadrupling of the number of cells). But things are not quite as simple as they seem. The apparent precision of vector is often not real, since many geographic phenomena simply cannot be located with high accuracy (see Section 6.3.2.2). So although raster data may look less attractive, they may be more honest to the

inherent quality of the data. Also, various methods exist for compressing raster data that can greatly reduce the capacity needed to store a given dataset (see Chapter 8). So the choice between raster and vector is often complex, as summarized in Table 3.3.

3.6.3 Representing Continuous Fields

While discrete objects lend themselves naturally to representation as points, lines, or areas using vector methods, it is less obvious how the continuous variation of a field can be expressed in a digital representation. In GIS six alternatives are commonly implemented (Figure 3.12):

A. Capturing the value of the variable at each of a grid of regularly spaced sample points (for example, elevations at 30-m spacing in a DEM).

B. Capturing the value of the field variable at each of a set of irregularly spaced sample points (for example, variation in surface temperature captured at weather stations).

C. Capturing a single value of the variable for a regularly shaped cell (for example, values of reflected radiation in a remotely sensed scene).

D. Capturing a single value of the variable over an irregularly shaped area (for example, vegetation cover class or the name of a parcel's owner).

E. Capturing the linear variation of the field variable over an irregularly shaped triangle (for example, elevation captured in a triangulated irregular network or TIN, Section 8.2.3.4).

F. Capturing the isolines of a surface, as digitized lines (for example, digitized contour lines representing surface elevation).

Each of these methods succeeds in compressing the potentially infinite amount of data in a continuous field to a finite amount, using one of the six options, two of which (A and C) are raster and four (B, D, E, and F) are vector. Of the vector methods one (B) uses points, two (D and E) use polygons, and one (F) uses lines to express the continuous spatial variation of the field in terms of a finite set of vector objects. But unlike the

Table 3.3 Relative advantages of raster and vector representation.

Issue	Raster	Vector
Volume of data	Depends on cell size	Depends on density of vertices
Sources of data	Remote sensing, imagery	Social and environmental data
Applications	Resources, environmental	Social, economic, administrative
Software	Raster GIS, image processing	Vector GIS, automated cartography
Resolution	Fixed	Variable

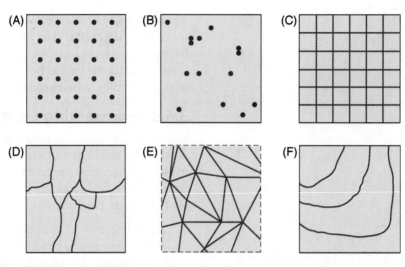

Figure 3.12 The six approximate representations of a field used in GIS. (A) Regularly spaced sample points. (B) Irregularly spaced sample points. (C) Rectangular cells. (D) Irregularly shaped polygons. (E) Irregular network of triangles, with linear variation over each triangle (the Triangulated Irregular Network or TIN model; the bounding box is shown dashed in this case because the unshown portions of complete triangles extend outside it). (F) Polylines representing contours (see the discussion of isopleth maps in Box 4.3) (Courtesy U.S. Geological Survey)

discrete object conceptualization, the objects used to represent a field are not real, but simply artifacts of the representation of something that is actually conceived as spatially continuous. The triangles of a TIN representation (E), for example, exist only in the digital representation and cannot be found on the ground, and neither can the lines of a contour representation (F).

3.7 The Paper Map

The paper map has long been a powerful and an effective means of communicating geographic information. In contrast to digital data, which use coding schemes such as ASCII, it is an instance of an *analog* representation, or a physical model in which the real world is scaled; in the case of the paper map, part of the world is scaled to fit the size of the paper. A key property of a paper map is its *scale* or *representative fraction*, defined as the ratio of distance on the map to distance on the Earth's surface. For example, a map with a scale of 1:24,000 reduces everything on the Earth to one-24,000th of its real size. This is a bit misleading because the Earth's surface is curved and a paper map is flat, so scale cannot be exactly constant.

> **A paper map is a source of data for geographic databases; an analog product from a GIS; and an effective communication tool.**

Maps have been so important, particularly prior to the development of digital technology, that many of the ideas associated with GIS are actually inherited directly from paper maps. For example, scale is often cited as a property of a digital database, even though the definition of scale makes no sense for digital data—ratio of distance *in the computer* to distance on the ground; how can there be distances in a computer? What is meant is a little more complicated: when a scale is quoted for a digital database, it is usually the scale of the map that formed the source of the data. So if a database is said to be at a scale of 1:24,000, one can safely assume that it was created from a paper map at that scale and includes representations of the features that are found on maps at that scale. Further discussion of scale can be found in Box 4.2 and in Chapter 6, where scale is important to the concept of uncertainty.

There is a close relationship between the contents of a map and the raster and vector representations discussed in the previous section. The U.S. Geological Survey, for example, distributes two digital versions of its topographic maps, one in raster form and one in vector form, and both attempt to capture the contents of the map as closely as possible. In the raster form, or *digital raster graphic* (DRG), the map is scanned at a very high density, using very small pixels, so that the raster looks very much like the original (Figure 3.13). The coding of each pixel simply records the color of the map picked up by the scanner, and the dataset includes all of the textual information surrounding the actual map.

In the vector form, or *digital line graph* (DLG), every geographic feature shown on the map is

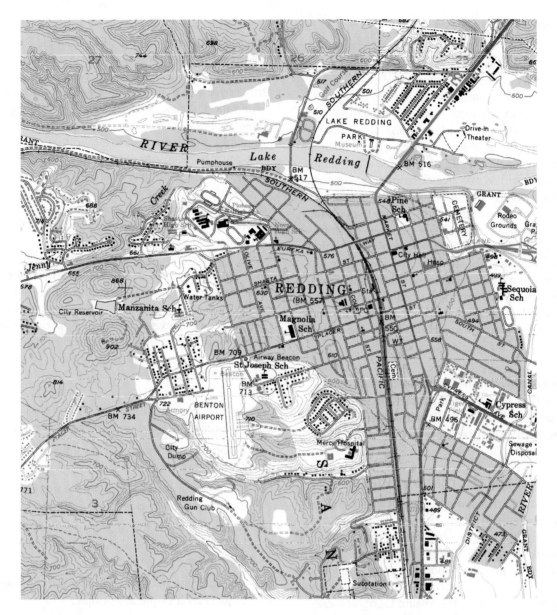

Figure 3.13 Part of a Digital Raster Graphic, a scan of a U.S Geological Survey 1:24 000 topographic map. (Courtesy USGS)

represented as a point, polyline, or polygon. The symbols used to represent point features on the map, such as the symbol for a windmill, are replaced in the digital data by points with associated attributes and must be regenerated when the data are displayed. Contours, which are shown on the map as lines of definite width, are replaced by polylines of no width and given attributes that record their elevations. In both cases, and especially in the vector case, there is a significant difference between the analog representation of the map and its digital equivalent. So it is quite misleading to think of the contents of a digital representation as a map and to think of a GIS as a container of digital maps. Digital representations can include information that would

be very difficult to show on maps. For example, they can represent the curved surface of the Earth, without the need for the distortions associated with flattening. They can represent changes, whereas maps must be static because it is very difficult to change their contents once they have been printed or drawn. Digital databases can represent all three spatial dimensions, including the vertical, whereas maps must always show two-dimensional views. So while the paper map is a useful metaphor for the contents of a geographic database, we must be careful not to let it limit our thinking about what is possible in the way of representation. This issue is pursued at greater length in Chapter 8, and map production is discussed in detail in Chapter 12.

3.8 Generalization

In Section 3.4 we saw how thinking about geographic information as a collection of atomic links—between a place, a time (not always, because many geographic facts are stated as if they were permanently true), and a property—led to an immediate problem because the potential number of such atomic facts is infinite. If seen in enough detail, the Earth's surface is unimaginably complex, and its effective description impossible. So instead, humans have devised numerous ways of simplifying their view of the world. Rather than making statements about each and every point, we describe entire areas, attributing uniform characteristics to them, even when areas are not strictly uniform; we identify features on the ground and describe their characteristics, again assuming them to be uniform; or we limit our descriptions to what exists at a finite number of sample points, hoping that these samples will be adequately representative of the whole (Section 4.4).

A geographic database cannot contain a perfect description; instead, its contents must be carefully selected to fit within the limited capacity of computer storage devices.

3.8.1 Generalization about Places

From this perspective some degree of generalization is almost inevitable in all geographic data. But cartographers often take a somewhat different approach, for which this observation is not necessarily true. Suppose we are tasked to prepare a map at a specific scale, say 1:25,000, using the standards laid down by a national mapping agency, such as the Institut Géographique National (IGN) of France. Every scale used by IGN has its associated rules of representation. For example, at a scale of 1:25,000 the rules specify that individual buildings will be shown only in specific circumstances, and similar rules apply to the 1:24,000 series of the U.S. Geological Survey. These rules are known by various names, including *terrain nominal* in the case of IGN, which translates roughly but not very helpfully to "nominal ground," and is perhaps better translated as "specification." From this perspective a map that represents the world by following the rules of a specification precisely can be perfectly accurate *with respect to the specification*, even though it is not a perfect representation of the full detail on the ground.

A map's specification defines how real features on the ground are selected for inclusion on the map.

Consider the representation of vegetation cover using the rules of a specification. For example, the rules might state that at a scale of 1:100,000, a vegetation cover map should not show areas of vegetation that cover less than 1 ha (hectare). But small areas of vegetation almost certainly exist, so deleting them inevitably results in information loss. But under the principle discussed above, a map that adheres to this rule must be accurate, *even though it differs substantively from the truth as observed on the ground.*

A GIS dataset's level of detail is one of its most important properties, as it determines both the degree to which the dataset approximates the real world and the dataset's complexity. In the interests of compressing data, it is often necessary to remove detail, fitting them into a storage device of limited capacity, processing them faster, or creating less confusing visualizations that emphasize general trends. Consequently, many methods have been devised for generalization, and several of the more important are discussed in this section.

McMaster and Shea identify the following types of generalization rules:

- *Simplification*, for example, by weeding out points in the outline of a polygon to create a simpler shape.

- *Smoothing*, or the replacement of sharp and complex forms by smoother ones.

- *Collapse*, or the replacement of an area object by a combination of point and line objects.

- *Aggregation*, or the replacement of a large number of distinct symbolized objects by a smaller number of new symbols.

- *Amalgamation*, or the replacement of several area objects by a single area object.

- *Merging*, or the replacement of several line objects by a smaller number of line objects.

- *Refinement*, or the replacement of a complex pattern of objects by a selection that preserves the pattern's general form.

- *Exaggeration*, or the relative enlargement of an object to preserve its characteristics when these would be lost if the object were shown to scale.

- *Enhancement*, through the alteration of the physical sizes and shapes of symbols.

- *Displacement*, or the moving of objects from their true positions to preserve their visibility and distinctiveness.

The differences between these types of rules are much easier to understand visually, and Figure 3.14 reproduces McMaster's and Shea's original

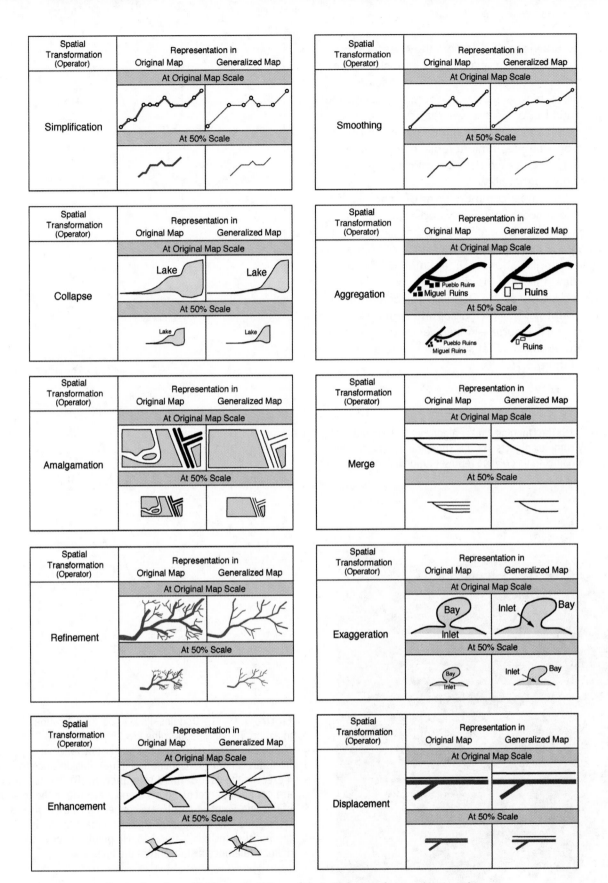

Figure 3.14 Illustrations from McMaster and Shea of their 10 forms of generalization. The original feature is shown at its original level of detail, and below it at 50% coarser scale. Each generalization technique resolves a specific problem of display at coarser scale and results in the acceptable version shown in the lower right.

illustrative drawings. In addition, they describe two forms of generalization of attributes, as distinct from geometric forms of generalization. *Classification* generalization reclassifies the attributes of objects into a smaller number of classes, whereas *symbolization* generalization changes the assignment of symbols to objects. For example, it might replace an elaborate symbol including the words "Mixed Forest" with a color identifying that class.

One of the most common forms of generalization in GIS is the process known as weeding, or the simplification of the representation of a line represented as a polyline. The process is an instance of McMaster and Shea's simplification. Standard methods exist in GIS for doing this, and the most common by far is the method known as the Douglas–Poiker algorithm after its inventors, David Douglas and Tom Poiker. The operation of the Douglas–Poiker weeding algorithm upon features such as that shown in Figure 3.15 is shown in Figure 3.16.

Figure 3.15 The Douglas–Poiker algorithm is designed to simplify complex objects like this shoreline by reducing the number of points in its polyline representation. (© iStockphoto)

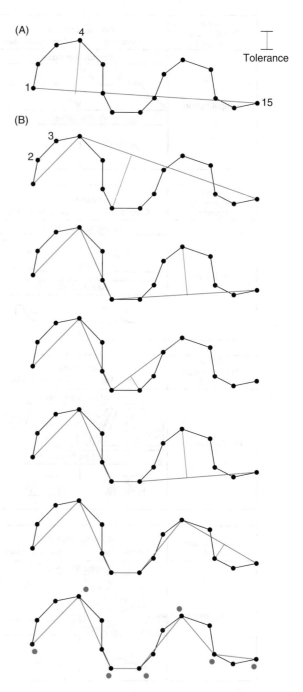

Figure 3.16 The Douglas–Poiker line simplification algorithm in action. The original polyline has 15 points. In (A) Points 1 and 15 are connected (red), and the furthest distance of any point from this connection is identified (blue). This distance to Point 4 exceeds the user-defined tolerance. In (B) Points 1 and 4 are connected (green). Points 2 and 3 are within the tolerance of this line. Points 4 and 15 are connected, and the process is repeated. In the final step 7 points remain (identified with green disks), including 1 and 15. No points are beyond the user-defined tolerance distance from the line.

Weeding is the process of simplifying a line or an area by reducing the number of points in its representation.

Note that the algorithm relies entirely on the assumption that the line is represented as a polyline—in other words, as a series of straight-line segments. GIS increasingly supports other representations, including arcs of circles, arcs of ellipses, and Bézier curves, but there is little consensus to date on appropriate methods for weeding or generalizing them, or on methods of analysis that can be applied to them.

3.8.2 Generalization about Properties

We saw in Sections 1.3 and 1.4 how the pace of scientific research had accelerated over the past two decades, in no small part because the Internet now provides a globally linked network of data warehouses and clearinghouses. In theory, the provenance of each of these various sources is known and is documented using metadata ("data about data": see Section 11.2). Many representations in GIS bring together multiple properties of places as composite *indicators* of conditions at particular locations, in order to fulfill particular needs. Thus, for example, the summary measure of social "deprivation" mapped for London in Figure 1.15 combines representations of conditions with respect to heath, employment, and housing, among other social phenomena. In a similar vein, multiple criteria might be combined in order to define "wilderness" in an environmental application (see Section 1.3.1). Problems arise if some of these constituent representations of places have been collected at scales, or using areal units, that are less appropriate to the purpose than others. Thus a standard soil map may not be appropriate to inform a gardener whether a land parcel is suitable for growing sunflowers, yet a terrain model of aspect and elevation may well be suitable for this purpose. We return to the procedures and problems for generalizing about the properties of places in Sections 4.4 and 6.3.2. These questions compound the fundamental problem that representations are necessarily incomplete (Section 3.4) and a representation may not be fit for its purpose if it is based upon inappropriate indicators of real world conditions (see Box 3.5).

3.8.3 Representation Using VGI

In Section 1.8 we discussed the rise of *volunteered geographic information* (VGI) and the ways in which it can facilitate the gathering of the widest range of properties of places at hitherto unimaginable levels of detail. In some cases, it may be possible to invoke automated procedures that are robust and open to scrutiny in order to ensure the acceptability of a representation for the widest range of purposes. But in the absence of knowledge about the proficiency and knowledge of every volunteer, the devil may be in the details. The wider "Wikification of GIS" entails few safeguards about the reliability of ways in which the properties of places are represented. VGI can be defined in broad or narrow terms, with the broad view including, for example, data that we "volunteer" to supermarket chains or airlines about our purchasing behavior in order to receive discounts or awards, rather than those supplied for purely altruistic purposes. From this broad perspective, here are some questions to ask about the generalizability of representations based on VGI:

1. Are all places equally accessible to volunteers, or is access to some areas difficult (e.g., physically remote areas) or constrained (e.g., gated communities)?

2. Are volunteers more likely to provide data on places or properties that interest them than those that do not? For example, social class bias among OpenStreetMap volunteers appears to have led to a lack of coverage of low-status, peripheral housing developments in the early stages of the endeavour.

3. Where volunteers supply information about themselves, is this representative of other nonvolunteers? Retail-store loyalty card data, for example, may not be representative of those who do not patronize the store chain, or those who opt out of loyalty programs.

4. Are volunteers at liberty to collect locationally sensitive data? For example, the Chinese government appears resistant to initiatives like OpenStreetMap, and other governments remain sensitive to record taking near military installations.

5. Should volunteers supply data that are open to malevolent use? This choice may appear straightforward in some circumstances (e.g., not revealing the locations of rare bird nesting sites to possible egg collectors) but in other instances is less clear-cut (e.g., mapping the geography of mobility-impaired individuals will help in developing a neighborhood fire evacuation strategy, but the information may also be of interest to criminals).

Denis Wood, Mapmaker

Denis Wood (Figure 3.17) is one of America's pre-eminent experts on the significance and meaning of maps. Wood loves maps and loves to talk about them, though he is skeptical about many of the uses to which they continue to be put. His bestselling book *The Power of Maps*—originally published in 1992—successively analyzes how maps work by serving interests, and describes how the selective nature of representation overwhelmingly serves the objectives of vested interests that use them to maintain or further distort the status quo. Some consider that GIS makes this process more transparent, while others suggest the opposite. Elsewhere in this chapter we very much take the former view, consistent with Denis's recent work (with John Krygier) that emphasizes the importance of good GIS design principles.

Figure 3.17 Denis Wood, critic of "re-presentation." (Courtesy of Denis Wood)

Denis has challenged our taken-for-granted ways of thinking about geographic representations, especially maps. During the 1980s and 1990s, he was among those committed to foregrounding the interests—political, economic, and other—that motivated even the most apparently dispassionate of representations. In 1986 he and John Fels published a seminal analysis of the way that the North Carolina State highway map served the mass of interests—economic, political, and racial—that gave that map its form. In 1992 and 1994 Denis curated a path-breaking exhibition for Smithsonian Institution museums in New York and Washington. This helped visitors to understand the interests informing Napoleon's mapping of the Middle East and the U.S. mapping of its West; the profound economic interests motivating topographic surveys; the interests of the automobile industry, big oil, and big rubber behind highway maps; and the interests, often personal, that arguably underlie *every* map.

During the 1990s Denis gradually shifted his attention from the interests shaping maps to the nature of cartographic representation itself, increasingly arguing that far from being "re-presentations," maps are actually systems of propositions that are rhetorically structured to make arguments about the nature of the world. It is through these propositions, his argument goes, that one worldview is advanced at the expense of another. This case is persuasively made in *The Natures of Maps*, that Denis also co-authored with John Fels, which explores the propositional logic of the map in the context of cartographic representation of the natural world.

Denis was born in 1945 in Cleveland, Ohio. He was educated at Case-Western Reserve and Clark Universities and taught landscape history, environmental psychology, and design for 23 years at North Carolina State University.

Check out Denis's Narrative Atlas of Boylan Heights at makingmaps.net.

---

6. Is it socially acceptable to make available *any* observations of uniquely identifiable individuals—especially if they might be undertaking behavior that might be construed as socially unacceptable)?

7. Is the information date-stamped, as an indicator of its provenance and current liability?

3.9 Conclusion

Representation, or more broadly *ontology*, is a fundamental issue in GIS, since it underlies all of our efforts to express useful information about the surface of the Earth in a digital computer. The fact that there are so many ways of doing this makes GIS at once complex and interesting, a point that will become much clearer on reading the technical chapter on data modeling (Chapter 8). But the broader issues of representation, including the distinction between field and object conceptualizations, underlie not only that chapter but many other issues as well, including uncertainty (Chapter 6) and analysis and modeling (Chapters 14 through 16).

Questions for Further Study

1. What fraction of the Earth's surface have you experienced in your lifetime? Make diagrams like that shown in Figure 3.1, at appropriate levels of detail, to show (a) where you have lived in your lifetime, (b) how you spent last weekend. How would you describe what is missing from each of these diagrams?

2. Table 3.3 summarized some of the relative merits of raster and vector representations. Expand on these, and use your enlarged table to assess the suitability of the two representational forms for a range of GIS applications.

3. The early explorers had limited ways of communicating what they saw, but many were very effective at it. Examine the published diaries, notebooks, or dispatches of one or two early explorers and look at the methods they used to communicate with others. What words did they use to describe unfamiliar landscapes, and how did they mix words with sketches?

4. With reference to Google Maps (maps.google. com) or another mapping site, identify the extent of your own neighborhood. Mark on it the discrete objects you are familiar with in the area. What features are hard to think of as discrete objects? For example, how will you divide up the various roadways in the neighborhood into discrete objects—where do they begin and end?

Further Reading

Chrisman, N.R. 2002. *Exploring Geographic Information Systems* (2nd ed.). New York: Wiley.

De Smith, M., Goodchild, M.F., and Longley, P.A. 2009 *Geospatial Analysis* (3rd ed.). Leicester, Troubador. See also www.spatialanalysisonline.com.

McMaster, R.B. and Shea, K.S. 1992. *Generalization in Digital Cartography*. Washington, DC: Association of American Geographers.

Wood, D. and Krygier, J. 2005. *Making Maps: A Visual Guide to Map Design for GIS*. New York: Guilford Press.

4 The Nature of Geographic Data

This chapter elaborates on the *spatial is special* theme by examining the nature of geographic data. It sets out the distinguishing characteristics of geographic data and suggests a range of guiding principles for working with them. Many geographic data are correctly thought of as sample observations, selected from the larger universe of possible observations that could be made. This chapter describes the main principles that govern scientific sampling and the principles that are invoked in order to infer information about the gaps between samples. When devising spatial sample designs, it is important to be aware of the nature of spatial variation, and here we learn how this is formalized and measured as spatial autocorrelation. Another key property of geographic information is the level of detail that is apparent at particular scales of analysis. The concept of fractals provides a solid theoretical foundation for understanding scale when building geographic representations.

LEARNING OBJECTIVES

After studying this chapter you will understand:

- How Tobler's First Law of Geography is formalized through the concept of spatial autocorrelation.

- The relationship between scale and the level of geographic detail in a representation.

- The principles of building representations around geographic samples.

- How the properties of smoothness and continuous variation can be used to characterize geographic variation.

- How fractals can be used to measure and simulate surface roughness.

4.1 Introduction

In Chapter 1 we identified the central motivation for scientific applications of GIS as the development of representations, not only of how the world *looks*, but also how it *works*. Chapter 3 established three governing principles that help us toward this goal, namely,

1. that the representations we build in GIS are of *unique places*;

2. that our representations of them are necessarily *selective* of reality, and hence *incomplete*;

3. that in building representations, it is useful to think of the world as either comprising continuously varying *fields* or as an empty space littered with crisp and well-defined *objects*

In this chapter we build on these principles to develop a fuller understanding of the ways in which the *nature of spatial variation* is represented in GIS. We do this by asserting three further principles:

4. that proximity effects are key to representing and understanding spatial variation, and to joining up incomplete representations of unique places;

5. that issues of geographic scale and level of detail are key to building appropriate representations of the world; and

6. that different measures of the world *covary*, and understanding the nature of covariation can help us to predict.

Implicit in all of this is one further principle that we will develop in Chapter 6: because almost all representations of the world are necessarily incomplete, they are *uncertain*. GIS is about representing spatial and temporal phenomena in the real world, and because the real world is complicated, this task is difficult and error prone. The real world provides an intriguing laboratory in which to examine phenomena, but is one in which it can be impossible to control for variation in all characteristics—be they relevant to landscape evolution, consumer behavior, urban growth, or whatever. In the terminology of Section 1.3.1, generalized *laws* governing spatial distributions and temporal dynamics are therefore most unlikely to work perfectly.

We choose to describe the seven points above as principles rather than laws because, like our discussion in Chapter 2, this chapter is grounded in empirical generalization about the real world. A more elevated discussion of the way that these principles

build into fundamental laws of GIScience has been published by Goodchild.

4.2 The Fundamental Problem Revisited

Consider for a moment a GIS-based representation of your own life history to date. It is infinitesimally small compared with the geographic extent and history of the world, but as we move to finer spatial and temporal scales than those shown in Figure 3.1, it nevertheless remains very intricate in detail. Viewed in aggregate, human behavior in space appears structured when we aggregate the outcomes of day-to-day (often repetitive) decisions about where to go, what to do, or how much time to spend doing it. Over the longer term, structure also arises out of (one-off) decisions about where to live, how to achieve career objectives, and how to balance work, leisure, and family pursuits.

It is helpful to distinguish between *controlled* and *uncontrolled* variation over time. Controlled variation in our lives oscillates around a steady state (daily, weekly) pattern, while uncontrolled variation (career changes, residential moves) does not. The same might be said, respectively, of seasonal change in climate and the phenomenon of global warming. When relating our own daily regimes and life histories, or indeed any short- or long-term *time series* of events, we are usually mindful of the contexts in which our decisions (to go to work, to change jobs, to marry) are made; "the past is the key to the present" aptly summarizes the effect of temporal context on our actions. The day-to-day operational context to our activities is very much determined by where we live and work. The longer-term strategic context may well be provided by where we were born, grew up, or went to college.

Our behavior in geographic space often reflects past patterns of behavior.

The relationship between consecutive events in *time* can be formalized in the concept of *temporal autocorrelation*. The analysis of time series data is in some senses straightforward, since the direction of causality is only one way: past events are sequentially related to the present and to the future. This chapter (and book) is principally concerned with spatial, rather than temporal, autocorrelation. Spatial autocorrelation shares some similarities with its temporal counterpart. Yet time moves in one direction only (forward), making temporal autocorrelation one-dimensional, while spatial events can potentially have consequences anywhere in two-dimensional or even three-dimensional space.

Explanation in time need only look to the past, but explanation in space must look in all directions simultaneously.

Assessment of spatial autocorrelation can be informed by knowledge of the degree and nature of *spatial heterogeneity*—the tendency of geographic places and regions to be different from each other. Everyone would recognize the extreme difference of landscapes between such regions as the Antarctic, the Nile Delta, the Sahara Desert, or the Amazon Basin, and many would recognize the more subtle differences between the Central Valley of California, the Northern Plain of China, and the valley of the Ganges in India. Heterogeneity occurs both in the way the landscape looks and in the way processes act on the landscape (see the form/process distinction of Section 1.3). As with change over time, variation in space may be controlled or uncontrolled: the spatial variation in some processes simply oscillates about an average (controlled variation), while other processes vary ever more the longer they are observed (uncontrolled variation). For example, controlled variation characterizes the operational environment of GIS applications in utility management (see Section 2.3.2), or the tactical environment of retail promotions (see Section 2.3.3), while longer-term processes such as global warming or deforestation may exhibit uncontrolled variation (see Section 2.3.5). As a general rule, spatial data exhibit an increasing range of values, hence increased heterogeneity, with increased distance. In this chapter we focus on the ways in which phenomena vary across space and on the general nature of geographic variation. Later, in Section 14.2.1, we return to the techniques for measuring spatial heterogeneity. Also, this requires us to move beyond thinking of GIS data as abstracted only from the continuous spatial distributions implied by Tobler's Law (Section 3.1) and from sequences of events over continuous time. Some events, such as the daily rhythm of the journey to work, are clearly incremental extensions of past practice, whereas others, such as residential relocation, constitute sudden breaks with the past. Similarly, landscapes of gently undulating terrain are best thought of as smooth and continuous, while others (such as the landscapes developed about fault systems, or mountain ranges) are best conceived as discretely bounded, jagged, and irregular. Smoothness and irregularity turn out to be among the most important distinguishing characteristics of geographic data.

Some geographic phenomena vary smoothly across space, while others can exhibit extreme irregularity, in violation of Tobler's Law.

Finally, it is highly likely that a representation of the real world that is suitable for predicting future change will need to incorporate information on how two or more factors *covary*. For example, planners seeking to justify improvements to a city's public transit system might wish to point out how house prices increase with proximity to existing rail stops. It is highly likely that patterns of spatial autocorrelation in one variable will, to a greater or lesser extent, be mirrored in another. However, while this is helpful in building representations of the real world, we will see in Section 15.5.1 that the property of spatial autocorrelation can frustrate our attempts to build inferential statistical models of the covariation of geographic phenomena.

Spatial autocorrelation helps us to build representations but frustrates our efforts to predict.

The nature of geographic variation, the scale at which uncontrolled variation occurs, and the way in which different geographic phenomena covary are all key to building effective representations of the real world. These principles are of practical importance and guide us toward answering questions such as: What is an appropriate scale or level of detail at which to build a representation for a particular application? How do I design my spatial sample? How do I generalize from my sample measurements? And what formal methods and techniques can I use to relate key spatial events and outcomes to one another?

Each of these questions is a facet of the fundamental problem of GIS, that is, of selecting what to leave in and what to take out of our digital representations of the real world (Section 3.4). The Tobler Law (Section 3.1) states that everything is related to everything else, but near things are more related than distant things, and this amounts to a succinct definition of spatial autocorrelation. An understanding of the nature of the spatial autocorrelation that characterizes a GIS application helps us to *deduce* how best to collect and assemble data for a representation, and also how best to develop inferences between events and occurrences. The concept of geographic *scale* or level of detail will be fundamental to observed measures of the likely strength and nature of autocorrelation in any given application. Together, the scale and spatial structure of a particular application suggest ways in which we should *sample* geographic reality and should *weight* sample observations in order to build our representation. We will return to the key concepts of scale, sampling, and weighting throughout much of this book.

4.3 Spatial Autocorrelation and Scale

In Chapter 3 (see Box 3.3) we classified attribute data into the nominal, ordinal, interval, ratio, and cyclic scales of measurement. Objects existing in space are described by locational (spatial) descriptors and are conventionally classified using the taxonomy shown in Box 4.1.

Spatial autocorrelation measures attempt to deal simultaneously with similarities in the location of spatial objects (Box 4.1) and their attributes (Box 3.3).

Technical Box 4.1

Types of Spatial Objects

We saw in Section 3.4 that geographic objects are classified according to their *topological dimension*, which provides a measure of the way they fill space. For present purposes we assume that dimensions are restricted to *integer* (whole number) values, though later (Section 4.7) we relax this constraint and consider geographic objects of noninteger (fractional, or *fractal*) dimension.

All geometric objects can be used to represent occurrences at absolute locations (*natural* objects), or they may be used to summarize spatial distributions (*artificial* objects).

A *point* has neither length nor breadth nor depth, and hence is said to be of dimension 0. Points may be used to indicate spatial occurrences or events and their spatial patterning. *Point pattern analysis* is used to identify whether occurrences or events are interrelated—as in analyzing the incidence of crime or in identifying whether patterns of disease infection might be related to environmental or social factors (see Section 14.3.3). The *centroid* of an area object is an artificial point reference, which is located so as to provide a summary measure of the location of the object (see Section 15.2.1).

Lines have length, but not breadth or depth, and hence are of dimension 1. They are used to represent linear entities such as roads, pipelines, and cables, which frequently build together into networks. They can also be used to measure distances between spatial objects, as in the measurement of intercentroid distance. In order to reduce the burden of data capture and storage, lines are often held in GIS in *generalized* form (see Section 3.8).

Area objects have the two dimensions of length and breadth, but not depth. They may be used to represent natural objects, such as agricultural fields, but are also commonly used to represent artificial aggregations, such as census tracts (see below). Areas may bound linear features and enclose points, and GIS functions can be used to identify whether a given area encloses a given point (Section 14.2.3).

Volume objects have length, breadth, and depth, and hence are of dimension 3. They are used to represent natural objects such as river basins, or artificial phenomena such as the population potential of shopping centers or the density of resident populations (Section 14.3.5).

Time is often considered to be the fourth dimension of spatial objects, although GIS remains poorly adapted to the modeling of temporal change.

The relationship between higher- and lower-dimension spatial objects is analogous to that between higher- and lower-order attribute data, in that lower-dimension objects can be derived from those of higher dimension but not vice versa.

Certain phenomena, such as population, may be held as natural or artificially imposed spatial object types. The chosen way of representing phenomena in GIS not only defines the apparent nature of geographic variation, but also the way in which geographic variation may be analyzed. Some objects, such as agricultural fields or digital terrain models, are represented in their natural state. Others are transformed from one spatial object class to another, as in the transformation of population data from individual points to census tract areas, for reasons of confidentiality or convention. Some high-order representations are created by interpolation between lower-order objects, as in the creation of digital elevation models (DEMs) from spot height data (see Box 8.1 and Section 9.2).

The classification of spatial phenomena into object types is dependent fundamentally on scale. For example, on a less detailed map of the world, New York is represented as a one-dimensional point. On a more detailed map such as a road atlas, it will be represented as a two-dimensional area. Yet if we visit the city, it is very much experienced as a three-dimensional entity, and virtual reality systems seek to represent it as such (see Section 13.4.4).

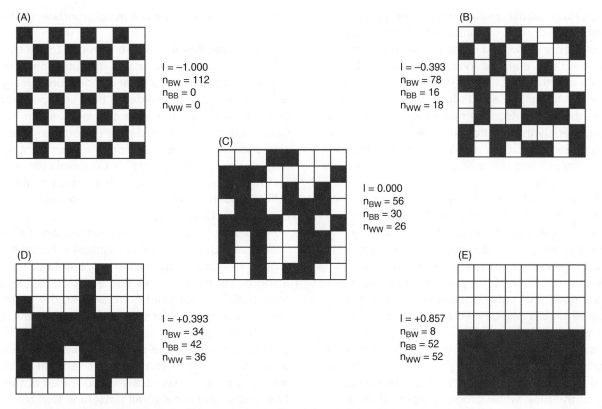

Figure 4.1 Field arrangements of blue and white cells exhibiting: (A) extreme negative spatial autocorrelation; (B) a dispersed arrangement; (C) spatial independence; (D) spatial clustering; and (E) extreme positive spatial autocorrelation. The values of the I statistic are calculated using the equation in Section 4.6. (Source: Goodchild 1986 CATMOG, GeoBooks, Norwich)

If features that are similar in location are also similar in attributes, then the pattern as a whole is said to exhibit *positive spatial autocorrelation*. Conversely, *negative spatial autocorrelation* is said to exist when features that are close together in space tend to be more dissimilar in attributes than features that are further apart (in opposition to Tobler's Law). Zero autocorrelation occurs when attributes are independent of location.

Figure 4.1 presents some simple field representations of a geographic variable in 64 cells that can each take one of two values, coded blue and white. Each of the five illustrations contains the same set of attributes, 32 white cells and 32 blue cells, yet the spatial arrangements are very different. Figure 4.1A presents the familiar chess board, and illustrates extreme negative spatial autocorrelation between neighboring cells. Figure 4.1E presents the opposite extreme of positive autocorrelation, when black and white cells cluster together in homogeneous regions. The other illustrations show arrangements that exhibit intermediate levels of autocorrelation. Figure 4.1C corresponds to spatial independence, or no autocorrelation; Figure 4.1B shows a relatively

dispersed arrangement; and Figure 4.1D, a relatively clustered one.

Spatial autocorrelation is determined by similarities in both position and attributes.

The patterns shown in Figure 4.1 are examples of a particular case of spatial autocorrelation. In terms of the classification developed in Chapter 3 (see Box 3.3), the attribute data are *nominal* (blue and white simply identify two different possibilities, with no implied order and no possibility of difference, or ratio) and their spatial distribution is conceived as a field, with a single value everywhere. The figure gives no clue as to the true dimensions of the area being represented. Usually, similarities in attribute values may be more precisely measured on higher-order measurement scales, enabling continuous measures of spatial variation (See Section 6.3.2.2 and Box 6.4 for a discussion of precision).

As we will see later in this chapter, the way in which we define what we mean by *neighboring* in investigating spatial arrangements may be more or less sophisticated. In considering the various arrangements shown in Figure 4.1, we have only considered the relationship between the attributes of a cell and those of its four *immediate* neighbors. But we could include a cell's four diagonal neighbors in the comparison (compare Figure 15.9), and more generally there is no reason we should not interpret Tobler's Law in terms of a gradual incremental attenuating effect of distance as we traverse successive cells.

We began this chapter by considering a time series analysis of events that are highly, even perfectly, repetitive in the short term. Activity patterns often exhibit strong positive temporal autocorrelation (where you were at this time last week or this time yesterday is likely to affect where you are now), but only if measures are made at the same time every day—that is, at the temporal scale of the daily interval. If, say, sample measurements were taken every 17 hours, measures of the temporal autocorrelation of your activity patterns would likely be much lower. Similarly, if the measures of the blue/white property were made at intervals that did not coincide with the dimensions of the squares of the chess boards in Figure 4.1, then the spatial autocorrelation measures would be different. Thus the issue of *sampling interval* is of direct importance in the measurement of spatial autocorrelation because spatial events and occurrences may or may not accommodate spatial structure.

Scale (see Box 4.2) is often integral to the trade-off between the level of spatial resolution and the degree of attribute detail that can be stored in a given application—as in the trade-off between spatial and spectral resolution in remote sensing (Section 9.2.1). Quattrochi and Goodchild have undertaken an extensive discussion of these and other meanings of scale (e.g., the degree of spectral or temporal coarseness) and their implications. The scale at which data from different sources are usually made available is discussed in Chapter 9.

A particular instance of the importance of scale is illustrated using a mosaic of squares in Figure 4.2. Figure 4.2A is a coarse-scale representation of attributes in nine squares and a pattern of negative spatial autocorrelation. However, the pattern is self-replicating at finer scales, and in Figure 4.4B, a finer-scale representation reveals that the smallest blue cells replicate the pattern of the whole area in a recursive manner. The pattern of spatial autocorrelation at the coarser scale is replicated at the finer scale, and the overall pattern is said to exhibit the property of *self-similarity*. Self-similar structure is characteristic of natural as well as social systems: for example, a rock may resemble the physical form of the mountain or coastline from which it was broken (Figure 4.3), small coastal features may resemble larger bays and inlets in structure and form,

Technical Box (4.2)

The Many Meanings of Scale

The concept of *scale* is fundamental to GIS, but unfortunately the word has acquired too many meanings in the course of time. Because they are to some extent contradictory, it is best to use other terms that have clearer meaning where appropriate.

Scale is in the details. Many scientists use scale in the sense of spatial resolution, or the level of spatial detail in data. Data are fine-scaled (or are at a fine level of granularity) if they include records of small objects, and coarse-scaled (coarse-grained) if they do not.

Scale is about extent. Scientists also use scale to talk about the geographic extent or scope of a project: a large-scale project covers a large area, and a small-scale project covers a small area. Scale can also refer to other aspects of the project's scope, including the cost or the number of people involved.

The scale of a map. Geographic data are often obtained from maps and often displayed in map form. Cartographers use the term *scale* to refer to a map's *representative fraction* (the ratio of distance on the map to distance on the ground; see Section 3.7). Unfortunately, this leads to confusion (and often bemusement) over the meaning of *large* and *small* with respect to scale. To a cartographer a large scale corresponds to a large representative fraction, in other words to plenty of geographic detail. This is exactly the opposite of what an average scientist understands by a large-scale study. In this book we have tried to avoid this problem by using the terms *coarse* and *fine* instead.

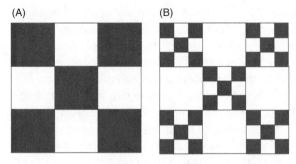

(A)   (B)

Figure 4.2 A Sierpinski carpet at two levels of resolution: (A) coarse scale and (B) finer scale. In general, measures of spatial and temporal autocorrelation are scale dependent (see Box 4.2: this point is explored in connection with the chessboard example in Section 14.2.5).

and neighborhoods may be of similar population size and each offer similar ranges of retail facilities right across a metropolitan area. Self-similarity is a core concept of fractals, a topic introduced in Section 4.7.

Figure 4.3 Individual rocks may resemble the forms of larger structures, such as rock outcrops or eroded coastlines. (© Paulo Ferreira/iStockphoto)

4.4 Spatial Sampling

The quest to represent the myriad complexity of the real world requires us to abstract, or sample, events and occurrences from the universe of eligible elements of interest, which is known as the *sample frame*. A spatial sampling frame might be bounded by the extent of a field of interest or by the combined extent of a set of areal objects. We can think of sampling as the process of selecting points from a continuous field or, if the field has been represented as a mosaic of areal objects, of selecting some of these objects while discarding others. The procedure of selecting some elements rather than others from a sample frame can very much determine the quality of the representation that is built using them.

Scientific sampling requires that each element in the sample frame has a known and prespecified chance of selection. In some important senses, we can think of any geographic representation as a kind of sample, in that the elements of reality that are retained are abstracted from the real world in accordance with some overall design. This is the case in remote sensing, for example (see Sections 3.6.1 and 9.2.1), in which each pixel value takes a spatially averaged reflectance value calculated at the spatial resolution characteristic of the sensor. In many situations, we will need consciously to select some observations, and not others, in order to create a generalizable abstraction. This is because, as a general rule, the resources available to any given project do not stretch to measuring every single one of the elements (soil profiles, migrating animals, shoppers) that we know to make up our population of interest. And even if resources were available, science tells us that this would be wasteful, since procedures of *statistical inference* allow us to infer from samples to the populations from which they were drawn. We will return to the process of statistical inference in Section 15.5. Here, we will confine ourselves to the question, how do we ensure a good sample?

Geographic data are only as good as the sampling scheme used to create them.

Classical statistics often emphasizes the importance of randomness in sound sample design. The purest form, simple random sampling, is well known: each element in the sample frame is assigned a unique number, and a prespecified number of elements are selected using a random number generator. In the case of a spatial sample from continuous space, *x, y* coordinate pairs might be randomly sampled within the range of *x* and *y* values (see Section 5.8 for information on coordinate systems). Since each randomly selected element has a known and prespecified probability of selection, it is possible to make robust and defensible generalizations about the population from which the sample was drawn. A spatially random sample is shown in Figure 4.4A.

Random sampling is integral to probability theory, and this enables us to use the distribution of values in our sample to tell us something about the likely distribution of values in the parent population from which the sample was drawn. However, sheer bad luck can mean that randomly drawn elements are disproportionately concentrated among some parts of the population at the expense of others, particularly when the size of our sample is small relative to the population. For example, a survey of household incomes might happen to select households with unusually low incomes. Systematic spatial sampling aims to circumvent this problem and ensure greater evenness of coverage across the sample frame. This is achieved by identifying a regular sampling interval k (equal to the reciprocal of the sampling fraction N/n, where n is the required sample size and N is the size of the population) and proceeding to select every kth element. In spatial terms, the sampling interval of spatially systematic samples maps into a regularly spaced grid, as shown in Figure 4.4B.

The advantage this offers over simple random sampling may be two-edged, however, if the sampling interval and the spatial structure of the study area coincide, that is, if the sample frame exhibits *periodicity*. A sample survey of urban land use along streets originally surveyed under the U.S. Public Land Survey System (PLSS: Section 5.6) would be ill-advised to take a sampling interval of one mile, for example, for this was the interval at which blocks within townships were originally laid out, and urban structure is still likely to be repetitive about this original design. In such instances, there may be a consequent failure to detect the true extent of heterogeneity of population attributes. For example, it is extremely unlikely that the attributes of street intersection locations would be representative of land uses elsewhere in a block structure. A number of hybrid sample designs have been devised to get around the vulnerability of spatially systematic sample designs to periodicity and the danger that simple random sampling may generate freak samples. These include stratified random sampling to ensure evenness of coverage (Figure 4.4C) and periodic random changes in the grid width of a spatially systematic sample (Figure 4.4D), perhaps subject to minimum spacing intervals.

In certain circumstances, it may be more efficient to restrict measurement to a specified range of sites—because of the prohibitive costs of transport over large areas, for example. Clustered sample designs, such as that shown in Figure 4.4E, may be used to generalize about attributes if the cluster presents a microcosm of surrounding conditions. In fact, this provides a legitimate use of a comprehensive study of one area to say something about conditions beyond it—as long as the study area is known to be representative of the broader study region. For example, political opinion polls are often taken in shopping centers where shoppers can be deemed broadly representative of the population at large. However, instances where they provide a comprehensive detailed picture of spatial structure are likely to be the exception rather than the rule, and in practice increased sample sizes are often used to mitigate this fact.

Figure 4.4 Spatial sample designs: (A) simple random sampling; (B) stratified sampling; (C) stratified random sampling; (D) stratified sampling with random variation in grid spacing; (E) clustered sampling; (F) transect sampling; and (G) contour sampling.

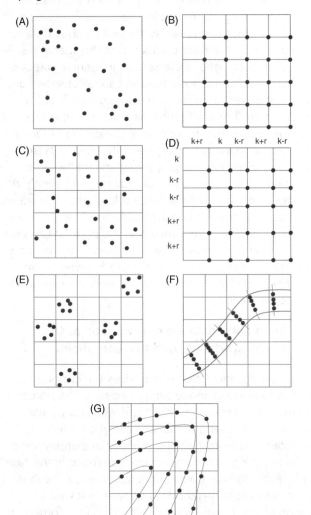

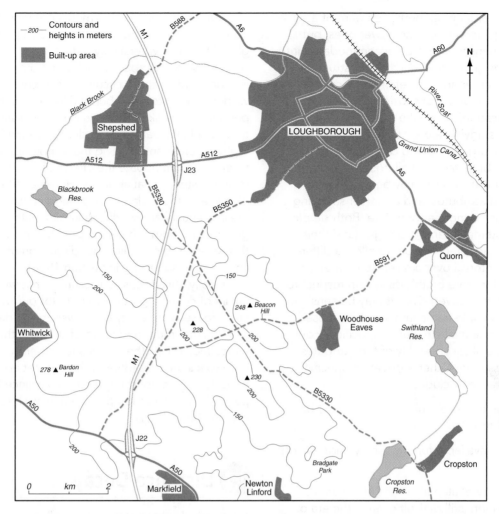

Figure 4.5 An example of physical terrain in which differential sampling would be advisable in order to construct a representation of elevation. (Source: M. Langford, University of Glamorgan)

Use of either simple random or spatially systematic sampling presumes that each observation is of equal importance, and hence of equal weight, in building a representation. As such, these sample designs are suitable for circumstances in which spatial structure is weak or nonexistent, or where (as in circumstances fully described by Tobler's Law) the attenuating effect of distance is constant in all directions. They are also suitable in circumstances where spatial structure is unknown. Yet in most practical applications, spatial structure is (to some extent at least) known, even if it cannot be wholly explained by Tobler's Law. These circumstances make it both more efficient and necessary to devise application-specific sample designs. This makes for improved quality of representation, with minimum resource costs of collecting data. Relevant sample designs include sampling along a transect, such as a soil profile (Figure 4.4F), or along a contour line (Figure 4.4G).

Consider the area of Leicestershire, UK, illustrated in Figure 4.5. It depicts a landscape in which the hilly

relief of an upland area falls away sharply toward a river's floodplain. In identifying the sample spot heights that we might measure and hold in a GIS to create a representation of this area, we would be well advised to sample a disproportionate number of observations in the upland part of the study area where the local variability of heights is greatest. In a socioeconomic context, imagine that you are required to identify the total repair cost of bringing all housing in a city up to a specified standard. (Such applications are common, for example, in forming bids for federal or central government funding.) A GIS that showed the time period in which different neighborhoods were developed (such as the Midwest settlements simulated in Figure 2.18) would provide a useful guide to effective use of sampling resources. Newer houses are all likely to be in more or less the same condition, while the repair costs of the older houses are likely to be much more variable and dependent on the attention that the occupants have lavished on them. As a general rule, the older neighborhoods warrant a higher sampling interval than the newer ones, but other considerations may also be accommodated into the sampling design as well—such as construction type (duplex versus apartment, etc.) and local geology (as an indicator of risk of subsidence).

In any application, where the events or phenomena that we are studying are spatially heterogeneous, we will require a large sample to capture the full variability of attribute values at all possible locations. Other parts of the study area may be much more homogeneous in attributes, and a sparser sampling interval may thus be more appropriate. Both simple random and systematic sample designs (and their variants) may be adapted in order to allow a differential sampling interval over a given study area (see Section 6.3.2.1 for more on this issue with respect to sampling vegetation cover). Thus it may be sensible to partition the sample frame into subareas, based on our knowledge of spatial structure—specifically our knowledge of the likely variability of the attributes that we are measuring. Other application-specific special circumstances include:

- whether source data are ubiquitous or must be specially collected;

- the resources available for any survey undertaking; and

- the accessibility of all parts of the study area to field observation (still difficult even in the era of ubiquitous availability of Global Positioning System receivers: Section 5.9).

Stratified sampling designs accommodate the unequal abundance of different phenomena on the Earth's surface.

It is very important to be aware that this discussion of sampling is appropriate to problems where there is a large hypothetical population of evenly distributed locations (elements, in the terminology of sampling theory, or atoms of information in the terminology of Section 3.4) and that each has a known and pre-specified probability of selection. Random selection of elements plays a part in each of the sample designs illustrated in Figure 4.4, albeit the probability of selecting an element may be greater for clearly defined subpopulations that lie along a contour line or across a soil transect, for example. In circumstances where spatial structure is either weak or is explicitly incorporated through clear definition of subpopulations, standard statistical theory provides a robust framework for inferring the attributes of the population from those of the sample.

But reality is somewhat messier. In most GIS applications, the population of elements (animals, glacial features, voters) may not be large, and its distribution across space may be far from random and independent. In these circumstances, conventional wisdom suggests a number of rules of thumb to compensate for the likely increase in error in estimating the true population value—as in clustered sampling, where slightly more than doubling the sample size is usually taken to accommodate the effects of spatial autocorrelation within a spatial cluster. However, the existence of spatial autocorrelation may fundamentally undermine the inferential framework and invalidate the process of generalizing from samples to populations. We examine this issue in more detail in our discussion of inference and hypothesis testing in Section 15.5.1.

Finally, this discussion assumes that we have the luxury of collecting our own data for our own particular purpose. The reality of analysis in our data-rich world is that more and more of the data that we use are collected by other parties for other purposes. In such cases the metadata of the dataset are crucially important in establishing their provenance for the particular investigation that we may wish to undertake (see Section 11.2.1).

4.5 Distance Decay

In selectively abstracting, or sampling, reality, judgment is required to fill in the gaps between the observations that make up a representation. This requires understanding of the likely attenuating effect of distance between the sample observations, and thus of the nature of geographic data (Figure 4.6). That is to say, we need to make an informed judgment about an appropriate *interpolation* function and how to *weight* adjacent observations. A literal interpretation of Tobler's Law implies a continuous, smooth, attenuating effect of distance on the attribute values of adjacent or contiguous spatial objects, or incremental variation in attribute values as we traverse a field. The polluting effect of a chemical spillage decreases in a predictable fashion with distance from the point source; aircraft noise decreases on a linear trend with distance from the flight path; and the number of visits from localities (suitably normalized by population) to a national park decreases at a regular rate as we traverse the counties that adjoin it. This section focuses on principles and introduces some of the functions that are used to describe attenuation effects over distance, or the nature of geographic variation. Section 14.3.6 discusses ways in which the principles

Figure 4.6 We require different ways of interpolating between points, as well as different sample designs, for representing mountains and forested hillsides. (© Rachel Turk/iStockphoto)

of distance decay are embodied in techniques of spatial interpolation.

The precise nature of the function used to represent the effects of distance is likely to vary between applications, and Figure 4.7 illustrates several hypothetical types. In mathematical terms, if i is a point for which we have a recorded measure of an attribute and j is a point with no recorded measurement, we use b as a parameter that determines the rate at which the weight w_{ij} assigned to point j declines with distance from i. A small b value produces a slower decrease than a large one. In most applications, the choice of distance attenuation function is the outcome of past experience, the fit of a particular application dataset, and convention. Figure 4.7A presents the simple case of linear distance decay, given by the expression:

$$w_{ij} = a - bd_{ij},$$

for $d_{ij} < a/b$ as might reflect the noise levels experienced across a transect perpendicular to an aircraft

flight path, for example. Figure 4.7B presents a negative power distance decay function, given by the expression:

$$w_{ij} = d_{ij}^{-b},$$

which some researchers have used to describe the decline in the density of resident population with distance from historic central business district (CBD) areas. Figure 4.7C illustrates a negative exponential statistical fit, given by the expression:

$$w_{ij} = e^{-bd_{ij}}$$

conventionally used in human geography to represent the decrease in retail store patronage with distance from it (e here denotes an exponential term, approximately equal to 2.71828, sometimes written exp).

Each of the attenuation functions illustrated in Figure 4.7 is idealized in that the effects of distance are presumed to be regular, continuous, and *isotropic* (uniform in every direction). This notion of smooth and continuous variation underpins many of the representational traditions in cartography, as in the creation of *isopleth* (or isoline) maps (see Box 4.3). It is also consistent with our experience of high school math, which is redolent of a world in which variation is continuous and best represented by interpolating smooth curves between everything. Yet even casual observation of the real world tells us that geographic variation is often far from smooth and continuous. The Earth's surface and geology, for example, are discontinuous at cliffs and fault lines, while the socioeconomic patterning of city neighborhoods can be similarly characterized by abrupt changes. Some illustrative issues pertaining to the catchment of a doctor's surgery general practice are apparent from Figure 4.9. A naïve GIS analysis might assume that a map of 10-, 20-, and 30-minute travel times to see the doctor depicts a series of equidistant concentric circles. On this basis

Figure 4.7 The attenuating effect of distance: (A) linear distance decay, $w_{ij} = a - bd_{ij}$; (B) negative power distance decay, $w_{ij} = d_{ij}^{-b}$; and (C) negative exponential distance decay, $w_{ij} = \exp(-bd_{ij})$.

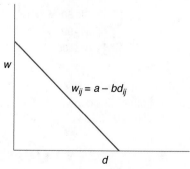
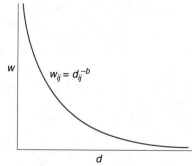
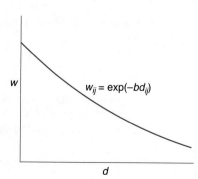

Isopleth and Choropleth Maps

Isopleth maps are used to visualize phenomena that are conceptualized as fields and measured on interval or ratio scales. An *isoline* connects points with equal attribute values, such as contour lines (equal height above sea level), *isohyets* (points of equal precipitation), *isochrones* (points of equal travel time), or *isodapanes* (points of equal transport cost). Figure 4.8 illustrates the procedures that are used to create a surface about a set of point measurements (Figure 4.8A), such as might be collected from rain gauges across a study region (see Section 14.3.6 for more technical detail on the process of spatial interpolation). A parsimonious number of user-defined values is identified to define the contour intervals (Figure 4.8B). The GIS then interpolates a contour between point observations of greater and lesser value (Figure 4.8C) using standard procedures of distance decay, and the other contours are then interpolated using the same procedure (Figure 4.8D).

Hue or shading can be added to improve user interpretability (Figure 4.8E).

Choropleth maps are constructed from values describing the properties of nonoverlapping areas, such as counties or census tracts. Each area is colored, shaded, or cross-hatched to symbolize the value of a specific variable, as in Figure 4.9. Two types of variables can be used, termed *spatially extensive* and *spatially intensive*. Spatially extensive variables are those whose values are true only of entire areas, such as total population or total number of children under 5 years of age. Spatially intensive variables are those that could potentially be true of every part of an area, if the area were homogeneous; examples include densities, rates, or proportions. Conceptually, a spatially intensive variable is a field, averaged over each area, whereas a spatially extensive variable is a field of density whose values are summed or integrated to obtain each area's value.

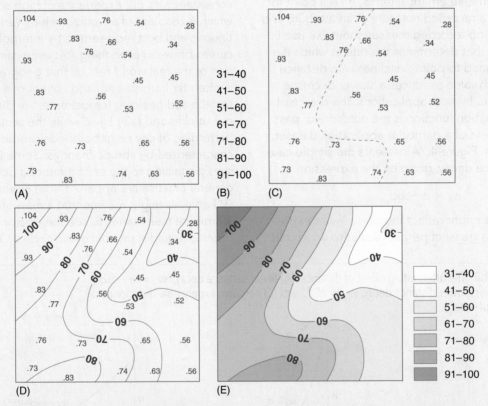

Figure 4.8 The creation of isopleth maps: (A) point attribute values; (B) user-defined classes; (C) interpolation of class boundary between points; (D) addition and labeling of other class boundaries; and (E) use of hue to enhance perception of trends (After Kraak and Ormeling 2003: 134).

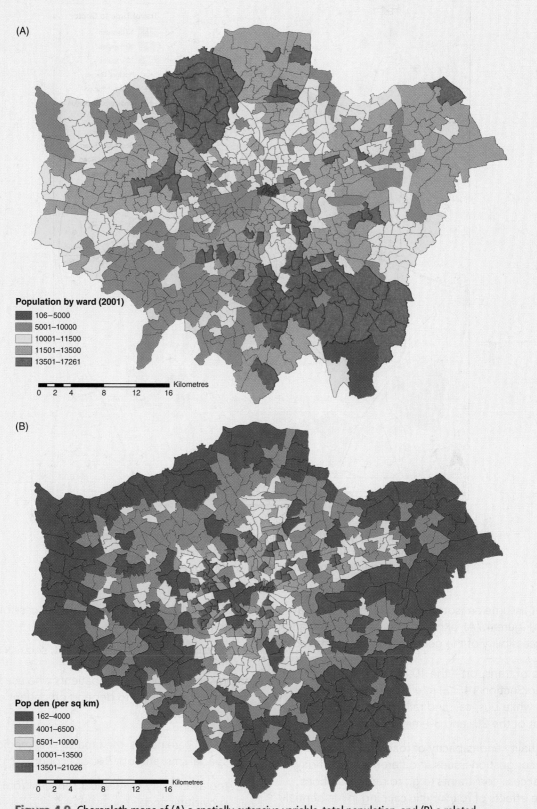

(A)

Population by ward (2001)

- 106–5000
- 5001–10000
- 10001–11500
- 11501–13500
- 13501–17261

Kilometres
0 2 4   8   12   16

(B)

Pop den (per sq km)

- 162–4000
- 4001–6500
- 6501–10000
- 10001–13500
- 13501–21026

Kilometres
0 2 4   8   12   16

Figure 4.9 Choropleth maps of (A) a spatially extensive variable, total population, and (B) a related but spatially intensive variable, population density. Many cartographers would argue that (A) is misleading and that spatially extensive variables should always be converted to spatially intensive form (as densities, ratios, or proportions) before being displayed as choropleth maps. (Courtesy Daryl Lloyd)

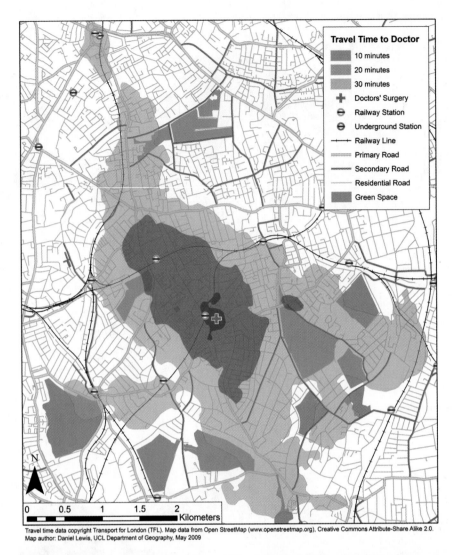

Figure 4.10 Map showing 10-, 20-, and 30-minute travel times to a doctor's surgery in South London. (Courtesy Daniel Lewis; travel time data © Transport for London; map data courtesy Open Streetmap www.openstreetmap.org).

we might assume an isotropic linear distance decay function (Figure 4.7A), although the GIS representation of the accessibility of the general practice is affected by

- mode of transport—the 10-minute travel time buffer (Section 14.3.2) in effect measures walk time, while bus, car, and rail modes increase the extent of the 20- and 30-minute buffers;

- the quality and capacity of road and rail infrastructure, congestion issues at different times of day, and access constraints (e.g., to railway stations, or the effects of road traffic calming in residential areas and one-way streets);

- availability of public transport—by road or rail;

- informal rights of way and the navigability of public open space (note that the public parks in

Figure 4.10 appear to impede accessibility rather than enhance it); and

- physical barriers to movement, such as rivers.

In practice, the number of patients who use the general practice is also likely to depend on patient characteristics, as measured by

- health care needs, their match with medical practice specialisms, and other socioeconomic characteristics;

- the availability of medical services from other local providers; and

- a demand constraint that requires the probabilities of attending all available practices at any point to sum to 1 (unless people opt out of the health care system).

4.6 Measuring Distance Effects as Spatial Autocorrelation

Understanding spatial structure is key to building real-world representations because it helps us to deduce a good sampling strategy and to use an appropriate means of interpolating between sampled points that is fit for purpose. Knowledge of the actual or likely nature of spatial autocorrelation can thus be used *deductively* in order to help build a spatial representation of the world. However, in many applications we do not understand enough about geographic variability, distance effects, and spatial structure to make reliable deductions. A further branch of spatial analysis thus emphasizes the *measurement* of spatial autocorrelation as an end in itself. This amounts to a more *inductive* approach to developing an understanding of the nature of a geographic dataset.

Induction reasons from data to build up understanding, while deduction begins with theory and principle as a basis for looking at data.

In Section 4.3 we saw that spatial autocorrelation measures the extent to which similarities in position match similarities in attributes. Methods of measuring spatial autocorrelation depend on the types of objects used as the basis of a representation, and as we saw in Section 4.2, the scale of attribute measurement is important too. Interpretation depends on how the objects relate to our conceptualization of the phenomena they represent. If the phenomenon of interest is conceived as a field, then spatial autocorrelation measures the smoothness of the field using data from the sample points, lines, or areas that represent the field. If the phenomenon of interest is conceived as a set of discrete objects, then spatial autocorrelation measures how the attribute values are distributed among the objects, distinguishing between arrangements that are clustered, random, and locally contrasting.

Figure 4.11 shows examples of each of the four object types, with associated attributes, chosen to represent situations in which a scientist might wish to measure spatial autocorrelation. The point data in Figure 4.11A comprise data on wellbores over an area of 30 km², and together provide information on the depth of an aquifer beneath the surface (the blue shading identifies those within a given threshold). We would expect values to exhibit strong spatial autocorrelation, with departures from this indicative

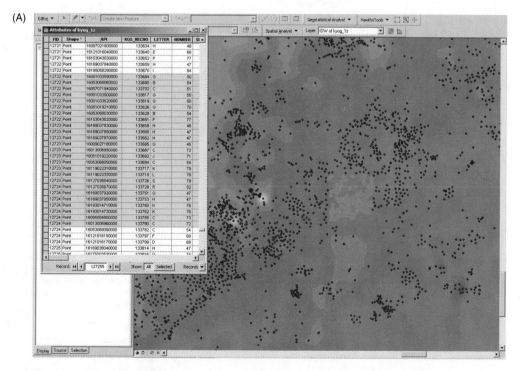

Figure 4.11 Situations in which a scientist might want to measure spatial autocorrelation: (A) point data (wells with attributes stored in a spreadsheet: linear extent of image 0.6 km); (A) Courtesy ESRI. *(continued)*

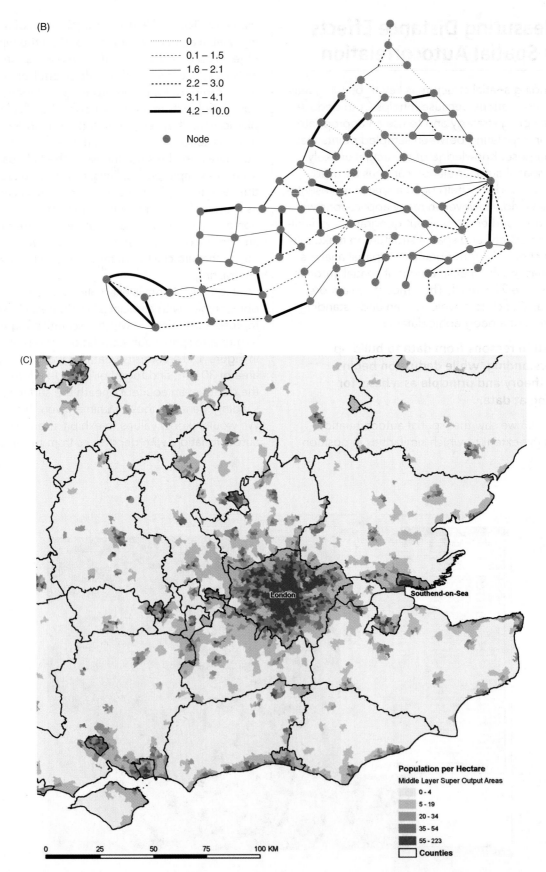

Figure 4.11 (*continued*) (B) line data (accident rates in the southwestern Ontario provincial highway network); (C) area data (percentage of population that are old age pensioners in Southeast England); and (C) Courtesy Daryl Lloyd;

(D)

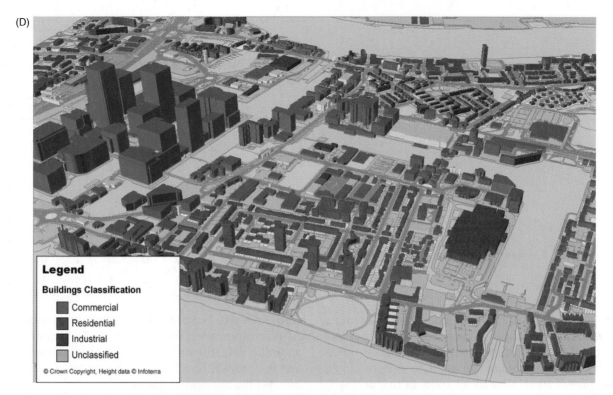

Figure 4.11 (*continued*) (D) volume data (elevation and volume of buildings in London: east to the top of the image; image extent 1.6 km north-south, 1.5 km east-west). (D) Courtesy Duncan Smith.

of changes in bedrock structure or form. The line data in Figure 4.11B present numbers of accidents for links of road over a lengthy survey period in the southwestern Ontario (Canada) provincial highway network. Low spatial autocorrelation in these statistics implies that local causative factors (such as badly laid out junctions) account for most accidents, whereas strong spatial autocorrelation would imply a more regional scale of variation, implying a link between accident rates and lifestyles, climate, or population density. The area data in Figure 4.11C

illustrate the settlement structure of South East England and invites analysis of the effectiveness of land use zoning (green belt) policies, for example. The volume data in Figure 4.11D allow some measure of the spatial autocorrelation of high-rise structures to be made, as part of a study of the relationship between the built form and economic function of London, UK. The way that spatial autocorrelation statistic might actually be calculated for the data used to construct Figure 4.11C is described in Box 4.4.

Technical Box (4.4)

Measuring Similarity between Neighbors

In the simple example shown in Figure 4.12, we compare neighboring values of spatial attributes by defining a weights matrix **W** in which each element w_{ij} measures the locational similarity of i and j (i identifies the row and j the column of the matrix). We use a simple measure of contiguity, coding $w_{ij} = 1$ if regions i and j are contiguous and $w_{ij} = 0$ otherwise. w_{ii} is set equal to 0 for all i. This is shown in Table 4.1.

The weights matrix provides a simple way of representing similarities between location and attribute values, in a region of contiguous areal objects. Autocorrelation is identified by the presence of neighboring cells or zones that take the same (binary) attribute value. More sophisticated measures of w_{ij} might include a decreasing function (such as one of those shown in Figure 4.7) of the

▶

Figure 4.12 A simple mosaic of zones.

Table 4.1 The weights matrix **W** derived from the zoning system shown in Figure 4.12.

	1	2	3	4	5	6	7	8
1	0	1	1	1	0	0	0	0
2	1	0	1	0	0	1	1	0
3	1	1	0	1	1	1	0	0
4	1	0	1	0	1	0	0	0
5	0	0	1	1	0	1	0	1
6	0	1	1	0	1	0	1	1
7	0	1	0	0	0	1	0	1
8	0	0	0	0	1	1	1	0

straight-line distance between points at the centers of zones, or the lengths of common boundaries. A range of different spatial metrics may also be used, such as existence of linkage by air, or a decreasing function of travel time by air, road, or rail, or the strength of linkages between individuals or firms on some (nonspatial) network.

The weights matrix makes it possible to develop measures of spatial autocorrelation using four of the attribute types (nominal, ordinal, interval, and ratio, but not cyclic) in Box 3.3 and the dimensioned classes of spatial objects in Box 4.1. Any measure of spatial autocorrelation seeks to compare a set of locational similarities w_{ij} (contained in a weights matrix) with a corresponding set of attribute similarities c_{ij}, combining them into a single index in the form of a cross-product:

$$\sum_i \sum_t c_{ij} w_{ij}$$

This expression is the total obtained by multiplying every cell in the **W** matrix with its corresponding entry in the **C** matrix and summing.

There are different ways of measuring the attribute similarities, c_{ij}, depending on whether they are measured on the nominal, ordinal, interval, or ratio scale. For nominal data, the usual approach is to set c_{ij} to 1 if i and j take the same attribute value, and

zero otherwise. For ordinal data, similarity is usually based on comparing the ranks of i and j. For interval and ratio data, the attribute of interest is denoted z_i, and the product $(z_i - \bar{z})(z_j - \bar{z})$ is calculated, where $\bar{z}$ denotes the average of the z's.

One of the most widely used spatial autocorrelation statistics for the case of area objects and interval scale attributes is the Moran Index. This is positive when nearby areas tend to be similar in attributes, negative when they tend to be more dissimilar than one might expect, and approximately zero when attribute values are arranged randomly and independently in space. It is given by the expression:

$$I = \left(n \sum_i \left[\sum_j w_{ij}(z_i - \bar{z})(z_j - \bar{z}) \right] \right) \Big/ \left(\sum_i \left(\sum_j w_{ij} \sum_j \right) \right) (z_i - \bar{z})^2$$

where n is the number of areal objects in the set. This brief exposition is provided at this point to emphasize the way in which spatial autocorrelation measures are able to accommodate attributes scaled as nominal, ordinal, interval, and ratio data, and to illustrate that there is flexibility in the nature of contiguity (or adjacency) relations that may be specified. Further techniques for measuring spatial autocorrelation are reviewed in connection with spatial interpolation in Section 14.3.6.

Spatial autocorrelation measures tell us about the interrelatedness of phenomena across space, one attribute at a time. Its measurement is key to formalizing and understanding many geographic problems, and it is central to locational analysis in

Geography (see Box 4.5). Another important facet to the nature of geographic data is the tendency for relationships to exist between different phenomena at the same place—between the values of two or more different fields, between two or more attributes of a

Peter Haggett, Spatial Analyst and Geographer

Our understanding of the nature of geographic data has been transformed by geographer Peter Haggett (Figure 4.13). Between 1965 and 1975, Peter systematically laid out the conceptual foundations for the modern science of Geography and provided a major stimulus to the development of the nomothetic approach to the subject (see Section 1.3.1). His *Locational Analysis in Human Geography* (first published in 1965) was a clarion call to a new generation of scholars concerned to develop systematic and formalized digital representation of geographical systems (Section 3.2), and also laid the foundations for subsequent collaborations with geomorphologist Richard Chorley and geostatistician Andy Cliff. Throughout, Peter's writing and analysis have retained the distinctive hallmarks of thoroughness, breadth, and attention to detail.

Figure 4.13 Peter Haggett. (Courtesy Peter Haggett)

Moreover, analytical rigor in representation has not accrued at the expense of real-world relevance. Peter's pioneering work on spatial structure and spatial diffusion has laid the foundation for a lengthy program of work with the World Health Organization, entailing the application of spatial analytical methods to the epidemiology of disease and to the diffusion of measles, influenza, and AIDS in particular. Peter has received many accolades as Europe's premier spatial analyst, and his work has been key to the emergence of geographic information science.

Peter was born in 1933 in the rural English county of Somerset, and he traces his interest in Geography to walking and cycling around the county during his childhood. Perhaps in contrast to the seemingly frenetic pace of developments in GIScience today, he has often emphasized the importance of direct experience of the real world at a measured pace. Peter taught in Cambridge University, University College London, and Bristol University during his career (he remains an emeritus professor at Bristol), and inspired many of the current generation of GIScientists, including all of the authors of this book!

Reflecting on his career, Peter says:

It's curmudgeonly to have regrets when one has got so much pleasure from a life in Geography but I would love not to have missed, by maybe just a decade, the GIS Revolution. Technological change and conceptual understanding are often lagged. The massive revolution in the understanding of DNA and molecular genetics followed on a technical breakthrough (in X-ray crystallography) made a generation earlier. In my view, breakthroughs in geographical understanding are starting to follow on from what has been achieved through GIS. Pandemic modeling is an area where tracking epidemic spread in a real-time setting is being totally transformed by the power of GIS to handle hugely complex spatial movements. It is so satisfying to see by how much GIS is changing the game.

set of discrete objects, or between the attributes of overlapping discrete objects. The interrelatedness of the various properties of a location is an important aspect of the nature of geographic data and is key to understanding how the world works (Section 1.3). But it is also a property that defies conventional statistical analysis, since most such methods assume zero spatial autocorrelation of sampled observations—in direct contradiction to Tobler's Law. We discuss this problem again in Section 15.5.

4.7 Taming Geographic Monsters

Thus far in our discussion of the nature of geographic data, we have assumed that spatial variation is smooth and continuous, apart from when we encounter abrupt truncations and discrete shifts at boundaries. However, much spatial variation does not appear to possess these properties, but rather is jagged and apparently irregular. The processes that give rise to the form of a mountain range produce features that are spatially autocorrelated (for example, the highest peaks tend to be clustered), yet it would be wholly inappropriate to represent a mountainscape using smooth interpolation between peaks and valley troughs.

Jagged irregularity turns out to be a property that is also often observed across a range of scales, and detailed irregularity may resemble coarse irregularity in shape, structure, and form. We commented on this in Section 4.3 when we suggested that a rock broken off a mountain may, for reasons of lithology, represent the mountain in form; this property has been termed *self-similarity*. Urban geographers also recognize that cities and city systems are also self-similar in organization across a range of scales, and the ways in which this echoes many of the earlier ideas of Walter Christaller's Central Place Theory have been discussed by Batty and Longley. It is unlikely that idealized smooth curves and conventional mathematical functions will provide useful representations for self-similar, irregular spatial structures: at what scale, if any, does it become meaningful to approximate the San Andreas Fault system by a continuous curve? Urban geographers, for example, have long sought to represent the apparent decline in population density with distance from historic central business districts (CBDs) as a continuous curve (Figure 4.7B), yet the three-dimensional profiles of cities (Figure 4.11D) are clearly characterized by urban canyons between irregularly spaced high-rise buildings. Each of these phenomena is characterized by spatial trends (the largest faults, the largest mountains, or the largest skyscrapers each tend to be close to one another), but they are not contiguous and smoothly joined, and the kinds of surface functions shown in Figure 4.7 present inappropriate generalizations of their structure.

For many years, such features were considered geometrical monsters that defied intuition. More recently, however, a more general geometry of the irregular, termed *fractal geometry* by Benoît Mandelbrot, has come to provide a more appropriate and general means of summarizing the structure and character of spatial objects. Fractals can be thought of as geometric objects that are, literally, between Euclidean dimensions, as described in Box 4.6.

In a self-similar object, each part has the same structure as the whole.

Ideas from fractal geometry are important, and for many phenomena a measurement of fractal dimension is as important as measures of spatial autocorrelation, or of medians and modes in standard statistics. An important application of fractal concepts is discussed in Section 15.3.1, and we return again to the issue of length estimation in GIS in Section 14.3.1. Ascertaining the fractal dimension of an object involves identifying the scaling relation between its length or extent and the yardstick (or level of detail) that is used to measure it. Regression analysis, which can be thought of as putting a best fit line through a scatter of points, provides one (of many) means of establishing this relationship. If we return to the Maine coastline example in Figures 4.15 and 4.16, we might obtain scale-dependent coast-length estimates (L) of 13.6 (4 × 3.4), 14.1 (2 × 7.1), and 16.6 (1 × 16.6) units for the step lengths (r) used in Figures 4.15B, 4.15C, and 4.15D, respectively. (It is arbitrary whether the steps are measured in kilometres or miles.)

If we then plot the natural log of L (on the y-axis) against the natural log of r for these and other pairs of values, we will build up the scatterplot shown in Figure 4.14. If the points lie more or less on a straight line and we fit a trend (regression) line through it, the value of the slope (b) parameter is equal to $(1 - D)$, where D is the fractal dimension of the line. This method for analyzing the nature of geographic lines was originally developed by Lewis Fry Richardson (Box 4.7).

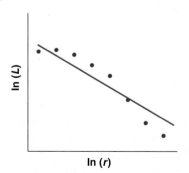

Figure 4.14 The relationship between recorded length (L) and step length (r).

The Strange Story of the Lengths of Geographic Objects

How long is the coastline of Maine (Figure 4.15)? (Benoît Mandelbrot, a French mathematician, originally posed this question in 1967 with regard to the coastline of Great Britain.) Consider the stretch of coastline shown in Figure 4.16A. With dividers set to measure 100-km intervals, we would take approximately 3.4 swings and record a length of 340 km (Figure 4.15B).

If we then halved the divider span so as to measure 50-km swings, we would take approximately 7.1 swings and the measured length would increase to 355 km (Figure 4.15C). If we halved the divider span once again to measure 25-km swings, we would take approximately 16.6 swings and the measured length would increase still further to 415 km (Figure 4.15D).

And so on until the divider span was so small that it picked up all of the detail on this particular representation of the coastline. But that would not be the end of the story.

If we were to resort instead to field measurement, using a tape measure or the Distance Measuring Instruments (DMIs) used by highway departments, the length would increase still further, as we picked up detail that even the most detailed maps do not seek to represent. If we were to use dividers, or even microscopic measuring devices, to measure every last grain of sand or earth particle, our recorded length measurement would stretch toward infinity, without apparent limit.

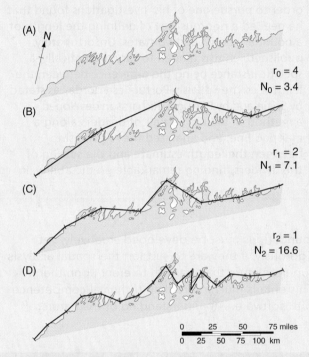

Figure 4.16 The coastline of Maine, at three levels of recursion: (A) the base curve of the coastline; (B) approximation using 100-km steps; (C) 50-km step approximation; and (D) 25-km step approximation.

Figure 4.15 Part of the Maine coastline. (© Anna Omeltchenko.iStockphoto)

4.8 Induction and Deduction and How It All Comes Together

The continual message of this chapter is that spatial is special—that geographic data have a unique nature. Tobler's Law presents an elementary general rule about spatial structure and provides a starting point for the measurement and simulation of spatially autocorrelated structures. This in turn assists us in devising appropriate spatial sampling schemes and creating improved representations, which tell us still more about the real world and how we might represent it.

Spatial data provide the foundations for operational and strategic applications of GIS, and these

Lewis Fry Richardson

Lewis Fry Richardson (1881–1953: Figure 4.17) was one of the founding fathers of the ideas of scaling and fractals. He was brought up a Quaker, and after earning a degree at Cambridge University went to work for the Meteorological Office, but his pacifist beliefs forced him to leave in 1920 when the Meteorological Office was militarized under the Air Ministry. His early work on how atmospheric turbulence is related at different scales established his scientific reputation. Later he became interested in the causes of war and human conflict, and in order to pursue one of his investigations found that he needed a rigorous way of defining the length of a boundary between two states. Unfortunately, published lengths tended to vary dramatically, a specific instance being the difference between the lengths of the Spanish–Portuguese border as stated by Spain and by Portugal. Richardson developed a method of walking a pair of dividers along a mapped line, and analyzed the relationship between the length estimate and the setting of the dividers, finding remarkable predictability. In

Figure 4.17 Lewis Fry Richardson. (Photo from wikipedia: http: 12/16/2009 120//en.wikipedia.org/wiki/File:Lewis_Fry_Richardson.png)

the 1960s Benoît Mandelbrot's concept of fractals finally provided the theoretical framework needed to understand this result.

foundations must be developed creatively, yet rigorously, if they are to support the spatial analysis superstructure that we wish to erect upon them. This entails much more than technical competence with software. An understanding of the nature of geographic data allows us to use induction (reasoning from observations) and deduction (reasoning from principles and theory) alongside each other to develop effective spatial representations that are safe to use.

Questions for Further Study

1. Many jurisdictions tout the number of miles of shoreline in their community—for example, Ottawa County, Ohio, claims 107 miles of Lake Erie shoreline. What does this mean, and how could you make it more meaningful?

2. With reference to Figure 4.9, list the design considerations that should be incorporated into GIS software in order to measure accessibility of (a) a neighborhood medical center to wheelchair-bound pedestrians; (b) a grocery store to high-income customers; and (c) all residential buildings in a small town from a single fire service station.

3. The apparatus of inference was developed by statisticians because they wanted to be able to reason from the results of experiments involving small samples to make conclusions about the results of much larger, hypothetical experiments—for example, in using samples to test the effects of drugs. Summarize the problems inherent in using this apparatus for geographic data, in your own words.

4. What important aspects of the nature of geographic data have not been covered in this chapter?

Further Reading

Batty, M. and Longley, P.A. 1994. *Fractal Cities: A Geometry of Form and Function*. London: Academic Press. Available for download from www.fractalcities.org.

De Smith, M., Goodchild, M.F., and Longley, P.A. 2009. *Geospatial Analysis: A Comprehensive Guide to Principles, Techniques and Software Tools*. Chapter 5: Data Exploration and Spatial Statistics. Leicester: Troubador. See also www.spatialanalysisonline.com.

Goodchild, M.F. 2004. The validity and usefulness of laws in geographic information science and geography. *Annals of the Association of American Geographers* 94(2): 300–303.

Kraak, M.J. and Ormeling, F.J. 2003. *Cartography: Visualization of Geospatial Data,* 2nd ed. Chapter 9. Harlow: Prentice Hall.

Mandelbrot, B.B. 1983. *The Fractal Geometry of Nature*. San Francisco: Freeman.

Quattrochi, D.A. and Goodchild, M.F. (eds.). 1996. *Scale in Remote Sensing and GIS*. Boca Raton, FL: Lewis Publishers.

5 Georeferencing

Geographic location is the element that distinguishes geographic information from all other types of information, so methods for specifying location on the Earth's surface are essential to the creation of useful geographic information. Many such techniques have been developed over the centuries, and this chapter provides a basic guide for GIS students—what you need to know about georeferencing to succeed in GIS. The first section lays out the principles of georeferencing, including the requirements that any effective system must satisfy. Subsequent sections discuss commonly used systems, starting with the ones closest to everyday human experience, including place-names and street addresses, and moving to the more accurate scientific methods that form the basis of geodesy and surveying, and the most recent methods developed for the Web. The final sections deal with issues that arise over conversions between georeferencing systems, with the Global Positioning System (GPS), with georeferencing of computers and cell phones, and with the concept of a gazetteer.

○ LEARNING OBJECTIVES

After Studying this chapter you will:

- Know the requirements for an effective system of georeferencing.

- Be familiar with the problems associated with place-names, street addresses, and other systems used every day by humans.

- Know how the Earth is measured and modeled for the purposes of positioning.

- Know the basic principles of map projections and the details of some commonly used projections.

- Know about conversion between different systems of georeferencing.

- Understand the principles behind GPS and learn some of its applications.

5.1 Introduction

Chapter 3 introduced the idea of an atomic element of geographic information: an atom made up of location, time (optionally), and attribute. To make GIS work there must be techniques for assigning values to all three aspects, in ways that are commonly understood by people who wish to communicate. Almost all the world agrees on a common calendar and time system, so there are only minor problems associated with communicating that element of the atom when it is needed (although different time zones, different names of the months in different languages, the annual switch to summer or Daylight Saving Time, and systems such as the classical Japanese convention of dating by the year of the emperor's reign all sometimes manage to confuse us).

Time is optional in a GIS, but location is not, so this chapter focuses on techniques for specifying location, and the problems and issues that arise. Locations are the basis for many of the benefits of GIS: the ability to map, to tie different kinds of information together because they refer to the same place, or to measure distances and areas. Without locations, data are said to be *nonspatial* or *aspatial* and would likely have no value at all within a geographic information system.

Time is an optional element in geographic information, but location is essential.

Several terms are commonly used to describe the act of assigning locations to atoms of information. We use the verbs *georeference*, *geolocate*, and *geocode*, and say that facts have been *georeferenced* or *geocoded*. We talk about *tagging* records with geographic locations, or about *locating* them. The term *georeference* will be used throughout this chapter.

The primary requirements of a georeference are (1) that it be *unique*, so that there is only one location associated with a given georeference, and therefore no confusion about the location that is referenced; and (2) that its meaning be *shared* among all of the people who wish to work with the information, including their geographic information systems. For example, the georeference 909 West Campus Lane, Goleta, California, USA, points to a single house—there is no other house anywhere on Earth with that address—and its meaning is shared sufficiently widely to allow mail to be delivered to the address from virtually anywhere on the planet. The address may not be meaningful to everyone living in China, but it will be meaningful to a sufficient number of people within China's postal service, so that a letter mailed from China to that address will likely be delivered successfully. Uniqueness and shared meaning are sufficient also to allow people to link different kinds of information based on common location: for example, a driving record that is georeferenced by street address can be linked to a record of purchasing.

To be as useful as possible, a georeference must be *persistent through time* because it would be very confusing if georeferences changed frequently, and very expensive to update all of the records that depend on them. This can be problematic when a georeferencing system serves more than one purpose or is used by more than one agency with different priorities. For example, a municipality may change its boundaries by incorporating more land, creating problems for mapping agencies and for researchers who wish to study the municipality through time. Street names sometimes change, and postal agencies sometimes revise postal codes. Changes even occur in the names of cities (Saigon to Ho Chi Minh City) or in their conventional transcriptions into the Roman alphabet (Peking to Beijing).

To be most useful, georeferences should stay constant through time.

Every georeference has an associated spatial resolution, equal to the size of the area that is assigned that georeference. A mailing address could be said to have a spatial resolution equal to the size of the mailbox, or perhaps to the area of the parcel of land or structure assigned that address. A U.S. state has a spatial resolution that varies from the size of Rhode Island to that of Alaska, and many other systems of georeferencing have similarly wide-ranging spatial resolutions.

Many systems of georeferencing are unique only within an area or *domain* of the Earth's surface. For example, many cities in the United States have the name Springfield (18 according to a recent edition of the Rand McNally Road Atlas; similarly, there are 9 places called Whitchurch in the 2003 AA Road Atlas of the United Kingdom). However, city name is unique within the domain of a U.S. state—a property that was engineered with the advent of the postal system in the Nineteenth Century—whereas in the UK city names are unique within counties. Today there is no danger that there are two Springfields in Massachusetts; a driver can therefore confidently ask for directions to "Springfield, Massachusetts" in the knowledge that there is no danger of being sent to the wrong Springfield. But people living in London, Ontario, Canada are well aware of the dangers of talking about "London" without specifying

Figure 5.1 Place-names are not necessarily unique at the global level. There are many Londons, for example, besides the largest and most prominent one in the UK. People living in other Londons must often add more information (e.g., London, *Ontario, Canada*) to resolve ambiguity. (Photolink/Photolibrary Group Limited)

the appropriate domain. Even in Toronto, Ontario a reference to "London" may be misinterpreted as a reference to the older (UK) London on a different continent rather than to the one 200 km away in the same province (Figure 5.1). Street name is unique in the United States within municipal domains, but not within larger domains such as county or state. There are 120 places on the Earth's surface with the same Universal Transverse Mercator coordinates (see Section 5.8.2), and a zone number and hemisphere must be added to make a reference unique in the global domain.

Although some georeferences are based on simple names, others are based on various kinds of *measurements* and are called *metric* georeferences. They include latitude and longitude and various kinds of coordinate systems, many of which are discussed in more detail later in this chapter, and are essential to the making of maps, the display of mapped information in GIS, and any kind of numerical analysis. One enormous advantage of such systems is that they provide the potential for infinitely fine spatial resolution: provided we have sufficiently accurate measuring devices and use enough decimal places, such systems will allow us to locate information to any level of accuracy. Another advantage is that from measurements of two or more locations it is possible to compute distances, a very important requirement of georeferencing in GIS.

Metric georeferences are much more useful because they allow maps to be made and distances to be calculated.

Other systems simply *order* locations. In most countries mailing addresses are ordered along streets, often using the odd integers for addresses on one side and the even integers for addresses on the other. This means that it is possible to say that 3000 State Street and 100 State Street are farther apart than 200 State Street and 100 State Street, allowing postal services to sort mail for easy delivery. In the western United States, it is often possible to infer estimates of the distance between two addresses on the same street by knowing that 100 addresses are assigned to each city block and that blocks are typically between 120 m and 160 m long.

This section has reviewed some of the general properties of georeferencing systems, and Table 5.1 shows some commonly used systems. The following sections discuss the specific properties of the systems that are most important in GIS applications.

5.2 Place-names

Giving names to places is the simplest form of georeferencing and was most likely the one first developed by early hunter-gatherer societies. Any distinctive feature on the landscape, such as a particularly old tree, can serve as a point of reference for two people who wish to share information, such as the existence of good game in the tree's vicinity. Human landscapes rapidly became littered with names, as people sought distinguishing labels to use in describing aspects of their surroundings and as other people adopted them. Today, of course, we have a complex system of naming oceans, continents, cities, mountains, rivers, and other prominent features. Each country maintains a system of authorized naming, often through national or state committees assigned with the task of standardizing geographic names. Nevertheless, multiple names are often attached to the same feature, for example when cultures try to preserve the names given to features by the original or local inhabitants (Mount Everest to many, but Chomolungma to many Tibetans), or when city names are different in different languages (Florence in English, Firenze in Italian).

Many commonly used place-names have meanings that vary between people and with the context in which they are used.

Language extends the power of place-names through words such as "between," which serves to refine

Table 5.1 Some commonly used systems of georeferencing.

System	Domain of uniqueness	Metric?	Example	Spatial resolution
Place-name	varies	no	London, Ontario, Canada	varies by feature type
Postal address	global	no, but ordered along streets in most countries	909 West Campus Lane, Goleta, California	size of one mailbox
Postal code	country	no	93117 (U.S. zip code); WC1E 6BT (UK Unit Postcode)	area occupied by a defined number of mailboxes
Telephone calling area	country	no	805	varies
Cadastral system	local authority	no	Parcel 01452954, City of Springfield, MA	area occupied by a single parcel of land
Public Land Survey System	Western Canada and United States only, unique to Prime Meridian	yes	Sec 5, Township 4 N, Range 6E	defined by level of subdivision
Latitude/ longitude	global	yes	119 degrees 45 minutes West, 34 degrees 40 minutes North	infinitely fine
Universal Transverse Mercator	zones six degrees of longitude wide, and N or S Hemisphere	yes	563146E, 4356732N	infinitely fine
State Plane Coordinates	United States only, unique to state and to zone within state	yes	55086.34E, 75210.76N	infinitely fine

references to location, or "near," which serves to broaden them. "Where State Street crosses Mission Creek" is an instance of combining two place-names to achieve greater refinement of location than either name could achieve individually. Even more powerful extensions come from combining place-names with directions and distances, as in "200 m north of the old tree" or "50 km west of Springfield."

But place-names are of limited use as georeferences. First, they often have very coarse spatial resolution. "Asia" covers over 43 million sq km, so the information that something is located "in Asia" is not very helpful in pinning down its location. Even Rhode Island, the smallest state of the United States, has a land area of over 2700 sq km. Second, only certain place-names are officially authorized by national or subnational agencies. Many more are recognized only locally, so their use is limited to communication between people in the local community. Place-names

may even be lost through time: although there are many contenders, we do not know with certainty where the "Camelot" described in the English legends of King Arthur was located, if indeed it ever existed.

The meaning of certain place-names can become lost through time.

The growing Web phenomenon of *user-generated content* is changing this situation, however, and creating an alternative to the traditional top-down system of naming. Individuals are now able to assign names to features quite independently of officialdom by using sites such as Wikimapia (www.wikimapia.org; see Figure 5.2). At time of writing, individuals worldwide had added descriptions of close to 12,000,000 features, from the largest cities to the smallest buildings. Descriptions can be in any language and a wide range

Figure 5.2 Wikimapia coverage of the University of California, Santa Barbara, and surrounding neighborhood. Each rectangle represents one entry, where a volunteer has provided a description and possible images and links to other information, all of which can be exposed by clicking on the rectangle. More rectangles will be displayed as the user zooms.

of scripts, and can include photographs and links to other sources of information.

Wikimapia's mantra is "Let's describe the whole world," and it is one example of a growing phenomenon that has been termed *volunteered geographic information* (VGI), and that draws individual citizens into the process of creating geographic knowledge. In essence, it echoes an earlier era before the establishment of national mapping agencies and authoritative committees, when place-names were created by explorers and local citizens. Many names fell by the wayside, but some were adopted and shown on maps. The process by which America was named, in 1507, was essentially of this nature: an act by an individual cartographer, Martin Waldseemüller, in St-Dié-des-Vosges in France. Waldseemüller had recently read letters from the Florentine explorer Amerigo Vespucci claiming credit for recognizing that the lands discovered to the west of Europe formed a previously unknown continent. Accordingly, he invented a name by feminizing Vespucci's first name, believing that all continents should have feminine names (Figure 5.3).

Copies of the map were distributed across Europe, and the name stuck.

5.3 Postal Addresses and Postal Codes

Postal addresses were introduced after the development of mail delivery in the Nineteenth Century. They rely on several assumptions:

- Every dwelling and office is a potential destination for mail.

- Dwellings and offices are arrayed along paths, roads, or streets, and are numbered sequentially.

- Paths, roads, and streets have names that are unique within local areas.

- Local areas have names that are unique within larger regions.

- Regions have names that are unique within countries.

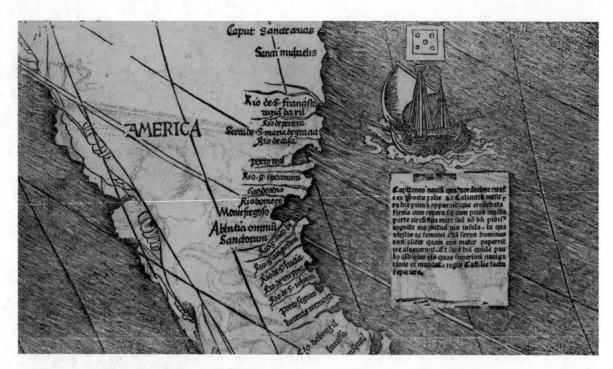

Figure 5.3 Detail of the Waldseemüller map of 1507, which for the first time showed the name the cartographer selected for the new continent. (Courtesy Library of Congress Map Collection)

If the assumptions are true, then mail address provides a unique identification for every dwelling and office on Earth.

Today, postal addresses are an almost universal means of locating many kinds of human activity: delivery of mail, place of residence, or place of business. They fail, of course, in locating anything that is not a potential destination for mail, including almost all kinds of natural features (Mount Everest does not have a postal address, and neither does Manzana Creek in Los Padres National Forest in California). They are not as useful when dwellings are not numbered consecutively along streets, as happens in some cultures (notably in Japan, where street numbering can reflect date of construction, not sequence along the street; it is temporal, rather than spatial) and in large building complexes like condominiums. Many GIS applications rely on the ability to locate activities by postal address and to convert addresses to some more universal system of georeferencing, such as latitude and longitude, for mapping and analysis.

Postal addresses work well to georeference dwellings and offices, but not natural features.

Postal codes were introduced in many countries in the late Twentieth Century in order to simplify the

sorting of mail. In the UK system, for example, the first characters of the code identify an Outward Code, and mail is initially sorted so that all mail directed to a single Local Delivery Office (LDO) is together. Incoming mail is accumulated in a local sorting station and sorted a second time by the last characters of the code so that it can be delivered by an LDO. Figure 5.4 shows a map of the Outward Codes for Southend-on-Sea (SS). The full six or seven characters of the postal code (e.g., SS5 4PJ) are unique to roughly 13 houses, a single large business, or a single building. Different countries are more or less insistent on completeness of postal codes; in the UK, for example, sorters will deliver mail with only an Outward Code since LDOs are required to know their local communities.

Postal codes have proven very useful for many purposes besides the sorting and delivery of mail. Although the area covered by a Canadian FSA, a U.S. ZIP code, or a UK postcode varies, and can be changed whenever the postal authorities want, it is sufficiently constant to be useful for mapping purposes, and many businesses routinely make maps of their customers by counting the numbers present in each postal code area, as well as by dividing by total population to get a picture of market penetration. Figure 5.5 shows an example of summarizing data by postal code. In Figure 5.5A it has been necessary to suppress the locations of postcodes where there are

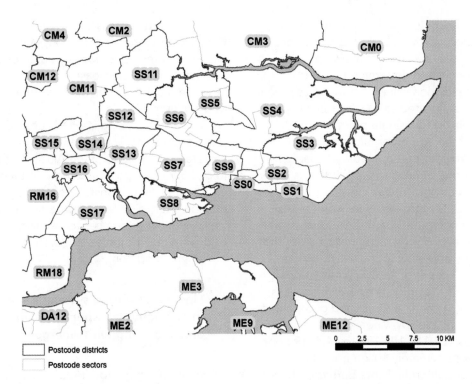

Postcode districts
Postcode sectors

Figure 5.4 Outward Codes for the Southend-on-Sea, UK, Local Delivery Offices (LDOs). Outward codes form the first two, three, or four characters of the UK postal code and are separated from the three characters that identify the Inward Codes. (Map courtesy of Maurizio Gibin)

fewer than five resident patients; it is not necessary to hide any data in the choropleth map shown in Figure 5.5B, since privacy is not violated at this coarser level of granularity. Most people know the postal code of their home, and in some instances postal codes have developed popular images (the ZIP code for Beverly Hills, California, 90210, became the title of a successful television series).

5.4 IP Addresses

Every device (computer, printer, etc.) connected to the Internet has a unique IP (Internet Protocol) address, such as 128.111.106.183, the address of the computer on which this text is being written. IP addresses are allocated to organizations, and the street address included in the registration record can

Figure 5.5 The use of Outward Code boundaries as a convenient basis for summarizing data. In this instance (A) points identifying the residential unit (Inward) postcode have been used to plot the locations of a doctor's patients, and (B) the Outward Code areas have been shaded according to the density of patients per square kilometer.

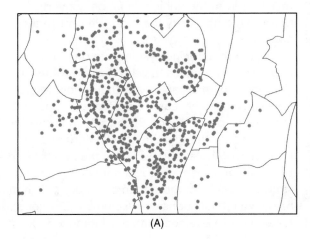

(A)

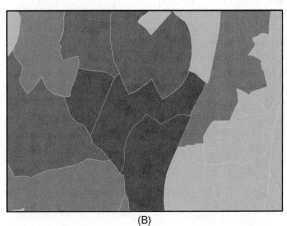

(B)

provide an approximate location. The IP address of the user's computer is provided whenever the computer is used to access a Web site, allowing the operators of major sites to determine the user's location. Spatial resolution will vary, however, since all of the IP addresses allocated to a university campus may be georeferenced to a single point on the campus, and all of the addresses allocated to an Internet Service Provider that operates a broadband service may be georeferenced to a point that is many kilometers from some customers.

Recently, the ability to determine even approximate locations for computers has led to some powerful applications. It allows search engines to order the results of search by proximity to the user's location, and it also permits sites such as Wikimapia to open centered on the user's location. Many Web services offer conversion between IP address and geographic coordinates. For example, www.networldmap.com resolves the address 128.111.106.183 to 34.4119 north, 119.7280 west, approximately 10 km from the office at the University of California, Santa Barbara, where the computer with this address is located. Wikimapia does rather better, opening centered on a location only 1 km away.

5.5 Linear Referencing Systems

A linear referencing system identifies location on a network by measuring distance from a defined point of reference along a defined path in the network. Figure 5.6 shows an example, an accident whose

Figure 5.6 Linear referencing—an incident's position is determined by measuring its distance (87 m) along one road (Birch Street) from a well-defined point (its intersection with Main Street).

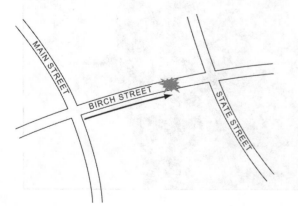

location is reported as being a measured distance from a street intersection, along a named street. Linear referencing is closely related to street address, but uses an explicit measurement of distance rather than the much less reliable surrogate of street address number.

Linear referencing is widely used in applications that depend on a linear network. This includes highways (e.g., Mile 1240 of the Alaska Highway), railroads (e.g., 25.9 miles from Paddington Station in London on the main line to Bristol, England), electrical transmission lines, pipelines, and canals. Highway agencies use linear references to define the locations of bridges, signs, potholes, and accidents, and to record pavement condition.

Linear referencing systems are widely used in managing transportation infrastructure and in dealing with emergencies.

Linear referencing provides a sufficient basis for georeferencing for some applications. Highway departments often base their records of accident locations on linear references, as well as their inventories of signs and bridges. (GIS has many applications in transportation that are known collectively as GIS-T, and in the developing field of intelligent transportation systems or ITS; see Section 2.3.4.) But for other applications it is important to be able to convert between linear references and other forms, such as latitude and longitude. For example, the OnStar system that is installed in many Cadillacs sold in the United States is designed to radio the position of a vehicle automatically as soon as it is involved in an accident. When the airbags deploy, a GPS receiver determines position, which is then relayed to a central dispatch office. Emergency response centers often use street addresses and linear referencing to define the locations of accidents, so the latitude and longitude received from the vehicle must be converted before an emergency team can be sent to the accident.

Linear referencing systems are often difficult to implement in practice in ways that are robust in all situations. In an urban area with frequent intersections, it is relatively easy to measure distance from the nearest one (e.g., on Birch St 87 m east of the intersection with Main Street). But in rural areas an incident may be a long way from the nearest intersection. Even in urban areas it is not uncommon for two streets to intersect more than once (e.g., Birch may have two intersections with Columbia Crescent), or for a street to intersect with itself (Figure 5.7).

Figure 5.7 The intersection of a street with itself (Calgary, Alberta, Canada), causing problems for linear referencing.

5.6 Cadasters and the U.S. Public Land Survey System

The *cadaster* is defined as the map of land ownership in an area, maintained for the purposes of taxing land or of creating a public record of ownership. The process of *subdivision* creates new parcels by legally subdividing existing ones.

Parcels of land in a cadaster are often uniquely identified, by number or by code, and are also reasonably persistent through time, thereby satisfying the requirements of a georeferencing system. Indeed, it has often been argued that the cadaster could form the universal basis of mapping (a *multipurpose cadaster*), with all other geographic information tied to it. But very few people know the identification code of their home parcel; thus use of the cadaster as a georeferencing system is limited largely to local officials, with one major exception.

The U.S. Public Land Survey System (PLSS) evolved out of the need to survey and distribute the vast land resources of the western United States, starting in the early Nineteenth Century, and expanded to become the dominant system of cadaster for all of the United States west of Ohio, and all of western Canada. Its essential simplicity and regularity make it useful for many purposes and understandable by the general public. Its geometric regularity also allows it to satisfy the requirement of a metric system of georeferencing because each georeference is defined by measured distances.

The Public Land Survey System defines land ownership over much of western North America and is a useful system of georeferencing.

To implement the PLSS in an area, a surveyor first laid out an accurate north-south line or *principal meridian*. An east-west *baseline* was then laid out perpendicular to this (Figure 5.8). Rows were then laid out 6 miles apart and parallel to the baseline, to become the *townships* of the system. Then blocks or *ranges* were laid out in 6 mile by 6 mile squares on either side of the principal meridian (see Figure 5.9). Each square is referenced by township number, range number, whether it is to the east or to the west, and the name of the principal meridian. Thirty-six *sections* 1 mile by 1 mile were laid out inside each township and numbered using a standard system (note how the numbers reverse in every other row). Each section was divided into four quarter-sections of 1/4 square mile, or 160 acres, the size of the nominal family farm or homestead in the original conception of the PLSS. The process can be continued by subdividing into four to obtain any level of spatial resolution.

The PLSS would be a wonderful system if the Earth were flat and if survey measurements were always exact. To account for the Earth's curvature, the squares are not perfectly 6 miles by 6 miles, and the rows must be offset frequently; errors in the original surveying complicate matters still further, particularly in rugged landscapes. Figure 5.9 shows the offsetting exaggerated for a small area. Nevertheless, the PLSS remains an efficient system and one with which many people in the Western USA and Western Canada are

Figure 5.8 Google Earth simulation of the view looking east along Baseline Road in Ontario, California—the road that follows the original survey baseline for Southern California laid out by Colonel Henry Washington in 1852. The monument marking the intersection between the baseline and the principal meridian is atop Mount San Bernardino, which appears on the horizon. (Image © 2009 131DigitalGlobe, Image USDAFarm Service Agency, Image County of SanBernardino)

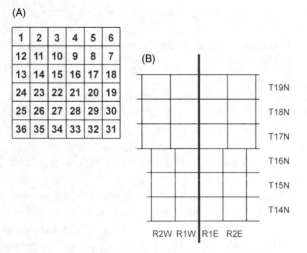

(A)

1	2	3	4	5	6
12	11	10	9	8	7
13	14	15	16	17	18
24	23	22	21	20	19
25	26	27	28	29	30
36	35	34	33	32	31

(B)

T19N
T18N
T17N
T16N
T15N
T14N

R2W R1W R1E R2E

Figure 5.9 Portion of the Township and Range system (Public Lands Survey System) widely used in the western United States as the basis of land ownership (B). Townships are laid out in 6 mile squares on either side of an accurately surveyed Principal Meridian. The offset shown between T16 N and T17 N is needed to accommodate the Earth's curvature (shown much exaggerated). The square mile sections within each township are numbered as shown in (A).

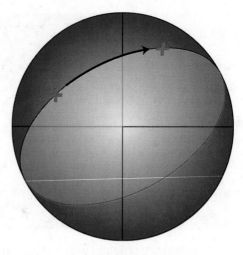

Figure 5.10 Definiton of longitude. The Earth is seen here from above the North Pole, looking along the axis, with the equator forming the outer circle. The location of Greenwich defines the Prime Meridian. The longitude of the point at the center of the red cross is determined by drawing a plane through it and the axis, and measuring the angle between this plane and the Prime Meridian.

familiar. It is often used to specify location, particularly in managing natural resources in the oil and gas industry and in mining, and in agriculture. Services have been built to convert PLSS locations automatically to and from latitude and longitude (Section 5.7).

5.7 Measuring the Earth: Latitude and Longitude

The most powerful systems of georeferencing are those that provide the potential for very fine spatial resolution, that allow distance to be computed between pairs of locations, and that support other forms of spatial analysis. The system of latitude and longitude is in many ways the most comprehensive and is often called the *geographic* system of coordinates, based on the Earth's rotation about its center of mass.

To define latitude and longitude, we first identify the *axis* of the Earth's rotation. The Earth's center of mass lies on the axis, and the plane through the center of mass perpendicular to the axis defines the *equator*. Slices through the Earth parallel to the axis, and perpendicular to the plane of the equator, define lines of constant longitude (Figure 5.10), rather like the segments of an orange. In 1884 a conference attended by delegates from 25 nations agreed that zero

longitude, the *prime meridian*, should be defined by a line marked on the ground at the Royal Observatory in Greenwich, England; the angle between this slice and any other slice defines the latter's measure of longitude. Each of the 360 degrees of longitude is divided into 60 minutes and each minute into 60 seconds. But it is more conventional to refer to longitude by degrees East or West, so longitude ranges from 180 degrees West to 180 degrees East of the prime meridian. Finally, because computers are designed to handle numbers ranging from very large and negative to very large and positive, we normally store longitude in computers as if West was negative and East was positive; and we store parts of degrees using decimals rather than minutes and seconds. A line of constant longitude is termed a *meridian*.

Longitude can be defined in this way for any rotating solid, no matter what its shape, because the axis of rotation and the center of mass are always defined. But the definition of latitude requires that we know something about the shape. The *geoid* is defined as a surface of equal gravity formed by the oceans at rest, and by an imaginary extension of this surface under the continents; it has a complex shape that is only approximately spherical. A much better approximation or *figure of the Earth* is the *ellipsoid of rotation*, the figure formed by taking a mathematical ellipse and rotating it about its shorter axis (Figure 5.11). The term *spheroid* is also commonly used.

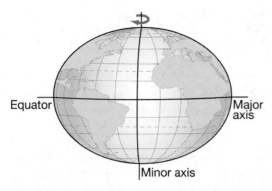

Figure 5.11 Definition of the ellipsoid, formed by rotating an ellipse about its minor axis (corresponding to the axis of the Earth's rotation).

The difference between the ellipsoid and the sphere is measured by its *flattening*, or the reduction in the minor axis relative to the major axis. Flattening is defined as

$$f = (a - b)/a$$

where a and b are the lengths of the major and minor axes, respectively (we usually refer to the *semi*-axes, or half the length of the axes, because these are comparable to radii). The actual flattening is about 1 part in 300.

The Earth is slightly flattened, such that the distance between the poles is about 1 part in 300 less than the diameter at the equator.

Much effort was expended over the past 200 years in finding ellipsoids that best approximated the shape of the Earth in particular countries, so that national mapping agencies could measure position and produce accurate maps. Early ellipsoids varied significantly in their basic parameters and were generally not centered on the Earth's center of mass. But the development of intercontinental ballistic missiles in the 1950s and the need to target them accurately, as well as new data available from satellites, drove the push to a single international standard. Without a single standard, the maps produced by different countries using different ellipsoids could never be made to fit together along their edges, and artificial steps and offsets were often necessary in moving from one country to another (navigation systems in aircraft would have to be corrected, for example). The ellipsoid known as WGS84 (the World Geodetic System of 1984) is now widely accepted, and North American mapping is being brought into conformity with it through the adoption of the virtually identical North American Datum of 1983 (NAD83). It specifies a semi-major axis (distance from the center to the equator) of 6378137 m, and a flattening of 1 part in 298.257; and its line of zero longitude passes about 100 m to the east of the Greenwich Observatory. But many other ellipsoids remain in use in certain parts of the world, and much older data still adhere to earlier standards, such as the North American Datum of 1927 (NAD27). For points in the United States the difference between two points with identical latitude and longitude, but determined according to the NAD27 and NAD83 datums, can be as much as 100 m. Thus GIS users sometimes need to convert between datums, and functions to do that are commonly available.

We can now define latitude. Figure 5.12 shows a line drawn through a point of interest perpendicular to the ellipsoid at that location. The angle made by this line with the plane of the equator is defined as the point's latitude and varies from 90 South to 90 North. Again, south latitudes are usually stored as negative numbers and north latitudes as positive. Latitude is often symbolized by the Greek letter phi (ϕ) and longitude by the Greek letter lambda (λ), so the respective ranges can be expressed in mathematical shorthand as: $-180 \leq \lambda \leq 180$; $-90 \leq \phi \leq 90$. A line of constant latitude is termed a *parallel*.

It is important to have a sense of what latitude and longitude mean in terms of distances on the surface. Ignoring the flattening, two points on the same north-south line of longitude and separated by one degree of latitude are 1/360 of the circumference of the Earth apart, or about 111 km apart. One minute of latitude corresponds to 1.86 km, and also defines one nautical mile, a unit of distance that is still commonly used in navigation. One second of latitude corresponds to about 30 m. But things are

Figure 5.12 Definition of the latitude of Point A, as the angle between the equator and a line drawn perpendicular to the ellipsoid.

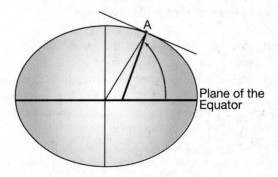

Newton, Descartes, and the Shape of the Earth

Newton's understanding of centrifugal forces led him to conclude that a planet such as the Earth rotating about its axis should bulge at the equator, where the diameter should therefore be greater than the distance between the poles. Descartes, on the other hand, argued that the reverse should be true—a greater distance between the poles than the diameter at the equator. Given the definition illustrated in Figure 5.12, lines of latitude should grow further apart as one moves away from the equator if Newton is right, and closer together if Descartes is right. In 1735 the French Academy of Sciences dispatched expeditions to Finland and Peru to make accurate deter- minations of the distance between lines of latitude. The results proved conclusively that Newton was right.

One of the members of the expedition to Peru, Jean Godin, married a Peruvian and promised to take her home to France via a difficult route over the Andes and down the Amazon, at a time when the Spanish authorities in Peru were at odds with the Portuguese authorities in Brazil. He left first, and the 20-year story of Isabel's epic and ultimately successful adventure to rejoin him is admirably told in Robert Whitaker's *The Mapmaker's Wife*.

more complicated in the east-west direction, and these figures only apply to east-west distances along the equator, where lines of longitude are furthest apart. Away from the equator the length of a line of latitude gets shorter and shorter, until it vanishes altogether at the poles. The degree of shortening is approximately equal to the cosine of latitude, or cos ϕ, which is 0.866 at 30 degrees North or South, 0.707 at 45 degrees, and 0.500 at 60 degrees. So a degree of longitude is only 55 km along the northern boundary of the Canadian province of Alberta (exactly 60 degrees North).

In GIS, latitude and longitude are often expressed as decimals of degrees, rather than degrees, minutes, and seconds. It is helpful to know that the 5th decimal place of degrees of latitude is about 1 m on the Earth's surface. In GIS it is very uncommon to know positions to greater accuracy, so any additional decimal places that may be displayed or recorded are probably beyond the limits of accuracy and therefore meaningless.

Lines of latitude and longitude are equally far apart only at the equator; toward the poles lines of longitude converge.

Given latitude and longitude, it is possible to deter- mine distance between any pair of points, not just pairs along lines of longitude or latitude. It is easiest to pretend for a moment that the Earth is spherical because the flattening of the ellipsoid makes the equations much more complex. But on a spheri- cal Earth the shortest path between two points is a *great circle*, or the arc formed if the Earth is sliced through the two points and through its center (Figure 5.13; an off-center slice creates a *small circle*). The length of this arc on a spherical Earth of radius R is given by

$$R \arccos[\sin\phi_1 \sin\phi_2 + \cos\phi_1 \cos\phi_2 \cos(\lambda_1 - \lambda_2)]$$

where the subscripts denote the two points. For example, the distance from a point on the equator at longitude 90 East (in the Indian Ocean between Sri Lanka and the Indonesian island of Sumatra) and the

Figure 5.13 The shortest distance between two points on the sphere is an arc of a great circle, defined by slicing the sphere through the two points and the center (all lines of longitude, and the equator, are great circles). The circle formed by a slice that does not pass through the center is a small circle (all lines of latitude except the equator are small circles).

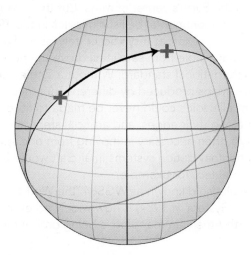

North Pole is found by evaluating the equation for $\phi_1 = 0$, $\lambda_1 = 90$, $\phi_2 = 90$, $\lambda_2 = 90$. It is best to work in radians (1 radian is 57.30 degrees, and 90 degrees is $\pi/2$ radians). The equation evaluates to $R \arccos 0$, or $R \pi/2$, or one quarter of the circumference of the Earth. Using a radius of 6378 km, this comes to 10,018 km, or close to 10,000 km (not surprisingly, since the French originally defined the meter in the late Eighteenth Century as one ten-millionth of the distance from the equator to the pole).

5.8 Projections and Coordinates

Latitude and longitude define location on the Earth's surface in terms of angles with respect to well-defined references: the prime meridian, the center of mass, and the axis of rotation. As such, they constitute the most comprehensive system of georeferencing and support a range of forms of analysis, including the calculation of distance between points, on the curved surface of the Earth. But many technologies for working with geographic data are inherently flat, including paper and printing, which evolved over many centuries long before the advent of digital geographic data and GIS. For various reasons, therefore, much work in GIS deals with a flattened or *projected* Earth, despite the price we pay in the distortions that are an inevitable consequence of flattening. Specifically, the Earth is often flattened because:

- Paper is flat, and paper is still used as a medium for inputting data to GIS by scanning or digitizing (see Section 9.3) and for outputting data in map or image form.

- Rasters (Section 8.2.2) are inherently flat, since it is impossible to cover a curved surface with equal squares without gaps or overlaps.

- Photographic film is flat, and film cameras are still used widely to take images of the Earth from aircraft to use in GIS.

- When the Earth is seen from space, the part in the center of the image has the most detail, and detail drops off rapidly, the back of the Earth being invisible; in order to see the whole Earth at once with approximately equal detail, it must be distorted in some way, and it is most convenient to make it flat.

The Cartesian coordinate system (Figure 5.14) assigns two coordinates to every point on a flat surface by measuring distances from an origin parallel to two axes drawn at right angles. We often talk of the two

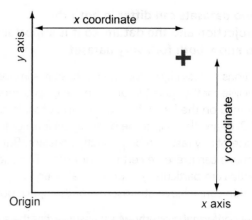

Figure 5.14 A Cartesian coordinate system, defining the location of the blue cross in terms of two measured distances from the origin, parallel to the two axes.

axes as x and y, and of the associated coordinates as the x- and y-coordinate, respectively. Because it is common to align the y-axis with North in geographic applications, the coordinates of a projection on a flat sheet are often termed *easting* and *northing*.

Although projections are not absolutely required, there are several good reasons for using them in GIS to flatten the Earth.

One way to think of a map projection, therefore, is that it transforms a position on the Earth's surface identified by latitude and longitude (ϕ, λ) into a position in Cartesian coordinates, (x, y). Every recognized map projection, of which there are many, can be represented as a pair of mathematical functions:

$$x = f(\phi, \lambda)$$
$$y = g(\phi, \lambda)$$

For example, the famous Mercator projection uses the functions:

$$x = \lambda$$
$$y = \ln \tan[\phi/2 + \pi/4]$$

where ln is the natural log function. The inverse transformations that map Cartesian coordinates back to latitude and longitude are also expressible as mathematical functions: in the Mercator case they are:

$$\lambda = x$$
$$\phi = 2 \arctan e^y - \pi/2$$

where e denotes the constant 2.71828. Many of these functions have been implemented in GIS, allowing users to work with virtually any recognized projection and datum, and to convert easily between them.

Two datasets can differ in both the projection and the datum, so it is important to know both for every dataset.

Projections necessarily distort the Earth, so it is impossible in principle for the scale (distance on the map compared to distance on the Earth; for a discussion of scale see Box 4.2) of any flat map to be perfectly uniform or for the pixel size of any raster to be perfectly constant. But projections can preserve certain properties. Two such properties are particularly important, although any projection can achieve at most one of them, not both:

- The *conformal* property, which ensures that the shapes of small features on the Earth's surface are preserved on the projection: in other words, that the scales of the projection in the *x*- and *y*-directions are always equal.

- The *equal area* property, which ensures that areas measured on the map are always in the same proportion to areas measured on the Earth's surface.

The conformal property is useful for navigation because a straight line drawn on the map has a constant bearing (the technical term for such a line is a *rhumb line* or *loxodrome*). The equal area property is useful for various kinds of analysis involving areas, such as the computation of the area of someone's property.

Besides their distortion properties, another common way to classify map projections is by analogy to a physical model of how positions on the map's flat surface are related to positions on the curved Earth. There are three major classes (Figure 5.15), but note that they do not cover all known projections:

- *Cylindrical* projections, which are analogous to wrapping a cylinder of paper around the Earth, projecting the Earth's features onto it, and then unwrapping the cylinder.

- *Azimuthal* or *planar* projections, which are analogous to touching the Earth with a sheet of flat paper.

- *Conic* projections, which are analogous to wrapping a sheet of paper around the Earth in a cone.

In each case, the projection's *aspect* defines the specific relationship, for example, whether the paper is wrapped around the equator or touches at a pole. Where the paper coincides with the surface the scale of the projection is 1, and where the paper is some distance outside the surface the projected feature will be larger than it is on the Earth. *Secant* projections attempt to minimize distortion by imagining the paper cutting through the surface, so that scale can be both greater and less than 1 (Figure 5.15; projections for which the paper touches the Earth and in which scale is always 1 or greater are called *tangent*).

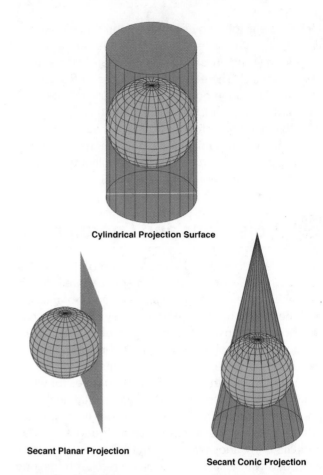

Cylindrical Projection Surface

Secant Planar Projection

Secant Conic Projection

Figure 5.15 The basis for three types of map projections—cylindrical, planar, and conic. In each case a sheet of paper is wrapped around the Earth, and positions of objects on the Earth's surface are projected onto the paper. The cylindrical projection is shown in the *tangent* case, with the paper touching the surface, but the planar and conic projections are shown in the *secant* case, where the paper cuts into the surface. (Reproduced by permission of Peter H. Dana)

All three types can have either conformal or equal area properties, but of course not both. Figure 5.16 presents examples of several common projections and shows how the lines of latitude and longitude map onto the projection, in a (distorted) grid known as a *graticule*.

The next sections describe several particularly important projections in detail, together with the coordinate systems that they produce. Each is important to GIS, and users are likely to come across them frequently. The map projection (and datum) used to make a dataset is sometimes not known to the user of the dataset, so it is helpful to know enough about map projections and coordinate systems to make intelligent guesses when trying to combine such a dataset with other data. Several excellent books on map projections are listed in Further Reading.

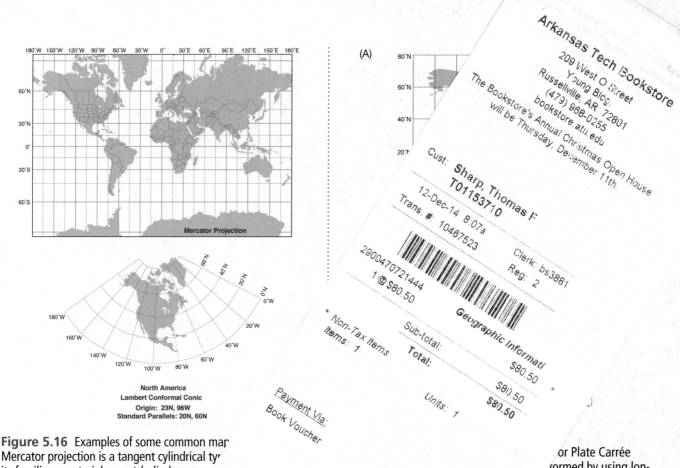

Figure 5.16 Examples of some common map [projections. The] Mercator projection is a tangent cylindrical ty[pe, shown here in] its familiar equatorial aspect (cylinder wrapp[ed around the equa-] tor). The Lambert Conformal Conic projecti[on is a secant conic] type. In this instance, the cone onto which the [surface was pro-] jected intersected the Earth along two lines of latit[ude, 20 North] and 60 North. (Reproduced by permission of Peter H. D[ana)]

[... or Plate Carrée ... formed by using lon-] [... arison of three familiar] [... Lambert Conformal Conic] [... hen the United States is] [... of the three to curve the] [... northern border on the 49th] [...ssion of Peter H. Dana)]

5.8.1 The Plate Carrée or Cylindrical Equidistant Projection

The simplest of all projections maps longitude as x and latitude as y; for that reason this map is also known informally as the *unprojected* projection. The result is a heavily distorted image of the Earth, with the poles smeared along the entire top and bottom edges of the map, and a very strangely shaped Antarctica. Nevertheless, it is the view that we most often see when images are created of the entire Earth from satellite data (for example, in illustrations of sea-surface temperature that show the El Niño or La Niña effects). The projection is not conformal (small shapes are distorted) and not equal area, though it does maintain the correct distance between every point and the equator. It is normally used only for the whole Earth, and maps of parts of the Earth, such as the United States or Canada, look distinctly odd in this projection. Figure 5.17 shows the projection applied to the world, and also presents a comparison of three

familiar projections of the United States: the Plate Carrée, Mercator, and Lambert Conformal Conic.

When longitude is assigned to x and latitude to y, a very odd-looking Earth results.

Serious problems can occur when doing analysis using this projection. Moreover, since most methods of analysis in GIS are designed to work with Cartesian coordinates rather than latitude and longitude, the same problems can arise in analysis when a dataset uses latitude and longitude, or so-called geographic coordinates. For example, a command to generate a circle of radius one unit in this projection will create a figure that is two degrees of latitude across in the north-south direction and two degrees of longitude across in the east-west direction. On the Earth's surface this figure is not a circle at all, and at high latitudes it is a very squashed ellipse.

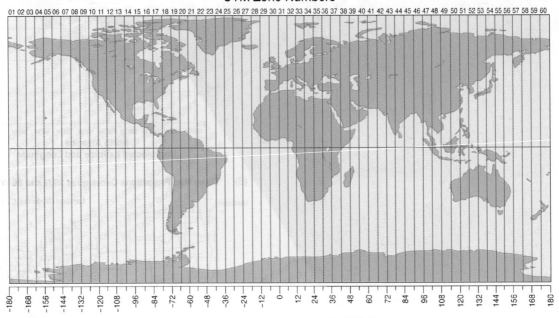

UTM Zone Numbers

01 02 03 04 05 06 07 08 09 10 11 12 13 14 15 16 17 18 19 20 21 22 23 24 25 26 27 28 29 30 31 32 33 34 35 36 37 38 39 40 41 42 43 44 45 46 47 48 49 50 51 52 53 54 55 56 57 58 59 60

Universal Transverse Mercator (UTM) System

Figure 5.18 The system of zones of the Universal Transverse Mercator system. The zones are identified at the top. Each zone is six degrees of longitude in width. (Reproduced by permission of Peter H. Dana)

It is wise to be careful when using a GIS to analyze data in latitude and longitude rather than in projected coordinates because serious distortions of distance, area, and other properties may result.

5.8.2 The Universal Transverse Mercator Projection

The UTM system is often found in military applications and in datasets with global or national coverage. It is based on the Mercator projection, but in the *transverse* rather than equatorial aspect, meaning that the projection is analogous to wrapping a cylinder around the poles rather than around the equator. There are 60 zones in the system, and each zone corresponds to a half cylinder wrapped along a particular line of longitude, each zone being 6 degrees wide. Thus Zone 1 applies to longitudes from 180 W to 174 W, with the half cylinder wrapped along 177 W; Zone 10 applies to longitudes from 126 W to 120 W centered on 123 W, and so on (Figure 5.18).

The UTM system is secant, with lines of scale 1 located some distance out on both sides of the central meridian. Because the projection is conformal, small features appear with the correct shape, and scale at each point is the same in all directions. Scale

is 0.9996 at the central meridian and at most 1.0004 at the edges of the zone, so the maximum distortion of distances using this projection is about 4/100 of 1%. Both parallels and meridians are curved on the projection, with the exception of the zone's central meridian and the equator. Figure 5.19 shows the major features of one zone.

The coordinates of a UTM zone are defined in meters and are set up such that the central meridian's easting is always 500,000 m (a *false* easting), so easting varies from near zero to near 1 million m. In the Northern Hemisphere the equator is the origin of northing, so a point at northing 5 million m is approximately 5000 km from the equator. In the Southern Hemisphere the equator is given a false northing of 10 million m and all other northings are less than this.

UTM coordinates are in meters, making it easy to make accurate calculations of short distances between points.

Because there are effectively 60 different projections in the UTM system, maps will not fit together across a zone boundary. Zones become so much of a problem at high latitudes that the UTM system is normally replaced with azimuthal projections centered on each pole (known as the UPS or Universal Polar Stereographic system) above 80 degrees latitude. The problem is

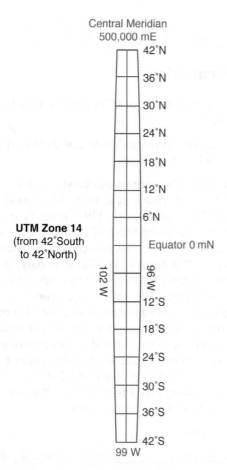

Central Meridian
500,000 mE

UTM Zone 14
(from 42°South
to 42°North)

102 W

99 W

Figure 5.19 Major features of UTM Zone 14 (from 102 W to 96 W). The central meridian is at 99 W. Scale factors vary from 0.9996 at the central meridian to 1.0004 at the zone boundaries. See text for details of the coordinate system. (Reproduced by permission of Peter H. Dana)

especially critical for cities that cross zone boundaries, such as Calgary, Alberta, Canada (which crosses the boundary at 114 W between Zone 11 and Zone 12). In such situations one zone can be extended to cover the entire city, but this results in distortions that are larger than normal. Another option is to define a special zone, with its own central meridian selected to pass directly through the city's center. Italy is split between Zones 32 and 33, and many Italian maps carry both sets of eastings and northings.

UTM coordinates are easy to recognize because they commonly consist of a six-digit integer followed by a seven-digit integer (and decimal places if precision is better than a meter), and sometimes include zone numbers and hemisphere codes. They are an excellent basis for analysis because distances can be calculated from them for points within the same zone with no more than 0.04% error. But they are complicated enough that their use is effectively limited to

professionals (the so-called spatially aware professionals or SAPs) except in applications where they can be hidden from the user. UTM grids are marked on many topographic maps, and many countries project their topographic maps using UTM. It is therefore easy to obtain UTM coordinates from maps for input to digital datasets, either by hand or automatically using scanning or digitizing (Section 9.3.2). In the United States the UTM system is the basis for the National Grid, a system designed to produce a single, unique code for any location in the country (Box 5.2).

5.8.3 State Plane Coordinates and Other Local Systems

Although the distortions of the UTM system are small, they are nevertheless too great for some purposes, particularly in accurate surveying. Zone boundaries also are a problem in many applications because they follow arbitrary lines of longitude rather than boundaries between jurisdictions. In the 1930s each U.S. state agreed to adopt its own projection and coordinate system, generally known as State Plane Coordinates (SPC), in order to support these high-accuracy applications. Projections were chosen to minimize distortion over the area of the state, so choices were often based on the state's shape. Some large states decided that distortions were still too great, and so they designed their SPCs with internal zones (for example, Texas has five zones based on the Lambert Conformal Conic projection (Figure 5.20), while Hawaii has five zones based on the Transverse Mercator projection). Many GIS have details of SPCs already stored, so it is easy to transform between them and UTM, or latitude and longitude. The system was revised in 1983 to accommodate the shift to the new North American Datum (NAD83).

All U.S. states have adopted their own specialized coordinate systems for applications such as surveying that require very high accuracy.

Many other countries have adopted coordinate systems of their own. For example, the UK uses a single projection and coordinate system known as the National Grid that is based on the Oblique Mercator projection and is marked on all topographic maps. Canada uses a uniform coordinate system based on the Lambert Conformal Conic projection, which has properties that are useful at mid- to high latitudes, for applications where the multiple zones of the UTM system would be problematic.

A National System of Georeferencing: The U.S. National Grid

Recent disasters such as Hurricane Katrina have drawn attention to the benefits of having a simple method of assigning a unique code to every location in the nation. Many other countries have such systems. For example, in Great Britain the National Grid is administered by the Ordnance Survey of Great Britain and is popular among hikers and other users of the outdoors. The U.S. National Grid was first proposed in 2000, adopted by the Federal Geographic Data Committee as a standard in 2001, and is now widely used in the emergency services.

Each National Grid code is composed as follows:

the two-digit UTM zone number

a letter denoting an 8-degree latitude range

a letter denoting a 100-km-wide swath oriented east-west

a letter denoting a 100-km-wide swath oriented north-south

up to five decimal digits of UTM easting, omitting the first digit

the same number of decimal digits of UTM northing, omitting the first two digits

For example, the address 909 West Campus Lane, Goleta, California is at latitude 34 degrees 24 minutes 42.7 seconds North, 119 degrees 52 minutes 14.4 seconds West using the NAD83 datum (the use of one decimal place implies an accuracy of about 3 m). It lies in UTM Zone 11 at 236150 m easting, 3811560 m northing (rounding to the nearest 10 m, which is roughly the accuracy of measurement in this case). The National Grid system uses a variable number of northing and easting digits to reflect accuracy, so that shorter codes correspond to less accurately determined positions. Because accuracy is about 10 m in this case we should use only digits 2 through 5 of the six-digit easting and 3 through 6 of the seven-digit northing, so the resulting code is:

11SKU36151156

Figure 5.20 The five State Plane Coordinate zones of Texas. Note that the zone boundaries are defined by counties, rather than parallels, for administrative simplicity.

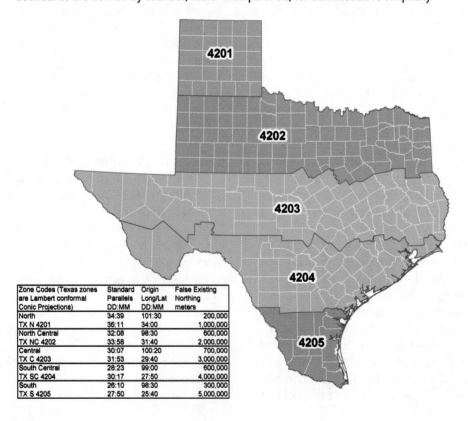

Zone Codes (Texas zones are Lambert conformal Conic Projections)	Standard Parallels DD:MM	Origin Long/Lat DD:MM	False Existing Northing meters
North	34:39	101:30	200,000
TX N 4201	36:11	34:00	1,000,000
North Central	32:08	98:30	600,000
TX NC 4202	33:58	31:40	2,000,000
Central	30:07	100:20	700,000
TX C 4203	31:53	29:40	3,000,000
South Central	28:23	99:00	600,000
TX SC 4204	30:17	27:50	4,000,000
South	26:10	98:30	300,000
TX S 4205	27:50	25:40	5,000,000

5.9 Measuring Latitude, Longitude, and Elevation: GPS

The Global Positioning System and its analogs (GLONASS in Russia and the Galileo system in Europe) have revolutionized the measurement of position, for the first time making it possible for people to know almost exactly where they are anywhere on the surface of the Earth. Previously, positions had to be established by a complex system of relative and absolute measurements. If one was near a point whose position was accurately known (a survey *monument*, for example), then position could be established through a series of accurate measurements of distances and directions starting from the monument. But if no monuments existed, then position had to be established through absolute measurements. Latitude is comparatively easy to measure, based on the elevation of the sun at its highest point (local noon), or on the locations of the sun, moon, or fixed stars at precisely known times. But longitude requires an accurate method of measuring time, and the lack of accurate clocks led to massively incorrect beliefs about positions during early navigation. For example, Columbus and his contemporary explorers had no means of measuring longitude, and believed that the Earth was much smaller than it is and that Asia was roughly as far west of Europe as the width of the Atlantic. The strength of this conviction is still reflected in the term we use for the islands of the Caribbean (the West Indies) and the first rapids on the St. Lawrence in Canada (Lachine, or China). The fascinating story of the accurate measurement of longitude is recounted by Sobel.

The GPS consists of a system of 24 satellites (plus some spares), each orbiting the Earth every 12 hours on distinct orbits at a height of 20,200 km and transmitting radio pulses at precisely timed intervals. To determine position, a receiver must make exact calculations from the signals, the known positions of the satellites, and the velocity of light. Positioning in three dimensions (latitude, longitude, and elevation) requires that at least four satellites are above the horizon, and accuracy depends on the number of such satellites and their positions (if elevation is not required, then only three satellites need be above the horizon). Several different versions of GPS exist, with distinct accuracies.

A simple GPS, such as one might buy in an electronics store for $100, or install as an optional addition to a laptop, cell phone, PDA (personal digital assistant), or vehicle, has an accuracy within 10 m. This accuracy will degrade in cities with tall buildings, or under trees, and GPS signals will be lost entirely under bridges or indoors. Differential GPS (DGPS) combines GPS signals from satellites with correction signals received via radio or telephone from base stations. Networks of such stations now exist, at precisely known locations, constantly broadcasting corrections; corrections are computed by comparing each known location to its apparent location determined from GPS. With DGPS correction, accuracies improve to 1 m or better. Even better accuracies are possible by using various sophisticated techniques or by remaining fixed and averaging measured locations over several hours.

GPS is very useful for recording ground control points when building GIS databases, for locating objects that move (for example, combine harvesters, tanks, cars, and shipping containers), and for direct capture of the locations of many types of fixed objects, such as utility assets, buildings, geological deposits, and sample points. Other applications of GPS are discussed in Chapter 11.

Some care is needed in using GPS to measure elevation. First, accuracies are typically poorer, and a position determined to 10 m in the horizontal may be no better than plus or minus 50 m in the vertical. Second, a variety of reference elevations or vertical datums are in common use in different parts of the world and by different agencies; for example, in the United States the topographic and hydrographic definitions of the vertical datum are significantly different.

5.10 Converting Georeferences

GIS are particularly powerful tools for converting between projections and coordinate systems because these transformations can be expressed as numerical operations. In fact, this ability was one of the most attractive features of early systems for handling digital geographic data and drove many early applications. But other conversions, for example, between place-names and geographic coordinates, are much more problematic. Yet they are essential operations. Almost everyone knows their mailing address and can identify travel destinations by name, but few are able to specify these locations in coordinates, or to interact with geographic information systems on that basis. GPS technology is attractive precisely because it allows its user to determine his or her latitude and longitude, or UTM coordinates, directly at the touch of a button.

Methods of converting between georeferences are important for:

- Converting lists of customer addresses to coordinates for mapping or analysis (the task known as *geocoding*; see Box 5.3).

- Combining datasets that use different systems of georeferencing.

Geocoding: Conversion of Street Addresses to Coordinates

Geocoding is the name commonly given to the process of converting street addresses to latitude and longitude, or some similarly universal coordinate system. It is widely used, as it allows any database containing addresses, such as a company mailing list or a set of medical records, to be input to a GIS and mapped. Geocoding requires a database containing records representing the geometry of street segments between consecutive intersections, and the address ranges on each side of each segment (a *street centerline database*; see Section 8.2.3.3). Addresses are geocoded by finding the appropriate street segment record and estimating a location based on

linear interpolation within the address range. For example, 950 West Broadway in Columbia, Missouri, lies on the side of the segment whose address range runs from 900 to 998, or 50/98 = 51.02% of the distance from the start of the segment to the end. The segment starts at 92.3503 West longitude, 38.9519 North latitude, and ends at 92.3527 West, 38.9522 North. Simple arithmetic gives the address location as 92.3515 West, 38.9521 North. Four decimal places suggests an accuracy of about 10 m, but the estimate also depends on the accuracy of the assumption that addresses are uniformly spaced, as well as on the accuracy of the street centerline database.

- Converting to projections that have desirable properties for analysis, for example, no distortion of area.

- Searching the Internet or other distributed data resources for data about specific locations.

- Positioning GIS map displays by recentering them on places of interest that are known by name (these last two are sometimes called *locator* services).

The oldest method of converting georeferences is the *gazetteer*, the name commonly given to the index in an atlas that relates place-names to latitude and longitude, and to relevant pages in the atlas where information about that place can be found. In this form the gazetteer is a useful locator service, but it works only in one direction as a conversion between georeferences (from place-name to latitude and longitude). Gazetteers have evolved substantially in the digital era, and it is now possible to obtain large databases of place-names and associated coordinates and to access services that allow such databases to be queried over the Internet (e.g., the Alexandria Digital Library gazetteer, www.alexandria.ucsb.edu; the U.S. Geographic Names Information System, www.geonames.usgs.gov). It is also possible to employ sophisticated software to detect place-names in text and to convert them to georeferences (e.g., www.metacarta.com).

5.11 Geotagging and Mashups

Online gazetteers, geocoding sites, and other services for converting georeferences have made it extremely easy to determine the geographic locations associated with many types of online data. It is now easy, for example, to take a mailing list containing street addresses, geocode them, and create maps. Services such as Google Earth, Google Maps, and their Yahoo! and Microsoft cousins provide the mapping capabilities and are readily invoked through their application programming interfaces (APIs). The term *mashup*, which derives originally from the popular-music industry, has been adopted as a way of describing the joining of two or more online services to create something that neither was able to do on its own. One of the best known mashups is www.housingmaps.com, which combines real-estate (property) listings from Craig's List (www.craigslist.org) with cartographic data from Google Maps (maps.google.com). A similar service for the UK is provided by www.nestoria.co.uk, and the www.londonprofiler.org site allows georeferenced property listings to be viewed against a range of thematic data. Today hundreds of thousands of such mashups have been created, many of them dealing with georeferenced data.

In many online services, such as Wikipedia (www.wikipedia.org), georeferences are embedded

Vincent Tao, Helping to Direct Microsoft's Virtual Earth Team

One of many graduates of Wuhan University who have gone on to successful careers in GIScience, Vincent Tao (Figure 5.21) obtained his Ph.D. from the University of Calgary in 1997. After several years there on the faculty, he moved to a Canada Research Chair at York University in 2001. His company GeoTango, which he launched in 2002, developed GlobeView, one of the earliest virtual globes, and SilverEye, the first software to build 3-D models from single images. GeoTango was acquired by Microsoft in 2005. Since then Tao has been director with the Microsoft Virtual Earth team, responsible for global product strategy and software development in local search and mapping, location-based services, and related technologies. He sees Virtual Earth developing as a powerful tool in many areas of human activity, from shopping and marketing to tourism and entertainment. Well known in the GIScience community, he has received many honors and awards and is a frequent speaker at conferences worldwide. In May 2009 he announced his departure from Microsoft to pursue opportunities in consulting.

Figure 5.21 Vincent Tao. (Photo Courtesy of Vincent Tao)

in text in the form of *geotags*, codes representing latitude and longitude in compressed form. Geotags make it easy to create mashups with mapping services (see, for example, www.geonames.org and the (www.maptube.org) service described in Box 13.3).

5.12 Georegistration

Converting between georeferences is often needed during the process of assembling a database, when two datasets use different and perhaps unknown coordinate systems. For example, two datasets were needed in a study of fire alarms in London, Ontario: the locations of each of several thousand alarms, and the boundaries of the tracts defined by the Canadian census. Linking the two datasets would allow the number of alarms to be determined in each tract and to be compared to statistics about the tract's housing and residents. The map of tract boundaries used UTM coordinates, but the map of alarms had been created from a street map using a digitizer (Section 9.3.2).

The standard approach in situations like this is to find a number of points that can serve as *registration* points or *tics*, using them to find equations that will convert coordinates. In this case it was convenient to use 10 major street intersections scattered over the city, since these could be readily located both in the tract boundaries and on the map of alarms, and to use UTM as the common system. Table 5.2 shows the coordinates of these intersections on the alarm map

(*x* and *y*) and in UTM. Note that the UTM coordinates in meters are rounded to the nearest hundred, given the limited spatial resolution of street intersections.

To convert the alarm map to UTM, we look for the simplest possible equations and often adopt an *affine transformation* of the form:

$$\text{UTM north} = a + bx + cy$$
$$\text{UTM east} = d + ex + fy$$

where *a* through *f* are values to be determined from the tic points. Standard methods exist in GIS for doing this, yielding the following values:

$$\text{UTM north} = 4744135 + 501.0x + 1222.2y$$
$$\text{UTM east} = 473668 + 1207.9x - 464.6y$$

Finally, these two equations are used to convert all of the fire alarm data points to UTM so that the analysis can proceed.

Figure 5.22 shows another example of georegistration. The highlighted area was the focus of a proposed project on the campus of the University of California, Santa Barbara, which was intended to restore the area to its natural wetland state. A number of wells were drilled to monitor groundwater, and accurately located using GPS. It was then necessary to merge these data with the campus database of detailed high-resolution imagery and ground elevations, which used a different coordinate system. Registration points were established on features, such as the corners of buildings and street lights, which could be easily located in both datasets. Note the difficulty in this case of registering

Table 5.2 Registration points, with coordinates in both systems.

Intersection	x	y	UTM north	UTM east
Oxford and Sanatorium	2.90	9.80	4757500	472700
Wonderland and Southdale	4.86	5.96	4753800	476500
Wharncliffe and Stanley	7.32	8.56	4758200	478600
Oxford and Wharncliffe	7.58	9.67	4759800	478400
Wellington and Southdale	8.24	4.90	4754300	481500
Highbury and Hamilton	11.32	6.67	4758200	484100
Trafalgar and Clarke	13.17	7.56	4759700	486200
Adelaide and Dundas	9.49	8.59	4759400	481100
Highbury and Fanshawe	11.50	12.68	4765400	481500
Richmond and Huron	8.28	10.73	4761500	478800

the lower part of the highlighted area, where there are no readily identifiable features to use for registration.

The figure shows the result, displayed in Google Earth. Note the obvious displacement between the highlighted area and the Google Earth imagery, but the apparent agreement between the highlighted area and the Google Earth roads. It seems clear that the roads and the highlighted area are registered much more accurately to the Earth than the Google Earth imagery, which is misplaced by more than 10 m.

Figure 5.22 Registration of a high-resolution image (highlighted) of part of the campus of the University of California, Santa Barbara, shown as a Google Earth mashup. (© 2009 Google, Image U.S. Geological Survey)

5.13 Summary

This chapter has looked in detail at the complex ways in which humans refer to specific locations on the planet and how they measure locations. Any form of geographic information must involve some kind of georeference, and so it is important to understand the common methods, together with their advantages and disadvantages. Many of the benefits of GIS rely on accurate georeferencing—the ability to link different items of information together through common geographic location; the ability to measure distances and areas on the Earth's surface, and to perform more complex forms of analysis; and the ability to communicate geographic information in forms that others can understand.

Georeferencing was introduced in early societies to deal with the need to describe locations. As humanity has progressed, we have found it more and more necessary to describe locations accurately, and over wider and wider domains, so that today our methods of georeferencing are able to locate phenomena unambiguously and accurately anywhere on the Earth's surface. Today, with modern methods of measurement, it is possible to direct another person to a point on the other side of the Earth to an accuracy of a few centimeters. This level of accuracy and referencing is regularly achieved in such areas as geophysics and civil engineering.

But georeferences can never be perfectly accurate, and it is always important to know something about spatial resolution. Questions of measurement accuracy are discussed at length in Chapter 6, and Section 6.2.3 deals with techniques for representation of phenomena that are inherently fuzzy, such that it is impossible to say with certainty whether a given point is inside or outside the georeference.

Questions for Further Study

1. Visit your local map library, and determine: (1) the projections and datums used by selected maps; (2) the coordinates of your house in several common georeferencing systems.

2. Summarize the arguments for and against a single global figure of the Earth, such as WGS84.

3. Access several online mapping services such as Google Maps, Google Earth, or Yahoo! Maps. What projection does each one use, how does it display map scale, and how does map scale vary over the map?

4. Chapter 14 discusses various forms of measurement in GIS. Review each of those methods and the issues involved in performing analysis on databases that use different map projections. Identify the map projections that would be best for measurement of (1) area, (2) length, and (3) shape.

Further Reading

Bugayevskiy, L.M. and Snyder, J.P. 1995. *Map Projections: A Reference Manual*. London: Taylor and Francis.

Kennedy, M. 1996. *The Global Positioning System and GIS: An introduction*. Chelsea, MI: Ann Arbor Press.

Maling, D.H. 1992. *Coordinate Systems and Map Projections*. 2nd ed. Oxford: Pergamon

Snyder, J.P. 1997. *Flattening the Earth: Two Thousand Years of Map Projections*. Chicago: University of Chicago Press.

Sobel, D. 1995. *Longitude: The True Story of a Lone Genius Who Solved the Greatest Scientific Problem of His Time*. New York: Walker.

Steede-Terry, K. 2000. *Integrating GIS and the Global Positioning System*. Redlands, CA: ESRI Press.

Whitaker, R.W. 2004. *The Mapmaker's Wife: A True Tale of Love, Murder, and Survival in the Amazon*. New York: Basic Books.

6 Uncertainty

Uncertainty in geographic representation arises because, of necessity, almost all representations of the world are incomplete. As a result, data in a GIS can be subject to measurement error, out of date, excessively generalized, or just plain wrong. This chapter identifies many of the sources of geographic uncertainty and the ways in which they operate in GIS-based representations. Uncertainty arises from the way that GIS users conceive of the world, how they measure and represent it, and how they analyze their representations of it. We investigate a number of conceptual issues in the creation and management of uncertainty, before reviewing the ways in which it may be measured using statistical and other methods. The propagation of uncertainty through geographical analysis is then considered. Uncertainty is an inevitable characteristic of GIS usage, and one that users must learn to live with. In these circumstances, it becomes clear that all decisions based on GIS are also subject to uncertainty.

○ LEARNING OBJECTIVES

After studying this chapter you will:

- Understand the concept of uncertainty and the ways in which it arises from imperfect representation of geographic phenomena.

- Be aware of the uncertainties introduced in the three stages (conception, measurement and representation, and analysis) of database creation and use.

- Understand the concepts of vagueness and ambiguity, and the uncertainties arising from the definition of key GIS attributes.

- Understand how and why scale of geographic measurement and analysis can both create and propagate uncertainty.

6.1 Introduction

GIS-based representations of the real world are used to reconcile science with practice, concepts with applications, and analytical methods with social context. Yet, almost always, such reconciliation is imperfect because, necessarily, representations of the world are incomplete (Section 3.4). In this chapter we will use *uncertainty* as an umbrella term to describe the problems that arise out of these imperfections. Occasionally, representations may approach perfect accuracy and precision (terms that we will define in Section 6.3.2.2)—as might be the case, for example, in the detailed site layout layer of a utility management system in which strenuous efforts are made to reconcile fine-scale multiple measurements of built environments. Yet perfect, or nearly perfect, representations of reality are the exception rather than the norm.

More often, the inherent complexity and detail of our world makes it virtually impossible to capture every single facet, at every possible scale, in a digital representation. (Neither is this usually desirable: see the discussion of spatial sampling in Section 4.4.) Furthermore, different individuals see the world in different ways, and in practice no single view is likely to be seen universally as the best or to enjoy uncontested status. In this chapter we discuss how the processes and procedures of abstraction create differences between the contents of our (geographic and attribute) database and real-world phenomena. Such differences are almost inevitable, and understanding them can help us to manage uncertainty and to live with it.

It is impossible to make a perfect representation of the world, so uncertainty about it is inevitable.

Various terms are used to describe differences between the real world and how it appears in a GIS, depending on the context. The concept of error in statistics arises in part from omission of some relevant aspects of a phenomenon—as in the failure to fully specify all of the predictor variables in a multiple regression model, for example. Similar problems arise when one or more variables are omitted from the calculation of a composite indicator—as, for example, in omitting road accessibility in an index of land value or omitting employment status from a measure of social deprivation (see Section 16.2.1 for a discussion of indicators). The established scientific notion of measurement *error* focuses on differences between observers or between measuring instruments. This raises issues of *accuracy*, which the Dutch geostatistician Gerard Heuvelink has defined as the difference between reality and *our* representation of reality. Although such differences are often principally addressed in formal mathematical terms, the use of the word *our* acknowledges the varying perspectives that different observers may take upon a complex, multiscale, and inherently uncertain world.

Yet even this established framework is too simple for understanding quality or the defining standards of geographic data. The terms *ambiguity* and *vagueness* (defined in Section 6.2.2) identify further considerations that need to be taken into account in assessing the *quality* of a GIS representation. Many geographic representations depend on inherently vague definitions and concepts. Quality is an important topic in GIS, and many attempts have been made to identify its basic dimensions. The U.S. Federal Geographic Data Committee's (FGDC's) various standards list five components of quality: attribute accuracy, positional accuracy, logical consistency, completeness, and lineage. Definitions and other details on each of these and several more can be found on the FGDC's Web pages (www.fgdc.gov). Error, inaccuracy, ambiguity, and vagueness all contribute to the notion of uncertainty in the broadest sense, and uncertainty may thus be defined as a measure of the user's understanding of the difference between the contents of a dataset and the real phenomena the data are believed to represent. This definition implies that phenomena are real, but includes the possibility that we are unable to describe them exactly. In GIS, the term *uncertainty* has come to be used as the catch-all term to describe situations in which the digital representation is simply incomplete and as a measure of the general quality of the representation.

Uncertainty accounts for the difference between the contents of a dataset and the phenomena that the data are supposed to represent.

The views outlined in the previous paragraph are themselves controversial and provide a rich ground for endless philosophical discussions. Some would argue that uncertainty can be inherent in phenomena themselves rather than just in their description. Others would argue for distinctions between *vagueness*, *uncertainty*, *fuzziness*, *imprecision*, *inaccuracy*, and many other terms that most people use as if they were essentially synonymous. Geographer Peter Fisher has provided a useful and wide-ranging discussion of these terms. We take the catch-all view here and leave these arguments to further study.

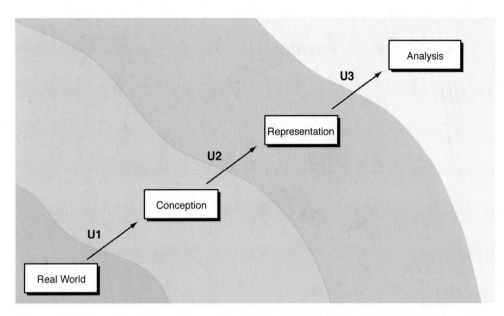

Figure 6.1 A conceptual view of uncertainty. The three filters, U1, U2, and U3, distort the way in which the complexity of the real world is conceived, represented, and analyzed in a cumulative way.

In this chapter, we will discuss some of the principal sources of uncertainty and some of the ways in which uncertainty degrades the quality of a spatial representation. The way in which we conceive of a geographic phenomenon very much prescribes the way in which we are likely to set about representing (or measuring) it. Representation, in turn, heavily conditions the ways in which it may be analyzed within a GIS. This chain sequence of events, in which *conception* prescribes *representation*, which in turn prescribes *analysis*, is a succinct way of summarizing much of the content of this chapter and is summarized in Figure 6.1. In this diagram, U1, U2, and U3 each denote *filters* that can be thought of as selectively distorting or transform the real world when it is stored and analyzed in GIS: a later chapter (Section 13.2.1) introduces a fourth filter that mediates interpretation of analysis and the ways in which feedback may be accommodated through improvements in representation.

6.2 U1: Uncertainty in the Conception of Geographic Phenomena

In Chapter 3 we defined an atom of geographic information as linking a descriptive property or *attribute*, a *place*, and a *time* (Section 3.4). We have acknowledged that what is left out of a representation may be important, but we have assumed that what is included in a representation is founded on clear conceptions of places and attributes. In fact, it often turns out that this is not the case, and the working definitions that are used to represent places or attributes may or may not be fit for purpose.

6.2.1 Conceptions of Place: Units of Analysis

The first component of an atom of geographic information is a *place*. Our discussion of Tobler's Law (Section 3.1) and of spatial autocorrelation (Section 4.6) established that geographic data handling is different from all other classes of nonspatial applications. A further characteristic that sets geographic information science apart from most every other science is that it is only rarely founded on *natural* units of analysis. What is the natural unit of measurement for a soil profile? What is the spatial extent of a *pocket* of high unemployment, or a *cluster* of cancer cases? How might we delimit the polluting effect of a coal-fired power station? The questions become still more difficult in bivariate (two variable) and multivariate (more than two variable) studies. At what scale is it appropriate to investigate any relationship between background radiation and the incidence of leukemia? Or to assess any relationship between labor-force qualifications and unemployment rates? Figure 6.2 shows some "hotspots" (local smoothed aggregations, designed to maintain the confidentiality of individual patient records) of the incidence of diabetes in the London Borough of Southwark: the raw data do indeed suggest local concentrations of the problem, but do not immediately suggest any areal basis to attempt interventions such as promoting healthier diets.

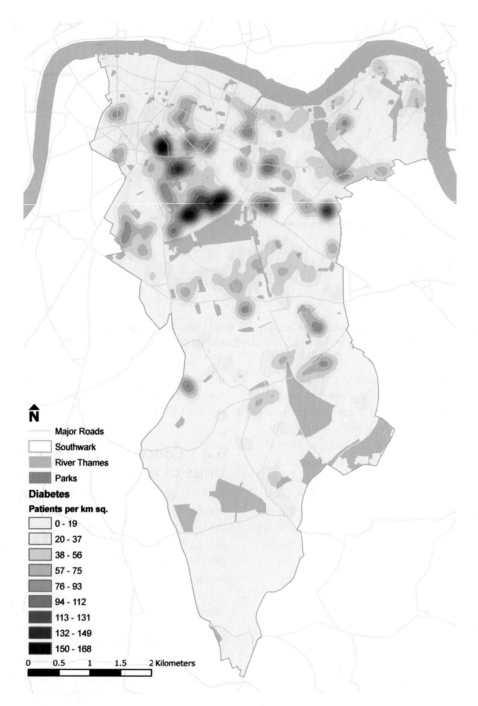

Figure 6.2 A map of local concentrations of diabetes in the London Borough of Southwark. (Courtesy Jakob Peterson and Maurizio Gibin)

In many cases there are no natural units for geographic analysis.

The discrete object view of geographic phenomena relies far more on the idea of natural units of analysis than the field view. As such, this problem is more likely to be manifest in vector GIS applications, such as those identified in Table 3.3. Things we manipulate, such as pencils, books, or screwdrivers, are obvious natural units. Biological organisms are almost always natural units of analysis, as are groupings such as households or families—though even here there are certainly difficult cases, such as the massive networks of fungal strands that are often claimed to be the largest living organisms on Earth, or extended families of human individuals. Most of the difficult cases fall into one of two categories—they are either instances of fields, where variation can be thought of

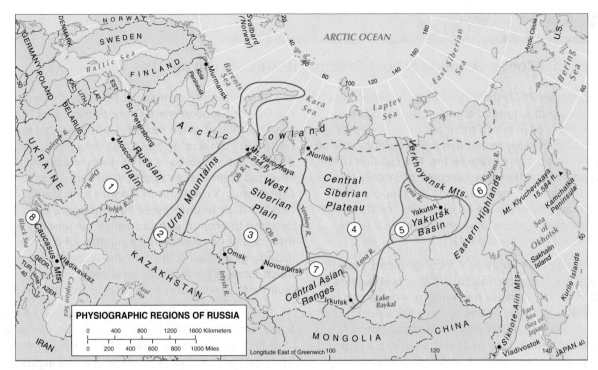

Figure 6.3 The regional geography of Russia. (Source de Blij H.J. and Muller P.O. 2000 *Geography: Realms, Regions and Concepts* (9th ed) New York: Wiley, p. 113)

as inherently continuous in space, or they are instances of poorly defined aggregations of discrete objects. In both of these cases it is up to the investigator to make the decisions about units of analysis, making the identification of the objects of analysis inherently subjective.

The absence of objective or uncontested definitions of place has by no means prevented the attempts of geographers to devise place classifications. The long-established regional geography tradition is fundamentally concerned with the delineation of zones characterized by internal homogeneity (with respect to climate, economic development, or agricultural land use, for example), within a zonal scheme that maximizes between-zone heterogeneity, such as the map illustrated in Figure 6.3. Regional geography is fundamentally about delineating *uniform* zones, and many employ multivariate statistical techniques such as cluster analysis to supplement, or post-rationalize, intuition.

> **Identification of homogeneous zones and spheres of influence lies at the heart of traditional regional geography as well as contemporary data analysis.**

Other geographers have tried to develop *functional* zonal schemes in which zone boundaries delineate the breakpoints between the spheres of influence of adjacent facilities or features—as in the definition of travel-to-work areas, for example, or the definition of a river catchment. Zones may be defined such that there is maximal interaction within zones and minimal interaction between zones. Any functional zoning system is likely to prove contentious, if spending public funds results in different outcomes for individuals in different locations. Nowhere is this more apparent than in the contentious domain of definiting and implementing of community school catchment areas (Box 6.1).

6.2.2 Conceptions of Attributes: Vagueness and Ambiguity

6.2.2.1 Vagueness

The frequent absence of objective geographic individual units means that, in practice, the labels that we assign to zones and the ways in which we draw zone boundaries are often only vague best guesses. What absolute or relative incidence of oak trees in a forested zone qualifies it for the label *oak woodland* (Figure 6.5)? Or, in a developing country context in which aerial photography rather than ground enumeration is used to estimate population size, what rate of incidence of dwellings identifies a zone of *dense* population? In each of these instances, it is

Functional Zones: Defining School Catchment Areas in Bristol, UK.

In September 2007, the City of Bristol opened an attractive public-funded school at Redland Green, a wealthy area in which the inadequacies of provision had previously resulted in many parents seeking alternatives in the private sector. The location of the new school was itself the outcome of intense local lobbying, and the geography of the school's area of primary responsibility (in effect, its "catchment") was somewhat odd in appearance relative to the school's location (Figure 6.4).

When public (in the U.S., public-funded, sense) schools are oversubscribed in Britain, local authorities frequently use distance measures to allocate places to children who live nearer to the school. Bristol City Council almost totally failed to understand and anticipate demand for what was, in effect, a new public service, and in the first year that the new school buildings were open, almost all of the places were allocated to prospective pupils living within 1.4 km of the school gates. Well under half of the school's catchment area was served by the new school.

Responding to complaints from local parents, the UK Local Government Ombudsman ruled that the Council had given parents an "unrealistic expectation" of securing a place at the school, and ordered the school to open its gates to appellants who had been denied access. While adhering to the Ombudsman's ruling, in the longer term the Council had to adapt to unforeseen circumstances by redefining the community for which the school was intended. It has considered a number of options, including redefining the catchment area to serve a much more geographically localized community of "winners" in the publicly funded schools lottery.

There is no such thing as a "natural" area for a school catchment. This being the case, GIS can assist in defining a functional zone that is fit for purpose by deriving measures of demand and fair access to the facility—using socioeconomic data such as numbers of children of school age, travel time, and so on. Bristol City Council failed to do this: it was able to use ArcGIS measures of distance for the

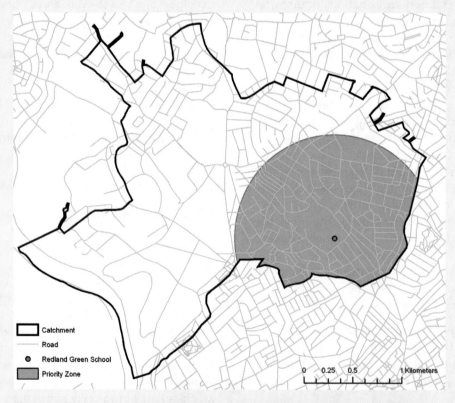

Figure 6.4 The original catchment area of Bristol's Redland Green School, and the subarea within which offers of places were originally made. (Courtesy Alex Singleton)

operational task of rationing places once the conceptual failure had become apparent, but this could not compensate for the failure to use any of the longer-term strategic management functions of GIS to anticipate demand for the new facility at the planning stage (see Chapter 18). The conception, hence definition, of the school's catchment area was wholly inadequate for the community function envisaged for the school when public funds were committed to it.

expedient to transform point-like events (individual trees or individual dwellings) into area objects, and pragmatic decisions must then be taken in order to create a working definition of a spatial distribution. These decisions rarely have any absolute validity, and they raise two important questions:

- Is the defining boundary of a zone crisp and well defined?
- Is the assignment of a particular label to a given zone robust and defensible?

Uncertainty can exist both in the positions of the boundaries of a zone and in its attributes.

The idiographic tradition in geography (see Section 1.3.1) has a long-held preoccupation with defining regions, and the vagaries inherent in tracing lines across maps can have important political or economic implications—as, for example with the drawing of the Dayton Agreement lines to partition Bosnia and Herzegovina in 1995, or in defining the geographical areas to which political decision making should be devolved. The data processing power of GIS may be harnessed through *geocomputation* (Section 16.1) to build regions from atoms of geographic information. Box 6.2 describes how the geography of Anglo-Saxon family names can be used to divide Great Britain into regions.

Figure 6.5 (Inset photo) A local assemblage of oak trees or part of an "oak woodland" on Ragged Boys Hill, in the New Forest, UK. The Ordnance Survey map of the area suggests that there are trees beyond the perimeter of the "woodland" area, while the tree symbology suggests that the woods are characterized by varying proportions of deciduous and coniferous trees. (Photo courtesy Jim Champion)

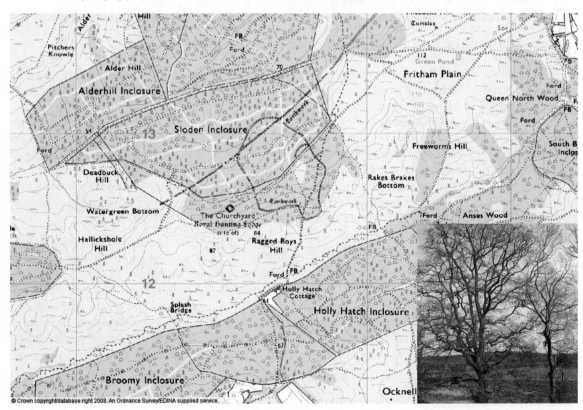

Many English language terms used to convey geographic information are inherently ambiguous.

These questions have statistical implications (can we put numbers on the confidence associated with boundaries or labels?), cartographic implications (how can we convey the meaning of vague boundaries and labels through appropriate symbols on maps and GIS displays?), and cognitive implications (do people subconsciously attempt to force things into categories and boundaries to satisfy a deep need to simplify the world?).

6.2.2.2 Ambiguity

Many objects are assigned different labels by different national or cultural groups, and such groups perceive space differently. Geographic prepositions in the English language such as *across*, *over*, and *in* (used in the Google Maps query in Figure 1.17) do not have simple correspondences with terms in other languages. Object names and the topological relations between them may thus be inherently *ambiguous*. Perception, behavior, language, and cognition all play a part in the conception of real-world entities and the relationships between them. GIS cannot provide the magic bullet of a value-neutral evidence base for decision making, yet this is not to say that GIS inevitably privileges one person's worldview over another's. Indeed, GIS can provide a formal framework for the reconciliation of different representations (see Sections 1.8 and 3.3).

Ambiguity also arises in the conception and construction of *indicators* (see also Section 16.2.1). *Direct* indicators are deemed to bear a clear correspondence with a mapped phenomenon. Detailed household income figures, for example, can provide a direct indicator of the likely geography of expenditure and demand for goods and services; tree diameter at breast height can be used to estimate stand value; and field nutrient measures can be used to estimate agronomic yield. *Indirect* indicators are used when the best available measure is a perceived surrogate link with the phenomenon of interest. Thus the incidence of central heating among households, or rates of multiple car ownership, might provide a surrogate for household income data if such data are not available, while local atmospheric measurements of nitrogen oxide can provide an indirect indicator of environmental health. Box 6.2 describes how people's names provide a direct indicator of their ethnicity and how this information may be used to improve on the detail, quality, and timeliness of ethnicity data collected in censuses of population.

Applications Box 6.2

Vagueness, Ambiguity, and the Geographies of Family Names

In Great Britain, family names ("surnames") entered common parlance in the Thirteenth Century. Most such names are toponyms (denoting landscape features or places, such as Castle or London), metonyms (denoting occupations, such as Smith or Baker) or diminutives (denoting family linkage, such as "William's son," Williamson or the abbreviated form Williams). Family names describing unique places remain concentrated near the places where they were first coined—you are 53 times more likely to meet someone called Rossall in Blackpool, England, for example, because Blackpool is very close to the settlement of that name. Other broader types of names exhibit strong regional concentrations—as with the widespread use of the diminutive "-s" suffix in Wales (e.g., Jones or Williams).

Analysis of family names tells us a lot about the enduring human geographies of places, for the very good reason that throughout history most people have not moved very far from the places where they were brought up. The concentrations of individual names and types of names in places provide us with an interesting indicator of the distinctiveness of places and populations. But using GIS, we can do much more than map a single family name or a single family name type (see Box 1.2). Using geocomputational clustering technique (see Section 16.1), we can identify the degree to which the mix of names in a particular place is distinctive.

Yet "distinctive" is a subjective, hence ambiguous, term. The lower the threshold that we adopt for defining a distinctive region, the more regions we will identify. Figure 6.6 shows how geographer James Cheshire has used names to partition Great Britain into between two and seven regions: note how the distinctiveness of Scottish and Welsh names dominates the first two maps, successively followed by the emergence

▶

of a "north-south divide" in England, the separation of London from the rest of the south, further partitioning of the north and finally the separation of the urban conurbations of northwest England.

Three regions are much better than vague best guesses, as they are rooted in the naming conventions of a bygone age. But why stop at seven regions? Check out James Cheshire's regional geographies of Britain at www.spatialanalysis.co.uk/surnames.

Naming conventions can also help us resolve the ambiguities inherent in mapping the local geographies of ethnic minority populations, which is important for a wide range of policy applications. Data on ethnicity are collected in many population censuses but are notoriously vulnerable to the ambiguities arising from the ways in which individuals assign themselves to ethnic groups. Partly for this reason, the categories that appear in the "ethnic group" question on the UK Census of Population are very broadbrush (Figure 6.7A). A different approach, which seeks neatly to circumvent these issues, begins with the adage that "a name is a statement." Recent research at University College London has used techniques of cluster analysis to classify names into no less than 165 cultural, ethnic, and linguistic groups. The resulting geocomputational

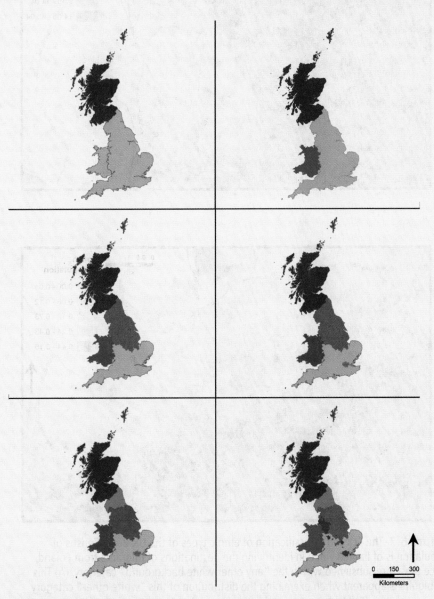

Figure 6.6 The use of family names to regionalize Great Britain. (Courtesy of James Cheshire)

classification is the result of inductive classification of over 600 million names and enables the crude aggregate groups of the UK Census to be compared with a truly multicultural atlas of the UK. Figures 6.7A and B compare the census classification with the names classification, called "Onomap."

Check whether the classification assigns your name to the correct group at www.onomap.org, and take a look at the multicultural atlas of London at www.londonprofiler.org.

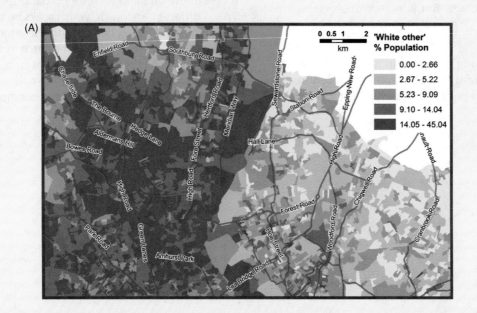

Figure 6.7 The "closed" classification of ethnic types of the UK 2001 Census of Population is of limited value in identifying the destinations of migrants from Poland, since these are subsumed within the "any other white background" category. (A) This ambiguity is apparent when examining the distribution of this "white other" category in west London. (B) Much more useful is the geography of residents with Polish names, as classified using the Onomap system. (Courtesy of James Cheshire and Ollie O'Brien)

Conception of the (direct or indirect) linkage between any indicator and the phenomenon of interest is subjective, hence ambiguous. Such measures will create errors of measurement if the correspondence between the two is imperfect, and these errors may be systematic. So, for example, differences in the conception of what hardship and deprivation entail can lead to specification of different composite indicators, while different geodemographic systems include different cocktails of census variables (Section 2.3.3). With regard to the natural environment, conception of critical defining properties of soils can lead to inherent ambiguity in their classification (see Section 6.2.3).

Ambiguity is introduced when imperfect indicators of phenomena are used instead of the phenomena themselves.

Fundamentally, GIS has upgraded our abilities to generalize about spatial distributions. Yet our abilities to do so may be constrained by the different taxonomies that are conceived and used by data-collecting organizations within our overall study area. A study of wetland classification in the United States found no fewer than six agencies engaged in mapping the same phenomena over the same geographic areas, and each with its own definitions of wetland types (see Section 1.3.1). If wetland maps are to be used in regulating the use of land, as they are in many areas, then uncertainty in mapping clearly exposes regulatory agencies to potentially damaging and costly lawsuits. How might soils data classified according to the UK national classification be assimilated within a pan-European soils map, which uses a classification honed to the full range and diversity of soils found across the European continent rather than those just on an assemblage of offshore island? How might different national geodemographic classifications be combined into a form suitable for a pan-European marketing exercise? These are all variants of the question:

How may mismatches between the categories of different classification schema be reconciled?

Differences in definitions are a major impediment to integration of geographic data over wide areas.

Like the process of pinning down the different nomenclatures developed in different cultural settings, the process of reconciling the semantics of different classification schema is an inherently *ambiguous* procedure. Ambiguity arises in data concatenation when we are unsure regarding the *meta-category* to which a particular class should be assigned.

6.2.3 Fuzzy Approaches to Attribute Classification

One way of resolving the assignment process is to adopt a probabilistic interpretation. If we take a statement like "the database indicates that this field contains wheat, but there is a 0.17 probability (or 17% chance) that it actually contains barley," there are at least two possible interpretations: (1) if 100 randomly chosen people were asked to make independent assessments of the field on the ground, 17 would determine that it contains barley and 83 would decide it contains wheat; or (2) of 100 similar fields in the database, 17 actually contained barley when checked on the ground and 83 contained wheat. Of the two we probably find the second more acceptable because the first implies that people cannot correctly determine the crop in the field.

But the important point is that, in conceptual terms, both of these interpretations are *frequentist* because they are based on the notion that the probability of a given outcome can be defined as the proportion of times the outcome occurs in some real or imagined experiment, when the number of tests is very large. While this interpretation is reasonable for classic statistical experiments, like tossing coins or drawing balls from an urn, the geographic situation is different—there is only one field with precisely these characteristics, and one observer, and in order to imagine a number of tests we have to invent more than one observer, or more than one field. (The problems of imagining larger populations for some geographic samples are discussed further in Section 15.5.)

In part because of this problem, many people prefer the *subjectivist* conception of probability—that it represents a judgment about relative likelihood that is not the result of any frequentist experiment, real or imagined. Subjective probability is similar in many ways to the concept of fuzzy sets, and the latter framework will be used here to emphasize the contrast with frequentist probability.

Suppose we are asked to examine an aerial photograph to determine whether a field contains wheat, and we decide that we are not sure. However, we are able to put a number on our degree of uncertainty by putting it on a scale from 0 to 1. The more certain we are, the higher the number. Thus we might say we are 0.90 sure it is wheat, and this would reflect a greater degree of certainty than 0.80. This degree of belonging to the class *wheat* is termed the *fuzzy membership*, and it is common, though not necessary, to limit memberships to the range 0 to 1. In effect, we have changed our view of membership in classes, and we have abandoned the notion that things must either belong to classes or not belong to them. In this new world, the boundaries of classes are no longer clean and crisp, and the set of things assigned to a set can be fuzzy.

In fuzzy logic, an object's degree of belonging to a class can be partial.

One of the major attractions of fuzzy sets is that they appear to let us deal with sets that are not precisely defined, and for which it is impossible to establish membership cleanly. Many such sets or classes are found in GIS applications, including land-use categories, neighborhood classifications, soil types, land cover classes, and vegetation types. Classes used for maps are often fuzzy, such that two people

asked to classify the same location might disagree, not because of measurement error, but because the classes themselves are not perfectly defined and because opinions vary. As such, mapping is often forced to stretch the rules of scientific repeatability, which require that two observers will always agree.

Box 6.3 shows a typical extract from the legend of a soil map, and it is easy to see how two people might disagree, even though both are experts with years of experience in soil classification. Figure 6.8 shows an example of mapping classes using the fuzzy methods developed by A-Xing Zhu of the University of Wisconsin-Madison, which take both remote-sensing images and the opinions of experts as inputs. There are three classes, and each map shows the fuzzy membership values in one class, ranging from 0 (darkest) to 1 (lightest). This figure also shows the result of converting to *crisp* categories, or *hardening*—to obtain Figure 6.8D, each pixel is colored according to the class with the highest membership value.

Figure 6.8 (A) Membership map for bare soils in the Upper Lake McDonald Basin, Glacier National Park. (B) Membership map for forest. (C) Membership map for alpine meadows. (D) Spatial distribution of the three cover types from hardening the membership maps. (Reproduced by permission of A-Xing Zhu)

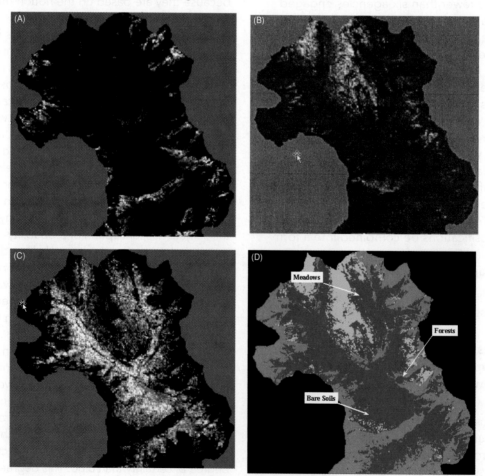

Fuzziness in Classification: Description of a Soil Class

The following is the description of the Limerick series of soils from New England (the type location is in Chittenden County, Vermont), as defined by the National Cooperative Soil Survey. Note the frequent use of vague terms such as "very," "moderate," "about," "typically," and "some." Because the definition is so loose, it is possible for many distinct soils to be lumped together in this one class—and two observers may easily disagree over whether a given soil belongs to the class, even though both are experts. The definition illustrates the extreme problems of defining soil classes with sufficient rigor to satisfy the criterion of scientific repeatability.

"The Limerick series consists of very deep, poorly drained soils on flood plains. They formed in loamy alluvium. Permeability is moderate. Slope ranges from 0 to 3 percent. Mean annual precipitation is about 34 inches and mean annual temperature is about 45 degrees F. Depth to bedrock is more than 60 inches. Reaction ranges from strongly acid to neutral in the surface layer and moderately acid to neutral in the substratum. Textures are typically silt loam or very fine sandy loam, but lenses of loamy very fine sand or very fine sand are present in some pedons. The weighted average of fine and coarser sands, in the particle-size control section, is less than 15 percent."

Fuzzy approaches are attractive because they capture the uncertainty that many of us feel about the assignment of places on the ground to specific categories. But researchers have struggled with the question of whether they are more *accurate*. In a sense, if we are uncertain about which class to choose, then it is more accurate to say so, in the form of a fuzzy membership, than to be forced into assigning a class without qualification. But that does not address the question of whether the fuzzy membership value is accurate. If Class A is not well defined, it is hard to see how one person's assignment of a fuzzy membership of 0.83 in Class A can be meaningful to another person, since there is no reason to believe that the two people share the same notions of what Class A means, or of what 0.83 means, as distinct from 0.91, or 0.74. So while fuzzy approaches make sense at an intuitive level, it is more difficult to see how they could be helpful in the process of communication of geographic knowledge from one person to another.

6.3 U2: Further Uncertainty in the Representation of Geographic Phenomena

As with the conception of uncertainty, it is helpful to consider the representation of uncertainty with regard to the components of geographic information—measures of places (locations), attributes, and time period (although we do not consider time in detail here). We consider both the mode of representing place (Section 6.3.1) and the accuracy and precision with which it can be measured (Section 6.3.3). We consider the measurement of attributes at the nominal, ordinal, interval, and ratio scales (see Box 3.3).

6.3.1 Representation of Place/Location

The conceptual models (fields and objects) that were introduced in Chapter 3 impose very different filters upon reality, and as a result, their usual corresponding representational models (raster and vector) are characterized by different uncertainties. The vector model enables a range of powerful analytical operations to be performed (see Chapters 14 through 16), yet it also requires a priori conceptualization of the nature and extent of geographic individuals and the ways in which they nest together into higher-order zones. The raster model defines individual elements as square cells, with boundaries that bear no relationship at all to natural features, but nevertheless provides a convenient and (usually) efficient structure for data handling within a GIS. However, in the absence of effective automated pattern recognition techniques, human interpretation is usually required to discriminate between real-world spatial entities as they appear in a rasterized image.

Although quite different representations of reality, both vector and raster data structures are attractive in their logical consistency, the ease with which they are able to handle spatial data, and (once the software is written) the ease with which they can be implemented in GIS. But neither can provide any substitute for robust conception of geographic units of analysis (Section 6.2). This said, however, the conceptual distinction between fields and discrete objects is often useful in dealing with uncertainty. Figure 6.9 shows a coastline, which is often conceptualized as a discrete line object. But suppose we recognize that its position is uncertain. For example, the coastline shown on a 1:2,000,000 map is a gross generalization, in which major liberties are taken, particularly in areas where the coast is highly indented and irregular. Consequently, the 1:2,000,000 version leaves substantial uncertainty about the true location of the shoreline. We might approach this by changing from a line to an area, and mapping the area where the actual coastline lies, as shown in the Figure. But another approach would be to reconceptualize the coastline as a field by mapping a variable whose value represents the probability that a point is land. This is shown in the figure as a raster representation. This would have far more information content, and consequently much more value in many applications. But at the same time it would be difficult to find an appropriate data

source for the representation—perhaps a fuzzy classification of an air photo, using one of an increasing number of techniques designed to produce representations of the uncertainty associated with objects discovered in images.

Uncertainty can be measured differently under field and discrete object views.

Indeed, far from offering quick fixes for eliminating or reducing uncertainty, the measurement process can actually increase it. Given that the vector and raster data models impose quite different filters on reality, it is unsurprising that they can each generate additional uncertainty in rather different ways. In field-based conceptualizations, such as those that underlie remotely sensed images expressed as rasters, spatial objects are not defined a priori. Instead, the classification of each cell into one or other category builds together into a representation. In remote sensing, when resolution is insufficient to detect all of the detail in geographic phenomena, the term *mixel* is often used to describe raster cells that contain more than one class of land—in other words, elements in which the outcome of statistical classification suggests the occurrence of multiple land cover categories. The total area of cells classified as mixed should decrease as the resolution of the satellite sensor increases, assuming the number of categories remains constant, yet a completely mixel-free classification is very unlikely at any level of resolution. Even where the Earth's surface is covered with perfectly homogeneous areas, such as agricultural fields growing uniform crops, the failure of real-world crop boundaries to line up with pixel edges ensures the presence of at least some mixels. Neither does finer-resolution imagery solve all problems: medium-resolution data (defined as pixel size of between 30 m × 30 m and 1000 m × 1000 m) are typically classified using between 3 and 7 bands, while fine-resolution data (pixel sizes 10 × 10 m or smaller) are typically classified using between 7 and 256 bands, and this can generate much greater heterogeneity of spectral values with attendant problems for classification algorithms.

A pixel whose area is divided among more than one class is termed a mixel.

The vector data structure, by contrast, defines spatial entities and specifies explicit topological relations (see Section 3.6) between them. Yet this often entails transformations of the inherent characteristics of spatial objects (Chapters 14 and 15). In conceptual terms, for example, while the true individual members of a population might each be defined as point-like

Figure 6.9 The contrast between discrete object (top) and field (bottom) conceptualizations of an uncertain coastline.

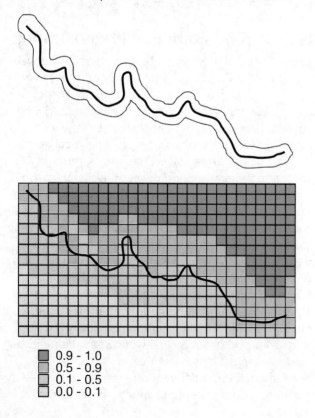

- 0.9 - 1.0
- 0.5 - 0.9
- 0.1 - 0.5
- 0.0 - 0.1

objects, they will often appear in a GIS dataset only as aggregate counts for apparently *uniform* zones. Such aggregation can be driven by the need to preserve the confidentiality of individual records, or simply by the need to limit data volume. Unlike the field conceptualization of spatial phenomena, this implies that there are good reasons for partitioning space in a particular way. In practice, partitioning is often made on grounds that are principally pragmatic, yet are rarely completely random (see Section 6.4). In much of socioeconomic GIS, for example, zones that are designed to preserve the anonymity of survey respondents may be largely ad hoc containers. Larger aggregations are often used for the simple reason that they permit comparisons of measures over time (see Box 6.3). They may also reflect the way that a cartographer or GIS interpolates a boundary between sampled points, as in the creation of isopleth maps (see Box 4.3).

6.3.2 Statistical Models of Uncertainty in Attribute Measures

Scientists have developed many widely used methods for describing errors in observations and measurements, and these methods may be applicable to GIS if we are willing to think of databases as collections of measurements. For example, a digital elevation model consists of a large number of measurements of the elevation of the Earth's surface. A map of land use is also in a sense a collection of measurements because observations of the land surface have resulted in the assignment of classes to locations. Both of these are examples of observed or measured attributes, but we can also think of location as a property that is measured.

> **A geographic database is a collection of measurements of phenomena on or near the Earth's surface.**

Here we consider errors in nominal class assignment, such as of types of land use and errors in continuous (interval or ratio) scales, such as elevation (see Section 3.4).

6.3.2.1 Nominal Case

The values of nominal data serve only to distinguish an instance of one class from an instance of another, or to identify an object uniquely (Section 3.4). If classes have an inherent ranking, they are described as ordinal data, but for purposes of simplicity the ordinal case will be treated here as if it were nominal. Consider a single observation of nominal data—for example, the observation that a single parcel of land is being used for agriculture (this might be designated by giving the parcel Class A as its value of the "Land-Use Class" attribute). For some reason, perhaps related to the quality of the aerial photography being used to build the database, the class may have been recorded falsely as Class G, Grassland. A certain proportion of parcels that are truly Agriculture might be similarly recorded as Grassland, and we can think of this in terms of a probability that parcels that are truly Agriculture are falsely recorded as Grassland.

Table 6.1 shows how this might work for all of the parcels in a database. Each parcel has a true class, defined by accurate observation in the field, and a recorded class as it appears in the database. The whole table is described as a *confusion matrix*, and instances of confusion matrices are commonly encountered in applications dominated by class data, such as classifications derived from remote sensing or aerial photography. The true class might be determined by ground check, which is inherently more accurate than classification of aerial photographs but much more expensive and time-consuming. Ideally, all of the observations in the confusion matrix should lie along the principal diagonal, in the cells that

Table 6.1 Example of a misclassification or confusion matrix. A grand total of 304 parcels have been checked. The rows of the table correspond to the land-use class of each parcel as recorded in the database, and the columns to the class as recorded in the field. The numbers appearing on the principal diagonal of the table (from top left to bottom right) reflect correct classification.

	A	B	C	D	E	Total
A	80	4	0	15	7	106
B	2	17	0	9	2	30
C	12	5	9	4	8	38
D	7	8	0	65	0	80
E	3	2	1	6	38	50
Total	104	36	10	99	55	304

correspond to agreement between true class and database class. But in practice certain classes are more easily confused than others, so certain cells off the diagonal will have substantial numbers of entries.

A useful way to think of the confusion matrix is as a set of rows, each defining a vector of values. The vector for any row i gives the proportions of cases in which what appears to be Class i is actually Class 1, 2, 3, and so on. Symbolically, this can be represented as a vector $\{p_1, p_2, \ldots, p_i, \ldots, p_n\}$, where n is the number of classes and p_i represents the proportion of cases for which what appears to be the class according to the database is actually Class i.

There are several ways of describing and summarizing the confusion matrix. If we focus on one row, then the table shows how a given class in the database falsely records what are actually different classes on the ground. For example, Row A shows that of 106 parcels recorded as Class A in the database, 80 were confirmed as Class A in the field, but 15 appeared to be truly Class D. The proportion of instances in the diagonal entries represents the proportion of correctly classified parcels, and the total of off-diagonal entries in the row is the proportion of entries in the database that appear to be of the row's class but are actually incorrectly classified. For example, there were only 9 instances of agreement between the database and the field in the case of Class D. If we look at the table's columns, the entries record the ways in which parcels that are truly of that class are actually recorded in the database. For example, of the 10 instances of Class C found in the field, 9 were recorded as such in the database and only 1 was misrecorded as Class E. The columns have been called the *producer's* perspective because the task of the producer of an accurate database is to minimize entries outside the diagonal cell in a given column; the rows have been called the *consumer's* perspective because they record what the contents of the database actually mean on the ground—in other words, the accuracy of the database's contents.

Users and producers of data look at misclassification in distinct ways.

For the table as a whole, the proportion of entries in diagonal cells is called the *percent correctly classified* (PCC) and is one possible way of summarizing the table. In this case 209/304 cases are on the diagonal, for a PCC of 68.8%. But this measure is misleading for at least two reasons. First, chance alone would produce some correct classifications, even in the worst circumstances, so it would be more meaningful if the scale were adjusted such that 0 represents chance. In this case, the number of chance hits on the diagonal

in a random assignment is 76.2 (the sum of the row total times the column total divided by the grand total for each of the five diagonal cells). So the actual number of diagonal hits, 209, should be compared to this number, not 0. The more useful index of success is the *kappa index*, defined as

$$\kappa = \frac{\sum_{i=1}^{n} c_{ii} - \sum_{i=0}^{n} c_{i.}\, c_{.i}/c_{..}}{c_{..} - \sum_{i=0}^{n} c_{i.}\, c_{.i}/c_{..}}$$

where c_{ij} denotes the entry in row i column j, the dots indicate summation (e.g., $c_{i.}$ is the summation over all columns for row i, that is, the row i total, and $c_{..}$ is the grand total), and n is the number of classes. The first term in the numerator is the sum of all the diagonal entries (entries for which the row number and the column number are the same). To compute PCC, we would simply divide this term by the grand total (the first term in the denominator). For kappa, both numerator and denominator are reduced by the same amount, an estimate of the number of hits (agreements between field and database) that would occur by chance. This involves taking each diagonal cell, multiplying the row total by the column total, and dividing by the grand total. The result is summed for each diagonal cell. In this case kappa evaluates to 58.3%, a much less optimistic assessment than PCC.

The second issue with both of these measures concerns the relative abundance of different classes. In the table, Class C is much less common than Class A. The confusion matrix is a useful way of summarizing the characteristics of nominal data, but to build it there must be some source of more accurate data. Commonly, this is obtained by ground observation, and in practice the confusion matrix is created by taking samples of more accurate data, by sending observers into the field to conduct spot checks. Clearly, it makes no sense to visit every parcel, and instead a sample is taken. Because some classes are more common than others, a random sample that made every parcel equally likely to be chosen would be inefficient because too many data would be gathered on common classes, and not enough on the relatively rare ones. So, instead, samples are usually chosen such that a roughly equal number of parcels are selected in each class. Of course, these decisions must be based on the class as recorded in the database, rather than the true class. This is an instance of sampling that is *stratified* by class (see Section 4.4).

Sampling for accuracy assessment should pay greater attention to the classes that are rarer on the ground.

Figure 6.10 An example of a vegetation cover map.

Parcels represent a relatively easy case, if it is reasonable to assume that the land-use class of a parcel is uniform over the parcel, and class is recorded as a single attribute of each parcel object. But as we noted in Sections 4.4 and 6.2.2.1, more difficult cases arise in sampling natural areas, for example, in the case of vegetation cover class, where parcel boundaries may not exist. Figure 6.10 shows a typical vegetation cover class map and is obviously highly generalized. If we were to apply the previous strategy, then we would test each area to see if its assigned vegetation cover class checks out on the ground. But unlike the parcel case, in this example the boundaries between areas are not fixed but are themselves part of the observation process, and we need to ask whether they are correctly located. Error in this case has two forms: misallocation of an area's class and mislocation of an area's boundaries. In some cases the boundary between two areas may be fixed because it coincides with a clearly defined line on the ground; but in other cases, the boundary's location is as much a matter of judgment as the allocation of an area's class.

Errors in land cover maps can occur in the locations of boundaries of areas, as well as in the classification of areas.

In such cases we need a different strategy that captures the influence both of mislocated boundaries and of misallocated classes. One way to deal with this is to think of error not in terms of classes assigned to areas, but in terms of classes assigned to points. In a raster dataset, the cells of the raster are a reasonable substitute for individual points. Instead of asking whether area classes are confused, and estimating errors by sampling areas, we ask whether the classes assigned to raster cells are confused, and we define the confusion matrix in terms of misclassified cells. This is often called *per-pixel* or *per-point* accuracy assessment, to distinguish it from the previous strategy of *per-polygon* accuracy assessment. As before, we would want to stratify by class, to make sure that relatively rare classes were sampled in the assessment.

6.3.2.2 Interval/Ratio Case

The second case addresses measurements that are made on interval or ratio scales. Here, error is best thought of not as a change of class but as a change of value, such that the observed value x' is equal to the true value x plus some distortion δx, where δx is hopefully small. δx might be either positive or negative, since errors are possible in both directions. For example, the measured and recorded elevation at some point might be equal to the true elevation, distorted by some small amount. If the average distortion is zero, so that positive and negative errors balance out, the observed values are said to be *unbiased*, and the average value will be true.

Error in measurement can produce a change of class, or a change of value, depending on the type of measurement.

Sometimes it is helpful to distinguish between *accuracy*, which has to do with the magnitude of δx, and *precision*. Unfortunately there are several ways of defining precision in this context, at least two of which are regularly encountered in GIS. Surveyors and others concerned with measuring instruments tend to define precision through the performance of an instrument in making repeated measurements of the same phenomenon. A measuring instrument is precise according to this definition if it repeatedly gives similar measurements, whether or not these are actually accurate. So a GPS receiver might make successive measurements of the same elevation, and if these are similar the instrument is said to be precise. Precision in this case can be measured by the variability among repeated measurements. But it is possible that all of the measurements are approximately 5 m too high, in which case the measurements are said to be biased, even though they are precise, and the instrument is said to be inaccurate. Figure 6.11 illustrates this meaning of precise and its relationship to accuracy. The other definition of precision is more common in science generally. It defines precision

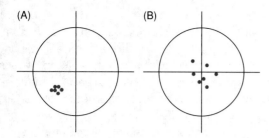

Figure 6.11 (A) Successive measurements have similar values (they are precise). (B) Precision is lower but accuracy is higher.

as the number of digits used to report a measurement, and again it is not necessarily related to accuracy. For example, a GPS receiver might measure elevation as 51.3456 m. But if the receiver is in reality only accurate to the nearest 10 cm, three of those digits are spurious, with no real meaning. So, although the precision is one ten-thousandth of a meter, the accuracy is only one-tenth of a meter. Box 6.4 summarizes the rules that are used to ensure that reported measurements do not mislead by appearing to have greater accuracy than they really do.

To most scientists, precision refers to the number of significant digits used to report a measurement, but it can also refer to a measurement's repeatability.

In the interval/ratio case, the magnitude of errors is described by the *root mean square error* (RMSE), defined as the square root of the average squared error, or:

$$\left[\sum \delta x^2 / n \right]^{1/2}$$

where the summation is over the values of δx for all of the n observations. The RMSE is similar in a number of ways to the standard deviation of observations in a sample. Although RMSE involves taking the square root of the average squared error, it is convenient to think of it as approximately equal to the average error in each observation, whether the error is positive or negative. The U.S. Geological Survey uses RMSE as its primary measure of the accuracy of elevations in digital elevation models, and published values range up to 7 m.

Although the RMSE can be thought of as capturing the magnitude of the average error, many errors will be greater than the RMSE and many will be less. It is useful, therefore, to know how errors are *distributed* in magnitude—how many are large, how many are small. Statisticians have developed a series of models of error distributions, of which the most common and most important is the Gaussian distribution, otherwise known as the error function, the "bell curve," or the Normal distribution. Figure 6.12 shows the curve's shape. The height of the curve at any value of x gives

---

Technical Box (6.4)

Good Practice in Reporting Measurements

Here are some simple rules that help to ensure that people receiving measurements from others are not misled by their apparently high precision.

1. The number of digits used to report a measurement should reflect the measurement's accuracy. For example, if a measurement is accurate to 1 m, then no decimal places should be reported. The measurement 14.4 m suggests accuracy to one-tenth of a meter, as does 14.0, but 14 suggests accuracy to 1 m.

2. Excess digits should be removed by rounding. Fractions above one-half should be rounded up, whereas fractions below one-half should be rounded down. The following examples reflect rounding to two decimal places:

14.57803 rounds to 14.58

14.57397 rounds to 14.57

14.57999 rounds to 14.58

14.57499 rounds to 14.57

3. These rules are not effective to the left of the decimal place; for example, they give no basis for knowing whether 1400 is accurate to the nearest unit or to the nearest hundred units.

4. If a number is known to be exactly an integer or whole number, then it is shown with no decimal point.

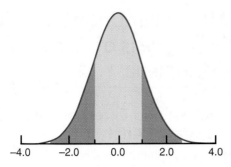

Figure 6.12 The Gaussian or Normal distribution. The lght blue area (between ±1 standard deviation) encloses 68% of the area under the curve, so 68% of observations will fall between these limits.

the relative abundance of observations with that value of x. The area under the curve between any two values of x gives the probability that observations will fall in that range. If observations are unbiased, then the mean error is zero (positive and negative errors cancel each other out), and the RMSE is also the distance from the center of the distribution (zero) to the points of inflection on either side, as shown in the figure.

Let us take the example of a 7 m RMSE on elevations in a USGS digital elevation model; if error follows the Gaussian distribution, this means that some errors will be more than 7 m in magnitude, whereas some will be less, and also that the relative abundance of errors of any given size is described by the curve shown. Sixty eight percent of errors will lie between −1.0 and +1.0 RMSEs, or −7 m and +7 m. In practice, many distributions of error do follow the Gaussian distribution, and there are good theoretical reasons why this should be so.

The Gaussian distribution is used to predict the relative abundances of different magnitudes of error.

To emphasize the mathematical formality of the Gaussian distribution, its equation is shown at the end of this paragraph. The symbol σ denotes the standard deviation, μ denotes the mean (in Figure 6.12 these values are 1 and 0, respectively), and exp is the exponential function, or "2.71828 to the power of" (also sometimes used to represent distance decay: see Section 4.5). Scientists believe that it applies very broadly and that many instances of measurement error adhere closely to the distribution because it is grounded in rigorous theory. It can be shown mathematically that the distribution arises whenever a large number of random factors contribute to error, and the effects of these factors combine additively—that is, a

given effect makes the same additive contribution to error whatever the specific values of the other factors. For example, error might be introduced in the use of a steel tape measure over a large number of measurements because some observers consistently pull the tape very taut, or hold it very straight, or fastidiously keep it horizontal, or keep it cool, and others do not. If the combined effects of these considerations always contribute the same amount of error (e.g., +1 cm, or −2 cm), then this contribution to error is said to be additive.

$$f(x) = \frac{1}{\sigma \sqrt{2\pi}} \exp\left[-\frac{(x - \mu)^2}{2\sigma^2} \right]$$

We can apply this idea to determine the inherent uncertainty in the locations of contours. The U.S. Geological Survey routinely evaluates the accuracies of its digital elevation models (DEMs) by comparing the elevations recorded in the database with those at the same locations in more accurate sources, for a sample of points. The differences are summarized in an RMSE, and in this example we will assume that errors have a Gaussian distribution with zero mean and a 7 m RMSE. Consider a measurement of 350 m. According to the error model, the truth might be as high as 360 m or as low as 340 m, and the relative frequencies of any particular error value are as predicted by the Gaussian distribution with a mean of zero and a standard deviation of 7. If we take error into account, using the Gaussian distribution with an RMSE of 7 m, it is no longer clear that a measurement of 350 m lies exactly on the 350 m contour. Instead, the truth might be 340 m, or 360 m, or 355 m.

6.3.3 Statistical Models of Uncertainty in Location Measures

In the case of measurements of position, it is possible for every coordinate to be subject to error. In the two-dimensional case, a measured position (x', y') would be subject to errors in both x and y; specifically, we might write $x' = x + \delta x$, $y' = y + \delta y$, and similarly in the three-dimensional case where all three coordinates are measured, $z' = z + \delta z$. The *bivariate Gaussian distribution* describes errors in the two horizontal dimensions, and it can be generalized to the three-dimensional case. Normally, we would expect the RMSEs of x and y to be the same, but z is often subject to errors of quite different magnitude, for example, in the case of determinations of position using GPS. The bivariate Gaussian distribution also allows for correlation between the errors in *x* and *y*, but normally there is little reason to expect correlations.

Because it involves two variables, the bivariate Gaussian distribution has somewhat different properties from the simple (univariate) Gaussian distribution. As shown in Figure 6.12, 68% of cases lie within one standard deviation for the univariate case. But in the bivariate case with equal standard errors in *x* and *y*, only 39% of cases lie within a circle of this radius. Similarly, 95% of cases lie within two standard deviations for the univariate distribution, but it is necessary to go to a circle of radius equal to 2.15 times the *x* or *y* standard deviations to enclose 90% of the bivariate distribution, and 2.45 times standard deviations for 95%.

National Map Accuracy Standards often prescribe the positional errors that are allowed in databases. For example, the 1947 U.S. National Map Accuracy Standard specified that 95% of errors should fall below 1/30 inch (0.85 mm) for maps at scales of 1:20,000 and finer (more detailed), and 1/50 inch (0.51 mm) for other maps (coarser, less detailed than 1:20,000). A convenient rule of thumb is that positions measured from maps are subject to errors of up to 0.5 mm at the scale of the map. Table 6.2 shows the distance on the ground corresponding to 0.5 mm for various common map scales.

> **A useful rule of thumb is that features on maps are positioned to an accuracy of about 0.5 mm.**

Table 6.2 Positions measured from maps should be accurate to about 0.5 mm on the map. Multiplying this by the scale of the map gives the corresponding distance on the ground.

Map scale	Ground distance corresponding to 0.5-mm map distance
1:1,250	62.5 cm
1:2,500	1.25 m
1:5,000	2.5 m
1:10,000	5 m
1:24,000	12 m
1:50,000	25 m
1:100,000	50 m
1:250,000	125 m
1:1,000,000	500 m
1:10,000,000	5 km

6.4 U3: Further Uncertainty in the Analysis of Geographic Phenomena

6.4.1 Internal and External Validation through Spatial Analysis

In Chapter 1 we defined a core remit of GIS as the resolution of scientific or decision-making problems through spatial analysis. Spatial analysis can be thought of as the process by which we turn raw spatial data into useful spatial information, and thus far have thought of the creation of spatial information as adding value to attribute data through selectivity or preparation for purpose (see Chapters 14 and 15). A further defining characteristic is that the results of spatial analysis change when the frame or extent of the space under investigation changes. This also implies that the frame can be divided into units of analysis that are clearly defined, yet we have seen (see Section 6.2.1) that there are likely to be few, if any, such units available to us. How can the outcome of spatial analysis be meaningful if it has such uncertain foundations?

Once again, this question has no easy answers, although we can begin to look for answers by anticipating possible errors of positioning, or the consequences of aggregating the subjects of analysis (such as individual people) into artificial geographic units of analysis (as when people are aggregated by census tracts, or disease incidences are aggregated by county). In so doing, we can illustrate how potential problems might arise, although we are unlikely to arrive at any definitive solutions—for the simple reason that the truth is inherently uncertain. The conception and representation of geographic phenomena may distort the outcome of spatial analysis by masking or accentuating apparent variation across space, or by restricting the nature and range of questions that can meaningfully be asked of a GIS.

> **Good GIS analysis cannot substitute for poor conceptions of geography or poor representation—but it can flag the likely consequences of both.**

We can deal with this risk in three ways. First, although the GIS analyst can only rarely tackle the *source* of uncertainty (analysts are rarely empowered to collect new, completely disaggregate data, for example), GIS can help to pinpoint the ways in which uncertainty is likely to *operate* (or *propagate*) within the GIS, and identify the likely degree of distortion arising from representational expedients.

Second, although we may have to work with areally aggregated data, GIS allows us to model within-zone spatial distributions, and this can ameliorate the worst effects of artificial zonation. Taken together, GIS allows us to gauge the effects of scale and aggregation through simulation of different possible outcomes. This is *internal validation* of the effects of scale, point placement, and spatial partitioning.

The third way that GIS can tackle uncertainty is by seeking to assess the quality of a representation with reference to other data sources, thus providing a means of *external validation* of the effects of zonal averaging. In today's advanced GIService economy (see Section 1.6.3), there may be other data sources that can be used to gauge the effects of aggregation on our analysis. In Section 13.2.1 we formally refine the basic scheme presented in Figure 6.1 to consider the role of geovisualization in evaluating representations, although some of the validation principles set out in this chapter also entail visual approaches.

GIS provides ways of validating our representations, sometimes with and sometimes without reference to external data sources.

6.4.2 Validation through Autocorrelation: The Spatial Structure of Errors

Understanding the spatial structure of errors is key to accommodating their likely effects on the results of GIS analysis, and hence estimates or measures of spatial autocorrelation can provide an important validation measure. This is because strong positive spatial autocorrelation will reduce the effects of uncertainty on estimates of properties such as slope or area. The cumulative effects of error, termed *error propagation*, can nevertheless produce impacts that are surprisingly large. Some of the examples in this section have been chosen to illustrate the substantial uncertainties that can be produced by apparently innocuous data errors.

Error propagation measures the impacts of uncertainty in data on the results of GIS operations.

The confusion matrix, or more specifically a single row of the matrix, along with the Gaussian distribution, provide convenient ways of describing the error present in a single observation of a nominal or interval/ratio measurement, respectively. When a GIS is used to respond to a simple query, such as "tell me the class of soil at this point," or "what is the elevation here?", then these methods are good ways of describing the uncertainty inherent in the response. For

example, a GIS might respond to the first query with the information "'Class A, with a 30% probability of Class C," and to the second query with the information "350 m, with an RMSE of 7 m." Notice how this makes it possible to describe nominal data as accurate to a percentage, but it makes no sense to describe a DEM, or any measurement on an interval/ratio scale, as accurate to a percentage. For example, we cannot meaningfully say that a DEM is "90% accurate."

However, many GIS operations involve more than the properties of single points, and this makes the analysis of error much more complex. For example, consider the query "how far is it from this point to that point?" Suppose the two points are both subject to error of position because their positions have been measured using GPS units with mean distance errors of 50 m. If the two measurements were taken some time apart, with different combinations of satellites above the horizon, it is likely that the errors are independent of each other, such that one error might be 50 m in the direction of North, and the other 50 m in the direction of South. Depending on the locations of the two points, the error in distance might be as high as 100 m. On the other hand, if the two measurements were made close together in time, with the same satellites above the horizon, it is likely that the two errors would be similar, perhaps 50 m North and 40 m North, leading to an error of only 10 m in determining distance. The difference between these two situations can be measured in terms of the degree of *spatial autocorrelation*, or the interdependence of errors at different points in space (see Section 4.6).

The spatial autocorrelation of errors can be as important as their magnitude in many GIS operations.

Spatial autocorrelation is also important in analyzing probable errors in nominal data. Reconsider the agricultural field discussed in Section 6.2.3 that is known to contain a single crop, perhaps barley. When seen from above, it is possible to confuse barley with other crops, so there may be error in the crop type assigned to points in the field. But since the field has only one crop, we know that such errors are likely to be strongly correlated. Spatial autocorrelation is almost always present in errors to some degree, but very few efforts have been made to measure it systematically. As a result, it is difficult to make good estimates of the uncertainties associated with many GIS operations.

The spatial structure or autocorrelation of errors is important in many ways. DEM data are often used to estimate the slope of terrain, and this is done by comparing elevations at points a short distance apart. For example, if the elevations at two points 10 m apart are 30 m and 35 m, respectively, the slope along the line between them is 5/10, or 0.5. (A somewhat more complex method is used in practice, to estimate slope at a point in the x- and y-directions in a DEM raster, by analyzing the elevations of nine points—the point itself and its eight neighbors. The equations in Section 15.3.1 detail the procedure.) Now consider the effects of errors in these two elevation measurements on the estimate of slope. Suppose the first point (elevation 30 m) is subject to a RMSE of 2 m, and consider possible true elevations of 28 m and 32 m. Similarly, the second point might have true elevations of 33 m and 37 m. We now have four possible combinations of values, and the corresponding estimates of slope range from (33 − 32)/10 = 0.1 to (37 − 28)/10 = 0.9. In other words, a relatively small amount of error in elevation can produce wildly varying slope estimates.

The spatial autocorrelation between errors in geographic databases helps to minimize their impacts on many GIS operations.

What saves us in this situation, and makes estimation of slope from DEMs a practical proposition at all, is spatial autocorrelation among the errors. In reality, although DEMs are subject to substantial errors in absolute elevation, neighboring points nevertheless tend to have similar errors, and errors tend to persist over quite large areas. Most of the sources of error

in the DEM production process tend to produce this kind of persistence of error over space, including errors due to misregistration of aerial photographs. In other words, errors in DEMs exhibit strong positive spatial autocorrelation.

Another important corollary of positive spatial autocorrelation can also be illustrated using DEMs. Suppose an area of low-lying land is predicted to be submerged by sea-level rise, and our task is to estimate the area of land affected (Figure 6.13). We are asked to do this using a DEM, which is known to have an RMSE of 2 m. Suppose the data points in the DEM are 30 m apart, and preliminary analysis shows that 100 points have elevations below the flood line. We might conclude that the area flooded is the area represented by these 100 points, or 900 × 100 sq m, or 9 hectares. But because of errors, it is possible that some of this area is actually above the flood line (we will ignore the possibility that other areas outside this may also be below the flood line, also because of errors), and it is possible that *all* of the area is above. Suppose the recorded elevation for each of the 100 points is 2 m below the flood line. This is one RMSE (recall that the RMSE is equal to 2 m) below the flood line, and the Gaussian distribution tells us that the chance that the true elevation is actually above the flood line is approximately 16% (see Figure 6.12).

But what is the chance that *all* 100 points are actually above the flood line? Here again the answer

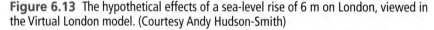

Figure 6.13 The hypothetical effects of a sea-level rise of 6 m on London, viewed in the Virtual London model. (Courtesy Andy Hudson-Smith)

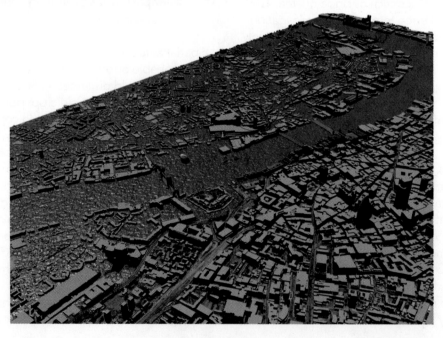

depends on the degree of spatial autocorrelation among the errors. If there is none, in other words if the error at each of the 100 points is independent of the errors at its neighbors, then the answer is $(0.16)^{100}$, or 1 chance in 1 followed by roughly 70 zeroes. But if there is strong positive spatial autocorrelation, so strong that all 100 points are subject to exactly the same error, then the answer is 0.16. One way to think about this is in terms of *degrees of freedom*. If the errors are independent, they can vary in 100 independent ways, depending on the error at each point. But if they are strongly spatially autocorrelated, the effective number of degrees of freedom is much less and may be as few as 1 if all errors behave in unison. Spatial autocorrelation has the effect of reducing the number of degrees of freedom in geographic data below what may be implied by the volume of information, in this case the number of points in the DEM.

Spatial autocorrelation acts to reduce the effective number of degrees of freedom in geographic data.

A further case concerns the accommodation of positional uncertainties in the location of the boundaries of a discrete object. Figure 6.14 shows a square approximately 100 m on each side. Suppose the square has been surveyed by determining the locations of its four corner points using GPS, and suppose the circumstances of the measurements are such that there is an RMSE of 1 m in both coordinates of all four points, and that errors are independent.

Suppose our task is to determine the area of the square. A GIS can do this easily, using a standard algorithm (see Figure 15.1). Computers are precise (in

the sense of Box 6.4) and capable of working to many significant digits, so the calculation might be reported by printing out a number to eight digits, such as 10,014.603 sq m, or even more. But the number of significant digits will have been determined by the precision of the machine, and not by the accuracy of the determination. Box 6.4 summarized some simple rules for ensuring that the precision used to report a measurement reflects as far as possible its accuracy, and clearly those rules will have been violated if the area is reported to eight digits. But what is the appropriate precision?

In this case we can determine exactly how positional accuracy affects the estimate of area. It turns out that area has an error distribution that is Gaussian, with a standard deviation (RMSE) which in our particular case is 200 sq m. In other words, each attempt to measure the area will give a different result, the variation between them having a standard deviation of 200 sq m. This means that the five rightmost digits in the estimate are spurious, including two digits to the left of the decimal point. So if we were to follow the rules of Box 6.4, we would print 10,000 rather than 10,014.603. (Note the problem with standard notation here, which does not let us omit digits to the left of the decimal point even if they are spurious, and so leaves some uncertainty about whether or not the tens and units digits are certain—and note also the danger that if the number is printed as an integer it may be interpreted as exactly the whole number). We can also turn the question around and ask how accurately the points would have to be measured to justify eight digits, and the answer is approximately 0.01 mm, far beyond the capabilities of normal surveying practice.

A useful way of visualizing spatial autocorrelation and interdependence is through animation. Each frame in the animation is a single possible map, or *realization* of the error process. If a point is subject to uncertainty, each realization will show the point in a different possible location, and a sequence of images will show the point shaking around its mean position. If two points have perfectly correlated positional errors, then they will appear to shake in unison, as if they were at the ends of a stiff rod. If errors are only partially correlated, then the system behaves as if the connecting rod were somewhat elastic.

The inherent difficulties in accommodating spatially autocorrelated errors have led many researchers to explore a more general strategy of simulation to evaluate the impacts of uncertainty on the results of spatial analysis. In essence, simulation requires

Figure 6.14 Error in the measurement of the area of a square 100 m on each side.

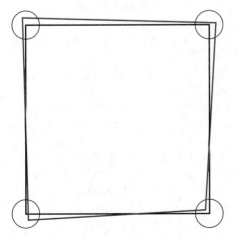

the generation of a series of realizations, as defined earlier. It is often called Monte Carlo simulation in reference to the realizations that occur when dice are tossed or cards are dealt in various games of chance. For example, we could simulate error in a single measurement from a DEM by generating a series of numbers with a mean equal to the measured elevation, and a standard deviation equal to the known RMSE and a Gaussian distribution. Simulation uses everything that is known about a situation, so if any additional information is available we would incorporate it in the simulation. We might assume that elevations must be whole numbers of meters, and we would simulate this by rounding the numbers obtained from the Gaussian distribution. With a mean of 350 m and an RMSE of 7 m the results of the simulation might, for example, be 341, 352, 356, 339, 349, 348, 355, 350, . . .

Simulation is an intuitively simple way of getting the uncertainty message across.

Because of spatial autocorrelation, it is impossible in most circumstances to think of databases as decomposable into component parts, each of which can be independently disturbed to create alternative realizations, as in the previous example. Instead, we have to think of the entire database as a realization and create alternative realizations of the database's contents that preserve spatial autocorrelation. Figure 6.15 shows an example, simulating the effects of uncertainty on a digital elevation model. Each of the three realizations is a complete map, and the simulation process has faithfully replicated the strong correlations present in errors across the DEM.

6.4.3 Validation through Investigating the Effects of Aggregation and Scale

We have already seen that a fundamental difference between geography and other scientific disciplines is that the definition of its objects of study is only rarely unambiguous and, in practice, rarely precedes our attempts to measure their characteristics. In socioeconomic GIS, these objects of study (geographic individuals) are usually aggregations, since the spaces that human individuals occupy are geographically unique, and confidentiality restrictions usually dictate that uniquely attributable information must be anonymized in some way. Even in natural-environment applications, the nature of sampling in the data collection process (Section 4.4) often makes it expedient to collect data pertaining to aggregations of one kind or another. Thus "geographic individuals" are likely to

be defined as areal aggregations of the units of study. Moreover, in cases where data are general-purpose series (as with census data, or natural-resource inventories), the zonal systems are unlikely to be determined with the end point of particular spatial analysis applications in mind.

As a consequence, we cannot be certain in ascribing even dominant characteristics *of* areas to true individuals or point locations *in* those areas. This source of uncertainty is known as the *ecological fallacy* and has long bedeviled the analysis of spatial distributions. (The opposite of ecological fallacy is atomistic fallacy, in which the individual is considered in isolation from his or her environment.) The ecological fallacy problem is a consequence of aggregation into the basic units of analysis and is illustrated in Figure 6.16.

Inappropriate inference from aggregate data about the characteristics of individuals is termed the ecological fallacy.

The likelihood of committing ecological fallacy in GIS analysis depends on the nature of the aggregation being used. If the members of a set of zones are all perfectly uniform and homogeneous (Section 6.2.1), then the only geographic variation (heterogeneity) that occurs will be between zones. However, nearly all zones are internally heterogeneous to some degree, and greater heterogeneity increases the likelihood and severity of the ecological fallacy problem. That said, it is important to be aware that there are few documented case studies that demonstrate the occurrence of ecological fallacy in practice, and many general-purpose zoning systems in socioeconomic GIS are designed to maximize within-zone homogeneity in the most salient population characteristics—as with the UK Census geography, for example.

The potential of *aggregation* to create problems in GIS analysis is compounded by problems arising from the different *scales* at which zonal systems may be defined. This was demonstrated more than half a century ago in a classic paper by Yule and Kendall, who used data for wheat and potato yields from the (then) 48 counties of England to demonstrate that correlation coefficients tend to increase with scale. They aggregated the 48-county data into zones so that there were first 24, then 12, then 6, and finally just 3 zones. Table 6.3 presents the range of their results, from near zero (no correlation) to over 0.99 (almost perfect positive correlation), although subsequent research has suggested that this range of values is atypical.

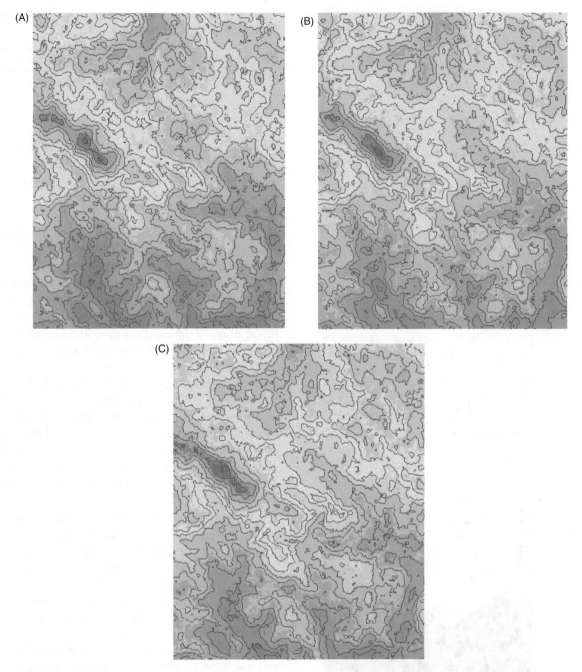

Figure 6.15 Three realizations of a model simulating the effects of error on a digital elevation model. (Reproduced by permission of Ashton Shortridge)

Relationships typically grow stronger when based on larger geographic units.

Scale turns out to be important because of the property of spatial autocorrelation outlined in Section 4.3. A succession of subsequent research papers has reaffirmed the existence of similar effects in multivariate analysis. However, rather discouragingly, scale effects in multivariate cases do not follow any consistent or predictable trends.

Further *aggregation* effects can arise in GIS analysis because there is a multitude of ways in which the mosaic of basic areal units (the "geographic individuals") can be assembled together into zones, and the requirement that zones be made up of spatially contiguous elements presents only a weak constraint on the huge combinatorial range. This gives rise to the related *aggregation* or *zonation* problem, in which different combinations of a given number of geographic individuals into coarser-scale areal units can yield widely different results.

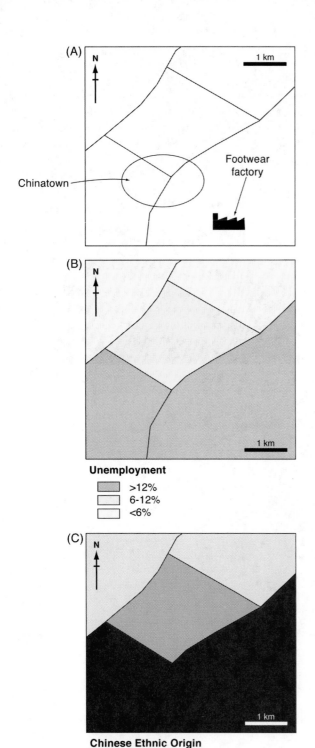

Table 6.3 1950 Yule and Kendall's data for wheat and potato yields from 48 English counties.

No. of geographic areas	Correlation
48	0.2189
24	0.2963
12	0.5757
6	0.7649
3	0.9902

In a classic 1984 study, geographer Stan Openshaw applied correlation and regression analysis to the attributes of a succession of zoning schemes. He demonstrated that the constellation of elemental zones within aggregated areal units could be used to manipulate the results of spatial analysis to a wide range of quite different prespecified outcomes. These numerical experiments have some sinister counterparts in the real world, the most notorious example of which is the political gerrymander of 1812 (see Section 15.1.2). Spatial distributions can be designed (with or without GIS) which are unrepresentative of the scale and configuration of real-world geographic phenomena; such outcomes may even emerge by chance. The outcome of multivariate spatial analysis is also similarly sensitive to the particular zonal scheme that is used.

Taken together, the effects of scale and aggregation are generally known as the *Modifiable Areal Unit Problem* (MAUP). The ecological fallacy and the MAUP have long been recognized as problems in applied spatial analysis, and, through the concept of spatial autocorrelation (Section 4.3), they are also understood to be related problems. Increased technical capacity for numerical processing and innovations in scientific visualization have refined the quantification and mapping of these measurement effects, and have also focused interest on the effects of within-area spatial distributions upon analysis.

6.4.4 Validation with Reference to External Sources: Data Integration and Shared Lineage

Goodchild and Longley use the term *concatenation* to describe the integration of two or more different data sources, such that the contents of each are accessible in the product. The polygon overlay operation that will be discussed in Section 14.2.4, and its field-view counterpart, is one simple form of concatenation. The term *conflation* is used to describe the range

Figure 6.16 The problem of ecological fallacy. (A) Before it closed down, the footwear factory drew its labor from its local neighborhood and a jurisdiction to the west. The closure caused high unemployment, but not amongst the service sector workers of Chinatown. Yet comparison of choropleth maps (B) and (C) indicates a spurious relationship between Chinese ethnicity and unemployment.

of functions that attempt to overcome differences between datasets, or to merge their contents (as with rubber-sheeting: see Section 9.3.2.3). Conflation thus attempts to replace two or more versions of the same information with a single version that reflects the pooling, or weighted averaging, of the sources. The individual items of information in a single geographic dataset often share lineage, in the sense that more than one item is affected by the same error. This happens, for example, when a map or photograph is registered poorly, since all of the data derived from it will have the same error. One indicator of shared lineage is the persistence of error—because all points derived from the same misregistration will be displaced by the same, or a similar, amount. Because neighboring points are more likely to share lineage than distant points, errors tend to exhibit strong positive spatial autocorrelation.

Conflation combines the information from two data sources into a single source.

When two datasets that share no common lineage are concatenated (for example, they have not been subject to the same misregistration), then the relative positions of objects inherit the absolute positional errors of both, even over the shortest distances. While the shapes of objects in each dataset may be accurate, the relative locations of pairs of neighboring objects may be wildly inaccurate when drawn from different datasets. The anecdotal history of GIS is full of examples of datasets that were perfectly adequate

for one application, but that failed completely when an application required that they be merged with some new dataset that had no common lineage. For example, merging GPS measurements of point positions with streets derived from the U.S. Bureau of the Census TIGER files may lead to surprises where points appear on the wrong sides of streets. If the absolute positional accuracy of a dataset is 50 m, as it is with parts of the TIGER database, points located less than 50 m from the nearest street will frequently appear to be misregistered.

Datasets with different lineages often reveal unsuspected errors when overlaid.

Figure 6.17 shows an example of the consequences of overlaying data with different lineages. In this case, two datasets of streets (shown in green and red) produced by different commercial vendors using their own process fail to match in position by amounts of up to 100 m; they also fail to match in the names of many streets, and even the existence of streets. The integrative functionality of GIS makes it an attractive possibility to generate multivariate indicators from diverse sources. Such data are likely to have been collected at a range of different scales, and for a range of areal units as diverse as census tracts, river catchments, land ownership parcels, travel-to-work areas, and market research surveys.

Established procedures of statistical inference can only be used to reason from representative samples to the populations from which they were drawn. Yet

Figure 6.17 Overlay of two street databases for part of Goleta, California.

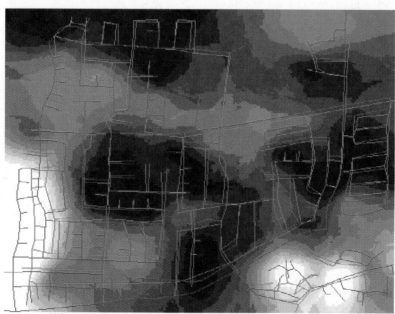

these procedures do not regulate the assignment of inferred values to other (usually smaller) zones, or their apportionment to ad hoc regions. There are emergent tensions within socioeconomic GIS, for there is a limit to the usefulness of inferences drawn from conventional, scientifically valid data sources that may be out-of-date, zonally coarse, and irrelevant to what is happening in modern societies. Yet the alternative of using new rich sources of crowd-sourced VGI (see Section 11.2) or marketing data may be profoundly unscientific in its inferential procedures.

6.4.5 Internal and External Validation; Induction and Deduction

Reformulation of the MAUP into a *geocomputational* (Section 16.1) approach to zone design amounts to inductive use of GIS to seek patterns through repeated scaling and aggregation experiments, alongside much better deductive *external* validation using any of the multitude of new datasets that are a hallmark of the information age. Neither of these approaches, used in isolation, is likely to resolve the uncertainties inherent in spatial analysis. Zone-design experiments are merely playing with the MAUP, and most of the new sources of external validation are unlikely to sustain full scientific scrutiny, particularly if they were assembled through nonrigorous survey designs.

> **There is no "solution" to the Modifiable Areal Unit Problem, but simulation of large numbers of alternative zoning schemes can gauge its likely effects.**

The conception and measurement of elemental zones, the geographic individuals, may be ad hoc, but it is rarely wholly random either. Can our recognition and understanding of the empirical effects of the MAUP help us to neutralize its effects? Not really. In measuring the distribution of all possible zonally averaged outcomes ("simple random zoning" in analogy to simple random sampling in Section 4.4), there is no tenable analogy with the established procedures of statistical inference and its concepts of precision and error. Even if there were, as we will see in Section 15.5, there are limits to the application of classic statistical inference to spatial data.

> **Zoning seems similar to sampling, but its effects are very different.**

The way forward seems to entail a twofold response: first, to use GIS to customize zoning schemes; and second to undertake validation of data with respect to external sources and focus clearer application-centered thinking upon the likely degree of within-zone heterogeneity that is concealed in our aggregated data. In this way, the MAUP will dissipate if GIS analysts understand the particular areal units that they wish to study.

There is also a sense here that resolution of the MAUP requires acknowledgment of the uniqueness of places. The time dimension is also important: the areal objects of study are ever-changing, and our perceptions of what constitutes an appropriate areal schema will change. Indeed, infusing the time dimension into GIS arguably creates the need for new overarching conceptions and paradigms (meta theories), exemplified by Michael Batty's work on complexity theory (Box 6.5). And finally, within the socioeconomic realm, the act of defining zones can also be self-validating if the allocation of individuals affects the interventions that are subsequently made, be they a mail-shot about a shopping opportunity or aid under an areal policy intervention. Spatial discrimination affects spatial behavior, and so the principles of zone design are of much more than academic interest.

6.5 Consolidation

Uncertainty is certainly much more than error. Just as the amount of available digital data and our abilities to process them have developed, so our understanding of the quality of digital depictions of reality has broadened. It is one of the supreme ironies of contemporary GIS that as we accrue more and better data and have more computational power at our disposal, we seem to become more uncertain about the quality of our digital representations and the adequacy of our areal units of analysis. Richness of representation and greater computational power serve to make us more aware of the range and variety of established uncertainties, and challenge us to integrate new ones.

The only way beyond this impasse is to continue advance hypotheses about the likely generalized structure of spatial data, but in a spirit of humility rather than of conviction. Hypothesis generation requires more than the brute force of high-power computing, and so progress requires greater a priori understanding about the structure in spatial as well as attribute data. There are some general rules to guide us here, and spatial autocorrelation measures provide further structural clues (Section 4.3). In Section 14.2.1 we discuss how context-sensitive spatial analysis techniques such as geographically weighted regression provide a bridge between general statistics and the case-study approach.

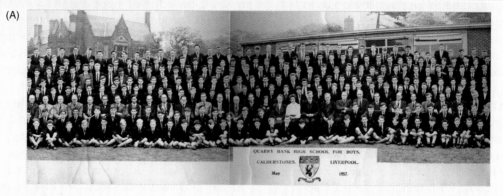

Figure 6.18 (A) Sometime schoolmate of Beatle John Lennon (Mike is in the front row position third from right: John Lennon is seventh boy from the left, three rows down).

Michael Batty

Michael Batty (Figure 6.18) is the most renowned and forward-thinking analytical planning researcher of his generation, and it is from this perspective that he has made enduring contributions to GIScience. A high school contemporary of Beatle John Lennon (Figure 6.18A and B), he studied Town and Country Planning at the University of Manchester prior to working in universities on both sides of the Atlantic—in Cardiff, London, Manchester, and Reading (all in the UK), Waterloo (Canada) and SUNY Buffalo (United States). Since 1995 he has directed the interdisciplinary Centre for Advanced Spatial Analysis (CASA: www.casa.ucl.ac.uk) at University College London. He was elected a Fellow of the British Academy in 2001, was awarded a CBE (Commander of the British Empire) for "services to geography," in 2004, and became the first ever professor of planning to be elected Fellow of the Royal Society in 2009.

Mike's 1976 book on *Urban Modelling* very much defined the agenda for quantitative analysis of city systems for the following 25 years, and his 1994 book *Fractal Cities* provides an early yet rich statement of ideas concerning the way in which cities grow and change. His 2005 monograph, *Cities and Complexity*, recasts much of this work within the emerging paradigm of complexity science: it posits that city systems are always far-from-equilibrium and hence that the action of processes governing their development is always uncertain, it follows that their evolution is inherently unpredictable and that uncertainty is inherent in their planning and management. Mike's related work has focused on the development of agent-based models in GIScience (see Section 16.2.2), and he has also explored the scaling laws that characterize city size distributions. Much of this work is also enriched by new ways of visualizing spatial complexity (see Chapter 13).

Figure 6.18 (B) Mike Batty, planner, geographer, GIScientist was a school contemporary of John Lennon at Liverpool's Quarry Bank High School. (Courtesy Mike Batty)

Figure 6.18 (C) Mike today.

Geocomputation helps too, by allowing us to gauge the sensitivity of outputs to inputs but, unaided, is unlikely to present unequivocal best solutions. The fathoming of uncertainty requires a combination of the cumulative development of a priori knowledge (we should expect scientific research to be cumulative in its findings), external validation of data sources, and inductive generalization in the fluid, eclectic data-handling environment that is contemporary GIS.

What does all this mean in practice? Here are some rules for how to live with uncertainty. First, since there can be no such thing as perfectly accurate GIS analysis, it is essential to acknowledge that uncertainty is inevitable. It is better to take a positive approach, by learning what one can about uncertainty, than to pretend that it does not exist. To behave otherwise is unconscionable and can also be very expensive in terms of lawsuits, bad decisions, and the unintended consequences of actions (see Section 18.5.1).

Second, GIS analysts often have to rely on others to provide data—historically through government-sponsored mapping programs (e.g., those of the U.S. Geological Survey or the Great Britain's Ordnance Survey) and commercial sources, and increasingly through volunteered geographic information (see Section 11.2). Data should never be taken as the truth; instead, it is essential to assemble all that is known about its quality and to use this knowledge to assess whether the data are fit for use. Metadata (Section 11.2.1) are designed specifically for this purpose and will often include assessments of quality. When these are not present, it is worth spending the extra effort to contact the creators of the data, or other people who have tried to use them, for advice on quality. Never trust data that have not been assessed for quality, or data from sources that do not have good reputations for quality.

Third, the uncertainties in the outputs of GIS analysis are often much greater than one might expect given knowledge of input uncertainties because many GIS processes are highly nonlinear. Yet some spatial processes dampen uncertainty rather than amplify it. Given this condition, it is important to gain some impression of the likely impacts of uncertain inputs to GIS upon outputs.

Fourth, we should rely on multiple sources of data whenever we can in order to facilitate external validation. It may be possible to obtain maps of an area at several different scales, for example, or to conflate several different open-source databases. Raster and vector datasets are often complementary (e.g., when combining a remotely sensed image with a topographic map). Digital elevation models can often be augmented with spot elevations, or GPS measurements.

38.74376% of all statistics are made up (including this one). Avoid spurious precision, and evaluate the provenance of the data that you use.

Finally, be honest and informative in reporting the results of GIS analysis. Recognize that uncertainty is never likely to be eliminated, but that it can be managed as part of good GIS practice. Input data sources may be presented with more apparent precision than is justified by their actual accuracy on the ground, and lines may have been drawn on maps with widths that reflect relative importance, rather than uncertainty of position. It is up to the users to redress this imbalance by finding ways of communicating what they know about accuracy, rather than relying on the GIS to do so. It is wise to include plenty of caveats into reported results, so that they reflect what we believe to be true, rather than a narrow and literal interpretation of what the GIS appears to be saying.

Questions for Further Study

1. What tools do GIS designers build into their products to help users deal with uncertainty? Take a look at your favorite GIS from this perspective. Does it allow you to associate metadata about data quality with datasets? Is there any support to accommodate propagation of uncertainty? How does it determine the number of significant digits when it prints numbers? What are the pros and cons of including such tools?

2. Using aggregate data for Iowa counties, Stan Openshaw found a strong positive correlation between the proportion of people over 65 and the proportion who were registered voters for the Republican Party. What if anything does this tell us about the tendency for older people to register as Republicans?

3. Where is the flattest place on Earth? Search the Web for appropriate documents and give reasons for your answer (Figure 6.19).

4. You are a senior retail analyst for Safemart, which is contemplating expansion from its home state to three others in the United States. Assess the relative merits of your own company's store loyalty card data (which you can assume are similar to those collected by any retail chain with which you are familiar) and of data from the most recent Census in planning this strategic initiative. Pay particular attention to issues of survey content, the representativeness of population characteristics, and problems of scale and aggregation. Suggest ways in which the two data sources might complement one another in an integrated analysis.

Figure 6.19 Salt "flats"? (Courtesy NASA)

Further Reading

Batty, M. 2008. The size, scale, and shape of cities. *Science* 319 (5864, February 8): 769–771.

Burrough, P.A. and Frank, A.U. (eds.). 1996. *Geographic Objects with Indeterminate Boundaries*. London: Taylor and Francis.

Fisher, P.F. 2005. Models of uncertainty in spatial data. In Longley, P.A., Goodchild, M.F., Maguire, D.J., and Rhind, D.W. (eds.). *Geographical Information Systems: Principles, Techniques, Management and Applications (Abridged Edition)*. New York: Wiley, pp. 191–205.

Goodchild, M.F. and Longley, P.A. 2005. The future of GIS and spatial analysis. In Longley, P.A., Goodchild, M.F., Maguire, D.J., and Rhind, D.W. (eds.). *Geographical Information Systems: Principles, Techniques, Management and Applications (Abridged Edition)*. New York: Wiley, pp. 567–580.

Heuvelink, G.B.M. 1998. *Error Propagation in Environmental Modelling with GIS*. London: Taylor and Francis.

De Smith, M., Goodchild, M.F., and Longley, P.A. 2009. *Geospatial Analysis* (3rd ed.). Leicester, Troubador. See also www.spatialanalysisonline.com.

Openshaw, S. and Alvanides, S. 2005. Applying geocomputation to the analysis of spatial distributions. In Longley, P.A., Goodchild, M.F., Maguire, D.J., and Rhind, D.W. (eds.). *Geographical Information Systems: Principles, Techniques, Management and Applications (Abridged Edition)*. New York: Wiley, pp. 267–282.

Zhang, J.X. and Goodchild, M.F. 2002. *Uncertainty in Geographical Information*. New York: Taylor and Francis.

3 TECHNIQUES

(Photo: © Doug Berry/iStockphoto)   **179**

7 GIS Software

GIS software forms the basis of the processing engine and is a vital component of an operational GIS. A GIS software system comprises an integrated collection of computer programs that implement geographic storage, processing, and display functions. The three key parts of any GIS software system are the user interface, the tools (functions), and the data manager. All three parts may be located on a single computer, or they may be spread over multiple machines in a departmental or an enterprise system configuration. Four main types of computer system architecture configurations are used to build operational GIS implementations: desktop, client server, centralized desktop, and centralized server. There are many different types of GIS software, and this chapter organizes the discussion around the main categories: desktop, Web mapping, server, virtual globes, developer, hand-held, and other. Software may be licensed using a commercial or open-source model. Examples of the main types of software products from the leading commercial developers include Autodesk AutoCAD, Bentley Map, ESRI ArcGIS, and Intergraph GeoMedia. There is also a significant and growing open-source GIS software community, of which Map Server is probably the best known example.

LEARNING OBJECTIVES

After studying this chapter you will be able to:

- Understand the architecture of GIS software systems, specifically

  - Organization by project, department, and enterprise

  - The three-tier architecture of software systems (graphical user interface, tools, and data access)

- Describe the process of GIS customization

- Describe the main types of commercial software:
  - Desktop
  - Web mapping
  - Server
  - Virtual globe
  - Developer
  - Hand-held
  - Other

- Outline the key characteristics of the main types of commercial GIS software products currently available.

7.1 Introduction

In Chapter 1, the four technical parts of a geographic information system were defined as the network, hardware, software, and data, which together functioned with reference to people and the procedural frameworks within which people operate (Section 1.5). This chapter is concerned with GIS software, the geographic storage, processing, and display engines of complete, working GIS. The functionality or capabilities of GIS software will be discussed later in this book (especially in Chapters 9–16). The focus here is on the different ways in which these capabilities are realized in GIS software products and implemented in operational GIS.

This chapter takes a fairly narrow view of GIS software, concentrating on systems with a range of generic capabilities to collect, store, manage, query, analyze, and present geographic information. It excludes atlases, simple graphics and mapping systems, route-finding software, simple location-based services, image-processing systems, and spatial extensions to database management systems (DBMS), which are not true GIS as defined here. The discussion is also restricted to GIS software products—well-defined collections of software and accompanying documentation, install scripts, and so on—that are subject to multiversioned release control (that is to say, the vendor releases software in a controlled way on a periodic basis). By definition, it excludes specific-purpose utilities, unsupported routines, and ephemeral codebases.

Earlier chapters, especially Chapter 3, introduced several fundamental computer concepts, including digital representations, data, and information. Two further concepts need to be introduced here. Programs are collections of instructions that are used to manipulate digital data in a computer. System software programs, such as a computer operating system, are used to support application software—the programs with which end users interact. Integrated collections of application programs are referred to as software packages or systems (or just software for short).

Software can be distributed to the market in several different ways. The dominant form of distribution is the sale of commercial off-the-shelf (COTS) software products on hard-copy media (DVD), although downloading over the Internet is becoming increasingly common and will soon replace distribution on hard-copy media. GIS software products of this type are developed with a view to providing users with a consistent and coherent model for interacting with geographic data. The product will usually comprise an integrated collection of software programs, an install script, online help files, sample data and maps, documentation, a license, and an associated Web site. Alternative distribution models that are becoming increasingly prevalent for smaller products include *shareware* (usually intended for sale after a trial period), *liteware* (shareware with some capabilities disabled), *freeware* (free software but with copyright restrictions), *public domain software* (free with no restrictions), and *open-source* software (the *source code* is provided and users agree not to limit the distribution of improvements). In recent years there has been an upsurge in interest in open-source software, and this is discussed in greater detail later in several sections.

GIS software packages provide a unified approach to working with geographic information.

GIS software vendors—the companies that design, develop, sell, and support major GIS software products—build on top of basic computer operating system capabilities such as security, file management, peripheral drivers (controllers), printing, and display management. GIS software is constructed on these foundations to provide a controlled environment for geographic information collection, management, analysis, display, and interpretation. The unified architecture and consistent approach to representing and working with geographic information in a GIS software package aim to provide users with a standardized approach to GIS.

In addition to the commercial-off-the-shelf GIS software products that have been the mainstay of GIS for many years, it is also important to acknowledge a growing movement that is creating public domain, open source, and free software. In the early days such software products provided only rather simple, poorly engineered tools with no user support. Today there are several high-quality, feature-rich software products (see, for example, www.freeGIS.org), and a number of products will be discussed in later sections.

7.2 The Evolution of GIS Software

In the formative GIS years, GIS software consisted simply of collections of computer routines that a skilled programmer could use to build an operational GIS. During this period, each and every GIS was unique in terms of its capabilities, and significant levels of resource were required to create a working system. As software engineering techniques advanced and the GIS market grew in the 1970s and 1980s, demand increased for higher-level applications with a standard user interface. In the late 1970s and early 1980s, the standard means of communicating with a GIS was to type in command lines and wait for a response from the system. User interaction with a GIS entailed typing instructions, for example, to draw a topographic map, query the attributes of a forest stand object, or summarize the length of highways in a project area. Essentially, a GIS software package was a toolbox of geoprocessing operators or commands that could be applied to datasets to create new derivative datasets. For example, three polygon-based data layers of the same geographic area—*Soil*, *Slope*, and *Vegetation*—could be combined using an overlay processing function (see Section 14.2.4) to create an *IntegratedTerrainUnit* dataset.

Two key developments in the late 1980s and 1990s made the software easier to use and more generic. First, command line interfaces were supplemented and eventually largely replaced by graphical user interfaces (GUIs). These menu-driven, form-based interfaces greatly simplify user interaction with a GIS (see Section 13.2). Second, a customization capability was added to allow specific-purpose applications to be created from the generic toolboxes. Software developers and advanced technical users could make calls using a high-level programming language (such as Visual Basic or Java) to published application programming interfaces (APIs) that exposed key functions. Together these stimulated enormous interest in GIS and led to much wider adoption and expansion into new areas. In particular, the ability to create custom application solutions allowed developers to build focused applications for end users in specific market areas. This led to the creation of GIS applications specifically tailored to the needs of major markets (e.g., government, utilities, military, and environment). New terms were developed to distinguish these subtypes of GIS software: planning information systems, automated mapping/facility management (AM/FM) systems, land information systems, and more recently, location-based services systems.

In the last few years a new method of software interaction has evolved that allows software systems to communicate over the Web using a Web services paradigm. A Web service is an application that exposes its functions via a well-defined published interface that can be accessed over the Web from another program or Web service. This new software interaction paradigm allows geographically distributed GIS functions to be linked together to create complete GIS applications. For example, a market analyst who wants to determine the suitability of a particular site for locating a new store can start a small browser-based program on his or her desktop computer that links to remote services over the Web that provide access to the latest population census and geodemographics data, as well as analytical models. Although these data and programs are remotely hosted and maintained, they can be used for site suitability analysis as though they were resident on the market analyst's desktop computer. Chapter 11 explores Web services in more depth in the context of distributed GIS.

The GIS software of today still embodies the same principles of an easy-to-use menu-driven interface and a customization capability, but can now be deployed on a desktop, on a mobile device, or distributed over the Web.

7.3 Architecture of GIS Software

7.3.1 Project, Departmental, and Enterprise GIS

Usually, GIS is first introduced into organizations in the context of a single, fixed-term project (Figure 7.1). The technical components (network, hardware, software, and data) of an operational GIS are assembled for the duration of the project, which may be from several months to a few years. Data are collected specifically for the project, and little thought is usually given to reuse of software, data, and human knowledge. In larger organizations, multiple projects may run one after another or even in parallel. The "one off" nature of the projects, coupled with an absence of organizational vision, often leads to duplication, as each project develops using different hardware, software, data, people, and procedures. Sharing data and experience is usually a low priority.

As interest in GIS grows, to save money and encourage sharing and resource reuse, several projects in the same department may be amalgamated. This often leads to the creation of common standards, development of a focused GIS team, and procurement of new GIS capabilities. Yet it is also quite common for different departments to have different GIS and standards.

As GIS becomes more pervasive, organizations learn more about it and begin to become dependent

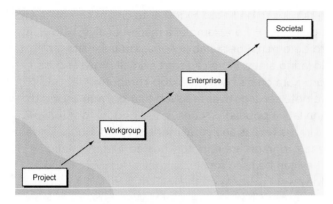

Figure 7.1 Types of GIS implementation.

on it. This leads to the realization that GIS is a useful way to organize many of the organization's assets, processes, and workflows. Through a process of natural growth, and possibly further major procurement (e.g., purchase of upgraded hardware, software, and data), GIS gradually becomes accepted as an important enterprise-wide information system. At this point, GIS standards are accepted across multiple departments, and resources to support and manage the GIS are often centrally funded and managed. A fourth type of societal implementation has been identified in which thousands or tens of thousands of users become engaged in GIS and connected over the Web. Web connections provide asynchronous, loosely coupled communication allowing messaging and data exchange. Google Earth, Microsoft Bing Maps, and ESRI's ArcGIS have been widely deployed over the Web and are being used to connect distributed users, provide widespread access to geographic information and processing, and establish geographic thinking and methods as a core part of organizational activities.

7.3.2 The Three-tier Architecture

From an information systems perspective a GIS has three key parts: the user interface, the tools, and the data management system (Figure 7.2). The user's interaction with the system is via a graphical user interface (GUI), an integrated collection of menus, tool bars, and other controls. The GUI provides organized access to the GIS tools. The tool set defines the capabilities or functions that the GIS software has available for processing geographic data. The data are stored as files, databases, or Web services and are organized by data management software. In standard information-system terminology this is a three-tier architecture, with the three tiers being called: presentation, business logic, and data server. Each of these

software tiers is required to perform different types of independent tasks. The presentation tier must be adept at collecting user inputs, rendering (displaying) data, and interacting with graphic objects. The business logic tier is responsible for performing all operations such as network routing, data overlay processing, and raster analysis. It is here also that the GIS data-model logic is implemented (see Section 7.3.3 for discussion of data models). The data-server tier must import and export data and service requests for subsets of data (queries) from a database or file system. In order to maximize system performance, it is useful to optimize hardware and operating systems settings differently for each type of task. For example, rendering maps requires large amounts of memory and fast CPU speeds, whereas database queries need fast disks and buses (interfaces between devices) for moving large amounts of data around. By placing each tier on a separate computer, some tasks can be performed in parallel, and greater overall system performance and scalability can be achieved.

Figure 7.2 Classical three-tier architecture of a GIS software system.

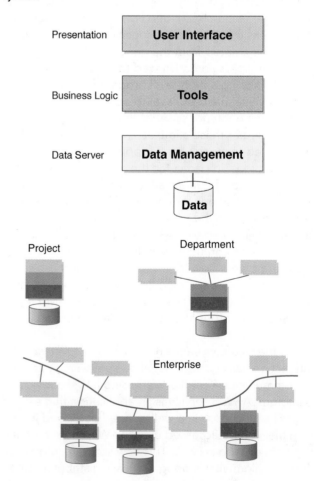

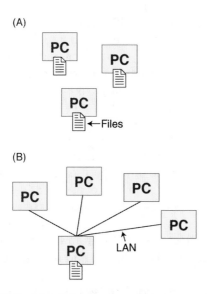

Figure 7.3 Desktop GIS software architecture used in project GIS: (A) standalone desktop GIS on PCs each with own local files; (B) Desktop GIS on PCs sharing files on a PC file server over a LAN.

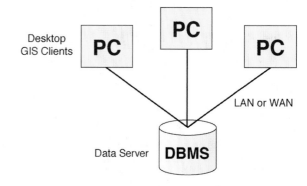

Figure 7.4 Client-server GIS: desktop GIS software and DBMS data server in a workgroup or departmental GIS configuration.

GIS software systems deal with user interfaces, tools, and data management.

Four types of computer system architecture configurations are used to build operational GIS implementations: desktop, client-server, centralized desktop, and centralized server. In the simplest desktop configuration, as used in single-user project GIS, the three software tiers are all installed on a single piece of hardware (most commonly a desktop PC) in the form of a desktop GIS software package, and users are usually unaware of their separate identity (Figure 7.3A). In a variation on this theme, data files are held on a PC (or sometimes a Windows, Linux, or Unix machine) centralized file server, but the data server functionality is still part of the desktop GIS. This means that the entire contents of any accessed file must be pulled across the network even if only a small amount of it is required (Figure 7.3B).

In larger and more advanced multiuser workgroup or departmental GIS, the three tiers can be installed on multiple machines to improve flexibility and performance (Figure 7.4). In this type of configuration, the users in a single department (for example, the planning or public works department in a typical local government organization) still interact with a desktop GIS GUI (presentation layer) on their desktop computer, which also contains all the business logic, but the database management software (data server layer) and data may be located on another machine connected over a network. This type of computing architecture is usually referred to as client-server because clients request data or processing services from servers that perform work to satisfy client requests. The data server has local processing capabilities and is able to query and process data and thus return part of the whole database to the client, which is much more efficient than sending back the whole dataset for client-side query. Clients and servers can communicate over a local area network (LAN), wide area network (WAN), or Internet network, but given the large amount of data communication between the client and server, faster local area networks are most widely used.

In a client-server GIS, clients request data or processing services from servers that perform work to satisfy client requests.

Both the desktop and client-server architecture configurations have significant amounts of functionality on the desktop, which is said to be a thick client (see also Section 7.3.4). In contrast, in the centralized desktop architecture configuration all the GUI and business logic is hosted on a centralized server, called an application (or middle tier) server (Figure 7.5). Typically, this is in the form of a desktop GIS software package. An additional piece of software is also installed on the application server (Citrix or Windows Terminal Server) that allows users on remote machines full access to the software over a LAN or WAN as though it were on the local desktop PC. Since the only application software that runs on the desktop PC is a small-client library, this is said to be a thin client. A data server (DBMS) is usually used for data management. This type of configuration is widely used in large departmental and enterprise applications where high-end capabilities such as advanced editing, mapping, and analysis are required.

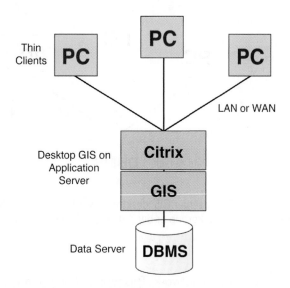

Figure 7.5 Centralized desktop GIS as used in advanced departmental and enterprise implementations. See Figure 7.2 for explanation of color-coding.

In a more common variation of the centralized desktop implementation, the business logic is implemented as a true server system and runs on a middle-tier server machine (Figure 7.6). In this configuration a range of thick and thin clients running on desktop PCs, Web browsers, and specialist devices (e.g., mobile hand-held devices) communicate with the middle-tier server over a wired or wireless network connection. In the case of thin-client access, the presentation tier (user interface) also runs on the server (although technically it is still presented on the desktop). A second server machine is usually employed to run data management software (DBMS). This type of implementation is the standard architecture for Web-based GIS and is common in enterprise

Figure 7.6 Centralized server GIS as used in advanced departmental and enterprise implementations. See Figure 7.2 for explanation of color-coding.

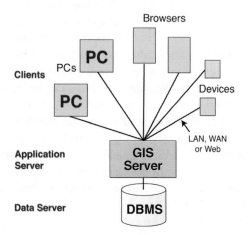

GIS. Large, enterprise GIS may involve more than 10 servers and hundreds or even thousands of clients that are widely dispersed geographically and nowadays are connected together using the Web. In recent years there has been a tendency to create very large server implementation farms, built from low-cost, standard consumer PC hardware, that are accessible over the Web. This new form of *cloud computing* architecture forms the platform for mass-participant Web-based services such as the e-mail and search systems from Google, Microsoft, and Yahoo, as well as various virtual globe implementations (see Section 7.6.4).

> **Server-based GIS is the standard architecture used in Web-based and enterprise GIS.**

Although organizations often standardize on either a project, a departmental, or an enterprise system, it is also common for large organizations to have all three configurations operating in parallel or as subparts of a full-scale system.

7.3.3 Software Data Models and Customization

In addition to the three-tier model, two further topics are relevant to an understanding of software architecture: data models and customization.

GIS data models will be discussed in detail in Chapter 8, and so the discussion here will be brief. From a software perspective, a data model defines how the real world is represented in a GIS. It will also affect the type of software tools (functions or operators) that are available, how they are organized, and their mode of operation. A software data model defines how the different tools are grouped together, how they can be used, and how they interact with data. Although such software facets are largely transparent to end users whose interaction with a GIS is via a user interface, they become very important to software developers that are interested in customizing or extending software. The software data model affects the capability, flexibility, and extensibility of GIS software systems.

Customization is the process of modifying GIS software to, for example, add new functionality to applications, embed GIS functions in other applications, or create specific-purpose applications. It can be as simple as deleting unwanted controls (for example, menu choices or buttons) from a GUI, or as sophisticated as adding a major new extension to a software package for such things as network analysis, high-quality cartographic production, or land parcel management.

To facilitate customization, GIS software products must provide access to the data model and expose capabilities to use, modify, and supplement existing functions. In the late 1980s when customization options were first added to GIS software products, each vendor had to provide a proprietary customization capability simply because no standard customization systems existed. Nowadays, with the widespread adoption of the .Net and Java frameworks, a number of industry-standard approaches and programming languages (such as Visual Basic, C/C++/C#, Java, and Python) are available for customizing GIS software systems.

Modern programming languages are one component of larger developer-oriented software packages called integrated development environments (IDEs). This term refers to the fact that the packages combine several software development tools, including a visual programming language, an editor, a debugger, and a profiler. Many of the so-called visual programming languages, such as C++/C#, Visual Basic, and Java, support the development of Windows-based GUIs containing forms, dialogs, buttons, and other controls. Program code can be entered and attached to the GUI elements using the integrated code editor. An interactive debugger will help identify syntactic problems in the code, for example, misspelled commands and missing instructions. Finally, there are also tools to support profiling programs. These show where resources are being consumed and how programs can be speeded up or improved in other ways.

Contemporary GIS typically use an industry-standard programming language like Visual Basic, C++/C#, or Java for customization.

To support customization using open, industry-standard IDEs, a GIS vendor must expose details of the software package's functionality. In modern GIS software systems the functionality is developed as software components—self-contained software modules that implement a coherent collection of functionality. A key feature of such software components is that they have well-defined application programming interfaces (APIs) that allow the functionality to be called by the programming tools in an IDE.

In recent years, three technology standards have emerged for defining and reusing software components. For building interactive desktop applications, Microsoft's .Net framework is the de facto standard for high-performance, interactive applications that use fine-grained components (that is, a large number of small functionality blocks). For server-centric GIS both .Net and the equivalent Java framework are widely deployed in operational GIS applications. Although both .Net and Java work very well for building fine-grained

client or server applications, they are less well suited for building applications that need to communicate over the Web. Because of the loosely coupled, comparatively slow, heterogeneous nature of Web networks and applications, fine-grained programming models do not work well. As a consequence, coarse-grained component and messaging systems have been built on top of the fine-grained .Net and Java applications using SOAP/XML (simple object access protocol/extensible markup language) and JavaScript/REST (representation state transfer protocol) that allow applications with Web services interfaces to interact over the Web. This higher-level approach makes it easier to create custom Web applications that use Web resources in a more efficient way by minimizing network traffic.

Components are important to software developers because they are the mechanism by which reusable, self-contained, software building blocks are created and used. They allow many programmers to work together to develop a large software system incrementally. The standard, open (published) format of components means that they can easily be assembled into larger systems. In addition, because the functionality within components is exposed through interfaces, developers can reuse them in many different ways, even supplementing or replacing functions if they so wish. Users also benefit from this approach because GIS software can evolve incrementally and support multiple third-party extensions. In the case of GIS products this includes, for example, tools for charting, reporting, and data table management.

7.3.4 GIS on the Desktop and on the Web

Today mainstream, high-end GIS users work primarily with software that runs either on the desktop or over the Web. In the desktop case, a PC (personal computer) is the main hardware platform and Microsoft Windows remains the dominant operating system (Figure 7.7 and Table 7.1). In the desktop paradigm, clients tend to be functionally rich and substantial in size, and are often referred to as thick or fat clients. Using the Windows standard facilitates interoperability (interaction) with other desktop applications, such as word processors, spreadsheets, and databases. As noted earlier (Section 7.3.2), most sophisticated and mature GIS workgroups have adopted the client-server implementation approach by adding either a thin or thick server application running on the Windows, Linux, or Unix operating system. The terms *thin* and *thick* are less widely used in the context of servers, but they mean essentially the same as when applied to clients. Thin servers perform relatively simple

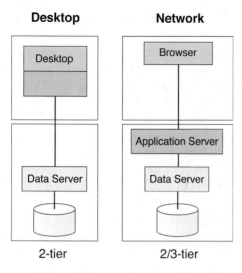

Desktop **Network**

2-tier   2/3-tier

Figure 7.7 Desktop and network GIS paradigms. See Figure 7.2 for explanation of color-coding.

tasks, such as serving data from files or databases, whereas thick servers also offer more extensive analytical capabilities such as geocoding, routing, mapping, and spatial analysis. In desktop GIS implementations, local area networks (LANs) and wide area networks (WANs) tend to be used for client-server communication. It is natural for developers to select Microsoft's .Net technology framework to build the underlying components making up these systems given the preponderance of the Windows operating system, although other component standards could also be used. The Windows-based client-server system architecture is a good platform for hosting interactive, high-performance GIS applications.

Table 7.1 Comparison of desktop and network GIS.

Feature	Desktop	Network
Client size	Thick	Thin
Client platform	Windows	Cross platform browser
Server size	Thin/thick	Thick
Server platform	Windows/Unix/ Linux	Windows/ Unix/Linux
Component standard	.Net	.Net/Java
Network	LAN/WAN	LAN/WAN/ Internet

Examples of applications well suited to this platform include those involving geographic data editing, map production, 2-D and 3-D visualization, spatial analysis, and modeling. It is currently the most practical platform for general-purpose systems because of its wide availability, good performance for a given price, and common usage in business, education, and government.

GIS users are standardizing their systems on the desktop and Web implementation models.

In the last few years there has been increasing interest in harnessing the power of the Web for GIS. Although desktop GIS have been and continue to be very successful, users are constantly looking for lower costs of ownership and improved access to geographic information. Network-based (sometimes called distributed) GIS allows previously inaccessible information resources to be made more widely available. A particularly interesting type of network GIS is Web-based GIS that uses the Web as a network. The network GIS model intrigues many organizations because it is based on centralized software and data management, which can dramatically reduce initial implementation and ongoing support and maintenance costs. It also provides the opportunity to link nodes of distributed user and processing resources using the medium of the Internet. The continued rise in network GIS will not signal the end of desktop GIS—indeed quite the reverse, since it is likely to stimulate the demand for content and professional GIS skills in geographic database automation and administration, and applications that are well suited to running on the desktop.

In contrast to desktop GIS, network GIS can use the cross-platform Web browser to host the viewer user interface. Currently, clients are typically very thin, often with simple display and query capabilities, although there is an increasing trend for them to become more functionally rich. Server-side functionality may be encapsulated on a single server, although in medium and large systems it is more common to have two servers, one containing the business logic (a middleware application server), the other the data manager (data server). The server applications typically contain all the business logic and are comparatively thick. The server applications may run on a Windows, Unix, or Linux platform.

Recently, there has been a move to combine the best elements of the desktop and network paradigms to create so-called rich clients. These are stored and managed on the server and dynamically downloaded to a client computer according to user demand. The business logic and presentation layers run on the server, with the client hardware simply used to render the data and manage use interaction. The new software capabilities in recent editions of the .Net and Java software development kits allow the development of applications with extensive user interaction that closely emulate the user experience of working with desktop software. Microsoft's Bing Maps 3D is an example of a rich client. When a user visits the Bing Web site, he or she is given the opportunity to download the Bing Maps 3D client. Upon installation, the client talks to the Bing Maps 3D server that holds vast databases of geographic information for the Earth and provides server-side query and analysis features.

7.4 Building GIS Software Systems

Widely used GIS software systems are built and released as GIS software products by GIS software development and product teams that may operate according to commercial or open-source models. Mature software products are subject to carefully planned versioned release cycles that incrementally enhance and extend the pool of capabilities. The key parts of a GIS software architecture—the user interface, business logic (tools), data manager, data model, and customization environment—were outlined in the previous section.

GIS product teams start with a formal design for a software system and then build each part or component separately before assembling the whole system. Typically, development will be iterative with creation of an initial prototype framework containing a small number of partially functioning parts, followed by increasing refinement of the system. Core GIS software systems are usually written in a modern programming language like Visual C++/C# or Java, with Visual Basic or another scripting language used for operations that do not involve significant amounts of computer processing like GUI interaction.

As standards for software development become more widely adopted, so the prospect of reusing software components becomes a reality. A key choice that then faces all software developers or customizers is whether to design a software system by buying in components, or to build it more or less from scratch. Both options have advantages: building components gives greater control over system capabilities and

enables specific-purpose optimization; buying components can save time and money.

A key GIS implementation issue is whether to buy a system or to build one.

A modern GIS software system comprises an integrated suite of software components of three basic types: a data management system for controlling access to data (Chapters 8, 9, and 10); a mapping system for display and interaction with maps and other geographic visualizations (Chapters 12 and 13); and a spatial analysis and modeling system for transforming geographic data using operators (Chapters 14, 15, and 16). The components for these parts may reside on the same computer or can be distributed widely (Chapter 11) over a network. The work of one of GIS's leading software product engineers is described in Biographical Box 7.1.

7.5 GIS Software Vendors

The GIS industry is fortunate to have a diverse range of significant and well-established software system vendors. Each of the major vendors brings its own heritage and experience, and has created unique elements and interpretations of core GIS capabilities in the form of software that is constantly evolving in response to ongoing IT changes and expanded user requirements horizons. The following sections briefly describe the main GIS software vendors. Daratech, an independent firm of consultants, publishes an annual report about the size of the GIS market, and in its most recent 2007 report the largest GIS vendors in terms of annual worldwide revenues were ESRI, Bentley, Autodesk, and Intergraph (Figure 7.9).

7.5.1 Autodesk

Autodesk is a large and well-known publicly traded company with headquarters in San Rafael, California. It is one of the world's leading digital design and content companies and serves customers in markets where design is critical to success: building, manufacturing, infrastructure, digital media, and location services. Autodesk is best known for its AutoCAD product family, which is used worldwide by more than 4 million customers. The company was founded more than 25 years ago and has grown to become a $2 billion entity employing over 7300 staff.

Clint Brown, Director of Software Products, ESRI

For over 25 years, Clint Brown (Figure 7.8) has worked at ESRI with Scott Morehouse, director of Software Engineering, to design, build, and ship ESRI's GIS software—from ArcInfo to ArcView and ArcGIS. His job has been to transform ESRI GIS software programs into useful GIS software products that could be applied by a wide range of customers and organizations to a host of problems. He has been one of the company's strongest GIS user advocates and has taken real pride in building useful software.

After obtaining an M.S. in Statistics in 1978 from Texas A&M, Clint worked with a team at the U.S. Fish and Wildlife Service in Fort Collins, Colorado, to develop a methodology known as the Habitat Evaluation Procedures (HEP). This was a new kind of "geographic accounting" methodology for evaluating habitat values for land involved in federally funded projects in the United States. This methodology was used to perform impact assessments and to manage project effects on fish and wildlife.

At the beginning of the 1980s, Clint went to Alaska and helped to apply this methodology using early GIS. His work supported congressionally mandated master planning of the U.S. National Wildlife Refuge System in Alaska. His work there included pioneering the use of GIS methods and databases for master planning of 16 national wildlife refuges covering more than 70 million acres of land and most of the coastal areas in Alaska. GIS was the only suitable methodology that could be used to evaluate these lands and to create useful assessments and management goals for these special areas.

Clint moved to ESRI in 1983 to work on ESRI software and the then new ArcInfo software. At the time,

Figure 7.8 Clint Brown, Software Product Director. (Courtesy Clint Brown)

ESRI had fewer than two dozen organizations using ArcInfo. Clint says: "in the early days of ESRI I set myself the goal of finding out what made users successful at applying ArcInfo to their specific situations. I thought that if ESRI could capture these methods in our product documentation, then many users could learn to be successful at using GIS for their own operations."

And then GIS's most successful software leadership partnership began between Clint and Scott Morehouse: Scott led the software development and Clint the nonprogramming aspects of ESRI's products—documentation, user support, testing, and capturing the practical applications of GIS methods. His job was and continues to be about making GIS work for many users.

Autodesk's product family for the GIS (or geospatial as they call it) marketplace features several main products: AutoCAD Map 3D, Autodesk MapGuide, and Autodesk Topobase. AutoCAD Map 3D is a desktop GIS software product that enables organizations to create, maintain, and visualize geographic data and is especially adept at bridging the gap between engineering and GIS teams, and the rest of the organization. Autodesk MapGuide is a server-based GIS for delivering spatial data and applications over the Web (see Box 7.4 for an extended discussion). Autodesk Topobase is an infrastructure design and management solution that extends AutoCAD Map 3D and Autodesk MapGuide by allowing spatial information to be stored in a database and shared across an organization.

7.5.2 Bentley

Bentley is a privately held developer of software solutions for the infrastructure life-cycle market that is headquartered in Exton, Pennsylvania. It has applications that help engineers, architects, contractors, governments, institutions, utilities, and owner-operators design, build, and operate productively, collaborate globally, and deliver infrastructure assets that perform sustainably. Bentley

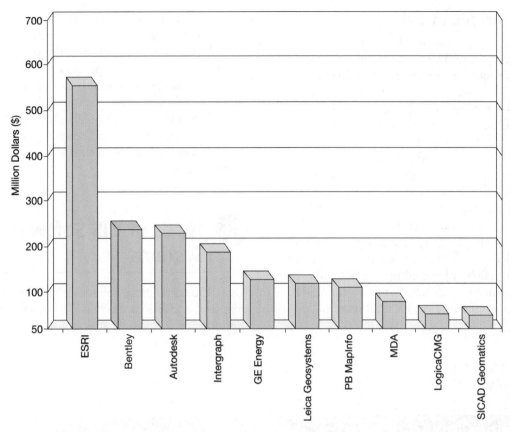

Figure 7.9 GIS software vendor market share (Courtesy Daratech).

has total annual revenues of $450 million and employs over 2700 staff worldwide; a little over $200 million in revenue is derived from Bentley's GIS business.

Bentley's flagship software product, MicroStation, supports a number of applications and serves the needs of a range of communities. Bentley Map is a GIS software system designed to address the needs of organizations that map, plan, design, build, and operate the world's infrastructure. It enhances the underlying MicroStation platform capabilities to power geospatial data creation, maintenance, and analysis. With Bentley Map, users can integrate data from a wide variety of sources into engineering and mapping workflows. Because Bentley Map is tightly integrated with MicroStation, it allows simultaneous manipulation of raster and vector data, which are standard features of the core platform. An interoperability environment makes it easy to work with many industry spatial data formats. Bentley's GIS solution can be implemented with any database connection supported by Micro-Station (e.g., Oracle Spatial or ArcGIS Server).

7.5.3 ESRI, Inc.

ESRI is a privately held company founded in 1969 by Jack and Laura Dangermond. Headquartered in

Redlands, California, ESRI employs over 4,500 people worldwide and has annual revenues of over $750 million. Today it serves more than 140,000 organizations and more than 1 million users. ESRI focuses solely on the GIS market, primarily as a software product company, but it also generates about a sixth of its revenue from project work such as advising clients on how to implement GIS. ESRI started building commercial software products in the late 1970s. Today ESRI's product strategy is centered on an integrated family of products called ArcGIS. The ArcGIS family is aimed at both end users (especially "professional" or highly technical users) and technical developers, and includes products that run on hand-held devices, desktop personal computers, servers, and over the Web. Most recently, the company has released an advanced Web-based GIS (ArcGIS Server) and has developed an online, hosted GIS called ArcGIS Online. ArcGIS Explorer is ESRI's virtual globe and is not unlike Google Earth and Microsoft Bing Maps.

ESRI is the classic high-end GIS vendor. It has a wide range of mainstream products covering all the main technical and industry markets. ESRI is a technically led geographic company focused squarely on the needs of hard-core GIS users. Box 7.1 highlights the work of one key ESRI employee, and Box 7.3 describes ESRI's ArcGIS product.

7.5.4 Intergraph, Inc.

Like ESRI, Intergraph was also founded in 1969 as a private company. The initial focus from their Huntsville, Alabama, offices was the development of computer graphics systems. After going public in 1981, Intergraph grew rapidly and diversified into a range of graphics areas including CAD and mapping software, consulting services, and hardware. In 2006 Intergraph returned to private ownership when it was purchased by TPG, Hellman & Friedman, and JMI Equity. After a series of reorganizations in the late 1990s and 2000s, Intergraph is today structured into two main operating units: Process, Power & Marine (PP&M) and Security, Government & Infrastructure (SG&I). The latter is the area that contains the main GIS activity of the company. Mapping and Geospatial Solutions accounts for more than $200 million of the annual Intergraph total revenue, which exceeds $720 million.

Intergraph has a large and diverse product line. From a GIS perspective the principal product family is GeoMedia, which spans the desktop and Web-based GIS markets across a range of application domains. Box 7.2 describes Intergraph GeoMedia. Other key Intergraph GIS products include G/Technology, which is designed to meet the geospatial resource management needs of utilities and communications companies, and ImageStation, a digital photogrammetric software suite.

7.6 Types of GIS Software Systems

Over 100 commercial software products claim to have mapping and GIS capabilities. The main categories of generic GIS software that dominate today are desk-

---

Applications Box 7.2

Desktop GIS: Intergraph GeoMedia

GeoMedia is an archetypal example of a mainstream commercial desktop GIS software product (Figure 7.10). First released in the late 1990s and built using a new codebase, it was created from the ground up to run on the Windows desktop operating system. Like other products in the desktop GIS category, it is primarily designed with the end user in mind (rather than technical users). It has a Windows-based graphical user interface and many tools for editing, querying, mapping, and spatial analysis. Data can be stored in proprietary GeoMedia files or in a DBMS such as Oracle, Microsoft Access, or SQL Server. GeoMedia enables data from multiple disparate databases to be brought into a single GIS environment for viewing, analysis, and presentation. The data are read and translated on the fly directly into memory. GeoMedia provides access to all major geospatial/CAD data file formats and to industry-standard relational databases.

GeoMedia is built as a collection of software object components. These underlying objects are exposed to developers who can customize the software using a programming language such as Visual Basic or C#.

GeoMedia offers a suite of analysis tools, including attribute and spatial query, buffer zones, spatial overlays, and thematic analysis. The product's layout composition tools provide the flexibility to design a range of types of maps that can be distributed on the Web, printed, or exported as files.

GeoMedia is the entry to a family of products. Several extensions (add-ons) offer additional functionality (e.g., image, grid and terrain analysis, and transaction management) and support for industry-specific workflows (e.g., transportation, parcel, and public works). Other members of the product family include GeoMedia Viewer (a free data viewer), GeoMedia Professional (offering high-end functionality), and Geo-Media WebMap (for Internet publishing).

All in all, the GeoMedia product family offers a wide range of capabilities for core GIS activities in the mainstream markets of government, education, and private companies. It is a modern and integrated product line with strengths in the areas of data access and user productivity. ▶

---

top, Web mapping, server, virtual globe, developer, and hand-held. The distinction between the Web mapping, server, and virtual globe categories is sometimes blurred. Web mapping systems are those that integrate software and data to create a unified online service; servers typically have more advanced functionality and can work with multiple data sources; and virtual globes provide 3-D visualization, query, and some analysis capabilities. In this section, the categories will be discussed followed by a brief summary of other types of software. Reviews of currently popular GIS software packages can be found in the various GIS magazines (see Box 1.4).

7.6.1 Desktop GIS Software

Since the mid-1990s, desktop GIS has been the mainstay of the majority of GIS implementations and the most widely used category of GIS software. Desktop GIS software owes its origins to the personal computer and the Microsoft Windows operating system. Initially,

the major GIS vendors ported their workstation or minicomputer GIS software to the PC, but subsequently redeveloped their software specifically for the PC platform. Desktop GIS software provides personal productivity tools for a wide variety of users across a broad cross section of industries. PCs are widely available, relatively inexpensive, and offer a large collection of user-oriented tools, including databases, word processors, and spreadsheets. The desktop GIS software category includes a range of options from simple viewers (such as ESRI ArcGIS ArcReader, Intergraph GeoMedia Viewer, Pitney Bowes MapInfo ProViewer, and open source Thuban) to desktop mapping and GIS software systems (such as Autodesk AutoCAD Map 3D, Clark Labs Idrisi, ESRI ArcGIS ArcView open source GRASS, GE Spatial Intelligence, Intergraph GeoMedia (see Box 7.2, Figure 7.10), the Manifold System (Figure 7.11), Pitney Bowes MapInfo Professional, and open source Quantum GIS, and at the high end, full-featured professional editor/analysis systems (such as ESRI ArcGIS ArcInfo (see Box 7.3, Figure 7.12)).

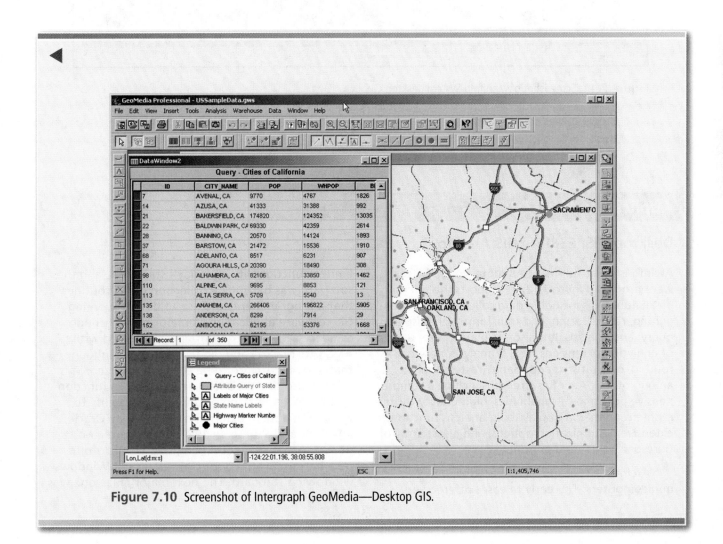

Figure 7.10 Screenshot of Intergraph GeoMedia—Desktop GIS.

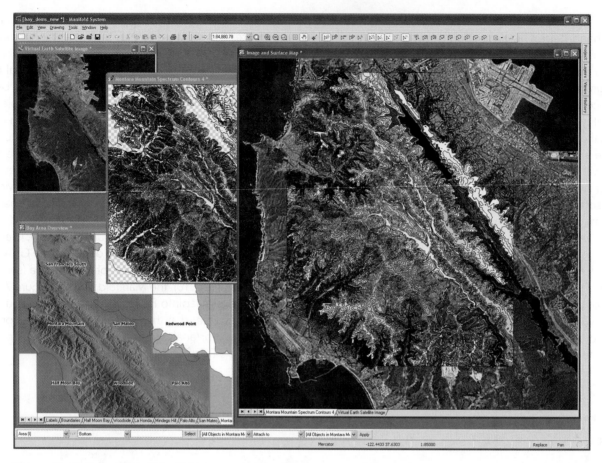

Figure 7.11 Screenshot of Manifold System desktop GIS.

Application Box 7.3

Desktop GIS: ESRI ArcGIS ArcInfo

ArcInfo is ESRI's full-featured professional GIS software product (Figure 7.12). It supports the full range of GIS functions including data collection and import; editing, restructuring, and transformation; display; query; and analysis. It is also the platform for a suite of analytical extensions for 3-D analysis, network routing, geostatistics and spatial (raster) analysis, among others. ArcInfo's strengths include a comprehensive portfolio of capabilities, high-quality cartography creation and display tools, analysis functions, extensive customization options, and a vast array of third-party tools and interfaces.

ArcInfo was originally released in 1981 on minicomputers. The early releases offered very limited functionality by today's standards, and the software was basically a collection of subroutines that a programmer could use to build a working GIS software application. A major breakthrough came in 1987 when ArcInfo 4 was released with AML (Arc Macro Language), a scripting language that allowed ArcInfo to be easily customized. This release also saw the software adapted to function on the Unix operating system and the ability to work with data in external databases like Oracle, Informix, and Sybase. In 1991, with the release of ArcInfo 6, ESRI again reengineered ArcInfo to take better advantage of Unix and the X-Windows windowing standard. The next major milestone was

▶

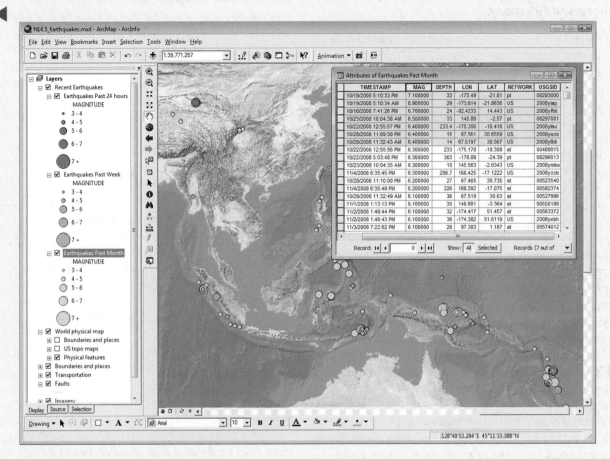

Figure 7.12 Screenshot of ESRI ArcGIS ArcInfo—Professional Desktop GIS.

the development of a menu-driven user interface in 1993 called ArcTools. This made the software considerably easier to use and also defined a standard for how developers could write ArcInfo-based applications. ArcInfo was ported to Windows NT at the 7.1 release in 1996. At about this time ESRI also made the decision to reengineer ArcInfo from first principles. This vision was realized in the form of ArcInfo 8, released in 1999.

ArcInfo 8 was quite unlike earlier versions of the software because it was designed from the outset as a collection of reusable, self-contained software components, based on Microsoft's COM standard. ESRI used these components to create an integrated suite of menu-driven, end-user applications: ArcMap—a map-centric application supporting integrated editing and viewing; ArcCatalog—a datacentric application for browsing and managing geographic data in files and databases; and ArcToolbox—a tool-oriented application for performing geoprocessing tasks such as proximity analysis, map overlay,

and data conversion. ArcInfo is customizable using any Microsoft .Net-compliant programming language such as Visual C++ and C#. The software is also notable because of the ability to store and manage all data (geographic and attribute) in standard commercial off-the-shelf DBMS (e.g., DB2, Informix, SQL Server, Oracle, and Postgres). In 2004, ESRI released Version 9, which builds on the foundations of Version 8. At this release ArcGIS desktop was supplemented by first a server and then online and mobile products. Another interesting aspect of ArcInfo is that for compatibility reasons since Release 8 ESRI has included a fully working version of the original ArcInfo workstation technology and applications. This has allowed ESRI users to migrate their existing databases and applications to the new version in their own time.

Version 10 of ArcGIS builds on the previous release, particularly through the addition of major new capabilities to use the software over the Web.

Desktop GIS are the mainstream work-horses of GIS today.

In the late 1990s, a number of vendors released free GIS viewers that are able to display and query popular file formats. Today, the GIS viewer has developed into a significant product subcategory. The rationale behind the development of these products from the commercial vendor point of view is that they help to establish market share and can create de facto standards for specific vendor terminology and data formats. A number of open-source products have also been developed to provide free and easy access to geographic information and tools. GIS users often work with viewers on a casual basis and use them in conjunction with more sophisticated GIS software products. GIS viewers have limited functional capabilities restricted to display, query, and simple mapping. They do not support editing, sophisticated analysis, modeling, or customization.

With their focus on data use rather than data creation, and their excellent tools for making maps, reports, and charts, desktop mapping and GIS software packages represent most people's experience of high-end GIS today. The successful systems have all adopted Microsoft standards for interoperability and user-interface style (although some open-source systems are based on Java). Users often see a desktop mapping and GIS package as simply a tool to enable them to do their full-time job faster, more easily, or more cheaply. Desktop mapping and GIS users work in planning, engineering, teaching, the army, marketing, and other similar professions; they are often not advanced technical users. Desktop GIS software prices typically range from $500 to $1,500 (these and other prices mentioned later typically have discounts for multiple purchases).

The term *professional* relates to the full-featured nature of this subcategory of software. The distinctive features of professional GIS include data collection and editing, database administration, advanced geoprocessing and analysis, and other specialist tools. Professional GIS offers a superset of the functionality of the systems in other classes. The people who use these systems typically are technically literate and think of themselves as GIS professionals (career GIS staff); they have degrees and, in many cases, advanced degrees in GIS or related disciplines. Prices for professional GIS are typically in the $5,000–15,000 range per user.

Professional GIS are high-end, fully functional systems.

7.6.2 Web Mapping

The last decade or so of GIS has been dominated by the desktop GIS architecture running on PCs. Already there are signs that the next decade will become dominated by Web and server GIS products. The defining characteristic of a Web mapping system is typically the quality of the base map. GIS on the Web has a comparatively long history that goes back to 1993 when researchers at Xerox Parc, California, created the first map server that could be accessed over the Internet. Since then several major milestones have been reached in developmental and organizational terms: the popularity of the MapQuest site and subsequent acquisition of the company by AOL for over $1.1 billion, the introduction of major mapping sites by Google, Yahoo, and Microsoft (among others), and the rise of computer mashups and neogeography (see Section 7.6.3 and Chapter 11) to name but a few.

Here the term *Web mapping* is taken to mean integrated Web-accessible software, a 2-D database (3-D base maps are covered later in Section 7.6.4, Virtual Globes) comprising one or more base maps, and an associated collection of services. Web access is provided via easily accessible, open interfaces. Typically, image tiles—each containing a fragment of a map—are returned in response to a user request for a map of a given area. Requests can be issued from, and responses processed and visualized in, standard Web browsers, and Web mapping sites have proven to be very popular (being consistently rated among the top 10 in terms of Web site traffic). In addition to providing maps, a range of other useful services are normally closely coupled, for example, with a gazetteer to find places of interest, multipoint driving directions, a choice of image and streetmap base maps, and the ability to overlay, or mashup, many other datasets.

Some of the success of Web mapping sites is also due to the fact that they can be easily accessed programmatically via well-defined application programming interfaces (APIs). The quintessential example here is the KML language and API for interacting with Google Maps. This has spawned a large number of add-ons and has allowed Google Maps to be integrated or mashed up with many other Web services. Google's MyMaps (www.maps.google.com) initiative provides a good window onto the world of mashups. Figure 7.13 shows videos from YouTube, pictures, and Wikipedia links on a Google Maps 2-D base map for Birmingham, UK.

OpenStreetMap is a relatively new and interesting entrant into this field. It is interesting because many of the data in the OpenStreetMap database have

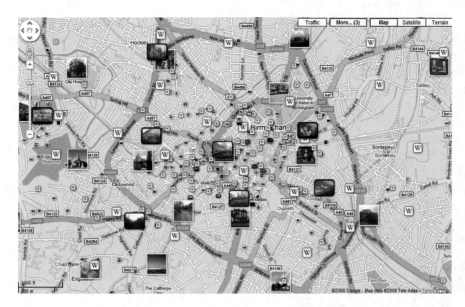

Figure 7.13 A geographic mashup of videos, pictures, and Wikipedia links on Google Maps.

been collected by volunteers and have made available using a copyright-free license. The software used to edit, access, and view the data is open source (including Mapnik and Osmarender for map rendering, JOSM and Potlatch for editing, and MySQL for data management: http://wiki.openstreetmap.org/wiki/Develop). In certain cities, OpenStreetMap is now beginning to rival and even exceed the coverage of commercial online base-map databases. Figure 7.14 shows a screenshot of an OpenStreetMap databases for Durham, UK (see also Box 13.1).

7.6.3 Server GIS

In simple terms, a server GIS is a GIS that runs on a computer server that can handle concurrent processing requests from a range of networked clients. Server GIS offers a wider range of GIS functions than Web

Figure 7.14 OpenStreetMap database for Durham, UK.

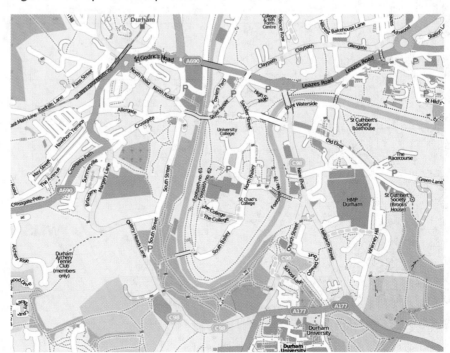

mapping systems, which focus solely on mapping and closely related services, and can work with any base map. Server GIS products have the potential for the largest user base and lowest cost per user of all GIS software system types. Stimulated by advances in server hardware and networks, the widespread availability of the Internet and market demand for greater access to geographic information, GIS software vendors have been quick to release server-based products that are accessible over the Web. Examples of server GIS include Autodesk MapGuide, ESRI ArcGIS Server, GE Spatial Application Server, Intergraph GeoMedia Webmap, MapServer (open source), and Pitney Bowes MapInfo MapXtreme. The cost of commercial server GIS products varies from around $5,000 to $25,000, for small to medium-sized systems, to well beyond for large multifunction, multiuser systems. Box 7.4 highlights Autodesk MapGuide.

Server GIS are growing in importance as capabilities increase and organization shift from desktop to network implementations.

The most popular open source GIS today is the MapServer Web mapping system (Figure 7.15). Originally developed at the University of Minnesota, it now has community ownership. The primary purpose of MapServer is to display and interact with dynamic maps over the Internet. It can work with hundreds of raster, vector, and database formats and can be accessed using popular scripting languages and development environments. It supports state-of-the-art functionality such as on-the-fly projections, high-quality map rendering, and map navigation and queries. MapServer has been used quite widely as the core of spatial data infrastructure projects

Figure 7.15 Architecture of MapServer. The central boxes are the software components that read data, render maps, and fulfill other client requests. A Web server (open-source Apache or Microsoft IIS) is used to communicate with browser clients. (Source: hhttp://mapserver.org/introduction.html)

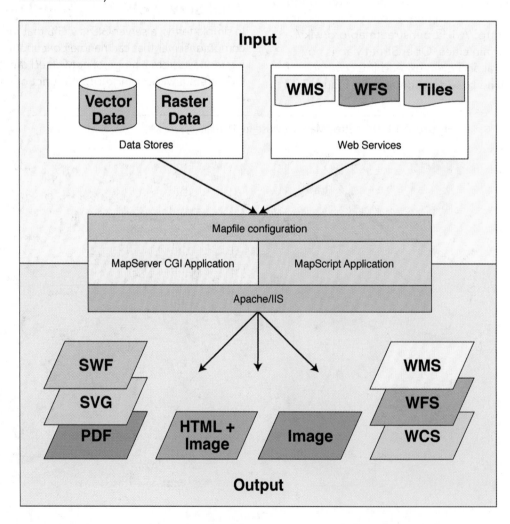

Web GIS: Autodesk MapGuide

In the late 1990s, a time when desktop GIS had become dominant, a small Canadian company called Argus released an Internet GIS product called MapGuide. Subsequently purchased by Autodesk, MapGuide marked the start of another chapter in the history of GIS software. The emphasis in MapGuide and other server GIS products is very much upon providing multiple users access to shared GIS resources over the Web.

MapGuide is an important innovation for the many users who have spent considerable amounts of time and money creating valuable databases, and who want to make them available to other users inside or outside their organization. Autodesk MapGuide allows users to leverage their existing GIS investment by publishing dynamic, intelligent maps

at the point at which they are most valuable—in the field, at the job site, or on the desks of colleagues, clients, and the public.

Autodesk, in partnership with the Open Source Geospatial Foundation, has released a version of MapGuide as an open-source project called MapGuide Open Source (see mapguide.osgeo.org). Figure 7.16 is a screenshot of an Autodesk MapGuide application that provides tourists and residents an easy way to locate streets and businesses in or around San Miguel de Allende, Guanajuato, Mexico. Using this application, community members are encouraged to include small local businesses that would not otherwise have an Internet presence.

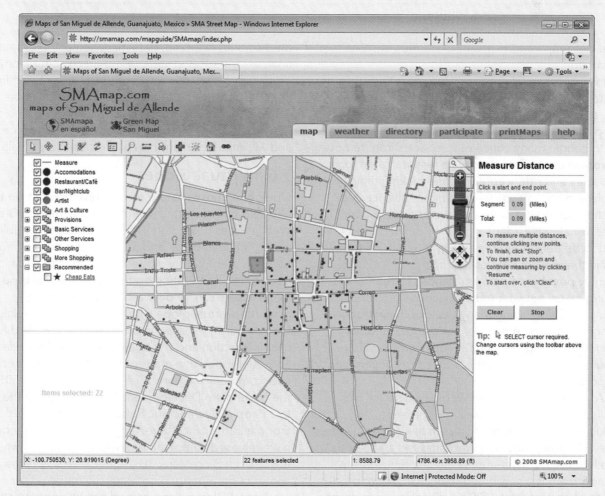

Figure 7.16 Screenshot of Autodesk MapGuide—Server GIS (www.smamap.com/mapguide/SMAmap/index.php).

There are three key components of MapGuide: the viewer—a relatively easy-to-use Web application with a browser-style interface; the author—a menu-driven authoring environment used to create and publish a site for client access; and the server—the administrative software that monitors site usage and manages requests from multiple clients and to external databases. MapGuide works directly with Web browsers and servers, and uses the HTTP (Internet protocol) for communication. It makes good use of standard Internet tools like HTML (hypertext markup language) and JavaScript for building client applications. Typical features of MapGuide sites include the display of raster and vector maps, map navigation (pan and zoom), geographic and attribute queries, buffering, report generation, and printing. Like other advanced Web-based Server GIS, MapGuide has tools for redlining (drawing on maps) and basic editing of geographic objects.

To date, MapGuide has been used most widely in existing mature GIS sites that want to publish their data internally or externally, and in new sites that want a way to publish dynamic maps quickly to a widely dispersed collection of users (for example, maps showing election results or transportation network status). Autodesk MapGuide can be used to serve maps (using the OGC Web Mapping Service (WMS) protocol) and features (using the OGC Web Feature Service (WFS) protocol). Configured as a mapping service, Autodesk MapGuide supports a client/server environment. It can retrieve geospatial data from WFS and WMS sites, enabling the use of data from other organizations that share their geospatial data. As a further service, Autodesk MapGuide allows organizations to share their data, in vector form, with authorized outside organizations.

to render maps and respond to queries for data download.

Initially, server GIS were nothing more than ports of desktop GIS products, but second-generation systems were subsequently built using a multiuser services-based architecture that allows them to run unattended and to handle many concurrent requests from remote networked users. These software systems often focus on display and query applications—making things simple and cost-effective—with more advanced applications being available at a higher price. Server GIS are especially well adapted for well-defined tasks that need to be performed repeatedly (ad hoc GIS tasks remain the preserve of desktop systems). Today, it is routinely possible to perform standard operations like making maps, network routing, geocoding, simple editing, and publishing many types of thematic and topographic data. A number of Web GIS services are indexed using geoportals—gateways to GIS resources—for example, the U.S. federal government geoportal geodata.gov.

A second generation of Web-based server GIS products is becoming increasingly prevalent. They exploit the unique characteristics of the Web and integrate GIS technology with Web browsers and servers. Initially, these new systems had limited functionality, but now there is a new breed of true GIS server products that offer complete GIS functionality in a multiuser, Web-based server environment (e.g., ArcGIS Server). These server GIS products have functions for distrib-

uted editing, mapping, data management, and spatial analysis, and support state of-the-art customization.

In conclusion, server GIS products are growing in importance. Their cost-effective nature, ability to be centrally managed, and focus on ease of use will help to disseminate geographic information even more widely and will introduce many new users to the field of GIS.

7.6.4 Virtual Globes

One of the most exciting recent developments in the GIS field has been the advent of virtual globes, 3-D Web services hosted on Web-based GIS that publish global 3-D databases and associated services for use over the Web. Virtual globes allow users to visualize geographic information on top of 3-D global base maps. Users can fly over a comparatively high-resolution virtual globe with thematic data overlaid. Google Earth, released in 2005, set the standard for user interaction and the quality and quantity of data, and established in a matter of months an engaged user community. Figure 7.17 shows information about the Millennium Development Goals (MDG) for the Lao People's Democratic Republic as represented in Google Earth. Since then, a number of other vendors have released comparable globes, not least of which is Microsoft Bing Maps 3D. Figure 7.18 shows the center of Paris in the Bing Maps 3D mode.

Virtual Globes have gained considerable traction in the wider GIS community because, comparatively-speaking, they are low cost (basic versions are free),

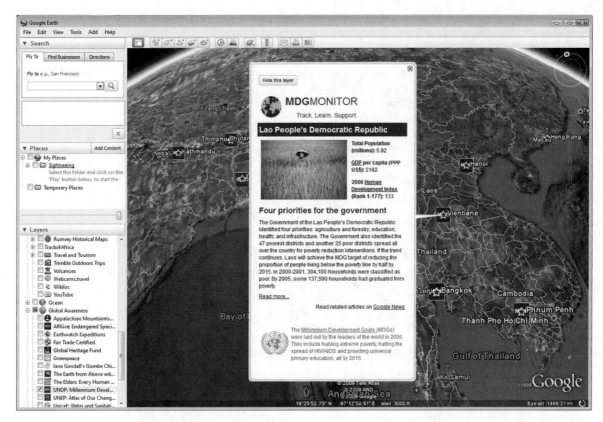

Figure 7.17 Google Earth (earth.google.com).

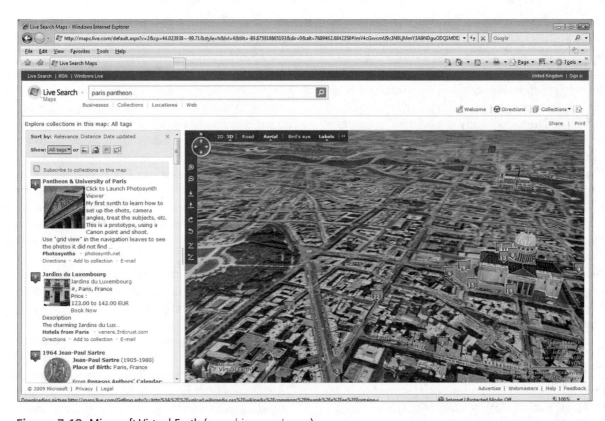

Figure 7.18 Microsoft Virtual Earth (www.bing.com/maps).

Figure 7.19 A Web application written using the ILOG JViews Map component toolkit.

they have high-quality image and vector databases for the Earth, it is simple to overlay user data on the globe, and they are easy to use. They have led to whole new subfields called neogeography (see Section 1.8) and volunteered geographic information (see Sections 1.8 and 11.2). Neogeography is the "new" geography that among other things includes the overlay or mashing up two or more sources of geographic information (for example webcams from Caltrans [California Department of Transportation] on top of a Yahoo base map). Volunteered geographic information focuses on the fact that humans are acting as sensors and are building and publishing content from the ground up, often using virtual globes as base maps. The non-authoritative and sometimes transient and dynamic nature of this information provides new geographic challenges and opportunities. These issues are explored in more depth in Chapter 11.

7.6.5 Developer GIS

With the advent of component-based software development (see Section 7.4), a number of GIS vendors have released collections of GIS software components

oriented toward the needs of developers. These are really tool kits of GIS functions (components) that a reasonably knowledgeable programmer can use to build a specific-purpose GIS application. They are of interest to developers because such components can be used to create highly customized and optimized applications that can either stand alone or can be embedded within other software systems. Typically, component GIS packages offer strong display and query capabilities, but only limited editing and analysis tools, mainly because there is greatest demand for products with such data exploration and visualization functionality.

Developer GIS products are collections of components used by developers to create focused applications.

Examples of component GIS products include ESRI ArcGIS Engine and Pitney Bowes MapInfo MapX. Most of the developer GIS products from mainstream vendors are built on top of Microsoft's .Net technology standards, but there are several cross-platform choices (e.g., ESRI ArcGIS Engine) and several Java-based toolkits (e.g., ObjectFX SpatialFX and ILOG JViews Maps—see Figure 7.19). The typical cost for a commercial developer GIS product is $1,000–5,000 for the developer kit and $100–$500 per deployed application. The people who use deployed applications may not even realize that they are using a GIS, because often the run-time deployment is embedded in other applications (e.g., customer care systems, routing systems, or interactive atlases).

There are a number of open-source developer GIS toolkits of which GeoTools and GDAL are two of the most widely used. GeoTools is a Java-based code library that provides OGC standards-compliant methods for the manipulation of geospatial data. It has been used in a number of GIS including GeoServer and is a key part of the GeoAPI set of tools for building GIS. GDAL/ODR is a software library for reading and writing raster and vector data formats. It is widely used in commercial and open-source GIS products.

7.6.6 Hand-held GIS

As hardware design and miniaturization have progressed dramatically over the past few years, so it has become possible to develop GIS software for mobile and personal use on hand-held systems. The development of low-cost, lightweight location positioning technologies (primarily based on the Global Positioning System; see Section 5.9) and wireless networking has further stimulated this market. With capabilities similar to the desktop systems of just a few years ago, these palm and pocket devices can support many display, query, and simple analytical applications, even on displays as small as 320 by 240 pixels (a quarter of the VGA (640 by 480) pixel screen resolution standard). An interesting characteristic of these systems is that all programs and data are held in local memory because of the lack of a hard disk. This provides fast access, but because of the cost of memory compared to disk systems, designers have had to develop compact data storage structures. ESRI's ArcPad product is one of a number of products in this space (Figure 7.20). It specializes in field data collection and mobile mapping. When linked to ESRI ArcGIS Server, it can act as an enterprise field client.

Hand-held GIS are lightweight systems designed for mobile and field use.

One of the most innovative areas at present is the development of hand-held software for high-end so-called smartphones. In spite of their compact size,

Figure 7.20 ESRI ArcPad running on a rugged hand-held field PC device.

they can deal with comparatively large amounts of data (8 GB and more) and surprisingly sophisticated software applications. The systems usually operate in a mixed connected/disconnected environment and so can make active use of data and software applications held on the server (see earlier discussion of server GIS in Section 7.6.3) and accessed over a wireless telephone network. Apple's iPhone, for example, provides access to Google Maps over a wireless connection (Figure 7.21).

Figure 7.21 The Apple iPhone showing Google Maps—an example of a hand-held GIS. (Courtesy Alex Singleton)

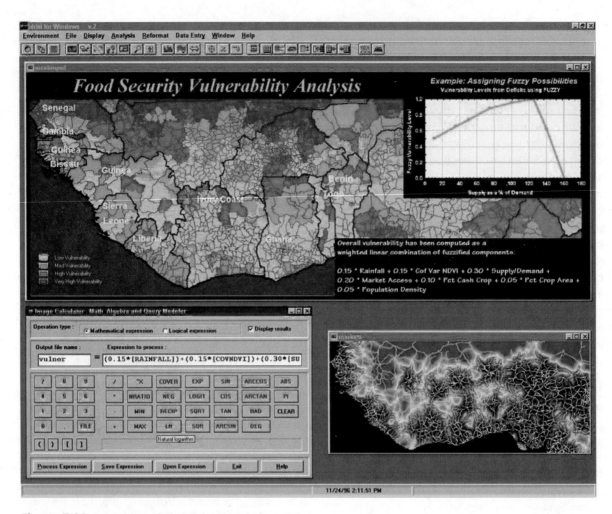

Figure 7.22 Screenshot of Clark Labs Idrisi desktop GIS.

7.6.7 Other Types of GIS Software

The previous section has focused on mainstream GIS software from the major commercial vendors. There are many other types of commercial and noncommercial software that provide valuable GIS capabilities. This section briefly reviews some of the main types of other GIS software. Raster-based GIS, as the name suggests, focus primarily on raster (image) data and raster analysis. Chapters 3 and 8 provide a discussion of the principles and techniques associated with raster and other data models, while Chapters 14 and 15 review their specific capabilities. Just as many vector-based systems have raster analysis extensions (for example, ESRI ArcGIS has Spatial Analyst, and Intergraph GeoMedia has Image and Grid), in recent years raster systems have added vector capabilities (for example, ERDAS Imagine and Clark Labs Idrisi now have vector capabilities built in (see Figure 7.22). As a result, the distinction between raster-based and other

software system categories is becoming increasingly blurred. The users of raster-based GIS are primarily interested in working with imagery and undertaking spatial analysis and modeling activities. The prices for raster-based GIS range from $500 to $10,000.

CAD (or Computer-Aided Design)-based GIS are systems that started life as CAD packages and then had GIS capabilities added on top. Typically, a CAD system is supplemented with database, spatial analysis, and cartography capabilities. Not surprisingly, these systems appeal mainly to users whose primary focus is in typical CAD application areas such as architecture, engineering, and construction, but who also want to use geographic information and geographic analysis in their projects. These systems are often used in data collection and mapping applications. The best known examples of CAD-based GIS are Autodesk Map 3D and Bentley Map (see Section 7.5). CAD-based GIS typically cost $3,000 to $5,000.

Many enterprise-wide GIS incorporate middleware (middle tier) GIS data and application servers. Their purpose is to manage multiple users accessing continuous geographic databases, which are stored and managed in commercial off-the-shelf (COTS) database management systems (DBMS). GIS middleware products offer centralized management of data, the ability to process data on a server (which delivers good performance for certain types of applications), and control over database editing and update (see Chapter 11 for further details). A number of GIS vendors have developed technology that fulfills this function. Examples of GIS application servers include Autodesk GIS Design Server, ESRI ArcGIS Server, and MapInfo SpatialWare. These systems typically cost $10,000 to $25,000 or more, depending on the number of users.

To assist in managing data in standard DBMS, some vendors—notably IBM, Microsoft, and Oracle—have developed technology to extend their DBMS servers so that they are able to store and process geographic information efficiently. Although not strictly GIS in their own right (because of the absence of editing, mapping, and analysis tools), they are included here for completeness. Box 10.1 provides an overview of Oracle's Spatial DBMS extension.

Some noteworthy examples of open-source software systems not already mentioned in earlier sections include GeoDA for spatial analysis and visualization (www.csiss.org/clearinghouse/GeoDa), gvSIG for desktop mapping and GIS (www.gvsig.gva.es/), and PostGIS for storing data in a DBMS (postgis. refractions.net).

The Open Geospatial Consortium (www. opengeospatial.org), although not itself a software vendor or software-development organization, has played a very important role in the evolution of GIS software in the last decade or so. OGC's role has been to encourage the development of standards that facilitate the sharing of geographic information and the interaction of geographic software. Perhaps the most significant progress has been in the area of interoperability of Web services, and a series of OGC standards now allow images, features, coverages, and metadata to be integrated together over the Web.

7.7 Conclusion

GIS software is a fundamental and critical part of any operational GIS. The software employed in a GIS project has a controlling impact on the type of studies that can be undertaken and the results that can be obtained. There are also far-reaching implications for user productivity and project costs. Today, there are many types of GIS software product to choose from and a number of ways to configure implementations. One of the exciting and at times unnerving characteristics of GIS software is its very rapid rate of development, not least in the areas of Web and open-source GIS. This trend seems set to continue as software developers push ahead with significant research and development efforts. The following chapters explore in more detail the functionality of GIS software and how it can be applied in real-world contexts.

Questions for Further Study

1. Design a GIS architecture that 25 users in three cities could use to create an inventory of recreation facilities.

2. Discuss the role of each of the three tiers of software architecture in an enterprise GIS implementation.

3. With reference to a large organization that is familiar to you, describe the ways in which its staff might use GIS, and evaluate the different types of GIS software systems that might be implemented to fulfill these needs.

4. Go to the Web sites of the main GIS software vendors and compare their product strategies with open-source GIS products. In what ways are they different?
   - Autodesk: www.autodesk.com
   - Bentley: www.bentley.com
   - ESRI: www.esri.com
   - Intergraph: www.intergraph.com

Further Reading

Coleman, D.J. 2005. GIS in networked environments. In Longley, P.A., Goodchild, M.F., Maguire, D.J., and Rhind, D.W. (eds.). *Geographical Information Systems: Principles, Techniques, Management and Applications (abridged edition)*. Hoboken, NJ: Wiley, 317–329.

Elshaw, Thrall S. and Thrall, G.I. 2005. Desktop GIS software. In Longley, P.A., Goodchild, M.F., Maguire, D.J., and Rhind, D.W. (eds.). *Geographical Information Systems: Principles, Techniques, Management and Applications (abridged edition)*. Hoboken, NJ: Wiley, 331–345.

Maguire, D.J. 2005. GIS customization. In Longley, P.A., Goodchild, M.F., Maguire, D.J., and Rhind, D.W. (eds.). *Geographical Information Systems: Principles, Techniques, Management and Applications (abridged edition)*. Hoboken, NJ: Wiley, 359–369.

Peng, Z-H. and Tsou, M-H. 2003. *Internet GIS: Distributed Geographic Information Services for the Internet and Wireless Networks*. Hoboken, NJ: Wiley.

Steiniger, S., and Bocher, E. 2009 An overview of current free and open source desktop GIS developments (www.geo.unizh.ch/~sstein).

8 Geographic Data Modeling

This chapter discusses the technical issues involved in modeling the real world in a GIS. It describes the process of data modeling and the various data models that have been used in GIS. A data model is a set of constructs for describing and representing parts of the real world in a digital computer system. Data models are vitally important to GIS because they control the way that data are stored and have a major impact on the type of analytical operations that can be performed. Early GIS were based on extended CAD, simple graphical, and image data models. In the 1980s and 1990s, the hybrid georelational model came to dominate GIS. In the last decade, major software systems have been developed on more advanced and standards-based geographic object models that include elements of all earlier models.

LEARNING OBJECTIVES

After studying this chapter you will be able to:

- Define what geographic data models are and discuss their importance in GIS.

- Understand how to undertake GIS data modeling.

- Outline the main geographic models used in GIS and their strengths and weaknesses.

- Understand key topology concepts and why topology is useful for data validation, analysis, and editing.

- Read data model notation.

- Describe how to model the world and create a useful geographic database.

8.1 Introduction

This chapter builds on the material on geographic representation presented in Chapter 3. By way of introduction, it should be noted that the terms *representation* and *model* overlap considerably (we will return to a more detailed discussion of models in Chapter 16). Representation is typically used in conceptual and scientific discussions, whereas model is used in practical and database circles and is preferred in this chapter. The focus here is on how geographic reality is modeled (abstracted or simplified) in GIS, with particular emphasis on the different types of data models that have been developed. A data model is an essential ingredient of any operational GIS and, as the discussion will show, has important implications for the types of operations that can be performed and the results that can be obtained.

8.1.1 Data Model Overview

The heart of any GIS is the data model, which is a set of constructs for representing objects and processes in the digital environment of the computer (Figure 8.1). People (GIS users) interact with operational GIS in order to undertake tasks such as making maps, querying databases, and performing site suitability analyses. Because the types of analyses that can be undertaken are strongly influenced by the way the real world is modeled, decisions about the type of data model to be adopted are vital to the success of a GIS project.

A data model is a set of constructs for describing and representing selected aspects of the real world in a computer.

Figure 8.1 The role of a data model in GIS.

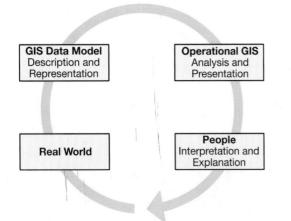

As described in Chapter 3, geographic reality is continuous and of seemingly infinite complexity, but computers are finite, comparatively simple, and can only work with digital data. Therefore, difficult choices have to be made about what things are modeled in a GIS and how they are represented. Because different types of people use GIS for different purposes, and the phenomena these people study have different characteristics, there is no single type of all-encompassing GIS data model that is best for all circumstances.

8.1.2 Levels of Data Model Abstraction

When representing the real world in a computer, it is helpful to think in terms of four different levels of abstraction (levels of generalization or simplification); these are shown in Figure 8.2. First, *reality* is made up of real-world phenomena (buildings, streets, wells, lakes, people, etc.) and includes all aspects that may or may not be perceived by individuals, or deemed relevant to a particular application. Second, the *conceptual model* is a human-oriented, often partially structured, model of selected objects and processes that are thought relevant to a particular problem domain. Third, the *logical model* is an implementation-oriented representation of reality that is often expressed in the form of diagrams and lists. Lastly, the *physical model* portrays the actual implementation in a GIS and often comprises tables stored as files or databases (see Chapter 10). Use of the term *physical* here is actually misleading because the models are not physical and only exist digitally in computers, but this is the generally accepted use of the term. Relating back to the discussion of uncertainty in Chapter 6 (Figure 6.1), the conceptual and logical models are found beyond the U1 filter, and the physical model is that on which analysis may be performed (beyond the U2 filter).

In data modeling, users and system developers participate in a process that successively engages with each of these levels. The first phase of modeling

Figure 8.2 Levels of GIS data model abstraction.

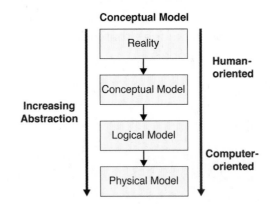

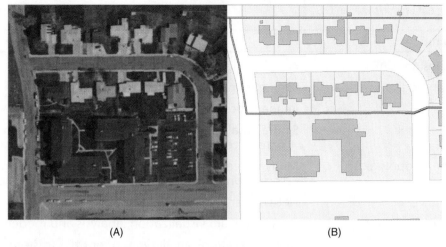

(A)   (B)

Figure 8.3 Different representational models of the same area in Colorado, USA: (A) aerial photograph; and (B) vector objects, some digitized from the photograph.

begins with the definition of the main types of objects to be represented in the GIS and concludes with a conceptual description of the main types of objects and relationships between them. Once this phase is complete, further work will lead to the creation of diagrams and lists describing the names of objects, their behavior, and the type of interaction between objects. This type of logical data model is very valuable for defining what a GIS will be able to do and the type of domain over which it will extend. Logical models are implementation independent, and can be created in any GIS with appropriate capabilities. The final data modeling phase involves creating a model showing how the objects under study can be digitally implemented in a GIS. Physical models describe the exact files or database tables used to store data, the relationships between object types, and the precise operations that can be performed. For more details about the practical steps involved in data modeling, see Sections 8.3 and 8.4.

A data model provides system developers and users with a common understanding and reference point. For developers, a data model is the means to represent an application domain in terms that may be translated into a design and then implemented in a system. For users, it provides a description of the structure of the system, independent of specific items of data or details of the particular application. A data model controls the types of things that a GIS can handle and the range of operations that can be performed on them.

The discussion of geographic representation in Chapter 3 introduced discrete objects and fields, the two fundamental conceptual models for representing real-world things geographically. In the same chapter the raster and vector logical models were also introduced. Figure 8.3 shows two representations of the same area in a GIS, one raster and the other vector. Notice the difference in the objects represented. The roads and buildings can be clearly seen on both figures, but the vector representation (Figure 8.3B) also shows the lot (property) boundaries and water mains and valves. Cars and other surface texture ephemera can be observed on the raster representation (Figure 8.3A). The next sections in this chapter focus on the logical and physical representation of raster, vector, and related models in GIS software systems.

8.2 GIS Data Models

In the past half-century, many GIS data models have been developed and deployed in GIS software systems. The key types of geographic data models and their main areas of application are listed in Table 8.1. All are based in some way on the conceptual discrete object/field and logical vector/raster geographic data models. All GIS software systems include a core data model that is built on one or more of these GIS data models. In practice, any modern comprehensive GIS supports at least some elements of all these models. As discussed earlier, the GIS software core system data model is the means to represent geographic aspects of the real world and defines the type of geographic operations that can be performed. It is the responsibility of the GIS implementation team to populate this generic model with information about a particular problem (e.g., utility outage management, military mapping, or natural resource planning). Some GIS software packages come with a fixed data model, while others have models that can be easily extended by adding new object types and relationships. Those that can easily be extended are better able to model

Table 8.1 Geographic data models used in GIS.

Data model	Example application
Computer-aided design (CAD)	Automating engineering design and drafting
Graphical (non–topological)	Simple mapping and graphic arts
Image	Image processing and simple grid analysis
Raster/grid	Spatial analysis and modeling, especially in environmental and natural resources applications
Vector/georelational topological	Many operations on geometric features in cartography, socioeconomic, and resource analysis and modeling
Network	Network analysis in transportation and utilities
Triangulated irregular network (TIN)	Surface/terrain visualization, analysis and modeling
Object	Many operations on all types of entities (raster/vector/TIN, etc.) in all types of applications

the richness of geographic domains, and in general are the easiest to use and the most productive systems.

When modeling the real world for representation inside a GIS, it is convenient to group entities of the same geometric type together (for example, all point entities such as lights, garbage cans, dumpsters, etc., might be stored together). A collection of entities of the same geometric type (dimensionality) is referred to as a class or layer. It should also be noted that the term *layer* is quite widely used in GIS as a general term for a specific dataset. It derived from the process of entering different types of data into a GIS from paper maps, which was undertaken one plate at a time. (All entities of the same type were represented in the same color and, using printing technology, were reproduced together on film or printing plates.) Grouping entities of the same geographic type together makes the storage of geographic databases more efficient (for further discussion, see Section 10.3). It also makes it much easier to implement rules for validating edit operations (for example, the addition of a new building or census administrative area) and for building relationships between entities. All of the data models discussed in this chapter use layers in some way to handle geographic entities.

A layer is a collection of geographic entities of the same geometric type (e.g., points, lines, or polygons). Grouped layers may combine layers of different geometric types.

8.2.1 CAD, Graphical, and Image GIS Data Models

The earliest GIS were based on very simple data models derived from work in the fields of CAD (computer-

aided design and drafting), computer cartography, and image analysis. In a CAD system, real-world entities are represented symbolically as simple point, line, and area vectors. This basic CAD data model never became widely popular in GIS because of three severe problems for most applications at geographic scales. First, because CAD models typically use local drawing coordinates instead of real-world coordinates for representing objects, they are of little use for map-centric applications. Second, because individual objects do not have unique identifiers, it is difficult to tag them with attributes. As the following discussion shows, this is a key requirement for GIS applications. Third, because CAD data models focus on graphical representation of objects, they cannot store details of any relationships between objects (e.g., topology or networks), yet this type of information is essential in many spatial analytical operations.

A second type of simple GIS geometry model was derived from work in the field of computer cartography. The main requirement for this field in the 1960s was the automated reproduction of paper topographic maps and the creation of simple thematic maps. Techniques were developed to digitize maps and store them in a computer for subsequent plotting and printing. All paper map entities were stored as points, lines, and areas, with annotation used for place-names. Like CAD systems, there was no requirement to tag objects with attributes or to work with object relationships.

At about the same time that CAD and computer cartography systems were being developed, a third type of data model emerged in the field of image processing. Because the main data sources for geographic image processing are scanned aerial photographs and digital satellite images, it was natural that

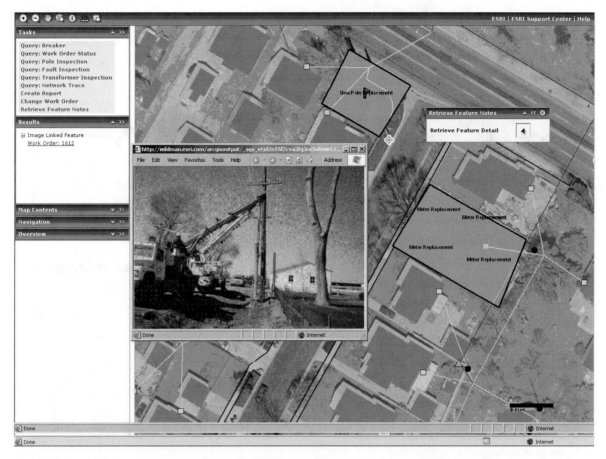

Figure 8.4 A photographic image used as a building object attribute in an electric facility system.

these systems would use rasters or grids to represent the patterning of real-world objects on the Earth's surface. The image data model is also well suited to working with pictures of real-world objects, such as photographs of water valves and scanned building floor plans that are held as attributes of geographically referenced entities in a database (Figure 8.4).

In spite of their many limitations, GIS still exist based on these simple data models, although the number is in significant decline. This is partly for historical reasons—the GIS may have been built before newer, more advanced models became available—but also because of lack of knowledge about the newer approaches described in this chapter.

8.2.2 Raster Data Model

The raster data model uses an array of cells, or pixels, to represent real-world objects (see Figure 3.9).

The cells can hold attribute values based on one of several encoding schemes, including categories, and integer and floating-point numbers (see Box 3.1 for details). In the simplest case a binary representation is used (for example, presence or absence of vegetation), but in more advanced cases floating-point values are preferred (for example, height of terrain above sea level in meters). In some systems, multiple attributes can be stored for each cell in a type of value attribute table where each column is an attribute and each row either a pixel or a pixel class (Figure 8.5).

Raster data are usually stored as an array of grid values, with metadata (data about data: see Section 11.2.1) about the array held in a file header. Typical metadata include the geographic coordinate of the upper-left corner of the grid, the cell size, the number of row and column elements, and the projection. The data array itself is usually stored as a compressed file or record in a database management system (see Section 10.3). Techniques for compressing rasters are described in Box 8.1.

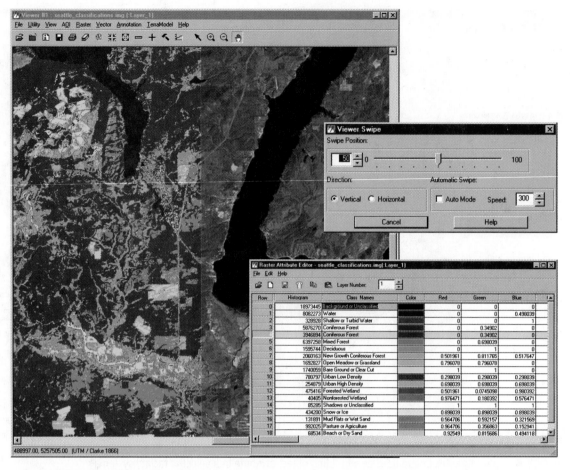

Figure 8.5 Raster data of the Olympic Peninsula, Washington State, with associated value attribute table. Bands 4,3,2 from Landsat 5 satellite with land cover classification overlaid. (Screenshot courtesy ERDAS Inc.; data courtesy USGS)

Technical Box (8.1)

Technical Raster Compression Techniques

Although the raster data model has many uses in GIS, one of the main operational problems associated with it is the sheer amount of raw data that must be stored. To improve storage efficiency, many types of raster compression technique have been developed such as run-length encoding, block encoding, wavelet compression, and quadtrees (see Section 10.7.2.2 for another use of quadtrees as a means of indexing geographic data). Table 8.2 presents a comparison of file sizes and compression rates for three compression techniques based on the image in Figure 8.6. It can be seen that even the comparatively simple run-length encoding technique compresses the file size by a factor of 5. The more sophisticated wavelet compression technique results in a compression rate of almost 40, reducing the file from 80.5 to 2.3 MB.

Table 8.2 Comparison of file sizes and compression rates for selected raster compression techniques (using image shown in Figure 8.6).

Compression technique	File size (MB)	Compression rate
Uncompressed original	80.5	–
Run-length	17.7	5.1
Wavelet	2.3	38.3

Run-length Encoding

Run-length encoding is perhaps the simplest compression method and is very widely used. It involves encoding adjacent row cells that have the same value, with a pair of values indicating the number of cells with the same value, and the actual value.

Block Encoding

Block encoding is a two-dimensional version of run-length encoding in which areas of common cell values are represented with a single value. An array is defined as a series of square blocks of the largest size possible. Recursively, the array is divided using blocks of smaller and smaller size. It is sometimes described as a quadtree data structure (see also Section 10.7.2.2).

Wavelet

Wavelet compression techniques invoke principles similar to those discussed in the treatment of fractals (see Section 4.7). They remove information by recursively examining patterns in datasets at different scales, always trying to reproduce a faithful representation of the original. A useful byproduct of this for geographic applications is that wavelet compressed raster layers can be quickly viewed at different scales with appropriate amounts of detail. The JPEG 2000 standard defines wavelet image compression.

Run-length and block encoding both result in lossless compression of raster layers; that is, a layer can be compressed and decompressed without degradation of information. In contrast, the wavelet compression technique is *lossy* since information is irrevocably discarded during compression. Although wavelet compression results in very high compression ratios, because information is lost its use is limited to applications that do not need to use the raw digital numbers for processing or analysis. It is not appropriate for compressing DEMs, for example, but many organizations use it to compress scanned maps and aerial photographs when access to the original data is not necessary.

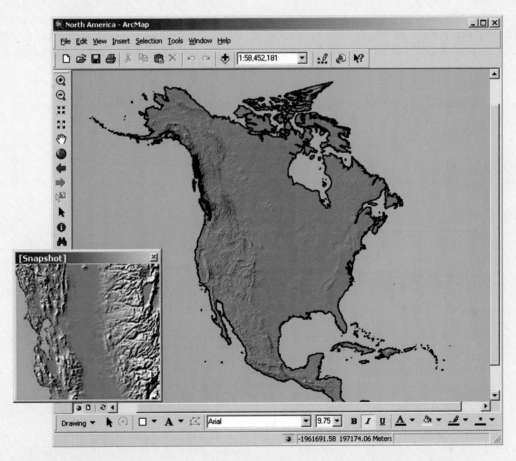

Figure 8.6 Shaded digital elevation model of North America used for comparison of image compression techniques in Table 8.2. Original image is 8,726 by 10,618 pixels, 8 bits per pixel. The inset shows part of the image at a zoom factor of 1,000 for the San Francisco Bay area.

Datasets encoded using the raster data model are particularly useful as a backdrop map display because they look like conventional maps and can communicate a lot of information quickly. They are also widely used for analytical applications such as disease dispersion modeling, surface-water flow analysis, and store location modeling.

8.2.3 Vector Data Model

The raster data model discussed earlier is most commonly associated with the field conceptual data model. The vector data model on the other hand is closely linked with the discrete object view. Both of these conceptual perspectives were introduced in Section 3.5. The vector data model is used in GIS because of the precise nature of its representation method, its storage efficiency, the quality of its cartographic output, and the wide availability of functional tools for operations like map projection, overlay processing, and cartographic analysis.

In the vector data model, each object in the real world is first classified into a geometric type: in the 2-D case point, line, or polygon (Figure 8.7). Points (e.g., wells, soil pits, and retail stores) are recoded as single coordinate pairs, lines (e.g., roads, streams, and geologic faults) as a series of ordered coordinate pairs (also called polylines—Section 3.6.2), and polygons (e.g., census tracts, soil areas, and oil license zones) as one or more line segments that close to form a polygon area. The coordinates that define the geometry of each object may have 2, 3, or 4 dimensions: 2 (x, y: row and column, or latitude and longitude), 3 (x, y, z: the addition of a height value), or 4 (x, y, z, m: the addition of another value to represent time or some other property—perhaps the offset of road signs from a road centerline, or an attribute).

For completeness, it should also be said that in some data models linear features can be represented not only as a series of ordered coordinates, but also as curves defined by a mathematical function (e.g., a spline or Bézier curve). These are particularly useful for representing built environment entities like road curbs and some buildings.

8.2.3.1 Simple Features

Geographic entities encoded using the vector data model are usually called features, and this will be the convention adopted here. Features of the same geometric type are stored in a geographic database as a feature class, or when speaking about the physical (database) representation the term *feature table* is preferred. Here each feature occupies a row, and

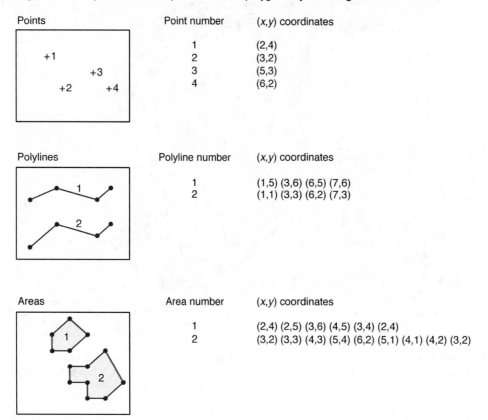

Figure 8.7 Representation of point, line, and polygon objects using the vector data model.

Points	Point number	(x,y) coordinates
	1	(2,4)
	2	(3,2)
	3	(5,3)
	4	(6,2)

Polylines	Polyline number	(x,y) coordinates
	1	(1,5) (3,6) (6,5) (7,6)
	2	(1,1) (3,3) (6,2) (7,3)

Areas	Area number	(x,y) coordinates
	1	(2,4) (2,5) (3,6) (4,5) (3,4) (2,4)
	2	(3,2) (3,3) (4,3) (5,4) (6,2) (5,1) (4,1) (4,2) (3,2)

each property of the feature occupies a column. GIS commonly deal with two types of feature: simple and topological. The structure of simple feature polyline and polygon datasets is sometimes called *spaghetti* because, like a plate of cooked spaghetti, lines (strands of spaghetti) and polygons (spaghetti hoops) can overlap and there are no relationships between any of the objects.

Features are vector objects of type point, polyline, or polygon.

Simple feature datasets are useful in GIS applications because they are easy to create and store, and because they can be retrieved and rendered on screen very quickly. However, because simple features lack more advanced data structure characteristics, such as topology (see next section), operations such as shortest-path network analysis and polygon adjacency cannot be performed without additional calculations.

8.2.3.2 Topological Features

Topological features are essentially simple features structured using topological rules. Topology is the mathematics and science of geometrical relationships. Topological relationships are nonmetric (qualitative) properties of geographic objects that remain constant when the geographic space of objects is distorted. For example, when a map is stretched, properties such as distance and angle change, whereas topological properties such as adjacency and containment do not change. Topological structuring of vector layers introduces some interesting and very useful properties, especially for polyline (also called 1-cell, arc, edge, line, and link) and polygon (also called 2-cell, area, and face) data. Topological structuring of line layers forces all line ends that are within a user-defined distance to be snapped together so that they are given exactly the same coordinate value. A node is placed wherever the ends of lines meet or cross. Following on from the earlier analogy, this type of data model is sometimes referred to as spaghetti with meatballs (the nodes being the meatballs on the spaghetti lines). Topology is important in GIS because of its role in data validation, modeling integrated feature behavior, editing, and query optimization.

Topology is the science and mathematics of relationships used to validate the geometry of vector entities, and for operations such as network tracing and tests of polygon adjacency.

Data validation   Much of the geographic data collected from basic digitizing, field data collection devices, photogrammetric workstations, and CAD systems comprise simple features of type point, polyline, or polygon, with limited structural intelligence (for example, no topology). Testing the topological integrity of a dataset is a useful way to validate the geometric quality of the data and to assess its suitability for geographic analysis. Some useful data validation topology tests include the following:

- Network connectivity—do all network elements connect to form a graph (i.e., are all the pipes in a water network joined to form a wastewater system)? Network elements that connect must be "snapped" together (that is, given the same coordinate value) at junctions (intersections).

- Line intersection—are there junctions at intersecting polylines, but not at crossing polylines? It is, for example, perfectly possible for roads to cross in planimetric 2-D view, but not intersect in 3-D (for example, at a bridge or underpass).

- Overlap—do adjacent polygons overlap? In many applications (e.g., land ownership), it is important to build databases free from overlaps and gaps so that ownership is unambiguous.

- Duplicate lines—are there multiple copies of network elements or polygons? Duplicate polylines often occur during data capture. During the topological creation process, it is necessary to detect and remove duplicate polylines to ensure that topology can be built for a dataset.

Modeling the integrated behavior of different feature types   In the real world, many objects share common locations and partial identities. For example, water distribution areas often coincide with water catchments, electric distribution areas often share common boundaries with building subdivisions, and multiple telecommunications fibers are often run down the same conduit. These situations can be modeled in a GIS database as either single objects with multiple geometry representations, or multiple objects with separate geometry integrated for editing, analysis, and representation. There are advantages and disadvantages to both approaches. Multiple objects with separate geometries are certainly easier to implement in commercially available databases and information systems. If one feature is moved during editing, then logically both features should move. This is achieved by storing both objects separately in the database, each with its own geometry, but integrating them inside the GIS editor application so that they are treated as single features. When the geometry of one is moved, the other automatically moves with it.

Editing productivity Topology improves editor productivity by simplifying the editing process and providing additional capabilities to manipulate feature geometries. Editing requires both topological data structuring and a set of topologically aware tools. The productivity of editors can be improved in several ways:

- Topology provides editors with the ability to manipulate common, shared polylines and nodes as single geometric objects to ensure that no differences are introduced into the common geometries.
- Rubberbanding is the process of moving a node, polyline, or polygon boundary and receiving interactive feedback on-screen about the location of all topologically connected geometry.
- Snapping (e.g., forcing the end of two polyline features to have the same coordinate) is a useful technique to both speed up editing and maintain a high standard of data quality.
- Auto-closure is the process of completing a polygon by snapping the last point to the first digitized point.
- Tracing is a type of network analysis technique that is used, especially in utility applications, to test the connectivity of linear features (e.g., is the newly designed service going to receive power?).

Optimizing queries Many GIS queries can be optimized by precomputing and storing information about topological relationships. Some common examples include:

- Network tracing (e.g., find all connected water pipes and fittings).
- Polygon adjacency (e.g., determine who owns the parcels adjoining those owned by a specific owner).
- Containment (e.g., find out which manholes lie within the pavement area of a given street).
- Intersection (e.g., examine which census tracts intersect with a set of health areas).

The remainder of this section focuses on a conceptual understanding of GIS topology, emphasizing the polygon case. The next section considers the network case; the relative merits and implementations of two approaches to GIS topology are discussed in Section 10.7.1. These two implementation approaches differ because in one case relationships are batch built and stored along with feature geometry, and in the other relationships are calculated interactively when they are needed.

Conceptually speaking, in a topologically structured polygon data layer each polygon is defined as a collection of polylines that in turn are made up of an ordered list of coordinates (vertices). Figure 8.8 shows an example of a polygon dataset comprising six polygons (including the "outside world": polygon 1). A number in a circle identifies a polygon. The lines that make up the polygons are shown in the polygon-polyline list. For example, polygon 2 can be assembled

Figure 8.8   A topologically structured polygon data layer. The polygons are made up of the polylines shown in the area polyline list. The lines are made up of the coordinates shown in the line coordinate list. (Source: after ESRI 1997)

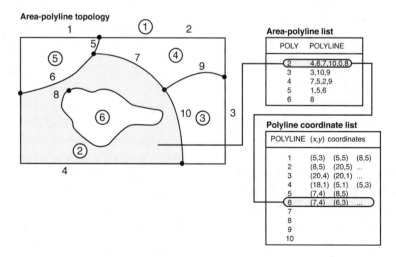

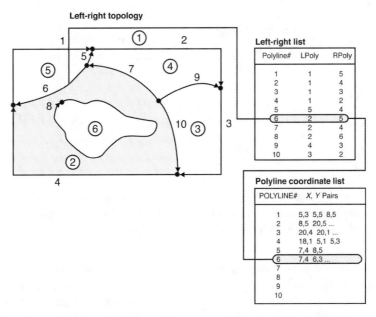

Figure 8.9 The contiguity of a topologically structured polygon data layer. For each polyline the left and right polygons are stored with the geometry data. (Source: after ESRI 1997)

from lines 4, 6, 7, 10, and 8. In this particular implementation example, the 0 before the 8 is used to indicate that line 8 actually defines an "island" inside polygon 2. The list of coordinates for each line is also shown in Figure 8.8. For example, line 6 begins and ends with coordinates 7,4 and 6,3—other coordinates have been omitted for brevity. A line may appear in the polygon-polyline list more than once (for example, line 6 is used in the definition of both polygons 2 and 5), but the actual coordinates for each polyline are only stored once in the polyline-coordinate list. Storing common boundaries between adjacent polygons avoids the potential problems of gaps (slivers) or overlaps between adjacent polygons. It has the added bonus that there are fewer coordinates in a topologically structured polygon feature layer compared with a simple feature layer representation of the same entities. The downside, however, is that drawing a polygon requires that multiple polylines must be retrieved from the database and then assembled into a boundary. This process can be time consuming when repeated for each polygon in a large dataset.

Planar enforcement is a very important property of topologically structured polygons. In simple terms, planar enforcement means that all the space on a map must be filled and that any point must fall in one polygon alone; that is, polygons must not overlap. Planar enforcement implies that the phenomenon being represented is conceptualized as a field.

The contiguity (adjacency) relationship between polygons is also defined during the process of topologi-

cal structuring. This information is used to define the polygons on the left- and right-hand side of each polyline, in the direction defined by the list of coordinates (Figure 8.9). In Figure 8.9, Polygon 2 is on the left of Polyline 6 and Polygon 5 is on the right (polygon identifiers are in circles). Thus from a simple look-up operation, we can deduce that Polygons 2 and 5 are adjacent.

Software systems based on the vector topological data model have become popular over the years. A special case of the vector topological model is the *georelational* model. In this model derivative, the feature geometries and associated topological information are stored in regular computer files, whereas the associated attribute information is held in relational database management system (RDBMS) tables. The GIS software maintains the intimate linkage between the geometry, topology, and attribute information. This hybrid data management solution was developed to take advantage of RDBMS to store and manipulate attribute information. Geometry and topology were not placed in RDBMS because, until a few years ago, RDBMS were unable to store and retrieve geographic data efficiently. Figure 8.10 is an example of a georelational model as implemented in ESRI's ArcInfo coverage polygon dataset. It shows file-based geometry and topology information linked to attributes in an RDBMS table. The ID (identifier) of the polygon, the label point, is linked (related or joined) to the ID column in the attribute table (see also Chapter 11). Thus, in this soils dataset Polygon 3 is soil B7, of Class 212, and its suitability is moderate.

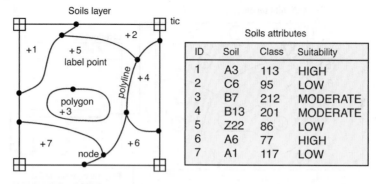

Figure 8.10 An example of a georelational polygon dataset. Each of the polygons is linked to a row in an RDBMS table. The table has multiple attributes, one in each column. (Source: after ESRI 1997)

The topological feature geographic data model has been extensively used in GIS applications over the last 20 years, especially in government and natural resources applications based on polygon representations (see Section 2.3.2). Typical government applications include cadastral management, tax assessment, land/property parcel management, land-use zoning, planning, and building control. In the areas of natural resources and environment, key applications include site suitability analysis, integrated land use modeling, license mapping, natural resource management, and conservation. The tax appraisal case study discussed in Section 2.3.2.2 is an example of a GIS based on the topological feature data model. The developers of this system chose this model because they wanted to avoid overlaps and gaps in tax parcels (polygons), to ensure that all parcel boundaries closed (were validated), and to store data in an efficient way. This is in spite of the fact that there is an overhead in creating and maintaining parcel topology, as well as degradation in draw and query performance for large databases.

8.2.3.3 Network Data Model

The network data model is a special type of the topological feature model. It is discussed here separately because it raises several new issues and has been widely applied in GIS studies.

Networks can be used to model the flow of goods and services. There are two primary types of networks: *radial* and *looped*. In radial or tree networks, flow always has an upstream and downstream direction. Stream and storm drainage systems are examples of radial networks. In looped networks, self-intersections are common occurrences. Water distribution networks are looped by design to ensure that service interruptions affect the fewest customers.

In GIS software systems, networks are modeled as points (for example, street intersections, fuses,

switches, water valves, and the confluence of stream reaches: usually referred to as nodes in topological models) and lines (for example, streets, transmission lines, pipes, and stream reaches). Network topological relationships define how lines connect with each other at nodes. For the purpose of network analysis, it is also useful to define rules about how flows can move through a network. For example, in a sewer network, flow is directional from a customer (source) to a treatment plant (sink), but in a pressurized gas network flow can be in any direction. The rate of flow is modeled as impedances (weights) on the nodes and lines. Figure 8.11 shows an example of a street network. The network comprises a collection of nodes (types of street intersection) and lines (types of street), as well as the topological relationships between them. The topological information makes it possible, for example, to trace the flow of traffic through the network and to examine the impact of street closures. An "impedance" defined on the intersections and streets determines the speed at which traffic flows. Typically, the rate of flow is proportional to the street speed limit and number of lanes, and the timing of stoplights at intersections. Although this example relates to streets, the same basic principles also apply to, for example, electric, water, and railroad networks.

In georelational implementations of the topological network feature model, the geometry and topology information is typically held in ordinary computer files and the attributes are stored in a linked database. The GIS software tools are responsible for creating and maintaining the topological information each time a change in the feature geometry takes place. In more modern object models the geometry, attributes, and topology may be stored together in a DBMS, or topology may be computed on the fly.

Many applications utilize networks. Prominent examples include calculating power load drops over an electricity network; routing emergency response

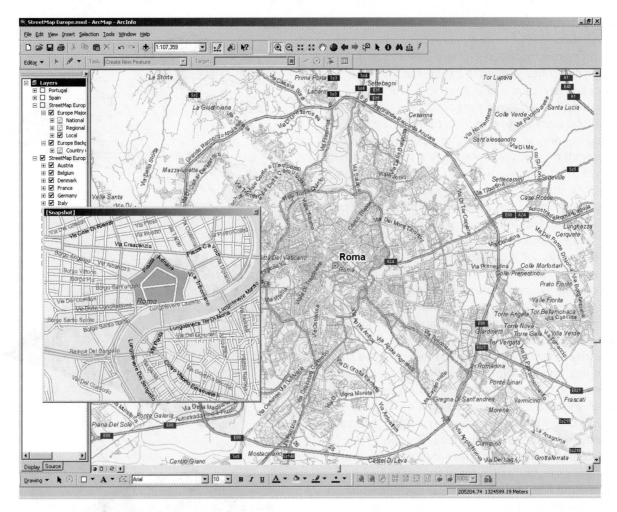

Figure 8.11 An example of a street network.

vehicles over a street network; optimizing the route of mail deliveries over a street network; and tracing pollution upstream to a source over a stream network.

Network data models are also used to support another data model variant called *linear referencing* (see Section 5.5). The basic principle of linear referencing is quite simple. Instead of recording the locations of geographic entities as explicit *x*, *y*, *z* coordinates, they are recorded as distances along a network (called a route system) from a point of origin. This is a very efficient way of storing information such as road pavement (surface) wear characteristics (e.g., the location of pot holes and degraded asphalt), geological seismic data (e.g., shockwave measurements at sensors along seismic lines), and pipeline corrosion data. An interesting aspect of this is that a two-dimensional network is reduced to a one-dimensional linear route list. The location of each entity (often called an event) is simply a distance along the route from the origin. Offsets are also often stored to indicate the distance from a network centerline. For example,

when recording the surface characteristics of a multi-carriageway road, several readings may be taken for each carriageway at the same linear distance along the route. The offset value will allow the data to be related to the correct carriageway. Dynamic segmentation is a special type of linear referencing. The term derives from the fact that event data values are held separately from the actual network route in database tables (still as linear distances and offsets) and then dynamically added to the route (segmented) each time the user queries the database. This approach is especially useful in situations in which the event data change frequently and need to be stored in a database due to access from other applications (e.g., traffic volumes or rate of pipe corrosion).

8.2.3.4 TIN Data Model

The geographic data models discussed so far have concentrated on one- and two-dimensional data. There are several ways to model three-dimensional data, such as terrain models, sales cost surfaces, and geologic strata. The term *2.5-D* is sometimes used to describe surface structures because they have dimensional properties between 2-D and 3-D. A true 3-D structure will contain multiple *z* values at the same *x*, *y* location and thus is able to model overhangs and tunnels, as well as support accurate volumetric calculations such as cut and fill (a term derived from civil-engineering applications that describes cutting earth from high areas and placing it in low areas to construct a flat surface, as is required in, for example, railroad construction). Both grids and triangulated irregular networks (TINs) are used to create and represent surfaces in GIS. A regular grid surface is really a type of raster dataset, as discussed earlier in Section 8.2.2. Each grid cell stores the height of the surface at a given location. The TIN structure, as the name suggests, represents a surface as contiguous nonoverlapping triangular elements (Figure 8.12). A TIN is created from a set of points with *x*, *y*, and *z* coordinate values. A key advantage of the TIN structure is that the density of sampled points, and therefore the size of triangles, can be adjusted to reflect the relief of the surface being modeled, with more points sampled in areas of variable relief (see Section 4.4). TIN surfaces can be created by performing what is called a Delaunay triangulation (Figure 8.13, Section 14.3.6.1). First, a convex hull is created for a dataset—the smallest convex polygon that contains the set of points. Next, straight lines that do not cross each other are drawn from interior points to points on the boundary of the convex hull and to each other. This divides the convex hull into a set of polygons, which are then divided into triangles by drawing more lines between vertices of the polygons.

A TIN is a topological data structure that manages information about the nodes comprising each triangle and the neighbors of each triangle. Figure 8.13 shows the topology of a simple TIN. As with other topological data structures, information about a TIN may be conveniently stored in a file or database table, or computed on the fly. TINs offer many advantages for surface analysis. First, they incorporate the original sample points, providing a useful check on the accuracy of the model. Second, the variable density of triangles means that a TIN is an efficient way of storing surface representations such as terrains that have substantial variations in topography. Third, the data structure makes it easy to calculate elevation, slope, aspect, and line-of-sight between points. The combination of these factors has led to the widespread use of the TIN data structure in applications such as volumetric calculations for roadway design, drainage studies for land development, and visualization of urban forms. Figure 8.14 shows two example

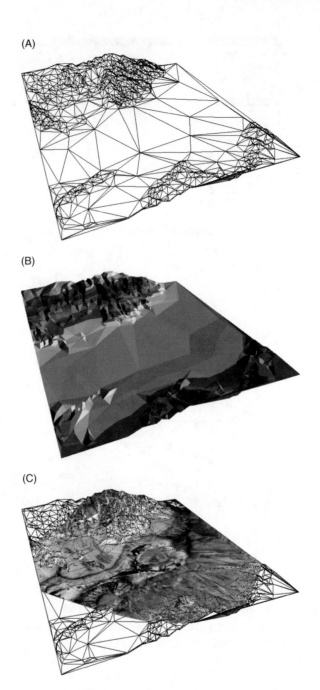

(A)

(B)

(C)

Figure 8.12 TIN surface of Death Valley, California: (A) "wireframe" showing all triangles; (B) shaded by elevation; and (C) draped with satellite image.

applications of TINs. Figure 8.14A is a shaded landslide risk TIN of the Pisa district in Italy, with building objects draped on top to give a sense of landscape. Figure 8.14B is a TIN of the Yangtse River, China, greatly exaggerated in the *z* dimension. It shows how TINs draped with images can provide photorealistic views of landscapes.

Like all 2.5-D and 3-D models, TINs are only as good as the input sample data (see Section 4.4). They are especially susceptible to extreme high and low

A TIN is a topologic data structure that manages information about the nodes that comprise each triangle and the neighbors to each triangle

Triangle	Node list	Neighbors
A	1, 2, 3	-, B, D
B	2, 4, 3	-, C, A
C	4, 8, 3	-, G, B
D	1, 3, 5	A, F, E
E	1, 5, 6	D, H, -
F	3, 7, 5	G, H, D
G	3, 8, 7	C, -, F
H	5, 7, 6	F, -, E

Triangles always have three nodes and usually have three neighboring triangles. Triangles on the periphery of the TIN can have one or two neighbors.

Figure 8.13 The topology of a TIN. (Source: after Zeiler 1999)

Figure 8.14 Examples of applications that use the TIN data model: (A) Landslide risk map for Pisa, Italy (Courtesy of Earth Science Department, University of Siena, Italy); (B) Yangtse River, China (Courtesy of Human Settlements Research Center, Tsinghua University, China).

(A)

(B)

values because there is no smoothing of the original data. Other limitations of TINs include their inability to deal with discontinuity of slope across triangle boundaries, the difficulty of calculating optimum routes, and the need to ensure that peaks, pits, ridges, and channels are captured if a drainage network TIN is to be accurate.

8.2.4 Object Data Model

All the geographic data models described so far are geometry-centric; that is, they model the world as collections of points, lines, areas, TINs, or rasters. Any operations to be performed on the geometry (and, in some cases, associated topology) are created as separate procedures (programs or scripts). Unfortunately, this approach can present several limitations for modeling geographic systems. All but the simplest of geographic systems contain many entities with large numbers of properties, complex relationships, and sophisticated behavior. Modeling such entities as simple geometry types is overly simplistic and does not easily support the sophisticated characteristics required for modern analysis. In addition, separating the state of an entity (attributes or properties defining what it *is*) from the behavior of an entity (methods defining what it *does*) makes software and database development tedious, time-consuming, and error prone. To try to address these problems, geographic object data models were developed; they allow the full richness of geographic systems to be modeled in an integrated way in a GIS.

The central focus of a GIS object data model is the collection of geographic objects and the

Object-oriented Concepts in GIS

An **object** is a self-contained package of information describing the characteristics and capabilities of an entity under study. An interaction between two objects is called a **relationship**. In a geographic object data model, the real world is modeled as a collection of objects and the relationships between the objects. Each entity in the real world to be included in the GIS is an object. A collection of objects of the same type is called a **class**. In fact, classes are a more central concept than objects from the implementation point of view because many of the object-oriented characteristics are built at the class level. A class can be thought of as a template for objects. When creating an object data model the designer specifies classes and the relationships between classes. Only when the data model is used to create a database are objects (instances or examples of classes) actually created.

Examples of objects include oil wells, soil bodies, stream catchments, and aircraft flight paths. In the case of an oil-well class, each oil-well object might include **properties** defining its **state**— annual production, owner name, date of construction, and type of geometry used for representation at a given scale (perhaps a point on a small-scale map and a polygon on a large-scale one). The oil-well class could have connectivity relationships with a pipeline class that represents the pipeline used to transfer oil to a refinery. There could also be a relationship defining the fact that each well must be located on a drilling platform. Finally, each oil-well object might also have **methods** defining the **behavior** or what it can do. Example behavior might include how objects draw themselves on a computer screen, how objects can be created and deleted, and editing rules about how oil wells snap to pipelines.

Three key facets of object data models make them especially good for modeling geographic systems: encapsulation, inheritance, and polymorphism.

Encapsulation describes the fact that each object packages together a description of its state and behavior. The state of an object can be thought of as its properties or attributes (e.g., for a forest object, it could be the dominant tree type, average tree age, and soil pH). The behavior is the sum of the methods or operations that can be performed on an object (for a forest object these could be create, delete, draw, query, split, and merge). For example, when splitting a forest polygon into two parts, perhaps following a part sale, it is useful to get the GIS to automatically calculate the areas of the two new parts. Combining the state and behavior of an object together in a single package is a natural way to think of geographic entities and a useful way to support the reuse of objects.

Inheritance is the ability to reuse some or all of the characteristics of one object in another object. For example, in a gas facility system a new type of gas valve could easily be created by overwriting or adding a few properties or methods to a similar existing type of valve. Inheritance provides an efficient way to create models of geographic systems by reusing objects and also a mechanism to extend models easily. New object classes can be built to reuse parts of one or more existing object classes and add some new unique properties and methods. The example described in Section 8.3 shows how inheritance and other object characteristics can be used in practice.

Polymorphism describes the process whereby each object has its own specific implementation for operations like draw, create, and delete. One example of the benefit of polymorphism is that a geographic database can have a generic object-creation component that issues requests to be processed in a specific way by each type of object class. A utility system's editor software can send a generic create request to all objects (e.g., gas pipes, valves, and service lines), each of which has specific create algorithms. If a new object class is added to the system (e.g., landbase), then this mechanism will work because the new class is responsible for implementing the create method. Polymorphism is essential for isolating parts of software as self-contained components (see Chapter 7).

relationships between the objects (see Box 8.2). Each geographic object is an integrated package of geometry, properties, and methods. In the object data model, geometry is treated like any other attribute of the object and not as its primary characteristic (although clearly from an applications perspective it is often the major property of interest). Geographic objects of the same type are grouped together as object classes, with individual objects in the class referred to as "instances." In many GIS software systems each object class is stored physically as a database table, with each row an object and each object property a column. The methods that apply are attached to the object instances when they are created in memory for use in the application.

An object is the basic atomic unit in an object data model and comprises all the properties that define the state of an object, together with the methods that define its behavior.

All geographic objects have some type of relationship to other objects in the same object class and, possibly, to objects in other object classes. Some of these relationships are inherent in the class definition (for example, some GIS remove overlapping polygons) while other interclass relationships are user-definable. Three types of relationships are commonly used in geographic object data models: topological, geographic, and general.

A class is a template for creating objects.

Generally, topological relationships are built into the class definition. For example, modeling real-world entities as a network class will cause network topology to be built for the nodes and lines participating in the network. Similarly, real-world entities modeled as topological polygon classes will be structured using the node–polyline model described in Section 8.2.3.2.

Geographic relationships between object classes are based on geographic operators (such as overlap, adjacency, inside, and touching) that determine the interaction between objects. In a model of an agricultural system, for example, it might be useful to ensure that all farm buildings are within a farm boundary using a test for geographic containment.

General relationships are useful to define other types of relationship between objects. In a tax assessment system, for example, it is advantageous to define a relationship between land parcels (polygons) and ownership data that is stored in an associated DBMS table. Similarly, an electric distribution system relating light poles (points) to text strings (called annotation) allows depiction of pole height and material of construction on a map display. This type of information is very valuable for creating work orders (requests for change) that alter the facilities. Establish-

ing relationships between objects in this way is useful because if one object is moved then the other will move as well, or if one is deleted then the other is also deleted. This makes maintaining databases much easier and safer.

In addition to supporting relationships between objects (strictly speaking, between object classes), object data models also allow several types of rules to be defined. Rules are a valuable means of maintaining database integrity during editing tasks. The most popular types of rules used in object data models are attribute, connectivity, relationship, and geographic.

Attribute rules are used to define the possible attribute values that can be entered for any object. Both range and coded-value attribute rules are widely employed. A range attribute rule defines the range of valid values that can be entered. Examples of range rules include the following: highway traffic speed must be in the range 25–70 miles per hour; forest compartment average tree height must be in the range 0–50 meters. Coded attribute rules are used for categorical data types. For example, land use must be of type commercial, residential, park, or other; or pipe material must be of type steel, copper, lead, or concrete.

Connectivity rules are based on the specification of valid combinations of features, derived from the geometry, topology, and attribute properties. For example, in an electric distribution system a 28.8-kV conductor can only connect to a 14.4-kV conductor via a transformer. Similarly, in a gas distribution system it should not be possible to add pipes with free ends (that is, with no fitting or cap) to a database.

Geographic rules define what happens to the properties of objects when an editor splits or merges them (Figure 8.15). In the case of a land parcel split following the sale of part of the parcel, it is useful to define rules to determine the impact on properties such as area, land-use code, and owner. In this example, the original parcel area value should be divided in proportion to the size of the two new parcels, the land-use code should be transferred to both parcels, and the owner name should remain for one parcel, but a new one should be added for the part that was sold off. In the case of a merge of two adjacent water pipes, decisions need to be made about what happens to attributes such as material, length, and corrosion rate. In this example, the two pipe materials should be the same, the lengths should be summed, and the new corrosion rate determined by a weighted average of both pipes.

(A)

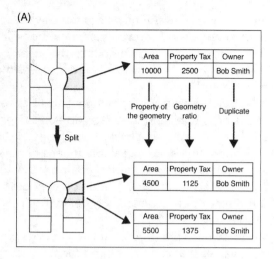

(B)

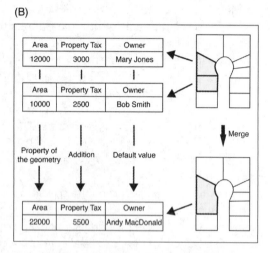

Figure 8.15 Example of split and merge rules for parcel objects: (A) split; and (B) merge. (Source: after MacDonald 1999)

8.3 Example of a Water-Facility Object Data Model

This section presents an example of a geographic object model and discusses how many of the concepts introduced earlier in this chapter are used in practice. The example selected is that of an urban water facility model. The types of issues raised in this example apply to all geographic object models, although of course the actual objects, object classes, and relationships under consideration will differ. The role of data modeling, as discussed in Section 8.1, is to represent the key aspects of the real world inside the digital computer for management, analysis, and display purposes.

Figure 8.16 is a diagram of part of a water distribution system, a type of pressurized network controlled by several devices. A pump is responsible for

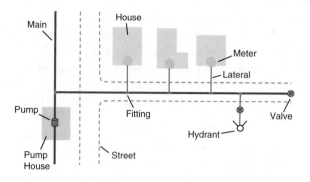

Figure 8.16 Water distribution system water-facility object types and geographic relationships.

moving water through pipes (mains and laterals) connected together by fittings. Meters measure the rate of water consumption at houses. Valves and hydrants control the flow of water.

The goal of the example object model is to support asset management, mapping, and network analysis applications. Based on this goal it is useful to classify the objects into two types: the landbase and the water facilities. Landbase is a general term for objects such as houses and streets that provide geographic context but are not used in network analysis. The landbase object types are Pump House, House, and Street. The water-facilities object types are: Main, Lateral (a smaller type of WaterLine), Fitting (water line connectors), Meter, Valve, and Hydrant. All of these object types need to be modeled as a network in order to support network analysis operations such as network isolation traces and flow prediction. A network isolation trace is used to find all parts of a network that are unconnected (isolated). By using the topological connectivity of the network and information about whether pipes and fittings support water flow, it is possible to determine connectivity. Flow prediction is used to estimate the flow of water through the network based on network connectivity and data about water availability and consumption. Figure 8.17 shows all the main object types and the

Figure 8.17 Water distribution system network.

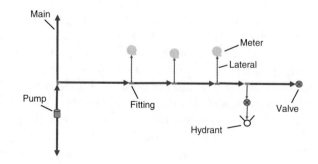

implicit geographic relationships to be incorporated into the model. The arrows indicate the direction of flow in the network. When digitizing this network using a GIS editor, it is useful to specify topological connectivity and attribute rules to control how objects can be connected (see Section 8.2.3.3). Before this network can be used for analysis, it will also be necessary to add flow impedances to each link (for example, pipe diameter).

Having identified the main object types, the next step is to decide how objects relate to each other and what is the most efficient way to implement them. Figure 8.18 shows one possible object model that uses the Unified Modeling Language (UML) to show objects and the relationships between them. Some additional color-coding has been added to help interpret the model. In UML models each box is an object class, and the lines define how one class reuses (inherits) part of the class above it in a hierarchy. Object class names in an italic font are abstract classes; those with regular font names are used to create (instantiate) actual object instances. Abstract classes do not have instances and exist for reasons of efficiency. It is sometimes useful to have a class that implements some capabilities once, so that several other classes can then be reused. For example, Main and Lateral are both types of *Line*, as is Street. Because Main and Lateral share several things in common—such as ConstructionMaterial, Diameter, and InstallDate properties, and connectivity and draw

behavior—it is efficient to implement these in a separate abstract class, called *WaterLine*. The triangles indicate that one class is a type of another class. For example, Pump House and House are types of *Building*, and Street and *WaterLine* are types of *Line*. The diamonds indicate composition. For example, a network is composed of a collection of *Line* and *Node* objects. In the water-facility object model, object classes without any geometry are colored pink. The Equipment and OperationsRecord object classes have their location determined by being associated with other objects (e.g., valves and mains). The Equipment and OperationsRecord classes are useful places to store properties common to many facilities, such as EquipmentID, InstallDate, ModelNumber, and SerialNumber.

Once this logical geographic object model has been created, it can be used to generate a physical data model. One way to do this is to create the model using a computer-aided software engineering (CASE) tool. A CASE tool is a software application that has graphical tools that draw and specify a logical model (Figure 8.19). A further advantage of a CASE tool is that physical models can be generated directly from the logical models, including all the database tables and much of the supporting code for implementing behavior. Once a database structure (schema) has been created, it can be populated with objects and the intended applications put into operation.

Figure 8.18 A water-facility object model.

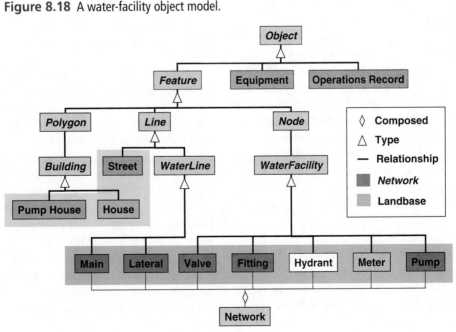

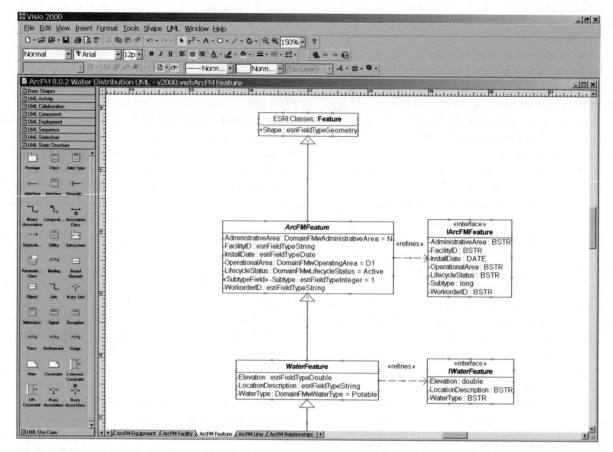

Figure 8.19 An example of a CASE tool (Microsoft Visio). The UML model is for a utility water system.

8.4 Geographic Data Modeling in Practice

Geographic analysis is only as good as the geographic database on which it is based, and a geographic database is only as good as the geographic data model from which it is derived. Geographic data modeling begins with a clear definition of the project goals and progresses through an understanding of user requirements, a definition of the objects and relationships, formulation of a logical model, and then creation of a physical model. These steps are a prelude to database creation and, finally, database use.

No step in data modeling is more important than understanding the purpose of the data-modeling exercise. This understanding can be gained by collecting user requirements from the main users. Initially, user requirements will be vague and ill-defined, but over time they will become clearer. Project goals and user requirements should be precisely specified in a list or narrative. Nicole Alexander's career and experience of GIS data modeling is highlighted in Box 8.3.

Formulation of a logical model necessitates identification of the objects and relationships to be modeled. Both the attributes and behavior of objects are required for an object model. A useful graphic tool for creating logical data models is a CASE tool, and a useful language for specifying models is UML. It is not essential that all objects and relationships be identified at the first attempt because logical models can be refined over time. The key objects and relationships for a water distribution system object model are shown in Figure 8.18.

Once an implementation-independent logical model has been created, this model can be turned into a system-dependent physical model. A physical model will result in an empty database schema—a collection

Nicole Alexander, Data Modeler

Nicole Alexander (Figure 8.20) started her geospatial career as a land surveyor working for Transmanche-Link (TML) on the Channel Tunnel project. The Channel Tunnel (le tunnel sous la Manche) is a 50-km undersea rail tunnel linking Folkestone, Kent in England to Coquelles, near Calais in France. Her interest in GIS began during the construction phase of the Channel Tunnel UK Terminal in Folkestone, where she was employed to perform detailed (as-built) surveys of the utilities, platforms, and other on-site structures. The data captured in these surveys were eventually overlaid in AutoCAD and used to determine whether to grant contractors a "Permit to Dig." Contractors required permission to excavate an area before installing or maintaining their plant to avoid contact with existing infrastructure. While working on the UK Terminal site, she entered the GIScience master's program at University College London, and for her thesis she designed a Facilities Management System for the UK Terminal site using ArcInfo GIS and Oracle DBMS.

After graduating from University College London, Nicole worked as a GIS consultant for Fujitsu-ICL Caribbean Limited on the Jamaica Land Titling Project. She also worked for Trimble Navigation as a GIS data-capture engineer, and a GIS specialist on their Mapping and GIS products. While at Trimble, she

Figure 8.20 Nicole Alexander. (Courtesy Nicole Alexander)

entered the doctoral program in Geography at the University of California, Santa Barbara.

Currently, she works at Oracle Corporation and at the Center for Geographic Analysis at Harvard University on many aspects of GIS data modeling. Nicole says: "GIS data modeling is important because it ultimately determines the efficiency with which specified tasks can be performed by a system. Therefore, in GIS data modeling it is necessary to have a clear understanding of the users' requirements and the knowledge of the application domain required to anticipate needs of the system."

of database tables and the relationships between them. Sometimes, for performance optimization reasons or because of changing requirements, it is necessary to alter the physical data model. Even at this relatively late stage in the process, flexibility is still necessary.

It is important to realize that there is no such thing as the correct geographic data model. Every problem can be represented with many possible data models. Each data model is designed with a specific purpose in mind and is suboptimal for other purposes. A classic dilemma is whether to define a general-purpose data model that has wide applicability, but that can, potentially, be complex and inefficient, or to focus on a narrower highly optimized model that will yield better performance. A small prototype can often help resolve some of these issues.

Geographic data modeling is both an art and a science. It requires a scientific understanding of the key geographic characteristics of real-world systems, including the state and behavior of objects, and the relationships between them. Geographic data models are of critical importance because they have a controlling influence over the type of data that can be represented and the operations that can be performed. As we have seen, object models are the best type of data model for representing rich object types and relationships in facility systems, whereas simple feature models are sufficient for elementary applications such as a map of the body. In a similar vein, so to speak, raster models are good for data represented as fields such as soils, vegetation, pollution, and population counts.

Questions for Further Study

1. Figure 8.21 is an oblique aerial photograph of part of the city of Kfar-Saba, Israel. Take 10 minutes to list all the object classes (including their attributes and behavior) and the relationships between the classes that you can see in this picture that would be appropriate for a city information system study.

2. Why is it useful to include the conceptual, logical, and physical levels in geographic data modeling?

3. Describe, with examples, five key differences between the topological vector and raster geographic data models. It may be useful to consult Figure 8.3 and Chapter 3.

4. Review the terms *encapsulation, inheritance,* and *polymorphism* and explain with geographic examples why they make object data models superior for representing geographic systems.

Figure 8.21 Oblique aerial view of Kfar-Saba, Israel. (Courtesy: Av-revaya Mordehay)

Further Reading

Arctur, D. and Zeiler, M. 2004. *Designing Geodatabases: Case Studies in GIS Data Modeling*. Redlands, CA: ESRI Press.

ESRI 1997. *Understanding GIS: The ArcInfo Method*. Redlands, CA: ESRI Press.

MacDonald, A. 1999. *Building a Geodatabase*. Redlands, CA: ESRI Press.

Worboys, M.F. and Duckham, M. 2004. *GIS: A Computing Perspective* (2nd ed.). Boca Raton, FL: CRC Press.

Zeiler, M. 1999. *Modeling Our World: The ESRI Guide to Geodatabase Design*. Redlands, CA: ESRI Press.

9 GIS Data Collection

Data collection is one of the most time-consuming and expensive, yet important, of GIS tasks. There are many diverse sources of geographic data, and many methods are available to enter them into a GIS. The two main methods of data collection are data capture and data transfer. It is useful to distinguish between primary (direct measurement) and secondary (derivation from other sources) data capture for both raster and vector data types. Data transfer involves importing digital data from other sources. Planning and executing an effective GIS data collection plan touches on many practical issues. This chapter reviews the main methods of GIS data capture and transfer and introduces key practical management issues.

LEARNING OBJECTIVES

After studying this chapter you will be able to:

- Describe data collection workflows.

- Understand the primary data capture techniques in remote sensing and surveying.

- Discuss secondary data capture techniques including scanning, digitizing, vectorization, photogrammetry, and COGO (a contraction of the term *coordinate geometry*) feature construction.

- Understand the principles of data transfer, sources of digital geographic data, and geographic data formats.

- Analyze practical issues associated with managing data capture projects.

9.1 Introduction

GIS can contain a wide variety of geographic data types originating from many diverse sources. Indeed, one of the key defining characteristics of GIS is their ability to integrate data about places and spaces from many different sources. Data collection activities for the purposes of organizing the material in this chapter are split into data capture (direct data input) and data transfer (input of data from other systems). From the perspective of creating geographic databases, it is convenient to classify raster and vector geographic data as primary and secondary (Table 9.1). Primary data sources are those collected in digital format specifically for use in a GIS project.

Typical examples of primary GIS sources include raster SPOT and Quickbird Earth satellite images, and vector building–survey measurements captured using a total survey station (see Section 9.2 below for discussion about these terms). Secondary sources are digital and analog datasets that may have been originally captured for another purpose and need to be converted into a suitable digital format for use in a GIS project. Typical secondary sources include raster-scanned color aerial photographs of urban areas, or United States Geological Survey (USGS) or Institut Géographique National (IGN, France) paper maps that can be scanned and vectorized. This classification scheme provides a useful organizing framework for this chapter, and, more importantly, it highlights the number of processing transformations that a dataset goes through, and therefore the opportunities for errors to be introduced.

However, the distinctions between primary and secondary, and raster and vector, are not always easy to determine. For example, are digital satellite remote-sensing data obtained on a DVD primary or secondary? Clearly, the commercial satellite sensor feeds do not run straight into GIS databases, but to ground stations where the data are preprocessed and then written onto digital media. Here they are considered primary because the data have usually undergone only minimal transformation since being collected by the satellite sensors and because the characteristics of the data make them suitable for direct use in GIS projects.

Primary geographic data sources are captured specifically for use in GIS by direct measurement. Secondary sources are reused from earlier studies or obtained from other systems.

Both primary and secondary geographic data may be obtained in either digital or analog format (see Section 3.7 for a definition of analog). Analog data must always be digitized before being added to a geographic database. Analog-to-digital transformation may involve scanning paper maps or photographs, optical character recognition (OCR) of text describing geographic object properties, or vectorization of selected features from an image. Depending on the format and characteristics of the digital data, considerable reformatting and restructuring may be required prior to importing into a GIS. Each of these transformations alters the original data and will introduce further uncertainty into the data (see Chapter 6 for discussion of uncertainty).

This chapter describes the data sources, techniques, and workflows involved in GIS data collection. The processes of data collection are also variously referred to as data capture, data automation, data conversion, data transfer, data translation, and digitizing. Although there are subtle differences between these terms, they essentially describe the same thing, namely, adding geographic data to a database. Here data capture refers to direct entry, and data transfer is the importing of existing digital data across a network connection (Internet, WAN, or LAN) or from physical media such as DVD or portable hard disk. This chapter focuses on the techniques of data collection; of equal, perhaps more, importance to a real-world GIS implementation are project management, cost, legal, and organization issues. These are covered briefly in Section 9.7 as a prelude to more detailed treatment in Chapters 17 through 19.

Table 9.1 Classification of geographic data for data collection purposes with examples of each type.

	Raster	Vector
Primary	Digital satellite remote-sensing images	GPS measurements
	Digital aerial photographs	Field survey measurements
Secondary	Scanned maps or photographs	Topographic maps
	Digital elevation models from topographic map contours	Toponymy (place-name) databases

Table 9.2 Breakdown of costs (in $1,000s) for two typical client-server GIS as estimated by the authors.

	10 seats		100 seats	
	$	%	$	%
Hardware	30	3.4	250	8.8
Software	25	2.8	150	5.3
Data	400	44.7	450	15.8
Staff	440	49.2	2,000	70.2
Total	895	100.0	2,850	100

Table 9.2 shows a breakdown of costs (in $1,000s) for two typical client-server GIS implementations: one with 10 seats (systems) and the other with 100. The hardware costs include desktop clients and servers only (i.e., not network infrastructure). The data costs assume the purchase of a landbase (e.g., streets, parcels, and landmarks) and digitizing assets such as pipes and fittings (water utility), conductors and devices (electrical utility), or land and property parcels (local government). Staff costs assume that all core GIS staff will be full-time, but that users will be part-time.

In the early days of GIS, when geographic data were very scarce, data collection was the main project task, and typically it consumed the majority of the available resources. Even today data collection still remains a time-consuming, tedious, and expensive process. Typically, it accounts for 15 to 50% of the total cost of a GIS project (Table 9.2). Data capture costs can in fact be much more significant because in many organizations (especially those that are government funded) staff costs are often assumed to be fixed and are not used in budget accounting. Furthermore, as the majority of data capture effort and expense tends to fall at the start of projects, data capture costs often receive greater scrutiny from senior managers. If staff costs are excluded from a GIS budget, then in cash expenditure terms data collection can be as much as 60 to 85% of costs.

Data capture costs can account for up to 85% of the cost of a GIS.

After an organization has completed basic data collection tasks, the focus of a GIS project moves on to data maintenance. Over the multiyear lifetime of a GIS project, data maintenance can turn out to be a far more complex and expensive activity than initial data collection. This is because of the high volume of update transactions in many systems (for example,

changes in land parcel ownership, maintenance work orders on a highway transport network, or logging military operational activities) and the need to manage multiuser access to operational databases. For more information about geographic data maintenance see Chapter 10.

9.1.1 Data Collection Workflow

In all but the simplest of projects, data collection involves a series of sequential stages (Figure 9.1). The workflow commences with planning, followed by preparation, digitizing/transfer (here taken to mean a range of primary and secondary techniques such as table digitizing, survey entry, scanning, and photogrammetry), editing and improvement and, finally, evaluation.

Planning is obviously important to any project, and data collection is no exception. It includes establishing user requirements, garnering resources (staff, hardware, and software), and developing a project plan. Preparation is especially important in data collection projects. It involves many tasks such as obtaining data, redrafting poor-quality map sources, editing scanned map images, and removing noise (unwanted data such as speckles on a scanned map image). It may also involve setting up appropriate GIS hardware and software systems to accept data. Digitizing and transfer are the stages where the majority of the effort will be expended. It is naive to think that data capture is really just digitizing, when in fact it involves very much more as discussed later in this chapter. Editing and improvement follows digitizing/transfer. This covers many techniques designed to validate data, as well as correct errors and improve quality. Evaluation, as the name suggests, is the process of identifying project successes and failures; these may be qualitative

Figure 9.1 Stages in data collection projects.

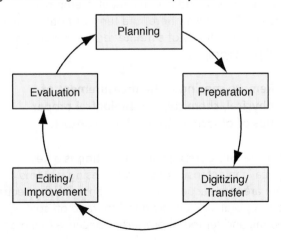

or quantitative. Since all large data projects involve multiple stages, this workflow is iterative, with earlier phases (especially a first, pilot, phase) helping to improve subsequent parts of the overall project.

9.2 Primary Geographic Data Capture

Primary geographic capture involves the direct measurement of objects. Digital data measurements may be input directly into the GIS database or may reside in a temporary file prior to input. Although the direct method is preferable as it minimizes the amount of time and the possibility of errors, close coupling of data collection devices and GIS databases is not always possible. Both raster and vector GIS primary data capture methods are available.

9.2.1 Raster Data Capture

The most popular form of primary raster data capture is remote sensing. Broadly speaking, remote sensing is a technique used to derive information about the physical, chemical, and biological properties of objects without direct physical contact. Information is derived from measurements of the amount of electromagnetic radiation reflected, emitted, or scattered from objects. A variety of sensors, operating throughout the electromagnetic spectrum from visible to microwave wavelengths, are commonly employed to obtain measurements (see Section 3.6.1). Passive sensors rely on reflected solar radiation or emitted terrestrial radiation; active sensors (such as synthetic aperture radar) generate their own source of electromagnetic radiation. The platforms on which these instruments are mounted are similarly diverse. Although Earth-orbiting satellites and fixed-wing aircraft are by far the most common, helicopters, balloons, masts, and booms are also employed (Figure 9.2). As used here, the term *remote sensing* subsumes the fields of satellite remote sensing and aerial photography.

Remote sensing is the measurement of physical, chemical, and biological properties of objects without direct contact.

From the GIS perspective, resolution is a key physical characteristic of remote sensing systems. There are three aspects to resolution: spatial, spectral, and temporal. All sensors need to trade off spatial, spectral, and temporal properties because of storage, processing, and bandwidth considerations. For further discussion of the important topic of resolution, see also Sections 3.4, 3.6.1, 6.3.2 and 16.1 and Box 4.2.

Three key aspects of resolution are spatial, spectral, and temporal.

Spatial resolution refers to the size of object that can be resolved, and the most usual measure is the pixel size. Satellite remote sensing systems typically provide data with pixel sizes in the range 0.5 m–1 km. The resolution of digital cameras used for capturing aerial photographs usually range from 0.01 m–5 m. Image (scene) sizes vary quite widely between sensors— typical ranges include 900 by 900 to 3,000 by 3,000 pixels. The total coverage of remote sensing images is usually in the range 9 by 9 km to 200 by 200 km.

Spectral resolution refers to the parts of the electromagnetic spectrum that are measured. Since different objects emit and reflect different types and amounts of radiation, selecting which part of the electromagnetic spectrum to measure is critical for each application area. Figure 9.3 shows the spectral signatures of water, green vegetation, and dry soil. Remote-sensing systems may capture data in one part of the spectrum (referred to as a single band) or simultaneously from several parts (multiband or multispectral). The radiation values are usually normalized and resampled to give a range of integers from 0–255 for each band (part of the electromagnetic spectrum measured), for each pixel, in each image. Until recently, remote-sensing satellites typically measured a small number of bands, in the visible part of the spectrum. More recently, a number of hyperspectral systems have come into operation that measure very large numbers of bands across a much wider part of the spectrum. Temporal resolution, or repeat cycle, describes the frequency with which images are collected for the same area. There are essentially two types of commercial remote-sensing satellite: Earth-orbiting and geostationary. Earth-orbiting satellites collect information about different parts of the Earth surface at regular intervals. To maximize utility, typically orbits are polar, at a fixed altitude and speed, and are sun synchronous.

The French SPOT (Système Probatoire d'Observation de la Terre) 5 satellite launched in 2002, for example, passes virtually over the poles at an altitude of 822 km, sensing the same location on the Earth surface every 2 to 3 days. The SPOT platform carries multiple sensors: a panchromatic sensor measuring radiation in the visible part of the electromagnetic spectrum at a spatial resolution of

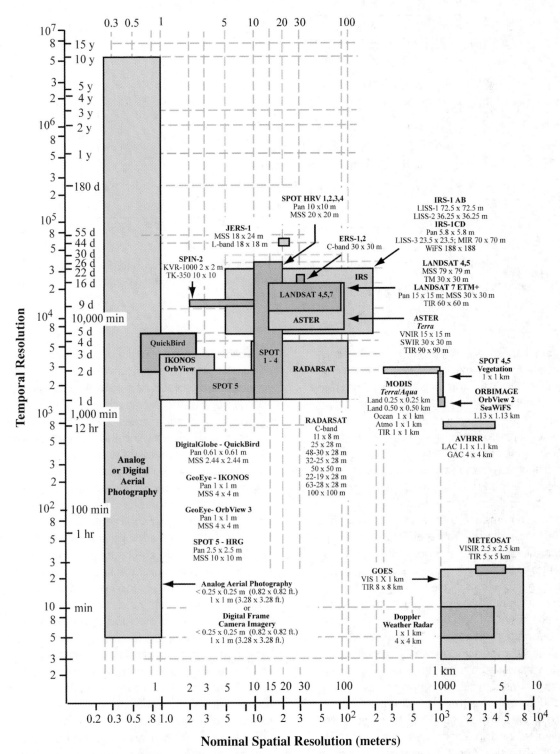

Figure 9.2 Spatial and temporal characteristics of commonly used Earth observation remote-sensing systems and their sensors. (Source: John Jensen)

2.5 by 2.5 m; a multispectral sensor measuring green, red, and reflected infrared radiation separately at a spatial resolution of 10 by 10 m; a shortwave near-infrared sensor with a resolution of 20 by 20 m; and a vegetation sensor measuring four bands at a spatial resolution of 1000 m. The SPOT system is also able to provide stereo images from which digital elevation models and 3-D measurements can be obtained. Each SPOT scene covers an area of about 60 by 60 km.

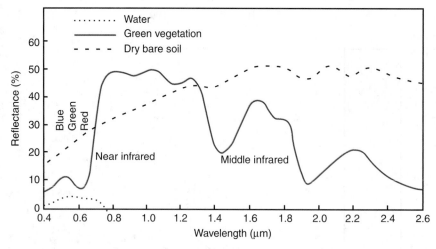

Figure 9.3 Typical reflectance signatures for water, green vegetation, and dry soil. (Source: After Jones C. 1997. *Geographic Information Systems and Computer Cartography*. Reading, MA: Addison-Wesley Longman)

Much of the discussion so far has focused on commercial satellite remote sensing systems. Of equal importance, especially in GIS projects that focus on smaller areas in great detail, is aerial photography. Although the data products resulting from remote sensing satellites and digital aerial photography systems are technically very similar (i.e., they are both images), there are some significant differences in the way data are captured and can, therefore, be analyzed and interpreted. One notable difference is that older aerial photography systems use analog optical cameras and then later rasterize the photographs (e.g., scanning a film negative) to create an image. Both the quality of the optics of the camera and the mechanics of the scanning process affect the spatial and spectral characteristics of the resulting images. Modern aerial photography systems now use digital cameras with on-board position-fixing units to collect digital imagery (although the term *aerial photography* still persists when the digital cameras are carried on planes). Most aerial photographs are collected on an ad hoc basis using cameras mounted in airplanes flying at low altitudes (3,000–9,000 m) and are either panchromatic (black and white) or color, although multispectral cameras/sensors operating in the nonvisible parts of the electromagnetic spectrum are also used. Aerial photographs are very suitable for detailed surveying and mapping projects.

An important feature of satellite and aerial photography systems is that they can provide stereo imagery from overlapping pairs of images. These images are used to create a 3-D analog or digital model from which 3-D coordinates, contours, and digital elevation models can be created (see Section 9.3.2.3).

Satellite and aerial photograph data offer a number of advantages for GIS projects. The consistency of the data and the availability of systematic global coverage make satellite data especially useful for large-area, very detailed projects (for example, mapping landforms and geology at the river catchment-area level) and for mapping inaccessible areas. The regular repeat cycles of commercial systems and the fact that they record radiation in many parts of the spectrum make such data especially suitable for assessing the condition of vegetation (for example, the moisture stress of wheat crops). Aerial photographs in particular are very useful for detailed surveying and mapping of, for example, urban areas and archaeological sites, especially those applications requiring 3-D data (see Chapter 12).

On the other hand, the spatial resolution of commercial satellites is too coarse for many detailed projects, and the data collection capability of many sensors is restricted by cloud cover. Some of this is changing, however, as the new generation of satellite sensors now provide data at 0.5-m spatial resolution and better, and radar data can be obtained that are not affected by cloud cover. The data volumes from both satellites and aerial cameras can be very large and create storage and processing problems for all but the most modern systems. The cost of data can also be prohibitive for a single project or organization.

Box 9.1 describes the work of Gilberto Câmara, a Brazilian who uses remote sensing technologies and techniques extensively in his work.

9.2.2 Vector Data Capture

Primary vector data capture is a major source of geographic data. The two main branches of vector data capture are ground surveying and GPS (which is

Gilberto Câmara, GI Engineer

Gilberto Câmara's (Figure 9.4) career has been focused on building a solid research and development base for GIS adoption and GIScience research in Brazil. He started designing GIS in the early 1980s, after earning a Master's degree in computer science at Brazil's National Institute for Space Research (INPE), where he has worked since 1980. Working with his team at INPE, he designed one of the first GIS for the PC environment in 1986. In the 1990s, he and his team built SPRING, one of the first GIS to adopt object-oriented techniques and ideas. Freely available on the Web since 1995, SPRING has a large user community in Latin America that uses it for teaching and research. SPRING has been downloaded by more than 100,000 users worldwide. From 2001 onward, he worked on TerraLib, an open-source GIS software library that extends object-relational DBMS technology to support spatiotemporal models, spatial analysis, spatial data mining, and image databases. A noteworthy application developed using TerraLib is TerraAmazon, Brazil's national database for surveying deforestation in Amazonia. Another relevant application is TerraME, a software environment for dynamic spatial modeling. TerraME supports scenario-building for future land change, thus enabling decision makers to understand the possible outcomes of public policies.

In his capacity as director of Brazil's INPE from 2006 to 2010, Gilberto has pushed for open access for geospatial data, and has been responsible for setting up open data policies for remote-sensing images in Brazil and abroad. He conducted the development of DETER, a system for real-time detection of deforestation in Amazonia, which *Science* described as "the envy

Figure 9.4 Gilberto Câmara. (Gilberto Câmara)

of the world" for tropical deforestation mapping. He has been one of the leaders on the design and development of the Brazilian series of remote sensing satellites, including the CBERS (China-Brazil Earth Resources Satellites) and the Amazonia programs.

Gilberto believes that GIS will be everywhere in the Twenty-First Century and will help shape a better world, ranging from more efficient transport to better scenario-building for fighting global environmental change. Following Fred Brooks, he maintains that "the scientist builds in order to study; the engineer studies in order to build." A good engineer studies the literature and chooses which scientific principles are relevant for his task. Thus, GI engineers need a critical understanding of the science produced in the field to produce the new generation of GIS.

covered in Section 5.9), although as more surveyors use GPS routinely, so the distinction between the two is becoming increasingly blurred. This is also leading a fundamental shift from measurement-based to coordinate-based ground surveying.

9.2.2.1 Surveying

Ground surveying is based on the principle that the 3-D location of any point can be determined by measuring angles and distances from other known points. Surveys begin from a benchmark point. If the coordinate system and location of this point are known, all subsequent points can be collected in this coordinate system. If they are unknown, then the survey will use a local or relative coordinate system (see Section 5.8).

Since all survey points are obtained from survey measurements, their known locations are always relative to other points. Any measurement errors need to be apportioned between multiple points in a survey. For example, when surveying a field boundary, if the last and first points are not identical in survey terms (within the tolerance employed in the survey), then errors need to be apportioned between all points that define the boundary (see Sections 6.3.2 and 6.4.2). As new measurements are obtained, these may change the locations of points.

Traditionally, surveyors used equipment such as transits and theodolites to measure angles, and tapes and chains to measure distances. Today these have been replaced by electro-optical devices

Figure 9.5 A tripod-mounted Leica Total Station. (Courtesy www.dgyeatman.co.uk/images/TPS1200_CanfordBridge.jpg) (DG Yeatman Surveying & Engineering Ltd)

called total stations that can measure both angles and distances to an accuracy of 1 mm (Figure 9.5). Total stations automatically log data, and the most sophisticated can create vector point, line, and area objects in the field, thus providing direct validation. The basic principles of surveying have changed very little in the past 100 years, although new technology has considerably improved accuracy and productivity. Two people are usually required to perform a survey, one to operate the total station and the other to hold a reflective prism that is placed at the object being measured. On some remote-controlled systems a single person can control both the total station and the prism.

Ground survey is a very time-consuming and expensive activity, but it is still the best way to obtain highly accurate point locations. Surveying is typically used for capturing buildings, land and property boundaries, manholes, and other objects that need to be located accurately. It is also used to obtain reference marks for use in other data capture projects. For example, detailed, fine-scale aerial photographs and satellite images are frequently georeferenced using points obtained from ground survey.

9.2.2.2 LiDAR

LiDAR (Light Detection And Ranging, also known as Airborne Laser Swath Mapping or ALSM) is a recent technology that employs a scanning laser rangefinder to produce accurate topographic surveys of great detail. A LiDAR scanner is an active remote-sensing instrument; that is, it transmits electromagnetic radiation and measures the radiation that is scattered back to a receiver after interacting with the Earth's atmosphere or objects on the surface. LiDAR

uses radiation in the ultraviolet, visible, or infrared region of the electromagnetic spectrum. The scanner is typically carried on a low-altitude aircraft that also has an inertial navigation system and a differential GPS to provide location. LiDAR scanners are capable of collecting extremely large quantities of very detailed information (i.e., scanning of the order of 30,000 points per second at an accuracy of around 15 cm). The data collected from a LiDAR scanner can be described as a point cloud, that is, a massive collection of independent points with x, y, z values. After initial data capture, extensive processing is usually required to remove tree canopies, buildings, and other unwanted features, and to correct errors in order to provide a "bare Earth" point dataset. The points in a LiDAR dataset are often rasterized to create a digital elevation model that is smaller in size and easier to work with in a GIS. Figure 9.6 presents a comparison of three datasets for the same area, one of which is derived from LiDAR.

9.3 Secondary Geographic Data Capture

Geographic data capture from secondary sources is the process of creating raster and vector files and databases from maps, photographs, and other hardcopy documents. Scanning is used to capture raster data. Heads-up digitizing, stereo-photogrammetry, and COGO data entry are the most widely used methods for capturing vector data.

9.3.1 Raster Data Capture Using Scanners

A scanner is a device that converts hardcopy analog media into digital images by scanning successive lines across a map or document and recording the amount of light reflected from a local data source (Figure 9.7). The differences in reflected light are normally scaled into bi-level black and white (1 bit per pixel) or multiple gray levels (8, 16, or 32 bits). Color scanners typically output data into 8-bit red, green, and blue color bands. The spatial resolution of scanners varies widely from as little as 200 dpi (8 dots per mm) to 2400 dpi (96 dots per mm) and beyond. Most GIS scanning is in the range 400–900 dpi (16–40 dots per mm). Depending on the type of scanner and the resolution required, it can take from 30 seconds to 30 minutes or more to scan a map.

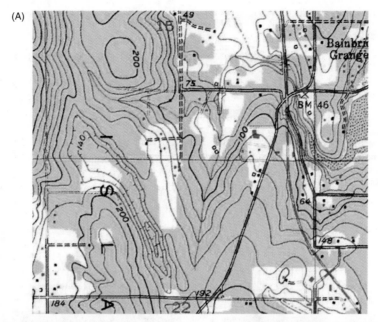

(A)

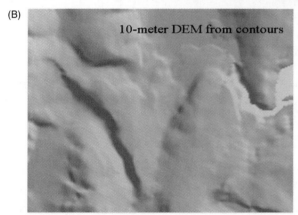

(B) 10-meter DEM from contours

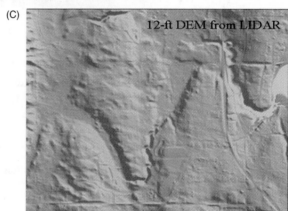

(C) 12-ft DEM from LIDAR

Figure 9.6 Comparison of three datasets for 1 square mile of Bainbridge Island, Washington State: (A) scanned USGS 1:24,000 topographic map sheet; (B) 10 m digital elevation model (DEM) derived from contours digitized from a map sheet; (C) 12 ft (365-cm) resolution DEM derived from a LiDAR survey. (From www.pugetsoundlidar.ess.washington.edu/About_LIDAR.htm)

Figure 9.7 A large-format roll-feed image scanner. (Courtesy Contex) (www.contex.com)

Scanned maps and documents are used extensively in GIS as background maps and data stores.

There are three main reasons to scan hardcopy media for use in GIS:

- Documents, such as building plans, CAD drawings, property deeds, and equipment photographs are scanned to reduce wear and tear, improve access, provide integrated database storage, and index them geographically (e.g., building plans can be attached to building objects in geographic space).

- Film and paper maps, aerial photographs, and images are scanned and georeferenced so that

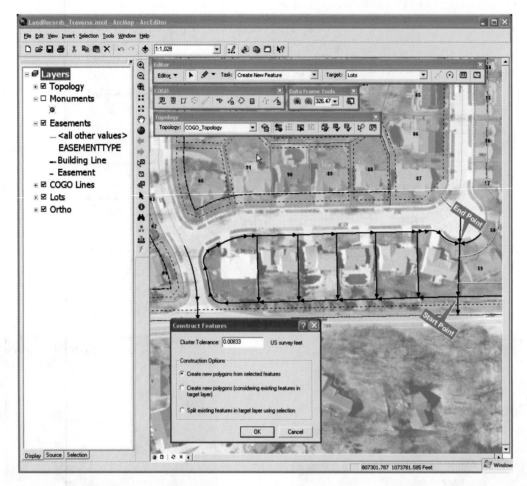

Figure 9.8 An example of raster background data (color aerial photography) underneath vector data (land parcels) that are being digitized on-screen.

they provide geographic context for other data (typically vector layers). This type of unintelligent image or background geographic wallpaper is very popular in systems that manage equipment and land and property assets (Figure 9.8).

- Maps, aerial photographs, and images are scanned prior to vectorization (see 9.3.2.1 below), and sometimes as a prelude to spatial analysis.

An 8 bit (256 gray level) 400 dpi (16 dots per mm) scanner is a good choice for minimum resolution for scanning maps to be used as a background GIS reference layer. For a color aerial photograph that is to be used for subsequent photo interpretation and analysis, a color (8 bit for each of three bands) 900 dpi (40 dots per mm) scanner provides a more appropriate minimum resolution. The quality of data output from a scanner is determined by the nature of the original source material, the quality of the scanning device, and the type of preparation prior to scanning (e.g., redrafting key features or removing unwanted marks will improve output quality).

9.3.2 Vector Data Capture

Secondary vector data capture involves digitizing vector objects from maps and other geographic data sources. The most popular methods are heads-up digitizing and vectorization, photogrammetry, and COGO data entry. Historically, manual digitizing from digitizing tables was the most popular way of secondary vector data capture, but its use has now almost entirely been replaced by the other more modern methods.

9.3.2.1 Heads-up Digitizing and Vectorization

One of the main reasons for scanning maps (see Section 9.3.1) is as a prelude to vectorization—the process of converting raster data into vector data. The simplest way to create vectors selectively from raster data is to digitize vector objects manually straight off a computer screen using a mouse or digitizing cursor. This method is called heads-up digitizing because the map is vertical and can be viewed without bending the head down. It is widely used for capturing, for example, land parcels, buildings, and utility assets.

Vectorization is the process of converting raster data into vector data. The opposite is called rasterization.

Any type of image derived from either primary sources, such as satellite images or aerial photographs, or secondary sources, such as maps or other documents, can be used in heads-up digitizing and vectorization. After loading a scanned image into a GIS database, it must be georeferenced (the term *georegistration* is also sometimes used synonymously: see Section 5.12) before digitizing can begin. This involves a geometric transformation process that uses well-known algorithms to convert image coordinates into database coordinates. The algorithms use both image and database coordinates from a minimum of three well-defined reference points. Once the transformation has been set up, any future coordinates digitized from the image will be in the database coordinate reference system.

Vertices defining point, line, and polygon objects are captured on-screen using point- or stream-digitizing methods. Point digitizing involves placing the screen cursor at the location for each object vertex and then clicking a button to record the location of the vertex. Stream-mode digitizing partially automates this process by collecting vertices automatically every time a distance or time threshold is crossed (e.g., every 0.02 inch (0.5 mm) or 0.25 second). Stream-mode digitizing is a much faster method, but it typically produces larger files with many redundant coordinates.

A faster and more consistent approach is to use software to perform automated vectorization in either batch or semi-interactive mode. Batch vectorization takes an entire raster file and converts it to vector objects in a single operation. Vector objects are created using software algorithms that build simple (spaghetti) line strings from the original pixel values. The lines can then be further processed to create topologically correct polygons (Figure 9.9). A typical map will take only a few minutes to vectorize using modern hardware and software systems. See Section 10.7 for further discussion of structuring geographic data.

Unfortunately, batch vectorization software is far from perfect, and postvectorization editing is usually required to clean up errors. To avoid large amounts of vector editing, it is useful to undertake a little raster editing of the original raster file prior to vectorization in order to remove unwanted noise that may affect the vectorization process. For example, text that overlaps lines should be deleted, and dashed lines are best converted into solid lines. Following vectorization, topological relationships are usually created for the vector objects. This process may also highlight some previously unnoticed errors that require additional editing.

Batch vectorization is best suited to simple bi-level maps of, for example, contours, streams, and highways. For more complicated images and maps and where selective vectorization is required (for example, digitizing electric conductors and devices, or water mains and fittings off topographic maps), interactive vectorization (also called semiautomatic vectorization, line following, or tracing) is preferred. In interactive vectorization, software is used to automate digitizing. The operator snaps the cursor to a pixel on the screen, indicates a direction for line following, and the software then automatically digitizes lines. Typically, many parameters can be tuned to control the density of points (level of generalization), the size of gaps (blank pixels in a line) that will be jumped, and whether to pause at junctions for operator intervention or always to trace in a specific direction (most systems require that all polygons are ordered either clockwise or counterclockwise). Interactive vectorization is still quite labor intensive, but generally it results in much greater productivity than manual or heads-up digitizing. It also produces

Figure 9.9 Batch vectorization of a scanned map: (A) original raster file; (B) vectorized polygons. Adjacent raster cells with the same attribute values are aggregated. Class boundaries are then created at the intersection between adjacent classes in the form of vector lines.

(A)

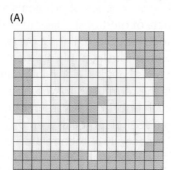

(B)

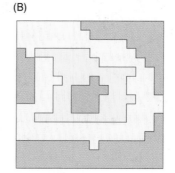

high-quality data, as software is able to represent lines more accurately and consistently than can humans. For these reasons specialized data capture groups much prefer vectorization to manual digitizing.

9.3.2.2 Measurement Error

Data capture, like all geographic workflows, is likely to generate errors. Because digitizing is a tedious and hence error-prone practice, it presents a source of measurement errors—as when the operator fails to position the cursor correctly or fails to record line segments. Figure 9.10 presents some examples of human errors that are commonly introduced in the

---

Figure 9.10 Examples of human errors in digitizing: (A) undershoots and overshoots; (B) invalid polygons; and (C) sliver polygons.

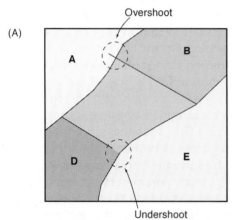

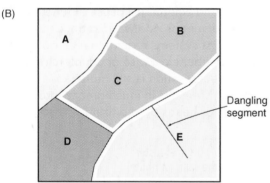

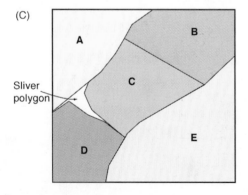

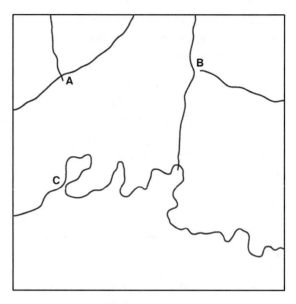

Figure 9.11 Error induced by data cleaning. If the tolerance level is set large enough to correct the errors at A and B, the loop at C will also (incorrectly) be closed.

digitizing process. These errors are: overshoots and undershoots where line intersections are inexact (Figure 9.10A); invalid polygons that are topologically inconsistent because of omission of one or more lines, or omission of attribute data (Figure 9.10B); and sliver polygons, in which multiple digitizing of the common boundary between adjacent polygons leads to the creation of additional polygons (Figure 9.10C).

Most GIS software packages include standard functions, which can be used to restore integrity and clean (or rather obscure, depending on your viewpoint!) obvious measurement errors. Such operations are best carried out immediately after digitizing, so that omissions may be easily rectified. Data-cleaning operations require sensitive setting of threshold values, or else damage can be done to real-world features, as Figure 9.11 shows.

Many errors in digitizing can be remedied by appropriately designed software.

Further classes of problems arise when the products of digitizing adjacent map sheets are merged together. Stretching of paper base maps, coupled with errors in rectifying them, give rise to the kinds of mismatches shown in Figure 9.12. Rubber sheeting is the term used to describe methods for removing such errors on the assumption that strong spatial autocorrelation exists among errors. If errors tend to be spatially autocorrelated up to a distance of x, say, then rubber sheeting will be successful at removing them, at least partially, provided control points can

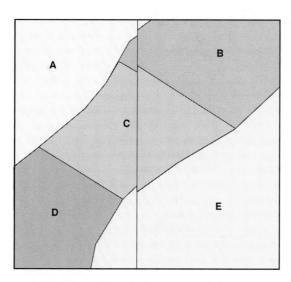

Figure 9.12 Mismatches of adjacent spatial data sources that require rubber sheeting.

be found that are spaced less than *x* apart. For the same reason, the shapes of features that are less than *x* across will tend to have little distortion, while very large shapes may be badly distorted. The results of calculating areas (Section 15.1.1), or other geometric operations that rely only on relative position, will be accurate as long as the areas are small, but will grow rapidly with feature size. Thus it is important for the user of a GIS to know which operations depend on relative position, and over what distance; and where absolute position is important (of course, the term *absolute* simply means relative to the Earth frame, defined by the equator and the Greenwich Meridian, or relative over a very long distance; see Section 5.7). Analogous procedures and problems characterize the rectification of raster datasets, be they scanned images of paper maps or satellite measurements of the curved Earth surface.

9.3.2.3 Photogrammetry

Photogrammetry is the science and technology of making measurements from pictures, aerial photographs, and images. Although in the strict sense it includes 2-D measurements taken from single aerial photographs, today in GIS it is almost exclusively concerned with capturing 2.5-D and 3-D measurements from models derived from stereopairs of photographs and images. In the case of aerial photographs, it is usual to have 60% overlap along each "flight" line and 30% overlap between "flight" lines. Similar layouts are used by remote-sensing satellites. The amount of overlap defines the area for which a 3-D model can be created.

Photogrammetry is used to capture measurements from photographs and other image sources.

To obtain true georeferenced Earth coordinates from a model, it is necessary to georeference photographs using control points (the procedure is essentially analogous to that described for digitizing in Section 9.3.2.1). Control points can be defined by ground survey, or nowadays more usually with GPS (see Section 9.2.2.1 for discussion of these techniques).

Measurements are captured from overlapping pairs of images using stereoplotters. These build a model and allow 3-D measurements to be captured, edited, stored, and plotted. Stereoplotters have undergone three major generations of development: analog (optical), analytic, and digital. Mechanical analog devices are seldom used today, whereas analytical (combined mechanical and digital) and digital (entirely computer-based) are much more common. It is likely that digital (softcopy) photogrammetry will eventually replace mechanical devices entirely.

There are many ways to view stereo models, including a split screen with a simple stereoscope and the use of special glasses to observe a red/green display or polarized light. To manipulate 3-D cursors in the *x*, *y*, and *z* planes, photogrammetry systems offer free-moving hand controllers, hand wheels and foot disks, and 3-D mice. The options for extracting vector objects from 3-D models are directly analogous to those available for manual digitizing as described earlier: namely, batch, interactive, and manual (Section 9.3.2.1). The obvious difference, however, is that there is a requirement for capturing *z* (elevation) values.

Figure 9.13 shows a typical workflow in digital photogrammetry derived from the work of Vincent Tao (see Box 5.4). There are three main parts to digital photogrammetry workflows: data input, processing, and product generation. Data can be obtained directly from sensors or by scanning secondary sources. Orientation and triangulation are fundamental photogrammetry processing tasks. Orientation is the process of creating a stereo model suitable for viewing and extracting 3-D vector coordinates that describe geographic objects. Triangulation (also called block adjustment) is used to assemble a collection of images into a single model so that accurate and consistent information can be obtained from large areas.

Photogrammetry workflows yield several important product outputs, including digital elevation models (DEMs), contours, orthoimages, vector features,

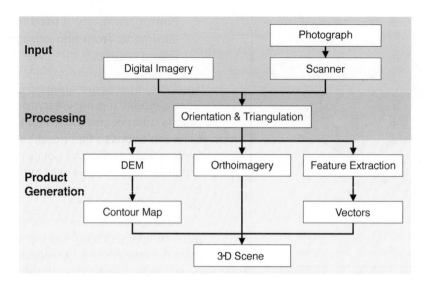

Figure 9.13 Typical photogrammetry workflow. (Reproduced by permission of GeoTec Media)

and 3-D scenes. DEMs—regular arrays of height values—are created by matching stereo image pairs together using a series of control points. Once a DEM has been created, it is relatively straightforward to derive contours using a choice of algorithms. Orthoimages are images corrected for variations in terrain using a DEM so as to appear as if every point was seen from vertically above. They have become popular because of their relatively low cost of creation (when compared with topographic maps) and ease of interpretation as base maps except

where tall buildings and other dramatic topographic features are present. They can also be used as accurate data sources for heads-up digitizing (see Section 9.3.2.1). Vector feature extraction is still an evolving field, and there are no widely applicable fully automated methods. The most successful methods use a combination of spectral analysis and spatial rules that define context, shape, proximity, and the like. Finally, 3-D scenes can be created by merging vector features with a DEM and an ortho-image (Figure 9.14).

Figure 9.14 Automatically created 3D city model of Oslo National Theatre. (Ludvig Emgård-C3 Technologies)

In summary, photogrammetry is a very cost-effective data capture technique that is sometimes the only practical method of obtaining detailed topographic data about an area of interest. Unfortunately, the complexity and high cost of equipment have restricted its use to primary data capture projects and specialist data capture organizations where very detailed information is required.

9.3.2.4 COGO Data Entry

COGO, which as noted earlier is a contraction of the term *coordinate geometry*, is a methodology for capturing and representing geographic data. COGO uses survey-style bearings and distances to define each part of an object in much the same way as described in Section 9.2.2. Figure 9.8 shows how land parcel features can be created using COGO tools and then formed into topologically correct polygons. Some examples of COGO object-construction tools are shown in Figure 9.15. The Construct Along tool creates a point along a curve using a distance along the curve. The Line Construct Angle Bisector tool constructs a line that bisects an angle defined by a from-point, through-point, to-point, and a length. The Construct Fillet tool creates a circular-arc tangent from two segments and a radius.

The COGO system is widely used in North America to represent land records and property parcels (also called lots). Coordinates can be obtained from COGO measurements by geometric transformation (i.e., bearings and distances are converted into x, y coordinates). Although COGO data obtained as part of a primary data capture activity are used in some projects, it is more often the case that secondary measurements are captured from hardcopy maps and documents. Source data may be in the form of legal descriptions, records of survey, tract (housing estate) maps, or similar documents.

> **COGO stands for coordinate geometry. It is a vector data structure and method of data entry.**

COGO data are very precise measurements and are often regarded as the only legally acceptable definition of land parcels. Measurements are usually very detailed, and data capture is often time-consuming. Furthermore, commonly occurring discrepancies in the data must be manually resolved by highly qualified individuals.

9.4 Obtaining Data from External Sources (Data Transfer)

One major decision that needs to be faced at the start of a GIS project is whether to build or buy part or all of a database. All of the preceding discussion has been concerned with techniques for building databases from primary and secondary sources. This section focuses on how to import or transfer data into a GIS that has been captured by others. Some datasets are freely available, but many of them are sold as commodities from a variety of outlets including, increasingly, Web sites.

There are many sources and types of geographic data. Space does not permit a comprehensive review of all geographic data sources here, but a small selection of key sources is listed in Table 9.3. In any case, the characteristics and availability of datasets are constantly changing, so those seeking an up-to-date list should consult one of the online sources described in this section. Section 18.4.1 also discusses the characteristics of geographic information and highlights several issues to bear in mind when using data collected by others.

The best way to find geographic data is to search the Internet. Several types of resources and technologies are available to assist searching, and these are

Figure 9.15 Example COGO construction tools used to represent geographic features.

Construct Along

distance (or ratio)

curve

Line Construct Angle Bisector

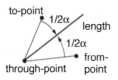

to-point

1/2α length

1/2α

from-point

through-point

Construct Fillet

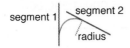

segment 1 segment 2

radius

Table 9.3 Examples of some digital data sources that can be imported into a GIS. NMOs = National Mapping Organizations, USGS = United States Geological Survey, NGA = U.S. National Geospatial-Intelligence Agency, NASA = National Aeronautics and Space Administration, DEM = Digital Elevation Model, EPA = U.S. Environmental Protection Agency, WWF = World Wide Fund for Nature, FEMA = Federal Emergency Management Agency, ESRI BIS = ESRI Business Information Solutions.

Type	Source	Details
Base Maps		
Geodetic framework	Many NMOs, e.g., USGS and Ordnance Survey	Definition of framework, map projections, and geodetic transformations
General topographic map data	NMOs and military agencies, e.g., NGA	Many types of data at detailed to medium scales
Elevation	NMOs, military agencies, and several commercial providers, e.g., USGS, SPOT Image, NASA	DEMs, contours at local, regional, and global levels
Transportation	National governments, and several commercial vendors, e.g., TeleAtlas and NAVTEQ	Highway/street centerline databases at national levels
Hydrology	NMOs and government agencies	National hydrological databases are available for many countries
Toponymy	NMOs, other government agencies, and commercial providers	Gazetteers of place-names at global and national levels
Satellite images	Commercial, government, and military providers, e.g., EROS Data Center, IRS, NASA, SPOT Image, i-cubed, Digital Globe, and GeoEye	See Figure 9.2 for further details
Aerial photographs	Many private and public agencies	Scales vary widely, typically from 1:500–1:20,000
Environmental		
Wetlands	National agencies, e.g., U.S. National Wetlands Inventory	Government wetlands inventory
Toxic release sites	National environmental protection agencies, e.g., EPA	Details of thousands of toxic sites
World eco-regions	Conservation agencies, e.g., WWF	Habitat types, threatened areas, biological distinctiveness
Flood zones	Many national and regional government agencies, e.g., FEMA	National flood-risk areas
Socioeconomic		
Population census	National governments, with value added by commercial providers	Typically, every 10 years with annual estimates
Lifestyle classifications	Private agencies (e.g., CACI and Experian)	Derived from population censuses and other socioeconomic data
Geodemographics	Private agencies (e.g., Claritas and ESRI BIS)	Many types of data at many scales and prices
Land and property ownership	National governments	Street, property, and cadastral data
Administrative areas	National governments	Obtained from maps at scales of 1:5,000–1:750,000

Table 9.4 Selected Web sites containing information about geographic data sources.

Source	URL	Description
AGI GIS Gigateway	www.gigateway.org.uk/	Indexed list of data sites
Geospatial One-Stop	www.geodata.gov	Geoportal providing metadata and direct access to over 15,000 datasets
MapMart	msrmaps.com	Extensive data and imagery provider
EROS Data Center	edc.usgs.gov/	U.S. government data archive
Terraserver	www.terraserver-usa.com/	Aerial imagery and topographic maps
National Geographic Society	www.nationalgeographic.com	Worldwide maps
GeoConnections	www.geoconnections.org	Canadian government's geographic data over the Web
EuroGeographics	www.eurogeographics.org	Coalition of European National Mapping Organizations offering topographic map data
GEOWorld Data Directory	www.geoplace.com	List of GIS data companies
The Data Depot	www.gisdatadepot.com	Extensive collection of mainly free geographic data

described in detail in Section 11.2. They include specialist geographic data catalogs and stores, as well as the sites of specific geographic data vendors (some Web sites are shown in Table 9.4). These sites provide access to information about the characteristics and availability of geographic data. Some also have facilities to purchase and download data directly. Probably the most useful resources for locating geographic data are the geolibraries and geoportals (see Section 11.2.2) that have been created as part of national and global spatial data infrastructure initiatives (SDI).

> **The best way to find geographic data is to search the Internet using one of the specialist geolibraries or SDI geographic data geoportals.**

A major challenge of using data obtained from the Web is evaluation of fitness for purpose. Too often inexperienced GIS practitioners download data from the Web and assume that its accuracy and licensing terms are adequate for use in a GIS project. It is essential that the suitability of all datasets be checked before being used. A good starting point is to examine the metadata records associated with the dataset (Section 11.2); these records should indicate age, provenance, projection, and a range of other relevant properties. Simple checks include overlay of the data on top of a base map of known and acceptable accuracy, and independent verification (e.g., by field

work or by comparison with other datasets) of the geometric and attribute properties of a representative sample of objects.

9.4.1 Geographic Data Formats

One of the biggest problems with data obtained from external sources is that they can be encoded in many different formats. There are so many different geographic data formats because no single format is appropriate for all tasks and applications. It is not possible to design a format that supports, for example, both fast rendering in police command and control systems, and sophisticated topological analysis in natural resource information systems: the two are mutually incompatible. Also, given the great diversity of geographic information, a single comprehensive format would simply be too large and cumbersome. The many different formats that are in use today have evolved in response to diverse user requirements.

Given the high cost of creating databases, many tools have been developed to move data between systems and to reuse data through open application programming interfaces (APIs). In the former case, the approach has been to develop software that is able to translate data (Figure 9.16), either by a direct read into memory or via an intermediate file format. In the latter case, software developers have created open interfaces to allow access to data.

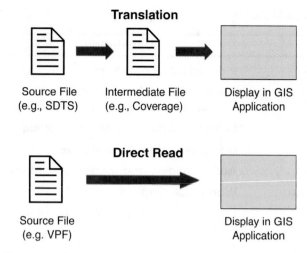

Translation

Source File
(e.g., SDTS)

Intermediate File
(e.g., Coverage)

Display in GIS
Application

Direct Read

Source File
(e.g. VPF)

Display in GIS
Application

Figure 9.16 Comparison of data access by translation and direct read.

Many GIS software systems are now able to read directly AutoCAD DWG and DXF, Microstation DGN, and ESRI Shapefile, VPF, and many image formats. Unfortunately, direct read support can only be easily provided for relatively simple product-oriented formats. Complex formats, such as SDTS (Spatial Data Transfer Standard), were designed for exchange purposes and require more advanced processing before they can be viewed (e.g., multipass read and feature assembly from several parts).

Data can be transferred between systems by direct read into memory or via an intermediate file format.

More than 25 organizations are involved in the standardization of various aspects of geographic data and geoprocessing; several of them are country and domain specific. At the global level, the ISO (International Standards Organization) is responsible for coordinating efforts through the work of technical committees TC 211 and 287. In Europe, CEN (Comité Européen de Normalisation) is engaged in geographic standardization. At the national level, there are many complementary bodies. One other standards-forming organization of particular note is OGC (Open Geospatial Consortium: www.opengeospatial.org), a group of vendors, academics, and users interested in the interoperability of geographic systems. To date, there have been promising OGC-coordinated efforts to standardize on simple feature access (simple geometric object types), metadata catalogs, and Web access (see Chapter 11 for further details).

The most efficient way to translate data between systems is usually via a common intermediate file format.

Having obtained a potentially useful source of geographic information, the next task is to import it into a GIS database. If the data are already in the native format of the target GIS software system, or the software has a direct read capability for the format in question, then this is a relatively straightforward task. If the data are not compatible with the target GIS software, then the alternatives are to ask the data supplier to convert the data to a compatible format, or to use a third-party translation software system, such as the Feature Manipulation Engine from Safe Software (www.safe.com lists over 225 supported geographic data formats) to convert the data. Geographic data translation software must address both syntactic and semantic translation issues. Syntactic translation involves converting specific digital symbols (letters and numbers) between systems, whereas semantic translation is concerned with converting the meaning inherent in geographic information. While syntactic translation is relatively simple to encode and decode, semantic translation is much more difficult and has seldom met with much success to date.

Although the task of translating geographic information between systems was described earlier as relatively straightforward, those that have tried this in practice will realize that things on the ground are seldom quite so simple. Any number of things can (and do!) go wrong, ranging from corrupted media to incomplete data files, incompatible versions of translators, and different interpretations of a format specification, to basic user error.

There are two basic strategies used for data translation: one is direct and the other uses a neutral intermediate format. For small systems that involve the translation of a small number of formats, the first is the simplest. Directly translating data back and forth between the internal structures of two systems requires two new translators (A to B, B to A). Adding two further systems will require 12 translators to share data between all systems (A to B, A to C, A to D, B to A, B to C, B to D, C to A, C to B, C to D, D to A, D to B, and D to C). A more efficient way of solving this problem is to use the concept of a data switchyard and a common intermediate file format. Systems now need only to translate to and from the common format. The four systems use only 8 translators instead of 12 (A to Neutral, B to Neutral, C to Neutral, D to Neutral, Neutral to A, Neutral to B, Neutral to C, and Neutral to D). The more systems there are, the more efficient the result. This is one of the key principles underlying the need for common file-interchange formats.

9.5 Capturing Attribute Data

All geographic objects have attributes of one type or another. Although attributes can be collected at the same time as vector geometry, it is usually more cost-effective to capture attributes separately. In part, this is because attribute data capture is a relatively simple task that can be undertaken by lower-cost clerical staff. It is also because attributes can be entered by direct data loggers, manual keyboard entry, optical character recognition (OCR), or, increasingly, voice recognition, methods that do not require expensive hardware and software systems. By far the most common method is direct keyboard data entry into a spreadsheet or database. For some projects, a custom data-entry form with built-in validation is preferred. On small projects single entry is used, but for larger, more complex projects data are entered twice and then compared as a validation check.

An essential requirement for separate data entry is a common identifier (also called a key, or object-id) that can be used to relate object geometry and attributes together following data capture (see Figure 8.10 for a diagrammatic explanation of relating geometry and attributes).

Metadata are a special type of nongeometric data that are increasingly being collected. Some metadata are derived automatically by the GIS software system (for example, length and area, extent of data layer, and count of features), but some must be explicitly collected (for example, owner name, quality estimate, and original source). Explicitly collected metadata can be entered in the same way as other attributes, as described earlier. For further information about metadata, see Section 11.2.

9.6 Citizen-centric Web-based Data Collection

New developments in Web technology have opened up a new vista of opportunities for distributed GIS data collection. A raft of new Web 2.0 technologies has enabled organizations and individual projects to use citizens to collect data very rapidly, across a wide variety of thematic and geographic areas that represent a wide spectrum of viewpoints. It is now very simple to create a Web site with a form-based interface to collect geographic data about many types of phenomena, events, and activities. Locations can be obtained by asking the user to digitize the location on a map, or to upload coordinates collected in any of the ways outlined earlier in this chapter. This type of approach to data collection by volunteers is discussed in Section 11.2. Box 9.2 presents an example of a Web 2.0, citizen-centric data collection application.

9.7 Managing a Data Collection Project

The subject of managing a GIS project is given extensive treatment in Chapters 17–19. The management of data capture projects is discussed briefly here, both because of its critical importance and because it involves several unique issues. That said, most of the general principles for any GIS project apply to data collection: the need for a clearly articulated plan, adequate resources, appropriate funding, and sufficient time.

In any data collection project there is a fundamental trade-off between quality, speed, and price. Collecting high-quality data quickly is possible, but it is also very expensive. If price is a key consideration, then lower-quality data can be collected over a longer period (Figure 9.18).

GIS data collection projects can be carried out intensively or over a longer period. A key decision facing managers of such projects is whether to pursue a strategy of incremental or very rapid collection. Incremental data collection involves breaking the data collection project into small manageable subprojects. This allows data collection to be undertaken with lower annual resource and funding levels (although total project resource requirements may be larger). It is a good approach for inexperienced organizations that are embarking on their first data collection project because they can learn and adapt as the project

Figure 9.18 Relationship between quality, speed, and price in data collection. (Source: after Hohl 1997)

E-Flora British Columbia, Citizen-centric Data Collection

The British Columbia, Canada E-Flora BC project is a good example of the way GIS data collection is changing to incorporate citizen-centric data input. E-Flora BC is an online, Web-accessible electronic atlas of plant species that allows interactive data collection, reporting, and mapping (www.eflora. bc.ca).

The public is encouraged to participate in an Invasive Alien Plant Program run by the Project by reporting suspected new occurrences of invasive plants using an interactive form-based Report-a-Weed tool. This tool (Figure 9.17) allows citizen-scientists to collect and enter pertinent information about invasive species, which is then delivered to an appropriate botanist for review and then entry into the organization's master database.

The Web has radically transformed the way that databases of this type are collected and maintained: no longer is this process the sole preserve of the official government organizations. This type of Web 2.0 data collection and public empowerment is especially good at handling local (in space and time) phenomena. There is a rapidly growing list of examples of Web projects that rely on volunteers to collect geographic information (see also Section 11.2 for further discussion).

(A)

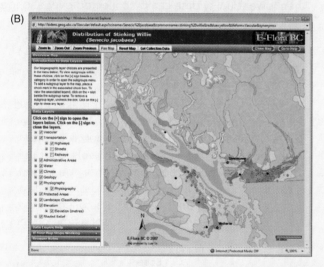

(B)

(C)

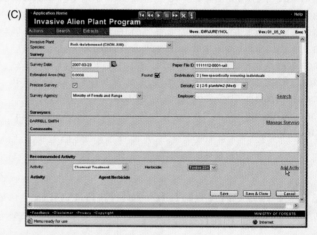

Figure 9.17 E-Flora information about *Senioco jacobaea* (the "stinking willie" or tansy ragwort): (A) photograph (© DND ASU Chilliwack) (B) distribution map; and (C) citizen-centric Web data input form. (Photo courtesy of Department of National Defense, ASU, Chilliwack)

proceeds. At the same time, these longer-term projects run the risk of employee turnover and burnout, as well as changing data, technology, and organizational priorities.

Whichever approach is preferred, a pilot project carried out on part of the study area and a selection of the data types can prove to be invaluable. A pilot project can identify problems in workflow, database design, personnel, and equipment. A pilot database can also be used to test equipment and to develop procedures for quality assurance. Many projects require a test database to carry out hardware and software acceptance tests, as well as to facilitate software customization. It is essential that project managers are prepared to discard all the data obtained during a pilot data collection project, so that the main phase can proceed unconstrained.

A further important decision is whether data collection should use in-house or external resources. It is now increasingly common to outsource geographic data collection to specialist companies that usually undertake the work in areas of the world with very low labor costs (e.g., India, Thailand, and Vietnam). Three factors influencing this decision are cost/schedule, quality, and long-term ramifications. Specialist external data collection agencies can often perform work faster, cheaper, and with higher quality than with in-house staff, but because of the need for real cash to pay external agencies this may not be possible. In the short term, project costs, quality, and time are the main considerations, but over time dependency on external groups may become a problem.

Questions for Further Study

1. Using the Web sites listed in Table 9.4 as a starting point, evaluate the suitability of free geographic data for your home region or country for use in a GIS project of your choice.

2. What are the advantages of batch vectorization over heads-up digitizing?

3. What quality assurance steps would you build into a data collection project designed to build a database of land parcels for tax assessment?

4. Why do so many geographic data formats exist? Which ones are most suitable for selling vector data?

Further Reading

Hohl, P. (ed.). 1997. *GIS Data Conversion: Strategies, Techniques and Management*. Santa Fe, NM: OnWord Press.

Jensen, J.R. 2007. *Remote Sensing of the Environment: An Earth Resource Perspective* (2nd ed.). Upper Saddle River, NJ: Prentice Hall.

Jones, C. 1997. *Geographic Information Systems and Computer Cartography*. Reading, MA: Addison-Wesley Longman.

Lillesand, T.M., Kiefer, R.W. and Chipman, R.W. 2004. *Remote Sensing and Image Interpretation* (5th ed.). Hoboken, NJ: Wiley.

Paine, D.P. and Kiser, J.D. 2003. *Aerial Photography and Image Interpretation* (2nd ed.). Hoboken, NJ: Wiley.

Walford, N. 2002. *Geographical Data: Characteristics and Sources*. Hoboken, NJ: Wiley

10 Creating and Maintaining Geographic Databases

All large operational GIS are built on the foundation of a geographic database. After the people who manage and operate a GIS, the database is arguably the most important part because of the costs of collection and maintenance, and because the database forms the basis of all query, analysis, and decision-making activities. Today, virtually all large GIS implementations store data in a database management system (DBMS), a specialist piece of software designed to handle multiuser access to an integrated set of data. Extending standard DBMS to store geographic data raises several interesting challenges. Databases need to be designed with great care and should be structured and indexed to provide efficient query and transaction performance. A comprehensive security and transactional access model is necessary to ensure that multiple users can access the database at the same time. Ongoing maintenance is also an essential, but very resource-intensive, activity.

LEARNING OBJECTIVES

After studying this chapter you will be able to:

- Understand the role of database management systems in GIS.

- Recognize Structured Query Language (SQL) statements.

- Understand the key geographic database data types and functions.

- Be familiar with the stages of geographic database design.

- Understand the key techniques for structuring geographic information, specifically creating topology and indexing.

- Understand the issues associated with multiuser editing and versioning.

10.1 Introduction

A database can be thought of as an integrated set of data on a particular subject. Geographic databases are simply databases containing geographic data for a particular area and subject. It is quite common to encounter the term *spatial* in the database world. As discussed in Section 1.1.1, spatial refers to data about any type of space at both geographic and nongeographic scales. A geographic database is a critical part of an operational GIS both because of the cost of creation and maintenance, and because of the impact of a geographic database on all analysis, modeling, and decision-making activities. Databases can be physically stored in files or by using specialist software programs called DBMS. Today, most large organizations use a combination of files and DBMS for storing data assets.

A database is an integrated set of data on a particular subject.

The database approach to storing geographic data offers a number of advantages over traditional file-based datasets.

- Collecting all data at a single location reduces redundancy.

- Maintenance costs decrease because of better organization and reduced data duplication.

- Applications become data independent so that multiple applications can use the same data and can evolve separately over time.

- User knowledge can be transferred between applications more easily because the database remains constant.

- Data sharing is facilitated, and a corporate view of data can be provided to all managers and users.

- Security and standards for data and data access can be established and enforced.

- DBMS are better able to manage large numbers of concurrent users working with vast amounts of data.

Using databases when compared to files also has some disadvantages.

- The cost of acquiring and maintaining DBMS software can be quite high.

- A DBMS adds complexity to the problem of managing data, especially in small projects.

- Single-user performance will often be better for files, especially for more complex data types and structures where specialist indexes and access algorithms can be implemented.

In recent years geographic databases have become increasingly large and complex (see Table 1.1). For example, a U.S. National Image Mosaic will be over 25 terabytes (TB) in size, a Landsat satellite global image mosaic at 15-m resolution is 6.5 TB, and the Ordnance Survey of Great Britain has approximately 450 million vector features in its MasterMap database covering all of Britain. This chapter describes how to create and maintain geographic databases, and presents the concepts, tools, and techniques that are available to manage geographic data in databases. Several other chapters provide additional information that is relevant to this discussion. In particular, the nature of geographic data and how to represent them in GIS were described in Chapters 3, 4, and 5, and data modeling and data collection were discussed in Chapters 8 and 9, respectively. Later chapters introduce the tools and techniques that are available to query, model, and analyze geographic databases (Chapters 14, 15, and 16). Finally, Chapters 17 through 19 discuss the important management issues associated with creating and maintaining geographic databases.

10.2 Database Management Systems

A DBMS is a software application designed to organize the efficient and effective storage and access of data.

Small, simple databases that are used by a small number of people can be stored on computer hard disks in standard files. However, large, more complex databases with many tens, hundreds, or thousands of users require specialist database management system (DBMS) software to ensure data integrity and longevity. A DBMS is a software application designed to organize the efficient and effective storage of and access to data. To carry out this function, DBMS provide a number of important capabilities. These are introduced briefly here and are discussed further in this and other chapters.

DBMS are successful because they are able to provide:

- *A data model.* As discussed in Chapter 8, a data model is a mechanism used to represent real-world objects digitally in a computer system. All DBMS include standard general-purpose core data models suitable for representing several object types (e.g., integer and floating-point numbers, dates, and text). Several DBMS now also support geographic (spatial) object types.

- *A data load capability.* DBMS provide tools to load data into databases. Simple tools are available to load standard supported data types (e.g., character, number, and date) in well-structured formats.

- *Indices.* An index is a data structure that is used to speed up searching. All databases include tools to index standard database data types.

- *A query language.* One of the major advantages of DBMS is that they support a standard data query/manipulation language called SQL (Structured/Standard Query Language).

- *Security.* A key characteristic of DBMS is that they provide controlled access to data. This includes the ability to restrict user access to just part of a database. For example, a casual GIS user might have read-only access to only part of a database, but a specialist user might have read and write (create, update, and delete) access to the entire database.

- *Controlled update.* Updates to databases are controlled through a transaction manager responsible for managing multiuser access and ensuring that updates affecting more than one part of the database are coordinated.

- *Backup and recovery.* It is important that the valuable data in a database are protected from system failure and incorrect (accidental or deliberate) update. Software utilities are provided to back up all or part of a database and to recover the database in the event of a problem.

- *Database administration tools.* The task of setting up the structure of a database (the schema), creating and maintaining indices, tuning to improve performance, backing up and recovering, and allocating user access rights is performed by a database administrator (DBA). A specialized collection of tools and a user interface are provided for this purpose.

- *Applications.* Modern DBMS are equipped with standard, general-purpose tools for creating, using, and maintaining databases. These include applications for designing databases (CASE [computer-aided software engineering] tools) and for building user interfaces for data access and presentations (e.g., forms and reports).

- *Application programming interfaces (APIs).* Although most DBMS have good general-purpose applications for standard use, most large, specialist applications will require further customization using a commercial off-the-shelf programming language and a DBMS programmable API.

This list of DBMS capabilities is very attractive to GIS users and so, not surprisingly, virtually all large GIS databases are managed using DBMS technology. Indeed, most GIS software vendors include DBMS software within their GIS software products, or provide an interface that supports very close coupling to a DBMS. For further discussion of this subject, see Chapter 7.

Today, virtually all large GIS use DBMS technology for data management.

10.2.1 Types of DBMS

DBMS can be classified according to the way they organize, store, and manipulate data. Three main types of DBMS have been used in GIS: relational (RDBMS), object (ODBMS), and object-relational (ORDBMS).

A relational database comprises a set of tables, each a two-dimensional list (or array) of records containing attributes about the objects under study. This apparently simple structure has proven to be remarkably flexible and useful in a wide range of application areas, such that historically over 95% of the data in DBMS have been stored in RDBMS. Most of the world's current DBMS are built on a foundation of core relational concepts.

Object database management systems (ODBMS) were initially designed to address several of the weaknesses of RDBMS. These include the inability to store complete objects directly in the database (both object state and behavior: see Box 8.2 for an overview of objects and object technology). Because RDBMS were focused primarily on business applications such as banking, human resource management, and stock control and inventory, they were never designed to deal with rich data types, such as geographic objects, sound, and video. A further difficulty is the poor performance of RDBMS for many types of geographic query. These problems are compounded by the difficulty of extending RDBMS to support geographic data types and processing functions, which obviously limits their adoption for geographic applications. ODBMS can store objects persistently (semipermanently on disk or other media) and provide object-oriented query tools. A number of commercial ODBMS have been developed including GemStone/S Object Server from GemStone Systems, Inc., Objectivity/DB from Objectivity, Inc., ObjectStore from Progress Software, and Versant from Versant Object Technology Corp.

In spite of the technical elegance of ODBMS, they have not proven to be as commercially successful as some people initially predicted. This is largely because of the massive installed base of RDBMS and the fact that RDBMS vendors have now added many of the important ODBMS capabilities to their standard RDBMS software systems to create hybrid object-relational DBMS (ORDBMS). An ORDBMS can be thought of as an RDBMS engine with some additional capabilities for dealing with objects. They can handle both the data describing what an object is (object attributes such as color, size, and age) and the behavior that determines what an object does (object methods or functions such as drawing instructions, query interfaces, and interpolation algorithms), and these can be managed and stored together as an integrated whole. Examples of ORDBMS software include IBM DB2 and Informix Dynamic Server, Microsoft SQL Server, and Oracle Corp. Oracle DBMS. Because ORDBMS and the underlying relational model are so important in GIS, these topics are discussed at length in Section 10.3.

A number of ORDBMS have now been extended to support geographic object types and functions through the addition of seven key capabilities (these are introduced here and discussed further later in this chapter):

- Query parser—the engines used to interpret queries by splitting them up and decoding them have been extended to deal with geographic types and functions.

- Query optimizer—software query optimizers have been enhanced so that they are able to handle geographic queries efficiently. Consider a query to find all potential users of a new brand of premier wine to be marketed to wealthy households from a network of retail stores. The objective is to select all households within 3 km of a store that have an income greater than $110,000. This could be carried out in two ways:
  1. Select all households with an income greater than $110,000; from this selected set, select all households within 3 km of a store.
  2. Select all households within 3 km of a store; from this selected set select all households with an income greater than $110,000.

  Selecting households with an income greater than $110,000 is an attribute query that can be performed very quickly. Selecting households within 3 km of a store is a geometric query that takes much longer. Executing the attribute query first (option 1 above) will result in fewer geometry query tests for store proximity, and therefore the whole query will be completed much more quickly.

- Query language—query languages have been improved to handle geographic types (e.g., points and areas) and functions (e.g., select areas that touch each other).

- Indexing services—standard unidimensional DBMS data index services have been extended to support multidimensional (i.e., x, y, z coordinates) geographic data types.

- Storage management—the large volumes of geographic records with different sizes (especially geometric and topological relationships) have been accommodated through specialized storage structures.

- Transaction services—standard DBMS are designed to handle short (subsecond) transactions, and these have been extended to deal with the long transactions common in many geographic applications.

- Replication—services for replicating databases have been extended to deal with geographic types and with problems of reconciling changes made by distributed users.

10.2.2 Geographic DBMS Extensions

A number of the major commercial DBMS vendors have released spatial database extensions to their standard ORDBMS products. IBM offers two solutions—DB2 Spatial Extender and Informix Spatial Datablade; Microsoft has released spatial capabilities in the core of SQLServer; and Oracle has spatial in the core of Oracle DBMS and a Spatial option that adds more advanced features (see Box 10.1). The open-source DBMS PostgreSQL has also been extended with spatial types and functions (PostGIS).

ORDBMS provide core support for geographic data types and functions.

Although these systems differ in technology, scope, and features, they all provide basic capabilities to store, manage, and query geographic objects. This is achieved by implementing the seven key database extensions described in the previous section. It is important to realize, however, that none of these is a complete GIS software system in itself. The focus of these extensions is data storage, retrieval, and management, and they lack the advanced capabilities for geographic editing, mapping, and analysis. Consequently, they must be used in conjunction with a conventional GIS software system except in the case of the simplest query-focused applications. Figure 10.1 shows how GIS and DBMS software can work together and some of the tasks best carried out by each system.

Oracle Spatial

Oracle Spatial is an extension to the Oracle DBMS that provides the foundation for the management of spatial (including geographic) data inside an Oracle database. The standard types and functions in Oracle (CHAR, DATE or INTEGER, etc.) are extended with geographic equivalents. Oracle Spatial supports three basic geometric forms:

- Points: points can represent locations such as buildings, fire hydrants, utility poles, oil rigs, box-cars, or roaming vehicles.

- Lines: lines can represent things like roads, rivers, utility lines, or fault lines.

- Polygons and complex polygons with holes: poly-gons can represent things like outlines of cities, districts, floodplains, or oil and gas fields. A poly-gon with a hole might geographically represent a parcel of land surrounding a patch of wetland.

These simple feature types can be aggregated to richer types using topology and linear referencing capabilities (see Section 8.2.3.3). In addition, Oracle Spatial can store and manage georaster (image) data. Oracle Spatial extends the Oracle DBMS query engine to support geographic queries. There is a set of spatial operators to perform: area-of-interest and spatial-join queries; length, area, and distance calcula-tions; buffer and union queries; and administrative tasks. The Oracle Spatial SQL used to create a table and populate it with a single record is shown in the following script. (The characters after the dash on each line are comments that describe the operations. The discussion of SQL syntax in Section 10.4 will help decode this program.)

```
-- Create a table for routes (highways).
CREATE TABLE lrs_routes (
 route_id  NUMBER PRIMARY KEY,
 route_name  VARCHAR2(32),
 route_geometry  MDSYS.SDO_GEOMETRY);
-- Populate table with just one route
   for this example.
INSERT INTO lrs_routes VALUES(
 1,
 'Route1',
```

```
MDSYS.SDO_GEOMETRY(
 3002, -- line string, 3 dimensions:
    X,Y,M
 NULL,
 NULL,
 MDSYS.SDO_ELEM_INFO_ARRAY(1,2,1), -- one
    line string, straight segments
MDSYS.SDO_ORDINATE_ARRAY(
 2,2,0, -- Starting point - Exit1; 0
    is measure from start.
 2,4,2, -- Exit2; 2 is measure from
    start.
 8,4,8, -- Exit3; 8 is measure from
    start.
 12,4,12, -- Exit4; 12 is measure from
    start.
 12,10,NULL, -- Not an exit; measure
    will be automatically calculated &
    filled.
 8,10,22, -- Exit5; 22 is measure from
    start.
 5,14,27) -- Ending point (Exit6); 27
    is measure from start.
 )
);
```

Geographic data in Oracle Spatial can be indexed using R-tree and quadtree indexing meth-ods (these terms are defined in Section 10.7.2). There are also capabilities for managing projections and coordinate systems, as well as long transactions (see discussion in Section 10.9.1). Finally, there are also some tools for elementary spatial data analysis (Chapters 14 and 15). Oracle Spatial can be used with all major GIS software products, and develop-ers can create specific-purpose applications that embed SQL commands for manipulating and querying data. Oracle has generated considerable interest among larger IT-focused organizations. IBM has approached this market in a similar way with its Spatial Extender for the DB2 DBMS and Spatial Datablade for Informix. Most recently Microsoft has added comparable spatial capabilities to its SQLServer DBMS.

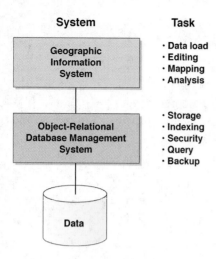

System

Geographic Information System

Task

• Data load
• Editing
• Mapping
• Analysis

Object-Relational Database Management System

• Storage
• Indexing
• Security
• Query
• Backup

Data

Figure 10.1 The roles of GIS and DBMS.

10.3 Storing Data in DBMS Tables

The lowest level of user interaction with a geographic database is usually the object class (also sometimes called a layer or feature class), which is an organized collection of data on a particular theme (e.g., all pipes in a water network, all soil polygons in a river basin, or all elevation values in a terrain surface). Object classes are stored in standard database tables. A table is a two-dimensional array of rows and columns. Each object class is stored as a single database table in a DBMS. Table rows contain objects (instances of object classes, e.g., data for a single pipe) and the columns contain object properties, or attributes as they are frequently called (Figure 10.2: see also Figure 8.10 as a conceptual example). The data stored at individual row, column intersections are usually referred to as values. Geographic database tables are distinguished from nongeographic tables by the presence of a geometry column (often called the shape column). To save space and improve performance, the actual coordinate values may be stored in a highly compressed binary form.

Relational databases are made up of tables. Each geographic class (layer) is stored as a table.

Tables are joined together using common row/column values or keys as they are known in the database world. Figure 10.2 shows parts of tables containing data about U.S. states. The STATES table (Figure 10.2A) contains the geometry (in the Shape field) and some basic attributes, an important one being a unique STATE FIPS (STATE_FIPS [Federal

Figure 10.2 Parts of GIS database tables for U.S. states: (A) STATES table; (B) POPULATION table (*continued*)

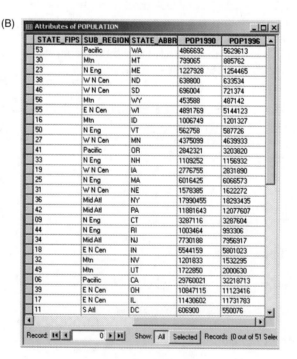

(A)

FID	Shape*	AREA	STATE_NAME	STATE_FIPS
41	Polygon	51715.656	Alabama	01
49	Polygon	576556.687	Alaska	02
35	Polygon	113711.523	Arizona	04
45	Polygon	52912.797	Arkansas	05
23	Polygon	157774.187	California	06
30	Polygon	104099.109	Colorado	08
17	Polygon	4976.434	Connecticut	09
27	Polygon	2054.506	Delaware	10
26	Polygon	66.063	District of Columbia	11
47	Polygon	55815.051	Florida	12
43	Polygon	58629.195	Georgia	13
48	Polygon	6381.435	Hawaii	15
7	Polygon	83340.594	Idaho	16
25	Polygon	56297.953	Illinois	17
20	Polygon	36399.516	Indiana	18
12	Polygon	56257.219	Iowa	19
32	Polygon	82195.437	Kansas	20
31	Polygon	40318.777	Kentucky	21
46	Polygon	45835.898	Louisiana	22
2	Polygon	32161.664	Maine	23
29	Polygon	9739.753	Maryland	24
13	Polygon	8172.482	Massachusetts	25
50	Polygon	57898.367	Michigan	26
9	Polygon	84517.469	Minnesota	27
42	Polygon	47618.723	Mississippi	28
34	Polygon	69831.625	Missouri	29
1	Polygon	147236.031	Montana	30

Record: 0   Show: All Selected Records (0 out of 51 Selected.)

(B)

STATE_FIPS	SUB_REGION	STATE_ABBR	POP1990	POP1996
53	Pacific	WA	4866692	5629613
30	Mtn	MT	799065	885762
23	N Eng	ME	1227928	1254465
38	W N Cen	ND	638800	633534
46	W N Cen	SD	696004	721374
56	Mtn	WY	453588	487142
55	E N Cen	WI	4891769	5144123
16	Mtn	ID	1006749	1201327
50	N Eng	VT	562758	587726
27	W N Cen	MN	4375099	4639933
41	Pacific	OR	2842321	3203820
33	N Eng	NH	1109252	1156932
19	W N Cen	IA	2776755	2831890
25	N Eng	MA	6016425	6066573
31	W N Cen	NE	1578385	1622272
36	Mid Atl	NY	17990455	18293435
42	Mid Atl	PA	11881643	12077607
09	N Eng	CT	3287116	3287604
44	N Eng	RI	1003464	993306
34	Mid Atl	NJ	7730188	7956917
18	E N Cen	IN	5544159	5801023
32	Mtn	NV	1201833	1532295
49	Mtn	UT	1722850	2000630
06	Pacific	CA	29760021	32218713
39	E N Cen	OH	10847115	11123416
17	E N Cen	IL	11430602	11731783
11	S Atl	DC	606900	550076

Record: 0   Show: All Selected Records (0 out of 51 Selec

FID	Shape*	AREA	STATE_NAME	STATE_FIPS	SUB_REGION	STATE_ABBR	POP1990	POP1996
0	Polygon	67286.875	Washington	53	Pacific	WA	4866692	5629613
1	Polygon	147236.031	Montana	30	Mtn	MT	799065	885762
2	Polygon	32161.664	Maine	23	N Eng	ME	1227928	1254465
3	Polygon	70810.156	North Dakota	38	W N Cen	ND	638800	633534
4	Polygon	77193.625	South Dakota	46	W N Cen	SD	696004	721374
5	Polygon	97799.492	Wyoming	56	Mtn	WY	453588	487142
6	Polygon	56088.066	Wisconsin	55	E N Cen	WI	4891769	5144123
7	Polygon	83340.594	Idaho	16	Mtn	ID	1006749	1201327
8	Polygon	9603.218	Vermont	50	N Eng	VT	562758	587726
9	Polygon	84517.469	Minnesota	27	W N Cen	MN	4375099	4639933
10	Polygon	97070.750	Oregon	41	Pacific	OR	2842321	3203820
11	Polygon	9259.514	New Hampshire	33	N Eng	NH	1109252	1156932
12	Polygon	56257.219	Iowa	19	W N Cen	IA	2776755	2831890
13	Polygon	8172.482	Massachusetts	25	N Eng	MA	6016425	6066573
14	Polygon	77328.336	Nebraska	31	W N Cen	NE	1578385	1622272
15	Polygon	48560.578	New York	36	Mid Atl	NY	17990455	18293435
16	Polygon	45359.238	Pennsylvania	42	Mid Atl	PA	11881643	12077607
17	Polygon	4976.434	Connecticut	09	N Eng	CT	3287116	3287604
18	Polygon	1044.850	Rhode Island	44	N Eng	RI	1003464	993306
19	Polygon	7507.302	New Jersey	34	Mid Atl	NJ	7730188	7956917
20	Polygon	36399.516	Indiana	18	E N Cen	IN	5544159	5801023
21	Polygon	110667.297	Nevada	32	Mtn	NV	1201833	1532295
22	Polygon	84870.187	Utah	49	Mtn	UT	1722850	2000630
23	Polygon	157774.187	California	06	Pacific	CA	29760021	32218713
24	Polygon	41192.863	Ohio	39	E N Cen	OH	10847115	11123416
25	Polygon	56297.953	Illinois	17	E N Cen	IL	11430602	11731783
26	Polygon	66.063	District of Columbia	11	S Atl	DC	606900	550076
27	Polygon	2054.506	Delaware	10	S Atl	DE	666168	724890

Record: 0   Show: All Selected Records (0 out of 51 Selected.)   Options ▾

Figure 10.2 (*continued*) (C) joined table—COMBINED STATES and POPULATION.

Information Processing Standard] code) identifier. The POPULATION table (Figure 10.2B) was created entirely independently, but also has a unique identifier column called STATE_FIPS. Using standard database tools, the two tables can be joined together based on the common STATE_FIPS identifier column (the key) to create a third table, COMBINED STATES and POPULATION (Figure 10.2C). Following the join, these can be treated as a single table for all GIS operations such as query, display, and analysis. There is additional discussion of relational joins and their generalization to spatial joins in Section 14.2.2.

Database tables can be joined together to create new views of the database.

In a groundbreaking description of the relational model that underlies the vast majority of the world's databases, in 1970 Ted Codd of IBM defined a series of rules for the efficient and effective design of database table structures. The heart of Codd's idea was that the best relational databases are made up of simple, stable tables that follow five principles:

1. There is only one value in each cell at the intersection of a row and column.

2. All values in a column are about the same subject.

3. Each row is unique.

4. There is no significance to the sequence of columns.

5. There is no significance to the sequence of rows.

Figure 10.3A shows a land parcel tax assessment database table that contradicts several of Codd's principles. Codd suggests a series of transformations called normal forms that successively improve the simplicity and stability, and reduce the redundancy of database tables (thus reducing the risk of editing conflicts) by splitting them into subtables that are rejoined at query time. Unfortunately, joining large tables is computationally expensive and can result in complex database designs that are difficult to maintain. For this reason, non-normalized table designs are often used in GIS.

Figure 10.3B is a cleansed version of 10.3A that has been entered into a GIS DBMS: there is now only one value in each cell (Date and AssessedValue are now separate columns); missing values have been added; an OBJECTID (unique system identifier) column has been added; and the potential confusion between Dave Widseler and D Widseler has been resolved. Figure 10.3C shows the same data after some normalization to make it suitable for use in a GIS tax assessment application. The database now

(A)

(B)

(C)

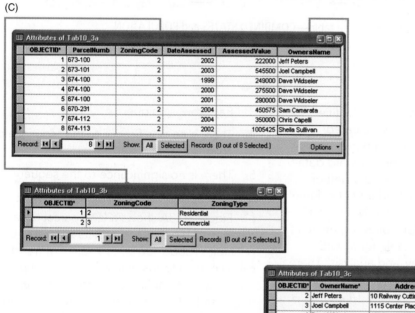

(D)

Figure 10.3 Tax assessment database: (A) raw data; (B) cleaned data in a GIS DBMS; (C) data partially normalized into three subtables; and (D) joined table.

consists of three tables that can be joined together using common keys. Figure 10.3C Attributes of Tab10_3b can be joined to Figure 10.3C Attributes of Tab10_3a using the common ZoningCode column, and Figure 10.3C **Attributes of Tab10_3c** can be joined using OwnersName to create Figure 10.3D. It is now possible to execute SQL queries against these joined tables as discussed in the next section.

10.4 SQL

The standard database query language adopted by virtually all mainstream databases is SQL (Structured or Standard Query Language: ISO Standard ISO/IEC 9075). There are many good background books and system implementation manuals on SQL, and so only brief details will be presented here. SQL may be used directly via an interactive command line interface; it may be compiled in a general-purpose programming language (e.g., C/C++/C#, Java, or Visual Basic); or it may be embedded in a graphical user interface (GUI). SQL is a set-based, rather than a procedural (e.g., Visual Basic) or object-oriented (e.g., Java or C#), programming language designed to retrieve sets (row and column combinations) of data from tables. There are three key types of SQL statements: DDL (data definition language), DML (data manipulation language), and DCL (data control language). The third major revision of SQL (SQL 3), which was released in 2004, defines spatial types and functions as part of a multimedia extension called SQL/MM.

The data in the database shown in Figure 10.3C may be queried to find parcels where the AssessedValue is greater than $300,000 and the ZoningType is Residential. This is an apparently simple query, but it requires three table joins to execute it. The SQL statements as implemented in the Microsoft Access DBMS are as follows:

```
SELECT Tab10_3a.ParcelNumb, Tab10_3c.Address,
   Tab10_3a.AssessedValue
FROM (Tab10_3b INNER JOIN Tab10_3a ON
   Tab10_3b.ZoningCode =
   Tab10_3a.ZoningCode) INNER JOIN Tab10_3c
   ON Tab10_3a.OwnersName =
   Tab10_3c.OwnerName
WHERE (((Tab10_3a.AssessedValue) >300000) AND
   ((Tab10_3b.ZoningType) ="Residential"));
```

The **SELECT** statement defines the columns to be displayed (the syntax is TableName.ColumnName). The **FROM** statement is used to identify and join the

Join3a3b3c : Select Query

ParcelNumb	Address	AssessedValue
673-101	1115 Center Place	545500
670-231	19 Big Bend Bld	450575
674-112	Hastings Barracks	350000
674-113	10034 Endin Mansions	1005425

Record: 1 of 4

Figure 10.4 Results of a SQL query against the tables in Figure 10.3C (see text for query and further explanation).

three tables (**INNER JOIN** is a type of join that signifies that only matching records in the two tables will be considered). The **WHERE** clause is used to select the rows from the columns using the constraints `(((Tab10_3a.AssessedValue) >300000) AND ((Tab10_3b.ZoningType) ="Residential"))`. The result of this query is shown in Figure 10.4. This triplet of **SELECT, FROM, WHERE** is the staple of SQL queries.

SQL is the standard database query language. Today it has geographic capabilities.

In SQL, data definition language statements are used to create, alter, and delete relational database structures. The **CREATE TABLE** command is used to define a table, the attributes it will contain, and the primary key (the column used to identify records uniquely). For example, the SQL statement to create a table to store data about Countries, with two columns (name and shape (geometry)) is as follows:

```
CREATE TABLE Countries (
name            VARCHAR(200) NOT NULL PRIMARY
   KEY,
shape           POLYGON NOT NULL
CONSTRAINT   spatial reference
CHECK         (SpatialReference(shape)  = 14)
)
```

This SQL statement defines several table parameters. The name column is of type **VARCHAR** (variable character) and can store values up to 200 characters. Name cannot be null (**NOT NULL**); that is, it must have a value, and it is defined as the **PRIMARY KEY**, which means that its entries must be unique. The shape column is of type **POLYGON**, and it is defined as **NOT NULL**. It has an additional spatial reference constraint (projection), meaning that a spatial reference is enforced for all shapes (Type 14—this will vary by system, but could be Universal Transverse Mercator (UTM)—see Section 5.8.2).

Data can be inserted into this table using the SQL INSERT command:

```
INSERT INTO Countries
(Name, Shape) VALUES ('Kenya', Polygon
   ('((x y, x y, x y, x y) ,2))
```

Actual coordinates would need to be substituted for the *x, y* values. Several additions of this type would result in a table like the following:

Name	Shape
Kenya	Polygon geometry
South Africa	Polygon geometry
Egypt	Polygon geometry

Data manipulation language statements are used to retrieve and manipulate data. Objects with a size greater than 11,000 can be retrieved from the countries table using a SELECT statement:

```
SELECT Countries.Name,
FROM Countries
WHERE Area(Countries.Shape) > 11000
```

In this system, the implementation area is computed automatically from the shape field using a DBMS function and does not need to be stored.

Data control language statements handle authorization access. The two main DCL keywords are GRANT and REVOKE, which authorize and rescind access privileges, respectively.

10.5 Geographic Database Types and Functions

Several attempts have been made to define a superset of geographic data types that can represent and process geographic data in databases. Unfortunately, space does not permit a review of them all. This discussion will instead focus on the practical aspects of this problem and will be based on the widely accepted ISO (International Standards Organization) and Open Geospatial Consortium (OGC) standards. The GIS community working under the auspices of ISO and OGC has defined the core geographic types and functions to be used in a DBMS and accessed using the SQL query language (see Section 10.4 for a discussion of SQL). The geometry types are shown in Figure 10.5. In this hierarchy, the Geometry class is the root class. It has an associated spatial reference (coordinate system and projection, for example, Lambert Azimuthal Equal Area). The Point, Curve, Surface, and GeometryCollection classes are all subtypes of Geometry. The other classes (boxes) and relationships (lines) show how geometries of one type are aggregated from others (e.g., a LineString is a collection of Points). For further explanation of how to interpret this object model diagram, see the discussion in Section 8.3.

According to this ISO/OGC standard, there are nine methods for testing spatial relationships between

Figure 10.5 Geometry class hierarchy. (Source: after OGC 1999, reproduced by permission of Open Geospatial Consortium, Inc.)

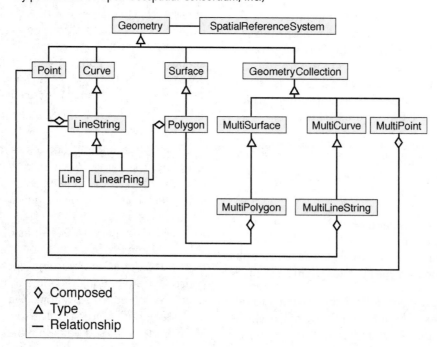

these geometric objects. Each method takes as input two geometries (collections of one or more geometric objects) and evaluates whether or not the relationship is true. Two examples of possible relations for all point, line, and area combinations are shown in Figure 10.6. In the case of the point–point Contain combination (northwest square in Figure 10.6A), two comparison geometry points (big circles) are contained in the set of base geometry points (small circles). In other words, the base geometry is a superset of the comparison geometry. In the case of line-area Touches combination (east square in Figure 10.6B), the two lines touch the area because they intersect the area boundary. The full set of Boolean operators to test the spatial relationships between geometries is.

- Equals—are the geometries the same?

- Disjoint—do the geometries share a common point?

- Intersects—do the geometries intersect?

- Touches—do the geometries intersect at their boundaries?

- Crosses—do the geometries overlap (can be geometries of different dimensions, for example, lines and polygons)?

- Within—is one geometry within another?

- Contains—does one geometry completely contain another?

- Overlaps—do the geometries overlap (must be geometries of the same dimension)?

- Relate—are there intersections between the interior, boundary, or exterior of the geometries?

  Seven methods support spatial analysis on these geometries. Four examples of these methods are shown in Figure 10.7.

- Distance—determines the shortest distance between any two points in two geometries (Section 14.3.1).

- Buffer—returns a geometry that represents all the points whose distance from the geometry is less than or equal to a user-defined distance (Section 14.3.2).

- ConvexHull—returns a geometry representing the convex hull of a geometry (a convex hull is the smallest polygon that can enclose another geometry without any concave areas).

- Intersection—returns a geometry that contains just the points common to both input geometries.

- Union—returns a geometry that contains all the points in both input geometries.

(A) **Contains**

Does the base geometry contain the comparison geometry?

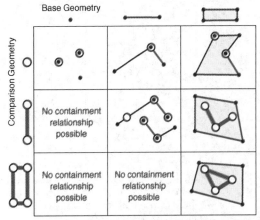

For the base geometry to contain the comparison geometry, it must be a superset of that geometry.

A geometry cannot contain another geometry of higher dimension.

(B) **Touches**

Does the base geometry touch the comparison geometry?

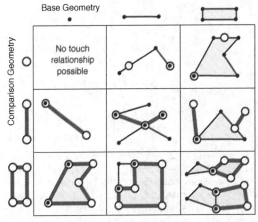

Two geometries touch when only their boundaries intersect.

Figure 10.6 Examples of possible relations for two geographic database operators: (A) Contains; and (B) Touches operators. (Source: after Zeiler 1999)

- Difference—returns a geometry containing the points that are different between the two geometries.

- SymDifference—returns a geometry containing the points that are in either of the input geometries, but not both.

(A) Buffer

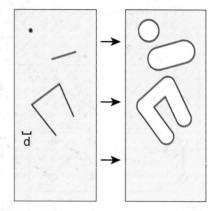

Given a geometry and a buffer distance, the buffer operator returns a polygon that covers all points whose distance from the geometry is less than or equal to the buffer distance.

(B) Convex Hull

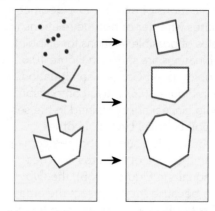

Given an input geometry, the convex hull operator returns a geometry that represents all points that are within all lines between all points in the input geometry.

A convex hull is the smallest polygon that wraps another geometry without any concave areas.

(C) Intersection

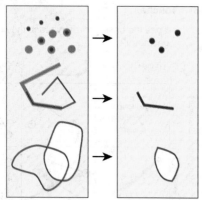

The intersect operator compares a base geometry (the object from which the operator is called) with another geometry of the same dimension and returns a geometry that contains the points that are in both the base geometry and the comparison geometry.

(D) Difference

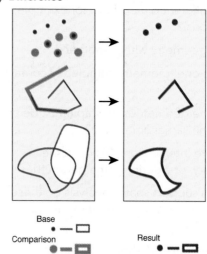

The difference operator returns a geometry that contains points that are in the base geometry and subtracts points that are in the comparison geometry.

Figure 10.7 Examples of spatial analysis methods on geometries: (A) Buffer; (B) Convex Hull; (C) Intersection; (D) Difference. (Source: after Zeiler 1999)

10.6 Geographic Database Design

This section is concerned with the technical aspects of logical and physical geographic database design. Chapter 8 provides an overview of these subjects, and Chapters 17 to 19 discuss the organizational, strategic, and business issues associated with designing and maintaining a geographic database.

10.6.1 The Database Design Process

All GIS and DBMS packages have their own core data model that defines the object types and relationships that can be used in an application (Figure 10.8). The DBMS package will define and implement a model for data types and access functions, such as those implemented in SQL and discussed in Section 10.4. DBMS are capable of dealing with simple geographic features and types (e.g., points, lines, areas, and sometimes also rasters) and relationships. A GIS can build on top of these simple feature types to create more advanced types and relationships (e.g., TINs, topologies, and feature-linked annotation geographic relationships: see Chapter 8 for a definition of these terms). The GIS types can be combined with domain data models that define specific object classes and relationships for specialist domains (e.g., water utilities, city parcel maps, and census geographies). Lastly, individual projects create specific physical data model instances that are populated with data (objects for the specified object classes). For example, a City Planning department may build a database of sewer lines that uses a water/wastewater (sewer) domain data model template that is built on top of core GIS and DBMS models. Figure 8.2 and Section 8.1.2 show three increasingly abstract stages in data modeling: conceptual, logical, and physical. The result of a data-modeling exercise is a physical database design. This design will include specification of all data types and relationships, as well as the actual database configuration required to store them.

Database design involves the creation of conceptual, logical, and physical models in the six practical steps that are shown in Figure 10.9.

Database design involves three key stages: conceptual, logical, and physical.

10.6.1.1 Conceptual Model

Model the User's View   This involves tasks such as identifying organizational functions (e.g., controlling

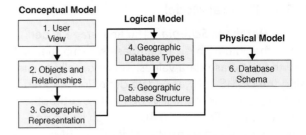

Figure 10.9 Stages in database design. (Source: after Zeiler 1999)

forestry resources, finding vacant land for new buildings, and maintaining highways), determining the data required to support these functions, and organizing the data into groups to facilitate data management. This information can be represented in many ways, for example, a report with accompanying tables is often used.

Define Objects and their Relationships   Once the functions of an organization have been defined, the object types (classes) and functions can be specified. The relationships between object types must also be described. This process usually benefits from the rigor of using object models and diagrams to describe a set of object classes and the relationships between them.

Select Geographic Representation   Choosing the types of geographic representation (discrete object or continuous field: see Section 3.6) will have profound implications for the way a database is used, and so it is a critical database design task. It is, of course, possible to change between representation types, but this is computationally expensive and results in loss of information.

10.6.1.2 Logical Model

Match to Geographic Database Types   This stage involves matching the object types to be studied to specific data types (point, line area, georaster, etc.) supported by the GIS that will be used to create and maintain the database. Because the data model of the GIS is usually independent of the actual storage mechanism (i.e., it could be implemented in Oracle, PostGIS, Microsoft Access, or a proprietary file system), this activity is defined as a logical modeling task.

Organize Geographic Database Structure   This stage includes tasks such as defining topological associations, specifying rules and relationships, and assigning coordinate systems.

Figure 10.8 Four levels of data model available for use in GIS projects.

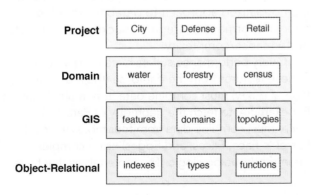

10.6.1.3 Physical Model

Define Database Schema The final stage is definition of the actual physical database schema that will hold the database data values. This is usually created using the DBMS software's data definition language. The most popular of these is SQL with geographic extensions (see Section 10.4), although some non-standard variants also exist in older GIS/DBMS.

10.7 Structuring Geographic Information

Once data have been captured in a geographic database according to a schema defined in a geographic data model, it is often desirable to perform some structuring and organization in order to support efficient query, analysis, and mapping. There are two main structuring techniques relevant to geographic databases: topology creation and indexing.

10.7.1 Topology Creation

The subject of topology was covered in Section 8.2.3.2 from a conceptual data-modeling perspective and is revisited here in the context of databases where the discussion focuses on the two main approaches to structuring and storing topology in a GIS DBMS.

Topology can be created for vector datasets using either batch or interactive techniques. Batch topology builders are required to handle CAD, survey, simple feature, and other unstructured vector data imported from non–topological systems. Creating topology is usually an iterative process because it is seldom possible to resolve all data problems during the first pass, and manual editing is required to make corrections. Some typical problems that may arise are shown in Figures 9.9, 9.10, and 9.12 and are discussed in Section 9.3.2.2. Interactive topology creation is performed dynamically at the time objects are added to a database using GIS editing software. For example, when adding water pipes using interactive vectorization tools (see Section 9.3.2.1), before each object is committed to the database topological connectivity can be checked to see if the object is valid (that is, it conforms to some pre–established database object connectivity rules).

Two database-oriented approaches have emerged in recent years for storing and managing topology: normalized and physical. The Normalized Model focuses on the storage of an arc-node data structure. It is said to be normalized because each object is decomposed into individual topological primitives for storage in a database and then subsequent reassembly when a query is posed. For example, polygon objects are assembled at query time by joining together tables containing the line segment geometries and topological primitives that define topological relationships (see Section 8.2.3.2 for a conceptual description of this process). In the Physical Model, topological primitives are not stored in the database, and the entire geometry is stored together for each object. Topological relationships are then computed on-the-fly whenever they are required by client applications.

Figure 10.10 is a simple example of a set of database tables that store a dataset according to the normalized topology model. The dataset (sketch in top left corner) comprises three feature classes (Parcels, Buildings, and Walls) and is implemented in three tables. In this example the three feature class tables have only one column (ID) and one row (a single instance of each feature in a feature class). The Nodes, Edges, and Faces tables store the points, lines, and polygons for the dataset and some of the topology (for each Edge the From-To connectivity and the Faces on the Left-Right in the direction of digitizing). Three other tables (Parcel × Face, Wall × Edge and Building × Face) store the cross-references for assembling Parcels, Buildings, and Walls from the topological primitives.

The normalized approach offers a number of advantages to GIS users. Many people find it comforting to see topological primitives actually stored in the database. This model has many similarities to the arc-node conceptual topology model (see Section 8.2.3.2), and so it is familiar to many users and easy to understand. The geometry is only stored once, thus minimizing database size and avoiding "double digitizing" slivers (Section 9.3.2.2). Finally, the normalized approach easily lends itself to access via a SQL API. Unfortunately, there are three main disadvantages associated with the normalized approach to database topology: query performance, integrity checking, and update performance/complexity.

Query performance suffers because queries to retrieve features from the database (the most common type of query) must combine data from multiple tables. For example, to fetch the geometry of Parcel P1, a query must combine data from four tables (Parcels, Parcel × Face, Faces, and Edges) using complex geometry/topology logic. The more tables that have to

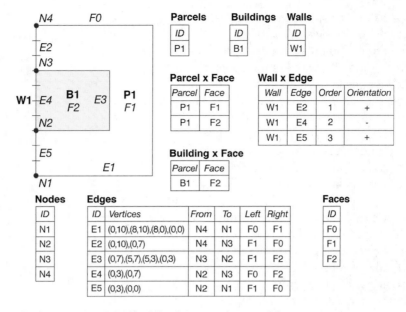

Figure 10.10 Normalized database topology model.

be visited, and especially the more that have to be joined, the longer it will take to process a query.

The standard referential integrity rules in DBMS are very simple and have no provision for complex topological relationships of the type defined here. There are many pitfalls associated with implementing topological structuring using DBMS techniques such as stored procedures (program code stored in a database), and in practice systems have resorted to implementing the business logic to manage things like topology external to the DBMS in a middle-tier application server or a set or server-side program code.

Updates are similarly problematic because changes to a single feature will have cascading effects on many tables. This raises attendant performance (especially scalability, that is, large numbers of concurrent queries) and integrity issues. Moreover, it is uncertain how multiuser updates will be handled that entail long transactions with design alternatives (see Sections 10.9 and 10.9.1 for coverage of these two important topics). For comparative purposes the same dataset used in the normalized model (Figure 10.10) is implemented using the physical model in Figure 10.11. In the physical model the three feature classes

Figure 10.11 Physical database topology model.

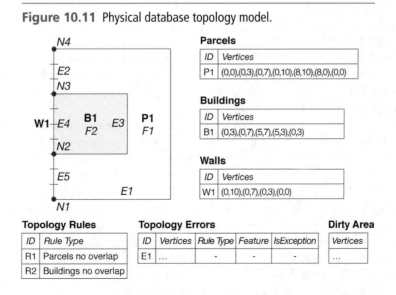

(Parcels, Buildings, and Walls) contain the same IDs, but differ significantly in that they also contain the geometry for each feature. The only other things required to be stored in the database are the specific set of topology rules that have been applied to the dataset (e.g., parcels should not overlap each other, and buildings should not overlap with each other), together with information about known errors (sometimes users defer topology cleanup and commit data with known errors to a database) and areas that have been edited but not yet validated (had their topology (re)built).

The physical model requires that an external client or middle-tier application server is responsible for validating the topological integrity of datasets. Topologically correct features are then stored in the database using a structure that is much simpler than the normalized model. When compared to the normalized model, the physical model offers two main advantages of simplicity and performance. Since all the geometry for each feature is stored in the same table column/row value, and there is no need to store topological primitives and cross-references, this is a very simple model. The biggest advantage and the reason for the appeal of this approach is query performance. Most DBMS queries do not need access to the topology primitives of features, and so the overhead of visiting and joining multiple tables is unnecessary. Even when topology is required, it is faster to retrieve feature geometries and recompute it outside the database than to retrieve geometry and topology from the database.

In summary, the normalized and physical topology models have both advantages and disadvantages. The normalized model is implemented in Oracle Spatial and can be accessed via a SQL API, making it easily available to a wide variety of users. The physical model is implemented in ESRI ArcGIS and offers fast update and query performance for high-end GIS applications. ESRI has also implemented a long transaction and versioning model based on the physical database topology model.

10.7.2 Indexing

Geographic databases tend to be very large and geographic queries computationally expensive. As a result, geographic queries, such as finding all the customers (points) within a store trade area (polygon), can take a very long time (perhaps 10 to 100 seconds or more for a 50 million customer database). The point has already been made in

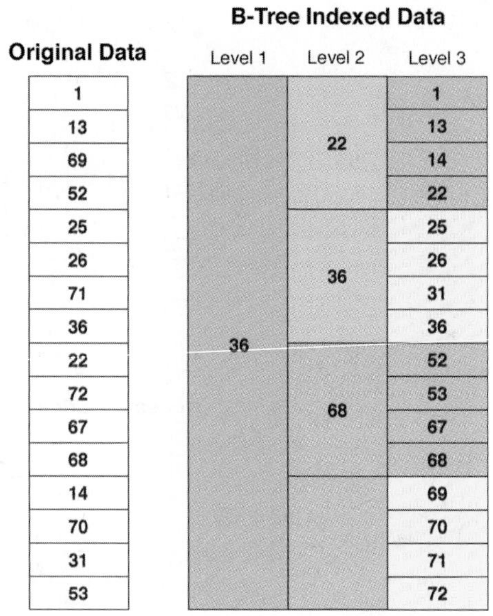

B-Tree Indexed Data

Original Data	Level 1	Level 2	Level 3
1			1
13		22	13
69			14
52			22
25			25
26		36	26
71			31
36	36		36
22			52
72		68	53
67			67
68			68
14			69
70			70
31			71
53			72

Figure 10.12 An example of a B-tree index.

Section 8.2.3.2 that topological structuring can help speed up certain types of queries such as adjacency and network connectivity. A second way to speed up queries is to index a database and use the index to find data records (database table rows) more quickly. A database index is logically similar to a book index; both are special organizations that speed up searching by allowing random instead of sequential access. A database index is, conceptually speaking, an ordered list derived from the data in a table. Using an index to find data reduces the number of computational tests that have to be performed to locate a given set of records. In DBMS jargon, indices avoid expensive full-table scans (reading every row in a table) by creating an index and storing it as a table column.

A database index is a special representation of information about objects that improves searching.

Figure 10.12 presents a simple example of the standard DBMS one-dimensional B-tree (balanced tree) index that is found in most major commercial DBMS. Without an index, a search of the original data in this table to guarantee finding any given value would involve 16 tests/read operations (one for each data point). The B-tree index orders the data and splits the ordered list into buckets of a given size (in this example it is four and then two), and the upper value for the bucket is stored (it is not necessary to store the uppermost value). To find a specific value, such

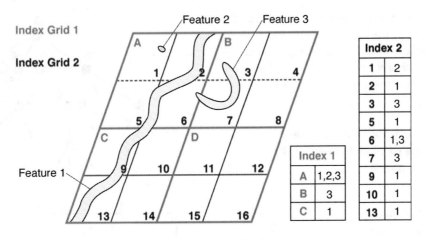

Figure 10.13 A multilevel grid geographic database index.

Index 2	
1	2
2	1
3	3
5	1
6	1,3
7	3
9	1
10	1
13	1

Index 1	
A	1,2,3
B	3
C	1

as 72, using the index involves a maximum of six tests: one at level 1 (less than or greater than 36), one at level 2 (less than or greater than 68), and a sequential read of four records at Level 3. The number of levels and buckets for each level can be optimized for each type of data. Typically, the larger the dataset, the more effective indices are in retrieval performance.

Unfortunately, creating and maintaining indices can be quite time-consuming; this is especially an issue when the data are very frequently updated. Since indices can occupy considerable amounts of disk space storage, requirements can be very demanding for large datasets. As a consequence, many different types of index have been developed to try and alleviate these problems for both geographic and nongeographic data. Some indices exploit specific characteristics of data to deliver optimum query performance, others are fast to update, and still others are robust across widely different data types.

The standard indices in DBMS, such as the B-tree, are one-dimensional and are very poor at indexing geographic objects. Many types of geographic indexing techniques have been developed, some of which are experimental and have been highly optimized for specific types of geographic data. Research shows that even a basic spatial index will yield very significant improvements in spatial data access and that further refinements often yield only marginal improvements at the costs of simplicity, as well as speed of generation and update. Three main methods of general practical importance have emerged in GIS: grid indices, quadtrees, and R-trees.

10.7.2.1 Grid Index

A grid index can be thought of as a regular mesh placed over a layer of geographic objects. Figure 10.13 shows a layer that has three features indexed using two grid levels. The highest (coarsest) grid (Index 1) splits the layer into four equal-sized cells. Cell A includes parts of Features 1, 2, and 3, Cell B includes a part of Feature 3, and Cell C has part of Feature 1. There are no features on Cell D. The same process is repeated for the second-level index (Index 2). A query to locate an object searches the indexed list first to find the object and then retrieves the object geometry or attributes for further analysis (e.g., tests for overlap, adjacency, or containment with other objects on the same or another layer). These two tests are often referred to as primary and secondary filters. Secondary filtering, which involves geometric processing, is much more computationally expensive. The performance of an index is clearly related to the relationship between grid and object size, and object density. If the grid size is too large relative to the size of object, too many objects will be retrieved by the primary filter, and therefore a lot of expensive secondary processing will be needed. If the grid size is too small, any large objects will be spread across many grid cells, which is inefficient for draw queries (queries made to the database for the purpose of displaying objects on a screen). For data layers that have a highly variable object density (for example, administrative areas tend to be smaller and more numerous in urban areas in order to equalize population counts), multiple levels can be used to optimize performance. Experience suggests that three grid levels are normally sufficient for good all-round performance.

Grid indices are one of the simplest and most robust indexing methods. They are fast to create and update and can handle a wide range of types and densities of data. For this reason they have been quite widely used in commercial GIS software systems.

Grid indices are easy to create, can deal with a wide range of object types, and offer good performance.

10.7.2.2 Quadtree Indices

Quadtree is a generic name for several kinds of index that are built by recursive division of space into quadrants. In many respects, quadtrees are a special type of grid index. The difference here is that in quadtrees space is always recursively split into four quadrants based on data density. Quadtrees are data structures used for both indexing and compressing geographic database layers, although the discussion here will relate only to indexing. The many types of quadtree can be classified according to the types of data that are indexed (points, lines, areas, surfaces, or rasters), the algorithm that is used to decompose (divide) the layer being indexed, and whether fixed or variable resolution decomposition is used.

In a point quadtree, space is divided successively into four rectangles based on the location of the points (Figure 10.14). The root of the tree corresponds to the region as a whole. The rectangular region is divided into four usually irregular parts based on the x, y coordinates of the first point. Successive points subdivide each new subregion into quadrants until all the points are indexed.

Region quadtrees are commonly used to index lines, areas, and rasters. The quadtree index is created by successively dividing a layer into quadrants. If a quadrant cell is not completely filled by an object, then it is subdivided again. Figure 10.15 is a quadtree of a woodland (red) and water (white) layer. Once a layer has been decomposed in this way, a linear index can be created using the search order shown in Figure 10.16. By reducing two-dimensional geographic data to a single linear dimension, a standard B-tree can be used to find data quickly.

Quadtrees have found favor in GIS software systems because of their applicability to many types of data (both raster and vector), their ease of implementation, and

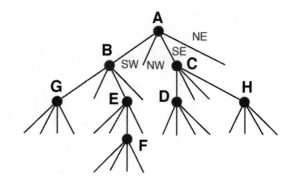

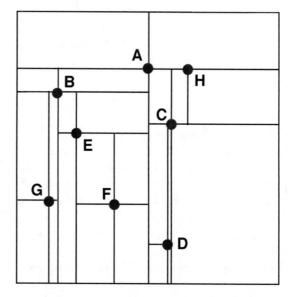

Figure 10.14 The point quadtree geographic database index. (Source: after van Oosterom 2005. Reproduced by permission of John Wiley & Sons, Inc.)

their relatively good improvements in search performance.

10.7.2.3 R-tree Indices

R-trees group objects using a rectangular approximation of their location called a minimum bounding rectangle (MBR) or minimum enclosing rectangle (see Box 10.2). Groups of point, line, or area objects are indexed based on their MBR. Objects are added to the index by choosing the MBR that would require the least expansion to accommodate each new object. If the object causes the MBR to be expanded beyond some preset parameter, then the MBR is split into two new MBRs. This may also cause the parent MBR to become too large, resulting in this also being split. The R-tree shown in Figure 10.17 has two levels. The lowest level contains three "leaf nodes"; the highest has one node with pointers to the MBR of the leaf nodes. The MBR is used to

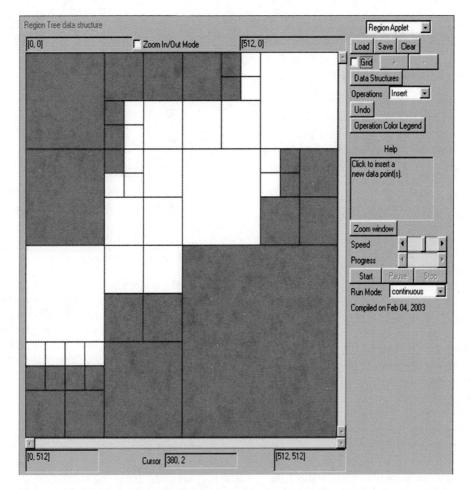

Figure 10.15 The region quadtree geographic database index. (Source: www.cs.umd. edu/~brabec/quadtree. Reproduced by permission of Hanan Samet and Frantisek Brabec)

reduce the number of objects that need to be examined in order to satisfy a query.

R-trees are popular methods of indexing geographic data because of their flexibility and excellent performance.

R-trees are suitable for a range of geographic object types and can be used for multidimen-

sional data. They offer good query performance, but the speed of update is not as fast as grids and quadtrees. The spatial datablade extensions to the IBM Informix DBMS and Oracle Spatial both use R-tree indices.

Figure 10.17 The R-tree geographic database index. (Source: After van Oosterom 2005. Reproduced by permission of John Wiley, Inc.)

Figure 10.16 Linear quadtree search order.

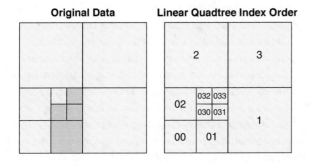

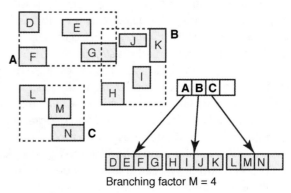

Minimum Bounding Rectangle

Minimum bounding rectangles (MBRs) are very useful structures that are widely implemented in GIS. An MBR essentially defines the smallest box whose sides are parallel to the axes of the coordinate system that encloses a set of one or more geographic objects. It is defined by the two coordinates at the bottom left (minimum x, minimum y) and the top right (maximum x, maximum y), as is shown in Figure 10.18.

MBRs can be used to generalize a set of data by replacing the geometry of each of the objects in the box with two pairs of coordinates defining the box. A second use is for fast searching. For example, all of the area objects in a database layer that are within a given study area can be found by performing an area on area contains test (see Figure 10.18) for each object and the study area boundary. If the area objects have complex boundaries (as is normally the case in GIS), this can be a very time-consuming task. A quicker approach is to split the task into two parts. First, screen out all the objects that are definitely in and definitely out by comparing their MBR. Because very few coordinate comparisons are required, this is very fast. Then use the full

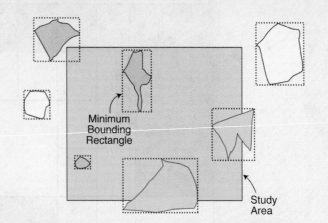

Figure 10.18 Area in area test using MBR. An MBR can be used to determine objects definitely in the study area (green) because of no overlap, definitely out (yellow), or possibly in (blue). Objects possibly in can then be analyzed further using their exact geometries. Note the red object that is actually completely outside, although the MBR suggests it may be partially within the study area.

geometry outline of the remaining area objects to determine containment. This is computationally expensive for areas with complex geometries.

10.8 Editing and Data Maintenance

Editing is the process of making changes to a geographic database by adding new objects or changing existing objects as part of data load or database update and maintenance operations. A database update is any change to the geometry and/or attribute of one or more objects, a change to an object relationship, or any change to the database schema. A general-purpose geographic database will require many tools for tasks such as geometry and attribute editing, database maintenance (e.g., system administration and tuning), creating and updating indices and topology, importing and exporting data, and georeferencing objects.

Contemporary GIS come equipped with an extensive array of tools for creating and editing geographic object geometries and attributes. These tools form workflow tasks that are exposed within the framework of a WYSIWYG (what you see is what you get) editing environment. The objects are displayed using map symbology usually in a projected coordinate system space and frequently on top of "background" layers such as aerial photographs or satellite images, or street centerline files. Object coordinates can be digitized into a geographic database using many methods including freehand digitizing on a digitizing table, on-screen heads-up vector digitizing by copying existing raster and/or vector sources, on-screen semiautomatic line following, automated feature recognition, and reading survey measurements from an instrument (e.g., GPS or Total Station) file (see Sections 5.9 and 9.2.2.1). The end result is always a layer of x, y coordinates with optional z and m (height and attribute) values. Similar tools also exist for loading and editing raster data.

Data entered into the editor must be stored persistently in a file system or database, and access to the database must be carefully managed to ensure continued security and quality. The mechanism for managing edits to a file or database is called a transaction. There are many challenging issues associated with implementing multiuser access to geographic data stored in a DBMS, as discussed in the next section.

10.9 Multiuser Editing of Continuous Databases

For many years, one of the most challenging problems in GIS data management was how to allow multiple users to edit the same continuous geographic database at the same time. In GIS applications the objects of interest (geographic features) do not exist in isolation and are usually closely related to surrounding objects. For example, a tax parcel will share a common boundary with an adjacent parcel, and changes in one will directly affect the other; similarly, connected road segments in a highway network need to be edited together to ensure continued connectivity. It is straightforward to provide multiple users with concurrent read and query access to a continuous shared database, but more difficult to deal with conflicts and avoid potential database corruption when multiple users want write (update) access. However, solutions to both of these problems have been implemented in mainstream GIS and DBMS. These solutions extend standard DBMS transaction models and provide a multiuser framework called versioning.

10.9.1 Transactions

A group of edits to a database, such as the addition of three new land parcels and changes to the attributes of a sewer line, is referred to as a "transaction." In order to protect the integrity of databases, transactions are atomic; that is, transactions are either completely committed to the database or they are rolled back (not committed at all). Many of the world's GIS and non-GIS databases are multiuser and transactional; that is, they have multiple users performing edit/update operations at the same time. For most types of database, transactions take a very short (subsecond) time. For example, in the case of a banking system, a transfer from a savings account to a checking account takes perhaps 0.001 second. It is important that the transaction is coordinated between the accounts and that it is atomic; otherwise one account might be debited and the other not credited. Multiuser access to banking and similar systems is handled simply by locking (preventing access to) affected database records (table rows) during the course of the transaction. Any attempt to write to the same record is simply postponed until the record lock is removed after the transaction is completed. Because banking transactions, like many other transactions, take only a very short amount of time, users never even notice whether a transaction is deferred.

A transaction is a group of changes that are made to a database as a coherent group. All the changes that form part of a transaction are either committed, or the database is rolled back to its initial state.

Although some geographic transactions have a short duration (short transactions), many extend to hours, weeks, and months, and are called long transactions. Consider, for example, the amount of time necessary to capture all the land parcels in a city subdivision (housing estate). This might take a few hours for an efficient operator working on a small subdivision, but an inexperienced operator working on a large subdivision might take days or weeks. This may cause three multiuser update problems. First, locking the whole or even part of a database to other updates for this length of time during a long transaction is unacceptable in many types of application, especially those involving frequent maintenance changes (e.g., utilities and land administration). Second, if a system failure occurs during the editing, work may be lost unless there is a procedure for storing updates in the database. Also, unless the data are stored in the database, they are not easily accessible to others who would like to use them.

10.9.2 Versioning

Short transactions use what is called a pessimistic locking concurrency strategy. That is, it is assumed that conflicts will occur in a multiuser database with concurrent users and that the only way to avoid database corruption is to lock out all but one user during an update operation. The term *pessimistic* is used because this is a very conservative strategy, assuming that update conflicts will occur and that they must be avoided at all cost. An alternative to pessimistic locking is optimistic versioning, which allows multiple users to update a database at the same time. Optimistic versioning is based on the assumption that conflicts are very unlikely to occur, but if they do occur then software can be used to resolve them.

The two strategies for providing multiuser access to geographic databases are pessimistic locking and optimistic versioning.

Versioning sets out to solve the long transaction and pessimistic locking concurrency problem that has been described earlier in this chapter. It also addresses a second key requirement peculiar to geographic databases—the need to support alternative representations of the same objects in the database. In some important geographic applications, it is a requirement to allow designers to create and maintain multiple object designs. For example, when designing a new housing subdivision, the water department manager may ask two designers to lay out alternative designs for a water system. The two designers would work concurrently to add objects to the same database layers, snapping to the same objects. At some point, they may wish to compare designs and perhaps create a third design based on parts of their two designs. While this design process is taking place, operational maintenance editors could be changing the same objects with which they are working. For example, maintenance updates resulting from new service connections or repairs to broken pipes will change the database and may affect the objects used in the new designs.

Figure 10.19 compares linear short transactions and branching long transactions as implemented in a versioned database. Within a versioned database, the different database versions are logical copies of their parents (base tables); that is, only the modifications (additions and deletions) are stored in the database (in version change tables). A query against a database version combines data from the base table with data in the version change tables. The process of creating two versions based on the same parent version is

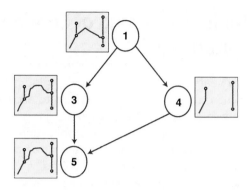

Figure 10.20 Version reconciliation. For Version 5 the user chooses via the GUI the geometry edit made in Version 3 instead of 1 or 4.

called branching. In Figure 10.19B, Version 4 is a branch from Version 2. Conversely, the process of combining two versions into one version is called merging. Figure 10.19B also illustrates the merging of Versions 6 and 7 into Version 8. A version can be updated at any time with any changes made in another version. Version reconciliation can be seen between Versions 3 and 5. Since the edits contained within Version 5 were reconciled with 3, only the edits in Versions 6 and 7 are considered when merging to create Version 8.

There are no database restrictions or locks placed on the operations performed on each version in the database. The versioning database schema isolates changes made during the edit process. With optimistic versioning, it is possible for conflicting edits to be made within two separate versions, although normal working practice will ensure that the vast majority of edits made will not result in any conflicts (Figure 10.20). In the event that conflicts are detected, the data management software will handle them either automatically or interactively. If interactive conflict resolution is chosen, the user is directed to each feature that is in conflict and must decide how to reconcile the conflict. The GUI will provide information about the conflict and display the objects in their various states. For example, if the geometry of an object has been edited in the two versions, the user can display the geometry as it was in any of its previous states and then select the geometry to be added to the new database state.

10.10 Conclusion

Database management systems are now a vital part of large modern operational GIS as well as related interdisciplinary fields such as computer

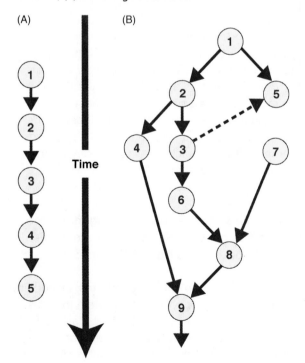

Figure 10.19 Database transactions: (A) linear short transactions; (B) branching version tree.

Shashi Shekhar, Computer Scientist

Shashi Shekhar (Figure 10.21) received a computer science and engineering (CSE) education from the Indian Institute of Technology, Kanpur (1981–1985) and the University of California, Berkeley (1985–1989). He is presently a McKnight Distinguished University Professor of CSE at the University of Minnesota. His knowledge of GIS started in the early 1990s from sponsored research projects on computational aspects of in-vehicle and Web-based navigation systems (USDOT), high-performance GIS for vehicle simulators (USDOD), Minnesota Mapserver (NASA), and the like. Soon, he realized there was a strong and growing demand for CSE advances for GIS, yet very few CSE scholars were dedicated to this promising area. Thus, in the mid-1990s, he decided to focus his research on understanding the structure of very large spatial computations (e.g., data analysis via spatial querying and spatial data mining). Illustrative contributions include: the Capacity Constrained Route Planner, an evacuation route planning algorithm, which is orders of magnitude faster than traditional linear programming-based methods; the Connectivity-Clustered Access Method, a min-cut graph-partitioning-based storage method, which outperforms geometry-based indices (e.g., R-tree family) in carrying out network computations; and the notion of "co-location" patterns in spatial datasets to provide a trade-off between computational scalability and spatial statistical rigor.

His textbook, *Spatial Databases: A Tour* (Shekhar and Chawla, 2003), describes object-relational database concepts for GIS at conceptual (e.g., entity-relationship models with pictograms), logical (e.g., SQL3, OGC simple feature types), and physical (e.g., R-trees) levels and introduces trends (e.g., spatial data mining). In a more recent project, *Encyclopedia of GIS*, he uncovers broader CSE issues in GIS, for example, open-

Figure 10.21 Shashi Shekhar. (Dr. Shashi Shekhar)

source and commercial software, standards, geosensor networks, the geospatial semantic Web, privacy, and indoor positioning.

Reflecting on the interplay between CSE and GIS, Shashi says:

> GIS problems often reveal limitations of current CSE methods, stimulating advances in CSE. At the same time, computational methods are becoming the third leg of science (e.g., GIS) complementing mathematical models and controlled experiments. In addition, scalability to large spatial datasets will often be expected from GIS in the post Google Earth era. As for the future, I believe that GIS work driven by stove-piped disciplines may be limited in solving complex problems facing our society, and will increasingly be replaced by interdisciplinary approaches leveraging ideas from and advancing multiple disciplines, e.g. GIS and CSE.

science and engineering (see Box 10.3). They bring with them standardized approaches for storing and, more importantly, accessing and manipulating geographic data using the SQL query language. GIS provide the necessary tools to load, edit, query, analyze, and display geographic data. DBMS require a database administrator

(DBA) to control database structure and security, and to tune the database to achieve maximum performance. Innovative work in the GIS field has extended standard DBMS to store and manage geographic data and has led to the development of long transactions and versioning that have application across several fields.

Questions for Further Study

1. Identify a geographic database with multiple layers and draw a diagram showing the tables and the relationships between them. Which are the primary keys, and which keys are used to join tables? Does the database have a good relational design?

2. What are the advantages and disadvantages of storing geographic data in a DBMS?

3. Is SQL a good language for querying geographic databases?

4. Why are there multiple methods of indexing geographic databases?

Further Reading

Date, C.J. 2005. *Database in Depth: Relational Theory for Practitioners*. Sebastopol, CA: O'Reilly Media, Inc.

Hoel, E., Menon, S., and Morehouse, S. 2003. Building a robust relational implementation of topology. In Hadzilacos, T., Manolopoulos, Y., Roddick, J.F., and Theodoridis, Y. (eds.). *Advances in Spatial and Temporal Databases. Proceedings of 8th International Symposium, SSTD 2003 Lecture Notes in Computer Science*, Vol. 2750.

OGC. 1999. *OpenGIS Simple Features Specification for SQL, Revision 1.1*. Available at www.opengis.org.

Samet, H. 2006. *The Foundations of Multidimensional and Metric Data Structure*. San Francisco: Morgan-Kaufmann.

Shekar, S., and Chawla, S. 2003. *Spatial Databases: A Tour*. Saddle River, NJ: Prentice Hall.

van Oosterom, P. 2005. Spatial access methods. In Longley, P.A., Goodchild, M.F., Maguire, D.J., and Rhind, D.W. (eds.) *Geographic Information Systems: Principles, Techniques, Applications and Management (abridged ed.)*. Hoboken, NJ: Wiley. 385–400

Zeiler, M. 1999. *Modeling Our World: The ESRI Guide to Geodatabase Design*. Redlands, CA: ESRI Press.

11 The GeoWeb

Until recently the only practical way to apply GIS to a problem was to assemble all of the necessary parts in one place, on the user's desktop. But today all of the parts—the data and the software—can be accessed remotely. The GeoWeb is a vision for the future, only partially realized to date, in which all of these parts are able to operate together, in effect turning the Internet into a massive GIS. The GeoWeb vision also includes real-time feeds of data from sensors and the ability to integrate these with other data. Rapidly developing technologies allow the user to move away from the desktop and hence to apply GIS anywhere. Limited GIServices are already available in common devices such as mobile phones and are increasingly being installed in vehicles. This chapter introduces the basic concepts of the GeoWeb, describes current capabilities, and looks to a future in which GIS is increasingly mobile and available everywhere.

After studying this chapter you will be able to:

- How the parts of GIS can be distributed instead of centralized.

- Geoportals, and the standards and protocols that allow remotely stored data to be discovered and accessed.

- The technologies that support real-time acquisition and distribution of geographic information.

- The service-oriented architectures and mashups that combine GIS services from different Web sites.

- The capabilities of mobile devices, including mobile phones and wearable computers.

- The concepts of augmented and virtual reality.

11.1 Introduction

Early computers were extremely expensive, forcing organizations like universities to provide computing services centrally, from a single site, and to require users to come to the computing center to access its services. As the cost of computers fell, from millions of dollars in the 1960s to hundreds of thousands in the late 1970s and now to less than a thousand, it became possible for departments, small groups, and finally individuals to own computers and to install them on their desktops. Today, of course, the average professional worker expects to have a computer in the office, and a substantial proportion of households in the industrialized countries have computers in the home. Computers are being embedded in familiar devices such as the car, entertainment systems, and mobile phones, suggesting a future in which computing will be everywhere and to an increasing extent invisible to the user.

In Section 1.5 we identified the six component parts of a GIS as its hardware, software, data, users, procedures, and network. This chapter describes how the network, to which almost all computers are now connected, has enabled a new vision of *distributed* GIS, in which the component parts no longer need to be co-located. New technologies are rapidly moving us to the point where it will be possible for a GIS project to be conducted not only on the desktop but anywhere the user chooses to be, using data located anywhere on the network, and using services provided by remote sites on the network.

In distributed GIS the six component parts may be at different locations.

In Chapter 7 we discussed some aspects of this new concept of distributed GIS. Many GIS vendors are now offering limited-functionality software that is capable of running on mobile devices, such as PDAs and tablet computers, which we have termed hand-held GIS (Section 7.6.6). Statistical agencies such as the U.S. Bureau of the Census or the U.S. Department of Agriculture are now widely using such devices for field data collection. Such agencies now routinely equip their field crews with PDAs, allowing them to record data in the field and to upload the results back in the office (Figure 11.1); the field crews of utility companies use similar systems to record the locations and status of transformers, poles, and switches (Figure 11.2). Vendors are also offering various forms of what we have termed server GIS (Section 7.6.3). These allow GIS users to access processing services at remote sites, using little more than a standard Web browser, thus avoiding the cost of installing a GIS locally.

Figure 11.1 Using a simple GIS in the field to collect data. The device on the pole is a GPS antenna, used to georeference data as they are collected. (AFP/Getty Images, Inc.)

Certain concepts need to be clarified at the outset. First, there are four distinct locations of significance to distributed GIS:

- The location of the user and the interface from which the user obtains GIS-created information, denoted by *U* (e.g., the PDA in Figure 11.2).
- The location of the data being accessed by the user, denoted by *D*. Traditionally, data had to be moved to the user's computer before being used, but new technology is allowing data to be accessed directly from data warehouses and archives.
- The location where the data are processed, denoted by *P*. In Section 1.5.3 we introduced the

Figure 11.2 Collecting geographic data in the field, an increasingly common task in many occupations. (© Sean Locke/iStockphoto)

concept of a GIService, a processing capability accessed at a remote site rather than provided locally by the user's desktop GIS.

- The area that is the focus of the GIS project, or the *subject* location, denoted by *S*. All GIS projects necessarily study some area, obtain data as a representation of the area, and apply GIS processes to those data.

In traditional GIS three of these locations—*U*, *D*, and *P*—are the same because both the data and processing occur at the user's desktop. The subject location could be anywhere in the world, depending on the project. But in distributed GIS there is no longer any need for *D* and *P* to be the same as *U*. Moreover, it is possible for the user to be located in the subject area *S*, and be able to see, touch, feel, and even smell it, rather than being in a distant office. The GIS might be held in the user's hand, or stuffed in a backpack, or mounted in a vehicle.

In distributed GIS the user location and the subject location can be the same.

Critical to distributed GIS are the standards and specifications that make it possible for devices, data, and processes to operate together, or *interoperate*. Some of these are universal, such as ASCII, the standard for the coding of characters (Box 3.1), and XML, the extensible markup language that is used by many services on the Web. Others are specific to GIS, and many of these standards have been developed through the Open Geospatial Consortium (OGC, www.opengeospatial.org), an organization set up to promote openness and interoperability in GIS. Among the many successes of OGC over the past decade are the simple feature specification, which standardizes many of the terms associated with primitive elements of GIS databases (polygons, polylines, points, etc.; see also Chapter 8); Geography Markup Language (GML), a version of XML that handles geographic features and enables open-format communication of geographic data; and specifications for Web services (Web Map Service, WMS; Web Feature Service, WFS; and Web Coverage Service, WCS) that allow a user's GIS software to request data automatically from remote servers.

Distributed GIS reinforces the notion that today's computer is not simply the device on the desk, with its hard drive, processor, and peripherals, but something more extended. The slogan "The Network Is the Computer" has provided a vision for at least one company—Sun Microsystems (www.sun.com)—and propels many major developments in computing. The term *cyberinfrastructure* describes a new approach to

Figure 11.3 The power of this home computer would be wasted at night when its owner is sleeping, so instead its power has been "harvested" by a remote server and used to process signals from radio telescopes as part of a search for extra-terrestrial intelligence.

the conduct of science, relying on high-speed networks, massive processors, and distributed networks of sensors and data archives (www.cise.nsf.gov/sci/reports/toc.cfm). Efforts are being made to integrate the world's computers to provide the kinds of massive computing power that are needed by projects such as SETI (the Search for Extra-Terrestrial Intelligence, www.seti.org), which processes terabytes of data per day in the search for anomalies that might indicate life elsewhere in the universe, and makes use of computer power wherever it can find it on the Internet (Figure 11.3). *Grid* computing is a generic term for such a fully integrated worldwide network of computers and data.

In the early days of GIS, all of the data were derived from paper maps and described features on the Earth's surface that were largely static and unchanging. But an increasing amount of real-time data is becoming available, thanks to GPS, and is being fed over the Internet to users. It is now routine, for example, to keep track of the progress of an arriving flight before going to the airport (Figure 11.4), or to make available real-time data on traffic congestion that are obtained by monitoring the speed of vehicles using GPS (see, for example, www.inrix.com). Real-time data are available from surveillance cameras, networks of environmental sensors, and an increasing range of small, cheap devices. RFID (radio frequency identification) allows the tracking of objects that have been implanted or tagged with small sensors, and is now widely used in retailing, livestock management, and building construction. Recently manufactured mobile phones carry RFID tags, as do new passports, items purchased from major retailers, vehicles using automatic highway toll gates, and even some oddly motivated individuals (Figure 11.5). All of this raises the prospect that at some point in the

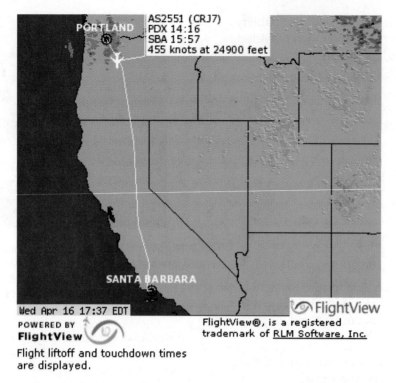

Flight liftoff and touchdown times are displayed.

FlightView®, is a registered trademark of RLM Software, Inc.

Figure 11.4 Tracking a flight: the real-time location of Alaska Airlines Flight 2551 from Portland to Santa Barbara, together with a base map and a depiction of current weather (light blue shows snow).

Figure 11.5 RFID (radio frequency identification) uses small tags to identify and track various kinds of objects. (Steven Puetzer/Photolibrary Group Limited)

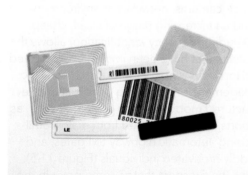

Deborah Estrin, Leading Researcher on Sensor Networks

Deborah Estrin (Figure 11.6) is professor of computer science at the University of California, Los Angeles, and founding director of the Center for Embedded Networked Sensing, which is funded by the U.S. National Science Foundation. The center conducts research on distributed systems that can sense and report environmental conditions, and also sponsors education programs to encourage awareness of the potential of such devices. While these observing systems are mostly fixed, most recently the center's research team has developed and studied mobile phones as the source of sensing information. These specially adapted phones can be used to create detailed maps of human activities and inferred exposures as people walk or bicycle through their local environments (Figure 11.7). Deborah received her Ph.D. from the Massachusetts Institute of Technology in 1985, and among many awards and honors has recently been elected to the National Academy of Engineering and the American Academy of Arts and Sciences.

Figure 11.6 Deborah Estrin, director of the Center for Embedded Network Sensing at the University of California, Los Angeles. (Deborah Estrin, UCLA)

(A)   (B)

Figure 11.7 Mobile phones can be powerful systems for: (A) capturing images; and (B) biometric data from volunteers. (Photo by Eyes of the World Media Group. Copyright 2008 Regents of the University of California)

future it will be possible to know where *everything* is, or at least everything that is important in some specific area of human activity.

In Chapter 5 we introduced the concept of a *mashup*, which describes the linking of Web sites to create new services that none of the component sites can provide alone, and has special significance when the linking is done through geographic location, when it is akin to overlay (see Section 14.2). Mashup, and the linking of services in general, is a key concept of the GeoWeb. Another is *service-oriented architecture* (SOA), the notion that any complex computer application can be decomposed into component parts, and that each of these parts can be provided by services that are distributed over the Internet. For example, suppose one is interested in developing an application to display current news stories, using a map interface that allows the user to zoom to any part of the world and to read about current events there. The service might combine a real-time feed of news stories, including text and images, from Web site A; sending the feed to a second site B that specializes in finding place-names in text and converting them to latitude and longitude; sending the results to a third site C that specializes in generating cartographic displays; and finally connecting the user with this third site. The locations of each of the three sites could be anywhere in the world that is connected to the Internet. SOA requires standards to ensure that the various services are interoperable, and mechanisms for searching out services that meet specific requirements. A more elaborate example of SOA is given in Section 11.4.1.

11.2 Distributing the Data

Since its popularization in the early 1990s, the Internet has had a tremendous and far-reaching impact on the accessibility of GIS data and on the ability of

GIS users to share datasets. As we saw in Chapter 9, a large and increasing number of Web sites offer GIS data for free, for sale, or for temporary use, and also provide services that allow users to search for datasets that satisfy certain requirements.

In the past few years many private citizens have become involved in the distributed creation and dissemination of geographic information, in a process known as *volunteered geographic information* (VGI). The notion that the content of the Web is increasingly created by its users has been termed Web 2.0 (introduced in Section 1.5.1), and today the Web is littered with hundreds of sites that support the creation of VGI. Section 5.2 described one prominent example, Wikimapia, in effect a place-name index created entirely by volunteers and with rich descriptive information. Another example is Open Street Map (OSM), one of many efforts around the world to enlist volunteers in the creation of open, free digital maps that is especially important in areas where such maps are not freely available. An extract of its coverage of São Paulo, Brazil (at time of writing) is shown in Figure 11.8. Volunteers provide GPS tracks, obtained by walking, cycling, or driving; these are then uploaded, integrated, and rendered into high-quality maps such as one might obtain from traditional national mapping agencies or commercial sources. VGI is becoming an important source of geographic information, particularly information that is difficult to obtain from any other source. On the other hand, its contributors have no authority, VGI sites do not generally conduct formal quality testing, and there is no equivalent of the trust that people place in traditional mapping sources (Section 18.5.5).

In effect, in a period of little more than 15 years we have gone from a situation in which geographic data were available only in the form of printed maps from map libraries and retailers, to one in which petabytes (Table 1.1) of information are available for download and use at electronic speed (about 1.5 million CDs would be required to store 1 petabyte). For example, the NASA-sponsored EOSDIS (Earth Observing System Data and Information System; spsosun.gsfc.nasa.gov/New_EOSDIS.html) archives and distributes the geographic data from the EOS series of satellites, acquiring new data at over a terabyte per day, with an accumulated total of more than 1 petabyte at this site alone.

Some GIS archives contain petabytes of data.

The vision of the GeoWeb goes well beyond the ability to access and retrieve remotely located data, however, because it includes the concepts of *search*, *discovery*, and *assessment*: in the world of distributed GIS, how do users search for data, discover their existence at remote sites, and assess their fitness for use? Four concepts are important in this respect: *object-level metadata*, *geolibraries*, *geoportals*, and *collection-level metadata*.

11.2.1 Object-Level Metadata

Strictly defined, metadata are data about data, and *object-level* metadata (OLM) describe the contents of a single dataset. We need information about data for many purposes, and OLM try to satisfy them all. First, we need OLM to automate the process of search and discovery over archives. In that sense OLM are similar to a library's catalog, which organizes the library's contents by author, title, and subject, and makes it easy for a user to find a

Figure 11.8 Open Street Map coverage of part of central São Paulo, Brazil. (© OpenStreetMap and contributors, CC-BY-SA)

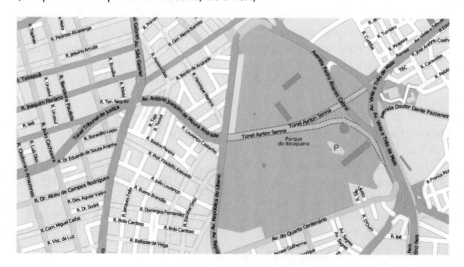

book. But OLM are potentially much more powerful, because a computer is more versatile than the traditional catalog in its potential for re-sorting items by a large number of properties, going well beyond author, title, and subject, and including geographic location. Second, we need OLM to determine whether a dataset, once discovered, will satisfy the user's requirements—in other words, to assess the fitness of a dataset for a given use. Does it have sufficient spatial resolution and acceptable quality? Such metadata may include comments provided by others who tried to use the data, or contact information for such previous users (users often comment that the most useful item of metadata is the phone number of the person who last tried to use the data). Third, OLM must provide the information needed to handle the dataset effectively. This may include technical specifications of format, or the names of software packages that are compatible with the data, along with information about the dataset's location and its volume. Finally, OLM may provide useful information on the dataset's contents. In the case of remotely sensed images, this may include the percentage of cloud obscuring the scene, or whether the scene contains particularly useful instances of specific phenomena, such as hurricanes.

Object-level metadata are formal descriptions of datasets that satisfy many different requirements.

OLM generalize and abstract the contents of datasets, and we would therefore expect the datasets to be smaller in volume than the data they describe. In reality, however, it is easy for the complete description of a dataset to generate a greater volume of information than the actual contents. OLM are also expensive to generate because they represent a level of understanding of the data that is difficult to assemble, and require a high level of professional expertise. Generation of OLM for a geographic dataset can easily take much longer than it takes to catalog a book, particularly if it is necessary to deal with technical issues such as the precise geographic coverage of the dataset, its projection and datum details (Chapter 5), and other properties that may not be easily accessible. Thus the cost of OLM generation, and the incentives that motivate people to provide OLM, are important issues.

For metadata to be useful, it is essential that they follow widely accepted standards. If two users are to be able to share a dataset, they must both understand the rules used to create its OLM, so that the custodian of the dataset can first create the description and so that the potential user can understand it. The most widely used standard for OLM is the U.S. Federal Geographic Data Committee's Content Standards for Digital Geospatial Metadata, or CSDGM, first published in 1993 and now the basis for many other standards worldwide. Box 11.2 lists some of its major features. As a *content standard*,

---

Technical Box (11.2)

Major Features of the U.S. Federal Geographic Data Committee's Content Standards for Digital Geospatial Metadata

1. Identification Information—basic information about the dataset.

2. Data Quality Information—a general assessment of the quality of the dataset.

3. Spatial Data Organization Information—the mechanism used to represent spatial information in the dataset.

4. Spatial Reference Information—the description of the reference frame for, and the means to encode, coordinates in the dataset.

5. Entity and Attribute Information—details about the information content of the dataset, including the entity types, their attributes, and the domains from which attribute values may be assigned.

6. Distribution Information—information about the distributor of and options for obtaining the dataset.

7. Metadata Reference Information—information on the currentness of the metadata information and the responsible party.

8. Citation Information—the recommended reference to be used for the dataset.

9. Time Period Information—information about the date and time of an event.

10. Contact Information—identity of, and means to communicate with, person(s) and organization(s) associated with the dataset.

CSDGM describes the items that should be in an OLM archive, but does not prescribe exactly how they should be formatted or structured. This allows developers to implement the standard in ways that suit their own software environments, but guarantees that one implementation will be understandable to another—in other words, that the implementations will be *interoperable*. For example, ESRI's ArcGIS provides two formats for OLM, one using the widely recognized XML standard and the other using ESRI's own format.

CSGDM was devised as a system for describing geographic datasets, and most of its elements make sense only for data that are accurately georeferenced and represent the spatial variation of phenomena over the Earth's surface. As such, its designers did not attempt to place CSGDM within any wider framework. But in the past decade a number of more broadly based efforts have also been directed at the metadata problem, and at the extension of traditional library cataloging in ways that make sense in the evolving world of digital technology.

One of the best known of these is the Dublin Core (see Box 11.3), which is the outcome of an effort to find the minimum set of properties needed to support search and discovery for datasets in general, not only geographic datasets. Dublin Core treats both space and time as instances of a single property, coverage, and unlike CSGDM does not lay down how such specific properties as spatial resolution, accuracy, projection, or datum should be described.

The principle of establishing a minimum set of properties is sharply distinct from the design of CSGDM, which was oriented more toward the capture of all knowable and potentially important properties of geographic datasets. Of direct relevance here is the problem of cost, and specifically the cost of capturing a full CSGDM metadata record. While many organizations have wanted to make their data more widely available, and have been driven to create OLM for their datasets, the cost of determining the full set of CSDGM elements is often highly discouraging. There is interest therefore in a concept of *light metadata*, a limited set of properties that is both comparatively

---

Technical Box (11.3)

The 15 Basic Elements of the Dublin Core Metadata Standard

1. TITLE. The name given to the resource by the CREATOR or PUBLISHER.

2. AUTHOR or CREATOR. The person(s) or organization(s) primarily responsible for the intellectual content of the resource.

3. SUBJECT or KEYWORDS. The topic of the resource, or keywords, phrases, or classification descriptors that describe the subject or content of the resource.

4. DESCRIPTION. A textual description of the content of the resource, including abstracts in the case of document-like objects or content description in the case of visual resources.

5. PUBLISHER. The entity responsible for making the resource available in its present form, such as a publisher, a university department, or a corporate entity.

6. OTHER CONTRIBUTORS. Person(s) or organization(s) in addition to those specified in the CREATOR element who have made significant intellectual contributions to the resource, but whose contribution is secondary to the individuals or entities specified in the CREATOR element.

7. DATE. The date the resource was made available in its present form.

8. RESOURCE TYPE. The category of the resource, such as home page, novel, poem, working paper, technical report, essay, or dictionary.

9. FORMAT. The data representation of the resource, such as text/html, ASCII, Postscript file, executable application, or JPEG image.

10. RESOURCE IDENTIFIER. String or number used to uniquely identify the resource.

11. SOURCE. The work, either print or electronic, from which this resource is delivered, if applicable.

12. LANGUAGE. Language(s) of the intellectual content of the resource.

13. RELATION. Relationship to other resources.

14. COVERAGE. The spatial locations and temporal durations characteristics of the resource.

15. RIGHTS MANAGEMENT. The content of this element is intended to be a link (a URL or other suitable URI as appropriate) to a copyright notice, a rights-management statement, or perhaps a server that would provide such information in a dynamic way.

cheap to capture and still useful to support search and discovery. Dublin Core represents this approach and thus sits at the opposite end of a spectrum from CSGDM. Every organization must somehow determine where its needs lie on this spectrum, which ranges from light and cheap to heavy and expensive.

Light, or stripped-down, OLM provide a short but useful description of a dataset that is cheaper to create.

11.2.2 Geolibraries and Geoportals

The use of digital technology to support search and discovery opens up many options that were not available in the earlier world of library catalogs and bookshelves. Books must be placed in a library on a permanent basis, and there is no possibility of reordering their sequence—but in a digital catalog it is possible to reorder the sequence of holdings in a collection almost instantaneously. So while a library's shelves are traditionally sorted by subject, it would be possible to re-sort them digitally by author name or title, or by any property in the OLM catalog. Similarly, the traditional card catalog allowed only three properties to be sorted—author, title, and subject—and discouraged sorting by multiple subjects. But the digital catalog can support any number of subjects.

Of particular relevance to GIS users is the possibility of sorting a collection by the coverage properties: location and time. Both the spatial and temporal dimensions are continuous, so it is impossible to capture them in a single property analogous to author that can then be sorted numerically or alphabetically. In a digital system this is not a serious problem, and it is straightforward to capture the coverage of a dataset and to allow the user to search for datasets that cover an area or time of interest defined by the user. Moreover, the properties of location and time are not limited to geographic datasets, since many types of information are associated with specific areas on the Earth's surface or with specific time periods. Searching based on location or time would enable users to find information about any place on the Earth's surface, or any time period—and to find reports, photographs, or even pieces of music, as long as they possessed geographical and temporal *footprints*.

The term *geolibrary* has been coined to describe digital libraries that can be searched for information about any user-defined geographic location. A U.S. National Research Council report (see Further Reading) describes the concept and its implementation, and many instances of geolibraries can be found on the Web.

A geolibrary can be searched for information about a specific geographic location.

Although geolibraries are very useful sources of data, it is often difficult for the user to predict *which* geolibrary is most likely to contain a given dataset, or the closest approximation to it. Rather than have to resort to trial and error, it would be much better if it were possible to search a single site for *all* datasets. This is the goal of a *geoportal*, which can be defined as a single point of entry to a distributed collection of geolibraries, with a single catalog, just as many networks of libraries provide a single *union catalog* to all of their holdings. There is a similarity here to the Web search engines such as Google, which provide a single point of entry for search over a large proportion of the entire Web. However, conventional search engines are not well designed for finding geographic datasets.

The Geospatial One-Stop is a good example of a geoportal, and in many ways it represents the state of the art in searching for geographic data. It was initiated by the U.S. government as a single point of entry to the holdings of government agencies (Figure 19.8), but it also catalogs many other geolibraries (though its coverage is essentially limited to the United States). A user visiting www.geodata.gov is presented with several ways of specifying requirements (Figure 11.9) and is provided with a list of datasets that appear to satisfy them, together with links to the geolibraries that contain the data. Mechanisms are provided for handling datasets that have usage restrictions and may require licensing or the payment of fees. The site receives tens of thousands of visitors per day and catalogs the contents of over a thousand geolibraries. It also supports automatic harvesting of catalog contents from geolibraries that are willing to contribute in this way.

Contemporary geoportals such as the Geospatial One-Stop support several forms of use. The most traditional is the download, when an entire dataset is simply transferred to the user's hard drive and then incorporated in some form of GIS-based analysis. But it is also possible to rely on many geoportals for simple GIS functions such as display, in which case the user needs no more than a Web browser. Finally, it may be possible to use a dataset "live" if the geoportal supports the appropriate standards. In this case the user's GIS behaves as if the geoportal's data were local, but sends requests over the Internet rather than to the local hard drive when specific data are needed in response to a GIS operation. In this mode there is no need to download an entire dataset, and standards

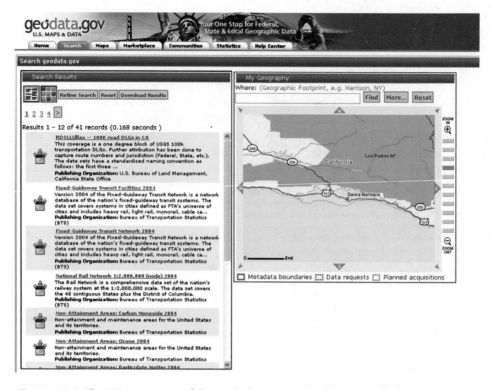

Figure 11.9 The U.S. Department of the Interior's Geospatial One-Stop, an example of a geoportal. In this case, the user has zoomed in to the area of Goleta, California, and requested data on transportation networks. A total of 41 "hits" were identified in this search, indicating 41 possible sources of suitable online information.

take care of such potential interoperability problems as differences of coordinate system or geographic coverage. Live data standards include the Open Geospatial Consortium's WMS, WFS, and WCS, as well as ESRI's IMS.

11.3 The Mobile User

The computer has become so much a part of our lives that for many people it is difficult to imagine life without it. We increasingly need computers to shop, to communicate with friends, to obtain the latest news, and to entertain ourselves. In the early days, the only place one could use computers was in a computing center, within a few meters of the central processor. Computing had extended to the office by the 1970s and to the home by the 1990s. The portable computers of the 1980s opened the possibility of computing in the garden, at the beach, or "on the road" in airports and airplanes. Wireless communication services such as WiFi (the wireless access technology based on the 802.11 family of standards) now allow broadband communication to the Internet from "hotspots" in many hotels, restaurants, airports, office buildings, and private homes (Figure 11.10). The range of mobile computing devices is also multiplying rapidly, from the relatively cumbersome but powerful laptop weighing several kilograms to the PDA, tablet computer, and the mobile phone weighing a hundred or so grams. Within a few years we will likely see the convergence of these devices into a single, low-weight, and powerful mobile personal device that acts as computer, storage device, and mobile phone.

In many ways the ultimate endpoint of this progression is the *wearable* computer, a device that is fully embedded in the user's clothing, goes everywhere, and provides ubiquitous computing service. Such devices are already obtainable, though in fairly cumbersome and expensive form. They include a small box worn on the belt and containing the processor and storage, and an output display clipped to the user's eyeglasses. Figure 11.11 shows such a device in use.

Computing has moved from the computer center and is now close to, even inside, the human body. The ultimate form of mobile computing is the wearable computer.

What possibilities do such systems create for GIS? Of greatest interest is the case in which the user is situated in the subject location; that is, *U* is contained within *S*, and the GIS is used to analyze immediate

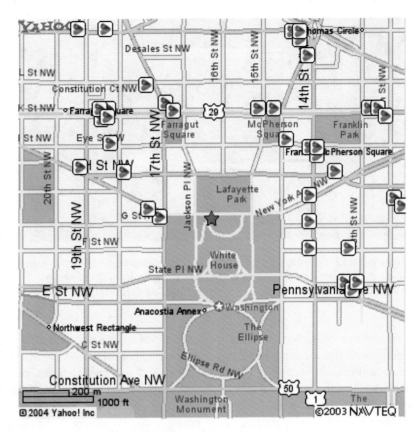

Figure 11.10 Map of WiFi (802.11) wireless broadband "hotspots" within 1 mile of the White House (1600 Pennsylvania Avenue, Washington, DC) and using the T-Mobile Internet provider.

Figure 11.11 A wearable computer in use. The outfit consists of a processor and storage unit hung on the user's waist belt; an output unit clipped to the eyeglasses with a screen approximately 1 cm across and VGA resolution; an input device in the hand; and a GPS antenna on the shoulder. The batteries are in a jacket pocket. (Courtesy Keith Clarke)

surroundings. A convenient way to think about the possibilities is by comparing *virtual* reality with *augmented* reality.

11.3.1 Virtual Reality and Augmented Reality

One of the great strengths of GIS is the window it provides on the world. A researcher in an office in Nairobi, Kenya, might use GIS to obtain and analyze data on the effects of salinization in the Murray Basin of Australia, combining images from satellites with base topographic data, data on roads, soils, and population distributions, all obtained from different sources via the Internet. In doing so, the researcher would build up a comprehensive picture of part of the world he or she might never have visited, all through the medium of digital geographic data and GIS, and might learn almost as much through this GIS window as from actually being there. The expense and time of traveling to Australia would be avoided, and the analysis could proceed almost instantaneously. Some aspects of the study area would be missing, of

Figure 11.12 View along East 64th Street in Manhattan generated in Google Maps using street-level photographs. (© 2008 Google-Imagery © Immersive Media, © 2008 DigitalGlobe, Bluesky, Sanborn, Map data © NAVTEQ TM)

course—aspects of culture, for example, that can best be experienced by meeting with the local people (see the discussion of representation in Section 3.1).

Research environments such as this are termed *virtual realities* because they replace what humans normally gather through their senses—sight, sound, touch, smell, and taste—by presenting information from a database. In most GIS applications, such as those discussed in Section 13.4.4, only one of the senses, sight, is used to create this virtual reality, or VR. In principle it is possible to record sounds and store them in GIS as attributes of features, but in practice very little use is made of any sensory channel other than vision. Moreover, in most GIS applications the view presented to the user is the view from *above*, even though our experience with looking at the world from this perspective is limited (for most of us, it is restricted to times when we requested a window seat in an airplane). As discussed in Section 1.8, GIS has been criticized for what has been termed the *God's-eye*, or a *privileged* view by some writers, on the basis that it distances the researcher from the real conditions experienced by people on the ground.

Virtual environments can place the user in distant locations.

More elaborate VR systems are capable of *immersing* the user by presenting the contents of a database in a three-dimensional environment using special eyeglasses, or by projecting information onto walls surrounding the user, and effectively *transporting* the user into the environment represented in the database. But even the standard personal computer is capable of creating remarkably close approximations to the physical appearance of the geographic landscape, and services such as Google Earth and Microsoft's Bing Maps have pushed the limits dramatically in recent years. Such *geobrowsers* or *virtual globes* create three-dimensional renderings of the Earth and rely on the powerful graphics capabilities now available in the standard office or home computer to support real-time movement—a "magic carpet ride" that even a child of 10 can learn to experience. Recent versions allow a smooth transition from the "God's-eye view" from above to street views assembled from hundreds of millions of photographs (Figure 11.12; see also Figure 2.8), and the 3-D extension of Microsoft's Bing Maps presents photorealistic exteriors of buildings (Figure 11.13).

Roger Downs, professor of geography at Pennsylvania State University, uses Johannes Vermeer's famous painting popularly known as *The Geographer* (Figure 11.14) to make an important point about virtual realities. On the table in front of the figure is a

Figure 11.13 Simulated view along East 64th Street in Manhattan generated in Microsoft's Bing Maps 3D, with three-dimensional buildings and exterior textures.

Figure 11.14 Johannes Vermeer's painting of 1669. (The Granger Collection, New York)

map, taken to represent the geographer's window on a part of the world that happens to be of interest. But the subject figure is shown looking out of the window, at the real world, perhaps because he needs the information he derives from his senses to understand the world as shown on the map. The idea of combining information from a database with information derived directly through the senses is termed *augmented reality*, or AR. In terms of the locations of computing discussed earlier, AR is clearly of most value when the location of the user *U* is contained within the subject area *S*, allowing the user to augment what can be seen directly with information retrieved about the same area from a database. This might include historic information, or predictions about the future, or information that is for some other reason invisible to the user.

AR can be used to augment or replace the work of senses that are impaired or absent. A team led by the late Reginald Golledge, professor of geography at the University of California, Santa Barbara (see Dedication), experimented with AR as a means of helping visually impaired people to perform the simple task of navigation. The system, which is worn by the user (Figure 11.15), includes a differential GPS for accurate positioning, a GIS database that includes very detailed information on the immediate environment, a compass to determine the position of the user's head, and a pair of earphones. The user gives the system

Figure 11.15 The system worn here by the late Reg Golledge (a leader of the development team) uses GIS and GPS to augment the senses of a visually impaired person navigating through a complex space such as a university campus. (Reproduced by permission of Reg Golledge)

information to identify the desired destination, and the system then generates sufficient information to replace the normal role of sight. This might be through verbal instructions, or through generation of stereo sounds that appear to come from the appropriate direction.

Augmented reality combines information from the database with information from the senses.

Steven Feiner, professor of computer science at Columbia University, has demonstrated another form of AR that superimposes historic images and other information directly on the user's field of view. For example, in Figures 11.16 and 11.17 a user wearing a head-mounted device coupled to a wearable computer is seeing both the Columbia University main library building and an image generated from a database showing the building that occupied the library's position prior to the university's move to this site in 1896: the Bloomingdale Insane Asylum.

11.3.2 Location-Based Services

"One of the four big trends in software is location-based applications."

Bill Gates, *Wireless 2000 Conference, March 2000*

A location-based service (LBS) is an information service provided by a device *that knows where it is* and is capable of *modifying* the information it provides based on that knowledge. Traditional computing devices, such as desktops or laptops, have no way of knowing where they are, and their functions are in no way changed when they are moved. But increasingly the essential information about a device's location is available, from a GPS or RFID tag incorporated in the device or from the device's IP address (Section 5.4), and is being used for a wide range of purposes.

A location-based service is provided by a computing device that knows where it is.

The simplest and most obvious form of locationally enabled device is the GPS receiver, and any computer that includes a GPS capability, such as a laptop with an added PCMCIA card, or a GPS-enhanced PDA, is capable of providing LBS. But the most ubiquitous LBS-capable device is the modern mobile phone. A

Figure 11.16 An augmented reality system developed at Columbia University by Professor Steve Feiner and his group. (Columbia University by Prof Steve Feiner and his group © 2001, Computer Graphics & User Interfaces Lab, Columbia University)

Figure 11.17 The user's field of view of Feiner's AR system, showing the Columbia University main library and the insane asylum that occupied the site in the 1800s. (Columbia University by Prof Steve Feiner and his group © 1999, T. Höllerer, S. Feiner & Pavlik Computer Graphics & User Interfaces Lab, Columbia University)

variety of methods exist for determining phone locations, including GPS embedded in the phone itself, measurements made by the mobile phone of signals received from towers, and measurements made by the towers of signals received from the phone. As long as the phone is on, the operator of the mobile phone network is able to pinpoint its location at least to the accuracy represented by the size of the cell, on the order of 10 km, and frequently much more accurately. Within a few years it is likely that the locations of the vast majority of active phones will be known to an accuracy of 10 m.

One of the strongest motives driving this process is emergency response. A large and growing proportion of emergency calls come from mobile phones, and while the location of each landline phone is likely to be recorded in a database available to the emergency responder, in a significant proportion of cases the user of a mobile phone is unable to report his or her current location to sufficient accuracy to enable effective response. Several well-publicized cases have drawn attention to the problem. The magazine *Popular Science*, for example, reported a case of a woman who lost control of her car in Florida in February 2001, skidding into a canal. Although she called 911 (the emergency number standard in the United States and Canada), she was unable to report her location accurately and died before the car was found.

One solution to the 911 problem is to install GPS in the vehicle, communicating location directly to the dispatcher. The Onstar system (www.onstar.com) is one such system, combining a GPS device and wireless communication. The system has been offered in the United States to Cadillac purchasers for a number of years and provides various advisory services in addition to emergency response, such as advice on local attractions and directions to local services. When the vehicle's airbags inflate, indicating a likely accident, the system automatically measures and radios its location to the dispatcher, who relays the necessary information to the emergency services.

Emergency services provide one of the strongest motivations for LBS.

There are many other examples of LBS that take advantage of locationally enabled mobile phones. A *yellow-page* service responds to a user who requests information on businesses that are close to his or her current location (Where is the nearest pizza restaurant? Where is the nearest hospital? Where is the nearest WiFi hotspot?) by sending a request that includes location to a suitable Web server. The response might consist of an ordered list presented on the mobile phone screen, or a simple map centered on the user's current location (Figure 11.10). A *trip planner* gives the user the ability to find an optimum driving route from the current location to some defined destination. Similar services are now being provided by public transport operators, and in some cases these services make use of GPS transponders on buses and trains to provide information on actual, as distinct from scheduled, arrival and departure times. Some social networking sites allow a mobile phone user to display a map showing the current locations of nearby friends (provided their mobile phones are active and they are also registered for this service). Undercover (www. playundercover.com) is an example of a *location-based game* that involves the actual locations of players moving around a real environment, while *geocaching* (see Section 13.2.1) is a type of orienteering in which contestants navigate their way to sites using GPS and conventional navigational means.

Direct determination of location, using GPS or measurement to or from towers, is only one basis on which a computing device might know its location, however. Other forms of LBS are provided by fixed devices and rely on the determination of location when the device was installed. For example, many *point-of-sale* systems that are used by retailers record the location of the sale, combining it with other information about the buyer obtained by accessing credit-card or store-affinity-card records (Figure 11.18). In exchange for the convenience or financial inducement of a card, the user effectively surrenders some degree of location privacy to the company whenever the card is used. One benefit is that it is possible for the company to analyze transactions, looking for patterns of purchase that are outside the card user's normal buying habits, perhaps because they occur in locations that the user does not normally frequent, or at anomalous times. Many of

Figure 11.18 Credit card in use. (© Cihan Taskin/iStockphoto)

us will have experienced the embarrassment of having a credit card transaction refused in an unfamiliar city because the techniques used by the company have flagged the transaction as an indicator that the card might have been stolen. In principle, a store-affinity card gives the company access to information about buying habits and locations, in return for a modest discount.

Location is revealed every time a credit, debit, or store-affinity card is used.

11.3.3 Issues in Mobile GIS

GIS in the field or "on the road" is very different from GIS in the office. First, the location of the user is important and is directly relevant to the application. It makes good sense to center maps on the user's location, to provide the capability for maps that show the view from the user's location rather than from above, and to offer maps that are oriented to the user's direction of travel, rather than north. Second, the field environment may make certain kinds of interaction impractical, or less desirable. In a moving vehicle, for example, it would be dangerous to present the driver with visual displays, unless perhaps these are directly superimposed on the field of view (on the windshield). Instead, such systems often provide instructions through computer-generated speech and use speech recognition to receive instructions. With wearable devices that provide output on minute screens attached to the user's eyeglasses, there is no prospect of conventional point-and-click with a mouse, so again voice communication may be more appropriate. On the other hand, many environments are noisy, creating problems for voice recognition.

One of the most important limitations to mobility remains the battery, and although great strides have been made in recent years, battery (or other wireless energy) technology has not advanced as rapidly as other components of mobile systems, such as processors and storage devices. Batteries typically account for the majority of a mobile system's weight and limit operating time severely.

The battery remains the major limitation to LBS and to mobile computing in general.

Although broadband wireless communication is possible using WiFi, connectivity remains a major issue. Wireless communication techniques tend to be:

- *Limited in spatial coverage.* WiFi hotspots are limited to a single building or perhaps a hundred

meters from the router; mobile phone-based techniques have wider coverage but lower bandwidth (communication speeds), and only the comparatively slow satellite-based systems approach global coverage.

- *Noisy.* Mobile phone and WiFi communications tend to "break up" at the edges of coverage areas, or as devices move between cells, leading to errors in communication.

- *Limited in temporal coverage.* A moving device is likely to lose signal from time to time, while some devices, particularly recorders installed in commercial vehicles and PDAs used for surveys, are unable to upload data until within range of a home system.

- *Insecure.* Wireless communications often lack adequate security to prevent unwanted intrusion. It is comparatively easy, for example, for someone to tap into a wireless session and obtain sensitive information.

Progress is being made on all of these fronts, but it will be some years before it becomes possible to "do GIS" as efficiently, effectively, and safely anywhere in the field as it currently is in the office.

11.4 Distributing the Software: GIServices

This final section addresses distributed processing, the notion that the actual operations of GIS might be provided from remote sites, rather than by the user's own computer. Despite the move to component-based software (Section 7.3.3), it is still true that almost all of the operations performed on a user's data are executed in the same computer, the one on the user's desk. Each copy of a popular GIS includes the same functions, which are replicated in every installation of that GIS around the world. When new versions are released, they must be copied and installed at each site, and because not all copies are replaced there is no guarantee that the GIS used by one user is identical to the GIS used by another, even though the vendor and product name are the same.

A GIService is defined as a program executed at a remote site, that performs some specific GIS task. The execution of the program is initiated remotely by the user, who may have supplied data, or may rely on data provided by the service, or both. A simple example of a GIService is that provided by wayfinding sites such as MapQuest (www.mapquest.com) or Yell.com (see Section 1.5.1.1). The user's current location and desired destination are provided to the service (the current location may be entered by the user or provided automatically from GPS), but the data representing the travel network are provided by the GIService. The results are obtained by solving for the shortest path between the origin and the destination (often a compromise between minimizing distance and minimizing expected travel time), a function that exists in many GIS but in this case is provided remotely by the GIService. Finally, the results are returned to the user in the form of driving instructions, a map, or both.

A GIService replaces a local GIS function with one provided remotely by a server.

In principle, any GIS function could be provided in this way, based on GIS server software (Section 7.6.3). In practice, however, certain functions tend to have attracted more attention than others. One obvious problem is commercial: how would a GIService pay for itself, would it charge for each transaction, and how would this compare to the normal sources of income for GIS vendors based on software sales? Some services are offered free and generate their revenue by sales of advertising space or by offering the service as an add-on to some other service. MapQuest and Yell are good examples, generating much of their revenue through direct advertising and through embedding their service in other Web sites, such as travel services. In general, however, the characteristics that make a GIS function suitable for offering as a service appear to be:

- Reliance on a database that must be updated frequently and is too expensive for the average user to acquire. Both geocoding (Box 5.3) and wayfinding services fall into this category, as do gazetteer services (Section 5.10).

- Reliance on GIS operations that are complex and can be performed better by a specialized service than by a generic GIS.

The number of available GIServices and GeoWeb services is growing steadily (see also Section 1.6), creating a need for directories and other mechanisms to help users find them, and standards and protocols for interacting with them. ArcGISOnline (www. arcgisonline.com; Figure 11.19) provides such a directory, in addition to its role as a geoportal to distributed data. The next section describes the state of the art in exploiting distributed GIServices.

11.4.1 Service-Oriented Architecture

When managing recovery in the aftermath of disasters, it is essential that managers have a complete picture of the situation: a map showing the locations and status of incidents, rescue vehicles and hospitals, traffic conditions, weather, and many other relevant assets and factors. A recent report of the U.S. National Research Council discusses the use of GIS and geographic information in all aspects of emergency management (see Further Reading).

Figure 11.19 ArcGISOnline provides a directory of remote GIServices. In this case a search for services related to transportation networks has identified fourteen maps and services.

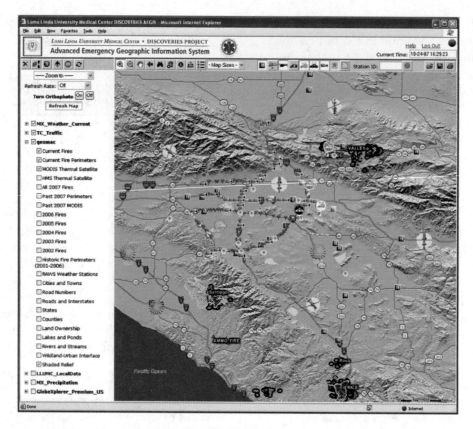

Figure 11.20 The Advanced Emergency GIS (AEGIS) in operation during the wildfire outbreak in Southern California, late 2007.

To illustrate the capabilities of current technology, particularly distributed GIS and the GeoWeb, the Loma Linda University Medical Center and ESRI recently teamed up to develop the Advanced Emergency GIS (AEGIS). It employs mashups and a service-oriented architecture to gather a wide variety of relevant data and to present it in readily understood form to emergency managers and other users. Figure 11.20 shows a typical synoptic visualization during the series of wildfires that hit Southern California in late 2007. The various icons signal the availability of information on traffic conditions (real-time video feeds from cameras on the freeway network), traffic incidents, hospitals (e.g., the number of available beds in each hospital's emergency room), helicopters, the current footprint of each fire, and much other useful information, any of which can be displayed by clicking the icon.

Figure 11.21 shows the architecture of the system. Roughly 10 servers maintained by the relevant agencies, such as the California Department of Transportation and the U.S. Geological Survey, provide real-time feeds of data. The server in the center polls each server at regular intervals, with a frequency depending on the rate of change of the respective server's data, which may range from seconds in the case of emergency vehicle locations to months in the case of base mapping. The format of data contributed by each server varies and in some cases requires the use of an additional service, such as geocoding, or special software to extract relevant information. The central server integrates all of the feeds into a composite mashup and distributes the result to users such as the one symbolized on the right. Several standards are involved, including SOAP (Simple Object Access Protocol), RSS (originally Rich Site Summary), and HTTP (Hypertext Transfer Protocol).

It would be dishonest to suggest that this kind of application is easy to develop. Although standards exist, there are a large number of them, and each requires its own specialized approach. In this example the worst case is the incident report feed from the California Department of Transportation, which requires analyzing the text Web pages posted by the agency using a specially developed program. Moreover, AEGIS relies on the correct operation of many servers and feeds, and is clearly vulnerable to power and network outages. However, there is no doubt that in future it will be easier to develop complex

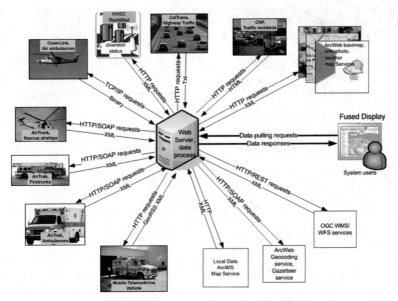

Figure 11.21 The service-oriented architecture of AEGIS.

SOA-based systems, given the obvious power and immediacy that is possible.

11.5 Prospects

Distributed GIS offers enormous benefits: it reduces duplication of effort, allows users to take advantage of remotely located data and services through simple devices, and provides ways of combining information gathered through the senses with information obtained from digital sources. Many issues continue to impede progress, however: complications resulting from the difficulties of interacting with devices in field settings; limitations placed on communication bandwidth and reliability; and limitations inherent in battery technology. Perhaps more problematic than

any of these at this time is the difficulty of imagining the full potential of distributed GIS. We are used to associating GIS with the desktop and conscious that we have not fully exploited its potential. So it is hard to imagine what might be possible when the GIS can be carried anywhere, and its information combined with the window on the world provided by our senses.

More broadly, the GeoWeb offers a vision of a future in which geographic location is central to a wide range of applications of information technology, and one in which individuals are able to access detailed information about the dynamic state of transportation, weather, and a host of other socially important domains. We are just beginning to see the potential of this vision and to recognize that while its benefits are often compelling, it raises invasion of privacy and surveillance issues that society as a whole will have to address.

Questions for Further Study

1. Design a location-based game based on GPS-enabled mobile phones.

2. Find a selection of geolibraries on the Web and identify their common characteristics. How do each of them allow the user (a) to specify locations of interest, (b) to browse through datasets that have similar characteristics, and (c) to examine the contents of data sets before acquiring them?

3. To what extent do citizens have a right to locational privacy? What laws and regulations control the use of locational data on individuals?

4. How many computers are there currently in your home, and how many items in your home have RFID tags? If you have a mobile phone, is it GPS-enabled? Do you have other GPS devices in the home?

Further Reading

National Research Council. 1999. *Distributed Geolibraries: Spatial Information Resources*. Washington, DC: National Academy Press.

National Research Council. 2007. *Successful Response Starts with a Map: Improving Geospatial Support for Disaster Management*. Washington, DC: National Academy Press.

Peng, Z.-H. and Tsou, M.-H. 2003. *Internet GIS: Distributed Geographic Information Services for the Internet and Wireless Networks*. Hoboken, N.J.: Wiley.

Scharl, A. and Tochtermann, K. (eds.). 2007. *The Geospatial Web: How Geobrowsers, Social Software and the Web 2.0 Are Shaping the Network Society*. Berlin: Springer.

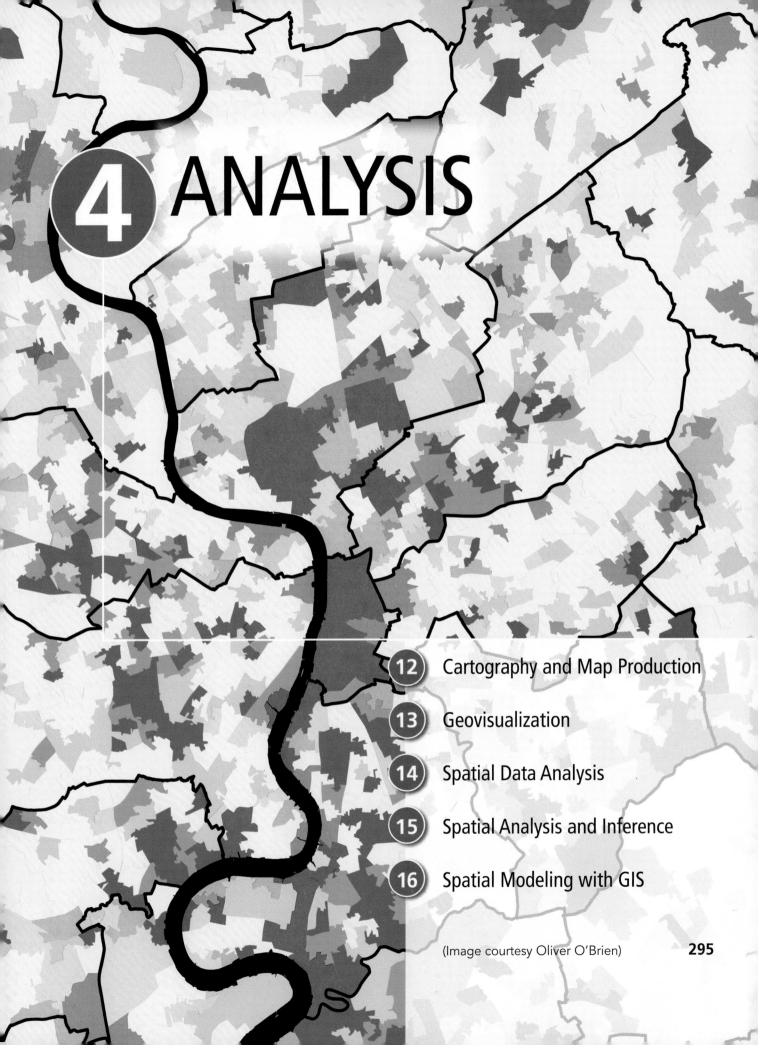

4 ANALYSIS

(Image courtesy Oliver O'Brien)

295

12 Cartography and Map Production

This chapter is the first of a series of chapters that examine GIS output; the next, Chapter 13, deals with the closely related but distinct subject of visualization. The current chapter reviews the nature of cartography and the ways that users interact with GIS in order to produce digital and hardcopy reference and thematic maps. Standard cartographic conventions and graphic symbology are discussed, as is the range of transformations used in map design. Map production is reviewed in the context of creating map series, as well as maps for specific applications. Some specialized types of mapping are introduced that are appropriate for particular applications areas.

LEARNING OBJECTIVES

After studying this chapter, you will understand:

- The nature of maps and cartography.

- Key map-design principles.

- The choices that are available to compose maps.

- The many types of map symbology.

- Concepts of map production flow lines.

12.1 Introduction

GIS output represents the pinnacle of many GIS projects. Since the purpose of information systems is to produce results, this aspect of GIS is vitally important to many managers, technicians, and scientists. Maps are a very effective way of summarizing and communicating the results of GIS operations to a wide audience. The importance of map output is further highlighted by the fact that many consumers of geographic information only interact with GIS through their use of map products.

For the purposes of organizing the discussion in this book, it is useful to distinguish between two types of GIS output: *formal maps,* created according to well-established cartographic conventions, that are used as a reference or communication product (e.g., a military mapping agency 1:250,000 scale topographic map, or a geological survey 1:50,000 scale paper map—see Box 12.3); and *transitory map and map-like visualizations,* used simply to display, analyze, edit, and query geographic information (e.g., results of a database query to retrieve areas with poor health indicators viewed on a desktop computer, or routing information displayed on a PDA device. Chapter 13 deals with many types of transitory maps. Both can exist in digital form on interactive display devices or in hardcopy form on paper and other media. In practice, this distinction is somewhat arbitrary and there is considerable overlap, but at the core the motivations, tools, and techniques for map production and map visualization are quite different. The current chapter will focus on maps and the next on visualization.

Cartography is concerned with the art, science, and techniques of making maps or charts. Conventionally, the term *map* is used for terrestrial areas (Figure 12.1)

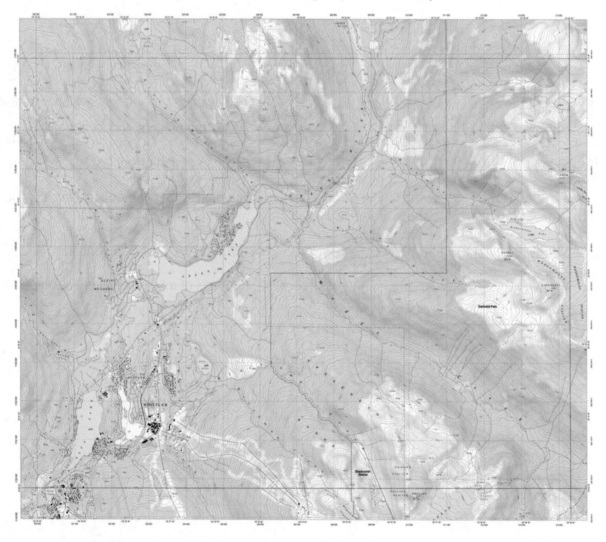

Figure 12.1 Terrestrial topographic map of Whistler, British Columbia, Canada. This is one of a collection of 7016 commercial maps at 1:20,000 scale covering the province. (Courtesy: ESRI)

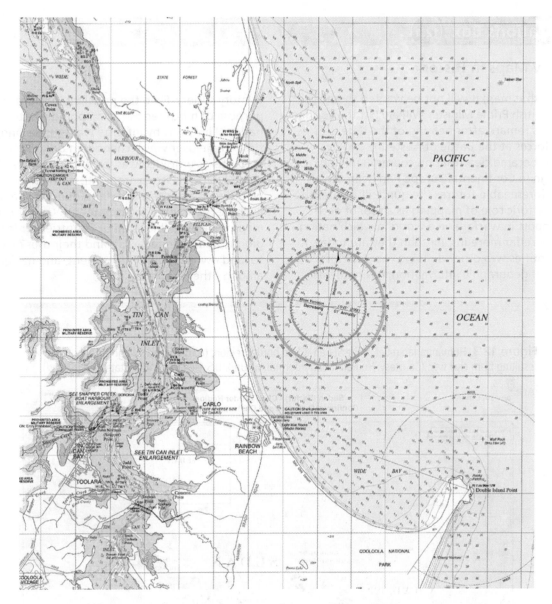

Figure 12.2 Marine chart of Great Sandy Strait (South), Queensland, Australia Boating Safety Chart. This chart conforms to international charting standards. (Courtesy: ESRI and Maritime Safety, Queensland)

and chart for marine areas (Figure 12.2), but they are both maps in the sense the word is used here. In statistical or analytical fields, charts provide the pictorial representation of statistical data, but this does not form part of the discussion in this chapter. Note, however, that statistical charts can be used on maps (e.g., Figure 12.3B). Cartography dates back thousands of years to a time before paper, but the main visual display principles were developed during the paper era; thus many of today's digital cartographers still use the terminology, conventions, and techniques from the paper era. Box 12.1 illustrates the importance of maps in a historic military context.

Maps are important communication and decision support tools.

Historically, the origins of many national mapping organizations can be traced to the need for mapping for "geographical campaigns" of infantry warfare, for colonial administration, and for defense. Today such organizations fulfill a far wider range of needs of many more user types (see Chapter 18). Although the military remains a heavy user of mapping, such territorial changes as arise out of today's conflicts reflect a more subtle interplay of economic, political, and historical considerations—though, of course,

Military Maps in History

"Roll up that map; it will not be wanted these ten years." British Prime Minister William Pitt the Younger made this remark after hearing of the defeat of British forces at the Battle of Austerlitz in 1805, where it became clear that his country's military campaign in Continental Europe had been thwarted for the foreseeable future. The quote illustrates the crucial historic role of mapping as a tool of decision support in warfare, in a world in which nation-states were far more insular than they are today. It also identifies two other defining characteristics of the use of geographic information in Nineteenth-Century society. First, the principal, straightforward purpose of much terrestrial mapping was to further national interests by infantry warfare. Second, the time frame over which changes in geographic activity patterns unfolded was, by today's standards, incredibly slow—Pitt envisaged that no British citizen would revisit this territory for a quarter of a (then) average lifetime! In the Nineteenth Century, printed maps were available in many countries for local, regional, and national areas, and their uses extended to land ownership, tax assessment, and navigation, among other activities.

Figure 12.3 Humanitarian relief in the Sudan–Chad border region: (A) damaged villages in July 2004 resulting from civil war. (Courtesy: Humanitarian Information Unit, U.S. Department of State) (*continued*)

(A)
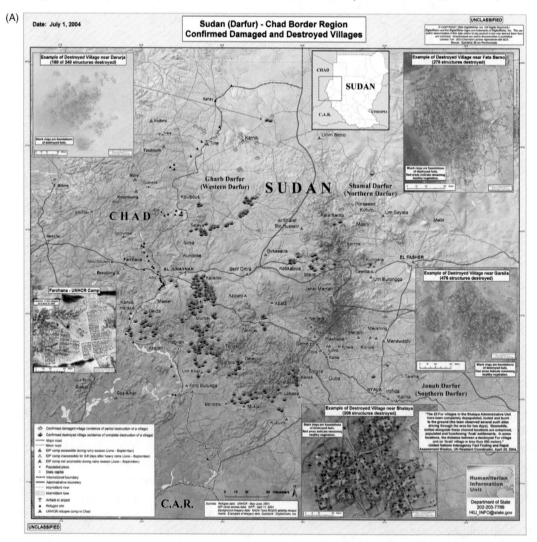

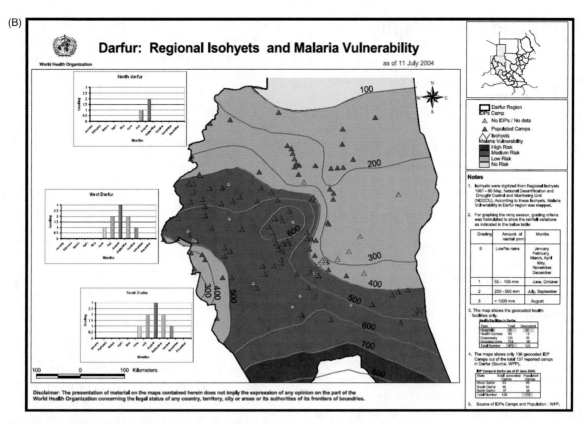

Figure 12.3 (*continued*) (B) malaria vulnerability (Courtesy World Health Organization)

the threat or actual deployment of force remains a pivotal consideration. Today, GIS-based terrestrial mapping underpins a wide range of activities, such as the support of humanitarian relief efforts (Figure 12.3) and the partitioning of territory through negotiation rather than force (e.g., GIS was used by senior decision makers in the partitioning of Bosnia in 1999). The time frame over which events unfold is also much more rapid: it is inconceivable to think of politicians, managers, and officials being able to neglect geographic space for months, weeks, or even days, never mind years.

Paper maps remain in widespread use because of their transportability, reliability, ease of use, and the routine application of printing technology that they entail. They are also amenable to conveying straightforward messages and supporting decision making. Yet our increasingly detailed understanding of the natural environment and the accelerating complexity of society mean that the messages that mapping can convey are increasingly sophisticated and, for the reasons set out in Chapter 6, uncertain. Greater democracy and accountability, coupled with the increased spatial reasoning abilities that better education brings, indicate that more people than ever feel motivated and able to contribute to all kinds of spatial policy. This makes the decision support role

immeasurably more challenging, varied, and demanding of visual media. Today's mapping must be capable of communicating an extensive array of messages and emulating the widest range of what-if scenarios (Section 16.1.1).

Both paper and digital maps have an important role to play in many economic, environmental, and social activities.

The visual medium of a given application must also be open to the widest community of users. Technology has led to the development of an enormous range of devices to bring mapping to the greatest range of users in the widest spectrum of decision environments. In-vehicle displays, palm-top devices, and wearable computers are all important in this regard (see Chapters 7 and 11 for examples). Most important of all, the innovation of the Internet makes societal representations of space a real possibility for the first time. That is to say, it is now very easy to create cartographic products that represent everything from consumer preferences about how they will vote in an election to indicators of social deprivation, and ethnicity. These ideas are explored more fully in the next chapter.

12.2 Maps and Cartography

There are many possible definitions of a map; here we use the term to describe digital or analog (softcopy or hardcopy) output from a GIS that shows geographic information using well-established cartographic conventions. A map is the final outcome of a series of GIS data processing steps (Figure 12.4) beginning with data collection, editing and maintenance (Chapter 9), extending through data management (Chapter 10) and analysis (Chapters 14–16), and concluding with a map. Each of these activities successively transforms a database of geographic information until it is in the form appropriate to display on a given technology.

Central to any GIS is the creation of a data model that defines the scope and capabilities of its operation (Chapter 8), and the management context in which it operates (Chapters 17–19). There are two basic types of map (Figure 12.5): reference maps, such as topographic maps from national mapping agencies (see also Figure 12.1), that convey general information; and thematic maps that depict specific geographic themes, such as population census statistics, soils, or climate zones (see Figures 12.13 and 12.14 for further examples).

Maps are both storage and communication mechanisms.

Maps fulfill two very useful functions, acting as both storage and communication mechanisms for geographic information. The old adage "a picture is worth a thousand words" connotes something of the efficiency of maps as a storage container. The modern equivalent of this saying is "a map is worth a million bytes." Before the advent of GIS, the paper map was the database, but now a map can be considered merely one of many possible products generated from a digital geographic database. Maps are also a mechanism to communicate information to viewers. They can present the results of analyses (e.g., the optimum site suitable for locating a new store, or analysis of the impact of an oil spill). They can communicate spatial relationships between phenomena across the same map, or between maps of the same or different areas. As such they can assist in the identification of spatial order and differentiation. Effective decision support requires that the message of the map be readily interpretable in the mind of the decision maker. A major function of a map is not simply to marshal and transmit known information about the world, but also to create or reinforce a particular message. The work of the leader of one important national mapping organization is discussed in Box 12.2.

Maps also have several limitations:

- Maps can be used to miscommunicate obfuscate accidentally or on purpose. For example, incorrect use of symbols can convey the wrong message to users by highlighting one type of feature at the expense of another (see Figure 12.13 for an example of different choropleth map classifications of the same data).

- Maps are a single realization of a spatial process. If we think for a moment about maps from a statistical perspective, then each map instance represents the outcome of a sampling trial and is therefore a single occurrence generated from all possible maps on the same subject for the same area (see Section 6.4.2). The significance of this is that other sample maps drawn from the same population would exhibit variations and, consequently, we need to be careful in drawing inferences from a single map sample. For example, a map of soil textures is derived by interpolating soil sample texture measurements (see Box 4.3). Repeated sampling of soils will show natural variation in the texture measurements.

- Maps are often created using complex rules, symbology, and conventions, and can be difficult to understand and interpret by the untrained viewer. This is particularly the case, for example, in multivariate statistical thematic mapping where the idiosyncrasies of classification schemes and color symbology can be challenging to comprehend.

Uncertainty pertains to maps just as it does to other geographic information.

Figure 12.4 GIS processing transformations needed to create a map.

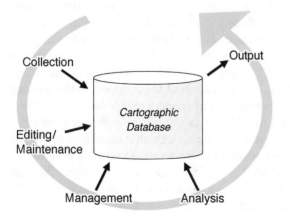

(A)

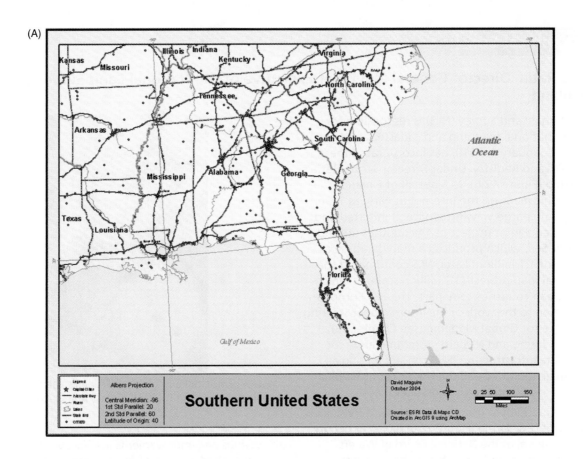

(B)

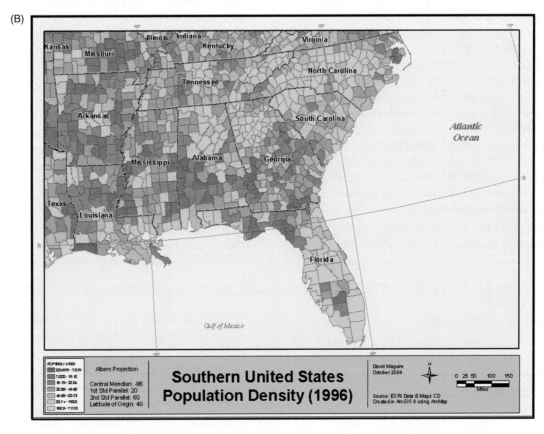

Figure 12.5 Two maps of the Southern United States: (A) a reference map showing topographic information; and (B) a thematic map showing population density.

Jarmo Ratia, Director General and Chief Executive of the National Land Survey (NLS) of Finland

The National Land Survey (NLS) is responsible for Finland's national mapping and cadaster agency and, since January 1, 2010, has incorporated the country's Land Registry. Under Director Jarmo Ratia's leadership (Figure 12.6), NLS put all of Finland's topographic maps on the Internet as early as 1996, as a free-of-charge viewing service. It initiated and coordinated a joint project of 14 countries around the Baltic Sea, which produced a map database at the scale of 1:1 million. Later, NLS coordinated the Euroglobal map project (1:1 million scale) in which 35 European countries took part. This is a European contribution to the work of the International Steering Committee for Global Mapping (see Chapter 19). Currently, Jarmo and NLS are promoting an Arctic spatial data infrastructure (SDI) concept among all eight Arctic countries—all the Nordic countries plus Russia, Canada, and the United States.

After graduating from the University of Helsinki as a Master of Law, Jarmo (Figure 12.6) served as deputy director general of the National Board of Waters and Environment for 8 years before becoming the director general of NLS in 1991. He also served as the permanent secretary of the Ministry of Agriculture and Forestry during Finland's presidency of the EU in 1999–2000 and as a member of the Supreme Administrative Court in 1985. He has been the chairman of the Finnish-Norwegian Border River Commission and the chairman of the Border Inspection Commissions of Finland–Sweden, Finland–Norway, and Finland–Russia. In 1999 he was given responsibilities for land administration issues in the Balkan region, especially in Kosovo, by the UN/ECE Secretariat in Geneva. In addition, he has been the president of CERCO (the group of 35 European NMOs). He reflects:

> We live in an era of a new paradigm—the knowledge-based society. As a governmental official, I understand how big administrative

Figure 12.6 Jarmo Ratia, head of Finland's national mapping organization (NMO) (Courtesy Antero Aaltonen)

savings and improvements in effectiveness can be reached when data are collected only once and transferred through a network to other potential users. I also know that GIS can organize and structure all information and human knowledge into easily understandable form.

> To maximize benefits, we need SDI. This is not only a national or regional exercise, but a real global one. As CEO of NLS I have seen how much work is needed to make data interoperable even inside one organization. It will be a huge task to create global or regional SDI (like INSPIRE), but it is well worth trying. Humanity is facing major supranational challenges like climate change and overpopulation. Of course SDI cannot change the climate but it is an indispensable tool for ensuring sustainable development.

12.2.1 Maps and Media

Without question, GIS has fundamentally changed cartography and the way we create, use, and think about maps (see Box 12.3). The digital cartography of GIS frees mapmakers from many of the constraints inherent in traditional (non-GIS) paper mapping (see also Section 3.7).

- The paper map is of fixed scale. Generalization procedures (Section 3.8) can be invoked in order

to maintain clarity during map creation. This detail is not recoverable, except by reference back to the data from which the map was compiled. The zoom facility of GIS can allow mapping to be viewed at a range of scales, and permit detail to be filtered out as appropriate at a given scale.

- The paper map is of fixed extent, and adjoining map sheets must be used if a single map sheet does not cover the entire area of interest.

Czech Geological Survey (CGS) Map Production

In 1994, the Czech Geological Survey (CGS) began extensive use of GIS technology to meet the increasing demand for digital information about the environment. GIS in the CGS focuses on the methods of spatial data processing, unification, and dissemination. Digital processing of geological maps and the development of GIS follows standardized procedures using common geological dictionaries and graphic elements. Recently, the main objective has been to create and implement a uniform geological data model and provide the public and the scientific community with easy access to geographic data, via a Web map server (www.geology.cz). CGS has a unique geographic information system containing more than 260,000 mapped geological objects from the entire Czech Republic. The fundamental part of this geographic database is the unified national geological index (legend), which consists of four main types of information—chronostratigraphical units, regional units, lithostratigraphical units, and lithological description of rocks. The database has been under revision since 1998, leading to the creation of a seamless digital geological map of the Czech Republic. This database has already been used for land-use planning by government and local authorities. The geological map of the Krkonose–Jizera Mountains shown in Figure 12.7 is a cartographic presentation of one part of the CGS database. The overview map in the bottom left corner shows the extent of all maps in the series.

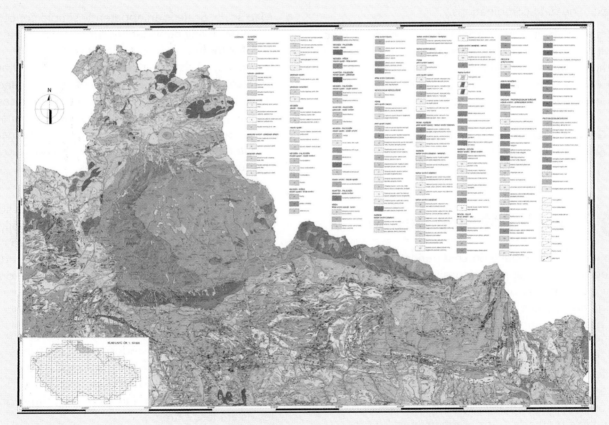

Figure 12.7 A 1:50,000 geological map of the Krkonose–Jizera Mountains, Czech Republic. (Courtesy ESRI)

(An unwritten law of paper map usage is that the most important map features always lie at the intersection of four paper map sheets!) GIS, by contrast, can provide a seamless medium for viewing space, and users are able to pan across wide swaths of territory.

- Most paper maps present a static view of the world, whereas conventional paper maps and charts are not adept at portraying dynamics. GIS-based representations are able to achieve this through animation.

- The paper map is flat and hence limited in the number of perspectives that it can offer on three-dimensional data. 3-D visualization is much more effective within GIS, which can support interactive pan and zoom operations (see Figure 9.14 for a 3-D view example).

- Paper maps provide a view of the world as essentially complete. GIS-based mapping allows the supplementation of base-map material with further data. Data layers can be turned on and off to examine data combinations.

- Paper maps provide a single, map-producer-centric, view of the world. GIS users are able to create their own, user-centric, map images in an interactive way. Side-by-side map comparison is also possible in GIS.

GIS is a flexible medium for the production of many types of maps.

12.3 Principles of Map Design

Map design is a creative process during which the cartographer, or mapmaker, tries to convey the message of the map's objective. The primary goals in map design are to share information, highlight patterns and processes, and illustrate results. A secondary objective is to create a pleasing and interesting picture, but this must not be at the expense of fidelity to reality and meeting the primary goals. Map design is quite a complex procedure requiring the simultaneous optimization of many variables and harmonization of multiple methods. Cartographers must be prepared to compromise and balance choices. It is difficult to define exactly what constitutes a good design, but the general consensus is that a good design is one that looks good, is simple and elegant, and, most importantly, leads to a map that is fit for the purpose.

Robinson et al. (1995) identify seven controls on the map-design process:

- *Purpose.* The purpose for which a map is being made will determine what is to be mapped and how the information is to be portrayed. Reference maps are multipurpose, whereas thematic maps tend to be single purpose (Figure 12.5). With the digital technology of GIS, it is easier to create maps, and many more are digital and interactive. As a consequence, today's maps are increasingly single purpose.

- *Reality.* The phenomena being mapped will usually impose some constraints on map design. For example, the orientation of the country— whether it be predominantly east-west (Russia) or north-south (Chile)—will determine layout in no small part.

- *Available data.* The specific characteristics of data (e.g., raster or vector, continuous or discrete, or point, line, or area) will affect the design. There are many different ways to symbolize map data of all types, as discussed in Section 12.3.2.

- *Map scale.* Scale is an apparently simple concept, but it has many ramifications for mapping (see Box 4.2 for further discussion). It will control the quality of data that can appear in a map frame, the size of symbols, the overlap of symbols, and much more. Although one of the early promises of digital cartography and GIS was scale-free databases that could be used to create multiple maps at different scales, this has never been realized because of technical complexities.

- *Audience.* Different audiences want different types of information on a map and expect to see information presented in different ways. Usually, executives (and small children!) are interested in summary information that can be assimilated quickly, whereas advanced users often want to see more information. Similarly, those with restricted eyesight find it easier to read bigger symbols.

- *Conditions of use.* The environment in which a map is to be used will impose significant constraints. Maps for outside use in poor or very bright light will need to be designed in ways different to maps for use indoors where the light levels are less extreme.

- *Technical limits.* The display medium, be it digital or hard copy, will impact the design process in several ways. For example, maps to be viewed in an Internet browser, where resolution and bandwidth are limited, should be simpler and based on fewer data than equivalents to be displayed on a desktop PC monitor.

12.3.1 Map Composition

Map composition is the process of creating a map comprising several closely interrelated elements (Figure 12.8):

- *Map body.* The principal focus of the map is the main map body, or in the case of comparative maps there will be two or more map bodies. It should be given space and use symbology appropriate to its significance.

- *Inset/overview map.* Inset and overview maps may be used to show, respectively, an area of the main map body in more detail (at a larger scale) and the general location or context of the main body.

- *Title.* One or more map titles are used to identify the map and to inform the reader about its content.

- *Legend.* This lists the items represented on the map and how they are symbolized. Many different layout designs are available, and there is a considerable body of information available about legend design.

- *Scale.* The map scale provides an indication of the size of objects and the distances between them. A paper map scale is a ratio, where one unit on the map represents some multiple of that value in the real world. The scale can be symbolized numerically (1:1,000), graphically (a scalebar), or texturally ("one inch equals 10,000 inches"). The scale is a representative fraction and so a 1:1,000 scale is larger (finer) than 1:100,000. A small (coarse) scale map displays a larger area than a large (fine) scale map, but with less detail. See also Box 4.2 for more details about scale.

- *Direction indicator.* The direction and orientation of a map can be conveyed in one of several ways, including grids, graticules, and directional symbols (usually north arrows). A grid is a network of parallel and perpendicular lines superimposed on a map (see Figure 12.1). A graticule is a network of longitude and latitude lines on a map that relates points on a map to their true location on the Earth (see Figure 12.5).

- *Map metadata.* Map compositions can contain many other types of information, including the map projection, date of creation, data sources, and authorship.

A key requirement for a good map is that all map elements are composed into a layout that has good visual balance. On large-scale maps, such as 1:10,000 national mapping agency topographic series, all of the contextual items (everything listed above except the

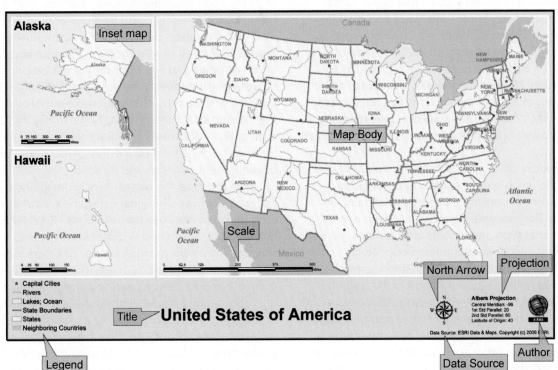

Figure 12.8 The principal components of a map composition layout.

map body) usually appear as marginal notations (or marginalia). In the case of map series or atlases (see Section 12.4), some of the common information may be in a separate document. On small- or medium-scale maps this information usually appears within the map border (Figure 12.7).

12.3.2 Map Symbolization

The data to be displayed on a map must be classified and represented using graphic symbols that conform to well-defined and accepted conventions. The choice of symbolization is critical to the usefulness of any map. Unfortunately, the seven controls on the design process listed in Section 12.3.1 also conspire to mean that there is not a single universal symbology model applicable everywhere, but rather one for each combination of factors. Again, we see that cartographic design is a compromise reached by simultaneously optimizing several factors.

Good mapping requires that spatial objects and their attributes can be readily interpreted in applications. In Chapter 3, attributes were classified as being measured on the nominal, ordinal, interval, ratio, or cyclic scales (see Box 3.3). Also, points, lines, and areas were described as types of discrete objects (see Section 3.5.1), and surfaces were discussed as a type of continuous field (see Section 3.5.2). We have already seen how attribute measures that we think of as continuous are actually discretized to levels of precision imposed by measurement or design (see Section 6.5). The representation of spatial objects is similarly imposed: cities might be captured as points, areas, mixtures of points, lines, and areas (as in a street map: Figure 12.19) or 3-D "walkthroughs" (see Section 13.4), depending on the base scale of a representation and the importance of city objects to the application. Measurement scales and spatial object types are thus one set of conventions that are used to abstract reality. Whether using GIS or paper, mapping may entail reclassification or transformation of attribute measures.

The process of mapping attributes frequently entails further problems of classification because many spatial attributes are inherently uncertain (Chapter 6). For example, in order to create a map of occupational type, individuals' occupations will be classified first into socioeconomic groups (e.g., "factory worker") and perhaps then into supergroups, such as "blue collar." At every stage in the aggregation process, we inevitably do injustice to many individuals who perform a mix of white and blue collar, intermediate and skilled functions, by lumping them into a single group (what social class is a frogman—a "professional" diver?). In practice, the validity and usefulness of an occupational classification will have become established over repeated applications, and the task of mapping is to convey thematic variation in as efficient a way as possible.

12.3.2.1 Attribute Representation and Transformation

Humans are good at interpreting visual data—much more so than interpreting numbers, for example—but conventions are still necessary to convey the message that the mapmaker wants the data to impart. Many of these conventions relate to the use of symbols (such as the way highway shields denote route numbers on many U.S. medium- and fine-scale maps; Figure 12.19) and colors (blue for rivers, green for forested areas, etc.), and have been developed over a period of hundreds of years. Mapping of different themes (such as vegetation cover, surface geology, and socioeconomic characteristics of human populations) has a more recent history. Here too, however, mapping conventions have developed, and sometimes they are specific to particular applications.

Attribute mapping entails the use of graphic symbols, which (in two dimensions) may be referenced by points (e.g., historic monuments and telecoms antennae), lines (e.g., roads and water pipes), or areas (e.g., forests and urban areas). Basic point, line, and area symbols are modified in different ways in order to communicate different types of information. The ways in which these modifications take place adhere to cognitive principles and the accumulated experience of application implementations. The nature of these modifications was first explored by Jacques Bertin in 1967 and was extended to the typology illustrated in Figure 12.9 by Alan MacEachren. The size and orientation of point and line symbols are varied principally to distinguish between the values of ordinal and interval/ratio data using graduated symbols (such as the proportional pie symbols shown in Figure 12.10). Figure 12.11 illustrates how orientation and color can be used to depict the properties of locations, such as ocean current strength and direction.

Hue refers to the use of color, principally to discriminate between nominal categories, as in agricultural or urban land-use maps (Figure 12.12). Different hues may be combined with different textures or shapes if there are a large number of categories in order to avoid difficulties of interpretation. The shape of map symbols can be used to communicate information about either a spatial attribute (e.g., a viewpoint or the start of a walking trail), or its spatial location (e.g., the location of a road or boundary of a particular type: Figure 12.12), or spatial relationships (e.g., the relationship between subsurface topography

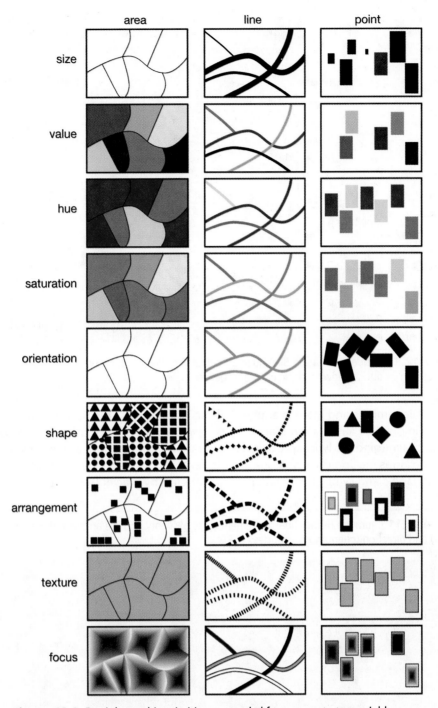

Figure 12.9 Bertin's graphic primitives, extended from seven to ten variables (the variable location is not depicted). Source: MacEachren 1994 (from *Visualization in Geographical Information Systems*, Hearnshaw H.M. and Unwin D.J. (eds.). Reproduced by permission of John Wiley & Sons, Ltd.)

and ocean currents). Arrangement, texture, and focus refer to within- and between-symbol properties that are used to signify pattern. A final graphic variable in the typologies of MacEachren and Bertin is location (not shown in Figure 12.9), which refers to the practice of offsetting the true coordinates of objects in order to improve map intelligibility, or changes in map projection. We discuss this in more detail later in this section. Some of the common ways in which these graphic variables are used to visualize spatial object types and attributes are shown in Table 12.1.

The selection of appropriate graphic variables to depict spatial locations and distributions presents one set of problems in mapping. A related task is how best to position symbols on the map, so as to optimize map interpretability. The representation of

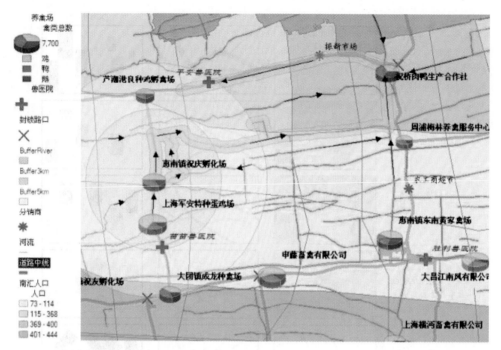

Figure 12.10 Incidence of Asian Bird Flu in Shanghai City. Pie size is proportional to the number of cases; red = duck, yellow = chicken, blue = goose.

nominal data by graphic symbols and icons is apparently trivial, although in practice automating placement presents some challenging analytical problems. Most GIS software packages include generic algorithms for positioning labels and symbols in relation to geographic objects. Point labels are positioned to avoid overlap by creating a window, or mask (often invisible to the user), around text or symbols. Linear features, such as rivers, roads, and contours, are often labeled by placing the text using a spline function to

Figure 12.11 Tauranga Harbor Tidal Movements, Bay of Plenty, NZ. The arrows indicate speed (color) and direction (orientation). (Courtesy ESRI)

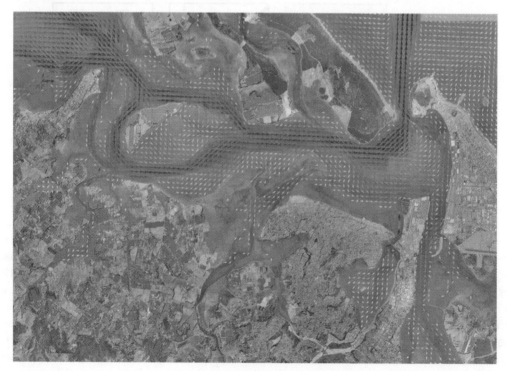

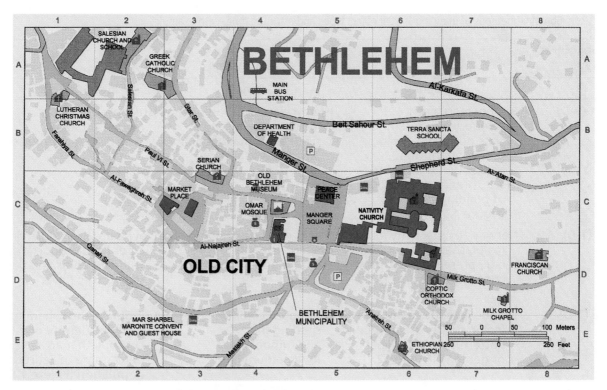

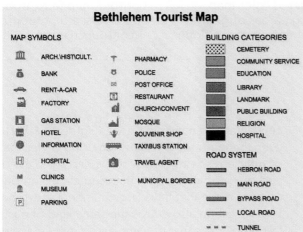

Figure 12.12 Use of hue (color) to discriminate between Bethlehem, West Bank, Israel urban land-use categories, and use of symbols to communicate location and other attribute information. (Courtesy ESRI)

give a smooth, even distribution, or distinguished by use of color. Area labels are assigned to central points (see Figure 12.8), using geometric algorithms similar to those used to calculate geometric centroids (Section 15.2.1). These generic algorithms are frequently customized to accommodate common conventions and rules for particular classes of application, such as topographic (Figure 12.1), utility, transportation, and seismic maps, for example. Generic and customized algorithms also include color conventions for map symbolization and lettering.

Ordinal attribute data are assigned to point, line, and area objects in the same rule-based manner, with the ordinal property of the data accommodated through use of a hierarchy of graphic variables (symbol and lettering sizes, types, colors, intensities, etc.). As a general rule, the typical user is unable to differentiate between more than seven (plus or minus two) ordinal categories, and this provides an upper limit on the normal extent of the hierarchy.

A wide range of conventions is used to visualize interval- and ratio-scale attribute data. Proportional

Table 12.1 Common methods of mapping spatial object types and attribute data, with examples.

Spatial object type	Attribute type		
	Nominal	Ordinal	Interval/Ratio
Point (0-D)	Symbol map (each category a different class of symbol—color, shape, orientation), and/or use of lettering: e.g., presence/absence of building type (Figure 12.12)	Hierarchy of symbols or lettering (color and size): e.g., small/medium/large depots	Graduated symbols (color and size): e.g., disease incidence (Figure 12.10)
Line (1-D)	Network connectivity map (color, shape, orientation): e.g., presence/absence of connection (Figure 12.20)	Graduated line symbology (color and size): e.g., road classifications (Figure 12.19)	Flow map with width or color lines proportional to flows (color and size): e.g., traffic flows
Area (2-D)	Unique category map (color, shape, orientation, pattern): e.g., soil types or geology (Figure 12.7)	Graduated color or shading map: e.g., timber yield low/medium/high	Continuous hue/shading, e.g., dot-density or choropleth map: e.g., percentage of retired population (Figure 12.5B)
Surface (2.5-D)	One color per category (color, shape, orientation, pattern), e.g., relief classes: mountain/valley	Ordered color map, e.g., areas of gentle/steep/very steep slopes	Contour map (e.g., isobars/isohyets: e.g., topography contours (Figure 12.3B)

circles and bar charts are often used to assign interval- or ratio-scale data to point locations (Figures 12.10 and 12.15). Variable line width (with increments that correspond to the precision of the interval measure) is a standard convention for representing continuous variation in flow diagrams.

A variety of ways can be used to ascribe interval- or ratio-scale attribute data to areal entities that are predefined. In practice, however, none is unproblematic. The standard method of depicting areal data is in zones (Figure 12.5B). However, as was discussed in Box 4.3, the choropleth map brings the dubious visual implication of within-zone uniformity of attribute value. Moreover, conventional choropleth mapping also allows any large (but possibly uninteresting) areas to dominate the map visually. A variant on the conventional choropleth map is the dot-density map, which uses points as a more aesthetically pleasing means of representing the relative density of zonally averaged data—but not as a means of depicting the precise locations of point events. Proportional circles provide one way around this problem; here the circle is scaled in proportion to the size of the quality being mapped and the circle can be centered on any

convenient point within a zone (Figure 12.10). However, there is a tension between using circles that are of sufficient size to convey the variability in the data and the problems of overlapping circles on busy areas of maps that have large numbers of symbols. Circle location also entails the same kind of positioning problem as that of name and symbol placement outlined earlier.

If the richness of the map presentation is to be equivalent to that of the representation from which it was derived, the intensity of color or shading should directly mirror the intensity or magnitude of attributes. The human eye is adept at discerning continuous variations in color and shading, and Waldo Tobler (see Section 3.1) has advanced the view that continuous scales present the best means of representing geographic variation. There is no natural ordering implied by use of different colors, and the common convention is to represent continuous variation on the red-green-blue (RGB) spectrum. In a similar fashion, difference in the hue, lightness, and saturation (HLS) of shading is used in color maps to represent continuous variation. International standards on intensity and shading have been formalized.

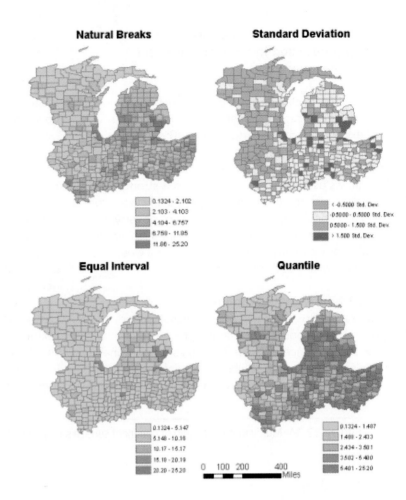

Natural Breaks

	0.1324 - 2.102
	2.103 - 4.103
	4.104 - 6.757
	6.758 - 11.85
	11.86 - 25.20

Standard Deviation

	< -0.5000 Std. Dev.
	-0.5000 - 0.5000 Std. Dev.
	0.5000 - 1.500 Std. Dev.
	> 1.500 Std. Dev.

Equal Interval

	0.1324 - 5.147
	5.148 - 10.16
	10.17 - 15.17
	15.18 - 20.19
	20.20 - 25.20

Quantile

	0.1324 - 1.467
	1.468 - 2.433
	2.434 - 3.501
	3.502 - 5.400
	5.401 - 25.20

Figure 12.13 Comparison of choropleth class definition schemes: natural breaks, quantile, equal interval, standard deviation. The data are Mobile Homes Density for North Central USA, 2004.

At least four basic classification schemes have been developed to divide interval and ratio data into categories (Figure 12.13):

1. *Natural (Jenks) breaks.* Here classes are defined according to apparently natural groupings of data values. The breaks may be imposed on the basis of break points that are known to be relevant to a particular application, such as fractions and multiples of mean income levels, or rainfall thresholds known to support different thresholds of vegetation ("arid," "semiarid," "temperate," etc.). This is "top down" or deductive assignment of breaks. Inductive ("bottom up") classification of data values may be carried out by using GIS software to look for relatively large jumps in data values.

2. *Quantile breaks.* Here each of a predetermined number of classes contains an equal number of observations. Quartile (four-category) classifications are widely used in statistical analysis, while quintile (five-category: Figure 12.13) classifications are well suited to the spatial display of

uniformly distributed data. Yet because the numeric size of each class is rigidly imposed, the result can be misleading. The placing of the boundaries may assign almost identical attributes to adjacent classes, or features with quite widely different values in the same class. The resulting visual distortion can be minimized by increasing the number of classes—assuming the user can assimilate the extra detail that this creates.

3. *Equal-interval breaks.* These are best applied if the data ranges are familiar to the user of the map, such as temperature bands or shaded relief contour maps.

4. *Standard deviation.* Such classifications show the distance of an observation from the mean. The GIS calculates the mean value and then generates class breaks in standard deviation measures above and below it. This would be an appropriate choice for a map of crimes showing variations above and below the mean for the whole map. Use of a two-color ramp helps to emphasize values above and below the mean (Figure 12.13).

Classification procedures are used in map production in order to ease user interpretation.

The selection of classification is very much the outcome of choice, convenience, and the accumulated experience of the cartographer. The automation of mapping in GIS has made it possible to evaluate different possible classifications. Looking at distributions allows us to see if the distribution is strongly skewed—which might justify using unequal class intervals in a particular application. A study of poverty, for example, could quite happily class millionaires along with all those earning over $50,000, as they would be equally irrelevant to the outcome.

12.3.2.2 Multivariate Mapping

Multivariate maps show two or more variables for comparative purposes. For example, Figure 12.10 shows the incidence of Asian Bird Flu as proportional symbols, and Figure 12.3B illustrates rainfall and malaria risk. Many maps are a compilation of composite maps based on a range of constituent indicators. Climate maps, for example, are compiled from direct measures such as amount and distribution of precipitation, diurnal temperature variation, humidity, and hours of sunshine, plus indirect measures such as vegetation coverage and type. It is unlikely there will ever be a perfect correspondence between each of these components, and historically it has been the role of the cartographer to integrate disparate data. In some instances, data may be averaged over zones that are devoid of any strong meaning, while the different components that make up a composite index may have been measured at a range of scales. Mapping can mask scale and aggregation problems where composite indicators are created at large scales using components that were only intended for use at small scales.

Figure 12.14 illustrates how multivariate data can be displayed on a shaded area map. In this soil texture map of Mexico, three variables are displayed simultaneously, as indicated in the legend, using color variations to display combinations of percent sand (base), silt (right), and clay (left).

Figure 12.14 Multivariate data map of dominant surface soil texture. (Courtesy ESRI)

12.4 Map Series

The discussion thus far has focused on the general principles of map design that apply equally to a single map or a collection of maps for the same area or a common theme being considered over multiple areas. The real power of GIS-based digital cartography is revealed when changes need to be made to maps or when collections of similar maps need to be created. Editing or copying a map composition to create similar maps of any combination of areas and data themes is relatively straightforward with a GIS.

Many organizations use GIS to create collections or series of topographic or thematic maps. Examples include a topographic map series to cover a state or country (e.g., USGS 1:24,000 quad sheets of the United States or Eurogeographics 1:250,000 maps of Europe), an atlas of reference maps (e.g., the *National Geographic Eighth Edition Atlas of the World)*, a series of geology maps for a country (see Box 12.3), land parcel maps within a jurisdiction, (Figure 12.16), and utility asset maps that might be bound into a map book. Map series by definition share a number of common elements (for example, projection, general layout, and symbology), and a number of techniques have been developed to automate the map-series production process.

GIS makes it much easier to create collections of maps with common characteristics from map templates.

Figure 12.4 showed the main GIS subcomponents and workflows used to create maps. In principle, this same process can be extended to build production flow lines that result in collections of maps such as an atlas, map series, or map book. Such a GIS requires considerable financial and human resource investments, both to create and maintain it, and many operatives will use it over a long period of time. In an ideal system, an organization would have a single "scale-free," continuous database, which is kept up to date by editing staff and is capable of feeding multiple flow lines at several scales, each with different content and symbology. Unfortunately, this vision remains elusive because of several significant scientific and technical issues. Nevertheless, considerable progress has been made in recent years toward this goal.

The heart of map production through GIS is a geographic database covering the area and data layers of interest (see Chapter 10 for background discussion about the basic principles of creating and maintaining geographic databases). This database is built using a data model that represents interesting aspects of the real world in the GIS (see Chapter 8 for

---

Applications Box 12.4

Multivariate Mapping of Groundwater Analyte Over Time

An increasing number of GIS users need to represent time-dependent (temporal) data on maps in a meaningful yet concise manner. The ability to analyze data for both spatial and temporal patterns in a single presentation provides a powerful incentive to develop tools for communication to map users and decision makers. Although there are many different methods to display scientific data to help discern either spatial or temporal trends, few visualization software packages allow for a single graphical presentation within a geographic context. Chemical concentration data for many different groundwater monitoring wells are widely available for one or more chemical or analyte samples over time. Figure 12.15 shows how this complex temporal data can be represented on a map so that the movement and change in concentration of the analyte can be

observed, both horizontally and vertically, and the potential for the analyte reaching the water table after many years can be assessed.

This map is innovative because it uses a clock diagram code for temporal and spatial visualization. The clock diagrams are analogous to the rose diagrams often used to depict wind direction in meteorological maps or strike direction in geologic applications. The clock diagram method consists of three steps: (1) sample location data, including depth, are compiled along with attribute data such as concentration and time; (2) all data are read into the software application and clock diagram graphics are created; and (3) clock diagram graphics are placed on the map at the location of the well where samples were collected. Although this method of representing temporal data through

▶

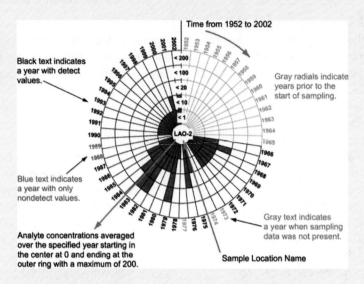

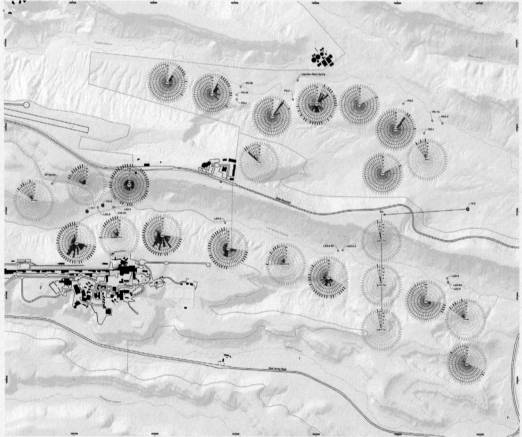

Figure 12.15 Representing temporal data as clock diagrams on maps. (Courtesy ESRI)

the use of clock diagrams is not entirely original, the use of multiple clock diagrams as GIS symbology to emphasize movement of sampled data over time is a new concept. The clock diagram code and method have proven to be an effective means of identifying temporal and spatial trends when analyzing ground-water data.

Figure 12.16 A page from the Town of Brookline, Massachusetts, Tax Assessor's map book. Each map book page uses a common template (as shown in the legend), but has a different map extent: (A) page 26; (B) legend. (Courtesy ESRI)

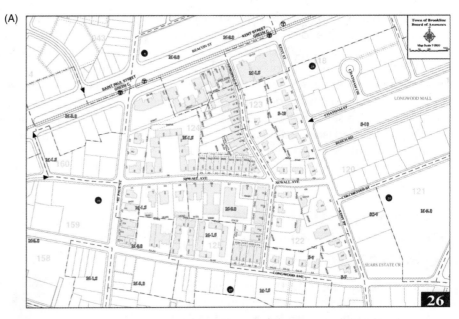

(A)

(B)

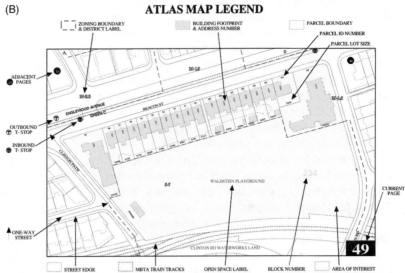

ATLAS MAP LEGEND

background discussion of GIS data models). Such a base cartographic data model is often referred to as a digital landscape model (DLM) because its role is to represent the landscape in the GIS as a collection of features that are independent of any map-product representation. Figure 12.17 is an extended version of

Figure 12.17 Key components and information flows in a GIS map-production system.

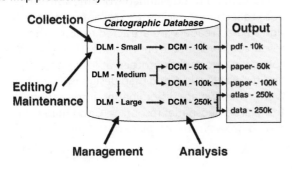

Figure 12.4 that shows the key components and flows in a map-production GIS.

The DLM continuous geographic database is usually stored in a database management system (DBMS) that is managed by the GIS. The database is created and maintained by editors that typically use sophisticated GIS desktop software applications that must be carefully tailored to enforce strict data integrity rules (e.g., geometric connectivity, attribute domains, and topology—see Section 10.7.1) to maintain database quality.

In an ideal world, it will be possible to create multiple different map products from this DLM database. Example products that might be generated by a civilian or military national mapping organization include a 1:10,000-scale topographic map in digital format (e.g., Adobe Portable Document Format (pdf)), a 1:50,000 topographic map sheet in paper format, a 1:100,000 paper map of major streets/highways, a 1:250,000 map suitable for inclusion in a digital or printed atlas, and a 1:250,000 digital database in GML (Geography Markup Language) digital format. In practice, however, for reasons of efficiency and because cartographic database generalization is still not fully automated, a series of intermediate data model layers are employed (see Section 3.8 for a discussion of generalization). The base, or fine-scale, DLM is generalized by both automated and manual means to create coarser-scale DLMs (Figure 12.17 shows two additional medium- and coarse-scale DLMs). From each of these DLMs one or more cartographic data models (digital cartographic models [DCMs]) can be created that derive cartographic representations from real-world features. For example, for a medium-scale DCM representation a road centerline network can be simplified to remove complex junctions and information about overpasses and bridges because they are not appropriate to maps at this scale.

Biographical Box 12.5

Cynthia Brewer, Cartographer

Cynthia Brewer (Figure 12.18) received her graduate degrees in geography from Michigan State University. She began her academic career at San Diego State University and is currently a professor at the Pennsylvania State University. Her early work examined use of color theory to solve cartography problems, such as accommodating simultaneous contrast effects on map-symbol perception, using color-order systems to guide map-color selection, and designing color symbols that colorblind people can read. Many of the lessons from these projects are now reflected in the recommendations offered by ColorBrewer.org, an online tool for selecting map-color schemes. Cindy advises on federal projects, such as national atlases produced by the U.S. Census Bureau and U.S. National Center for Health Statistics. She is currently working with the U.S. Geological Survey on topographic map design and researching multi-scale mapping. Connections between her research and teaching, and the goal of encouraging high-quality cartographic production using GIS tools, have inspired her to publish two books, entitled *Designing Better Maps* and *Designed Maps,* which offer guidance to mapmakers using GIS.

Reflecting on the present state of cartography, Cindy says:

> The look of maps has become an arena of intense competition as the many geospatial tools aimed at consumers offer their information with increasingly attractive and readable maps. GIS data that include names, redundant geometries, categories, and rank orders are essential to making the leap from analysis to competent mapping. In work on multi-scale cartography, we find that database features such as center lines for linear areal features (a line following the course of a wide river,

Figure 12.18 Cynthia Brewer. (Courtesy Cynthia Brewer)

> for example) are needed for labeling at large scales and then for collapsing to a single line at smaller scales to generalize features and reduce map clutter. Accurate topologies that allow flows to be modeled in hydrological and transport networks may be specified for analysis goals, but they are also important for allowing many small segments to be accurately dissolved to single lines to guide label placement and produce smooth multi-layer lines on maps. General categories and rank orders within feature types are used to distinguish less essential features, such as local roads and minor stream tributaries, so they can be symbolized with less prominent lines at larger scales and removed as a group at smaller scales. These basic characteristics for GIS data need to be planned from the start of data collection to facilitate excellent map design so map readers are attracted to using the resources developed from the extensive data collection efforts that proceed at local, state, and federal levels. Good map design isn't craft work when we are disseminating information from large geospatial resources. Good map design grows from complete data.

Once the necessary DCMs have been built, the process of map creation can proceed. Each individual map will be created in the manner described in Section 12.3, but since many similar maps will be required for each series or collection, some additional work is necessary to automate the map production process. Many similar maps can be created efficiently from a common map template that includes any material common to all maps (for example, inset/overview maps, titles, legends, scales, direction indicators, and map metadata; see Section 12.3.1 and Figure 12.8). Once a template has been created, it is simple to specify the individual map-sheet content (e.g., map data from DCM data layers, and specific title, and metadata information) for each separate sheet. The geographic extent of each map sheet is typically maintained in a separate database layer. This could be a regular grid (see inset to Figure 12.7), but commonly there is some overlap in sheets to accommodate irregular-shaped areas and to ensure that important features are not split at sheet edges. An automated batch process can then "stamp out" each sheet from the database as and when required. Finally, if the sheets are to form a map book or atlas, it is convenient to generate an index of names (for example, places or streets) that shows on which sheet they are located.

In Box 12.5 Cindy Brewer reflects on the current state of cartography and describes some of her work with national mapping organizations.

12.5 Applications

Obviously, a huge number and wide range of applications use maps extensively. The goal here is to highlight a few examples that raise some interesting cartographic issues.

The relative importance of representing space and attributes will vary within and between different applications, as will the ability to broker improved measures of spatial distributions through integration of ancillary sources. These tensions are not new. In a topographic map, for example, the width of a single-carriageway road may be exaggerated considerably (Figure 12.19). This is done in order to enhance the legibility of features that are central to general-purpose topographic mapping. In some instances, these prevailing conventions will have evolved over long periods of time, while in others the newfound capabilities of GIS entail a distinct break from the past. As a general rule, where accuracy and precision of georeferencing are important, the standard conventions of topographic mapping will be applied (e.g., Figure 12.1).

Utility applications use GIS that have come to be known as Automated Mapping and Facilities Management (AM/FM) systems. The prime objectives of such systems are asset management and the production of maps that can be used in work orders that drive asset-maintenance projects. Some of the maps produced

Figure 12.19 Topographic map showing street centerlines from a TeleAtlas database for the area around Chicago. The major roads are exaggerated with "cased" road symbols.

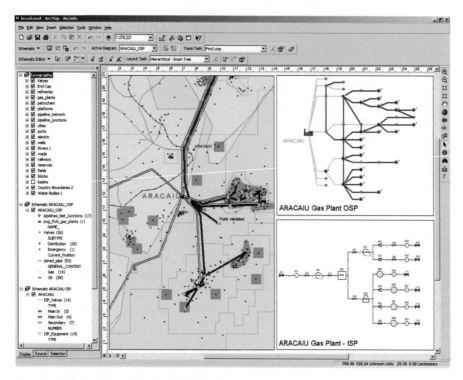

Figure 12.20 Schematic gas pipeline utility maps for Aracaiu, Brazil. Geographic view (left), Outside Plant schematic (top right), Inside Plant schematic (bottom right). (Data courtesy of IHS-Energy/Tobin)

from an AM/FM system use conventional cartographic representations, but for some maintenance operations schematic representations are preferred (Figure 12.20). A schematic map provides a view of the way in which a system functions, and is used for operational activities such as identifying faults during power outages and routine pipeline maintenance. Hybrid schematic and geographic maps, known as geoschematics, combine the synoptic view of the current state of the network as a schematic superimposed on a background map with real-world coordinates.

Transportation applications use a procedure known as linear referencing (Section 5.5) to visualize point (such as street furniture, such as road signs), linear (such as parking restrictions and road-surface quality measures), and continuous objects or events (such as speed limits). In linear referencing, two-dimensional geography is collapsed into one-dimensional linear space. Data are collected as linear measures from the start of a route (a path through a network). The linear measures are added at display time, usually dynamically, and segment the route into smaller sections (hence the term *dynamic segmentation*, which is sometimes used to describe this type of model). Figure 12.21 is a map of Montpelier, Vermont, that shows the results of several highway analysis studies (average daily traffic, roadway

and bridge condition, and frequent crash locations). The data were collected as linear events using several methods, including driving the roads with an in-vehicle sensor. For more information on linear referencing, see Sections 5.5 and 8.2.3.3.

We began this chapter with a discussion of the military uses of mapping, and such applications also have special cartographic conventions—as in the operational overlay maps used to communicate battle plans (Figure 12.22). On these maps, friendly, enemy, and neutral forces are shown in blue, red, and green, respectively. The location, size, and capabilities of military units are depicted with special multivariate symbols that have operational and tactical significance. Other subjects of significance, such as minefields, impassable vegetation, and direction of movement, also have special symbols. Animations of such maps can be used to show the progression of a battle, including future what-if scenarios.

12.6 Conclusion

Cartography is both an art and a science. The modern cartographer must be very familiar with the application of computer technology. The very nature of

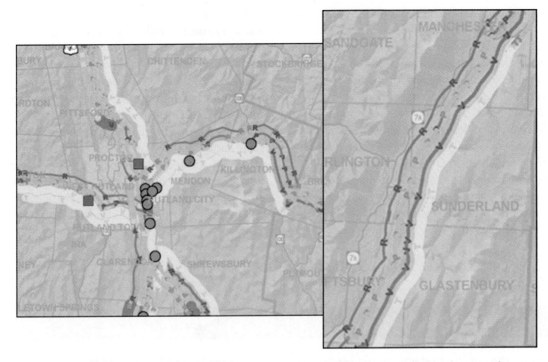

Figure 12.21 Linear-referencing transportation data for Montpelier, Vermont, displayed on top of street centerline files.

Figure 12.22 Military map showing overlay graphics on top of base-map data. Blue rectangles are friendly forces units, red diamonds are enemy force units, and black symbols are military operations—maneuvers, obstacles, phase lines, and unit boundaries. These "tactical graphics" depict the battle space and the tactical environment for the operational units.

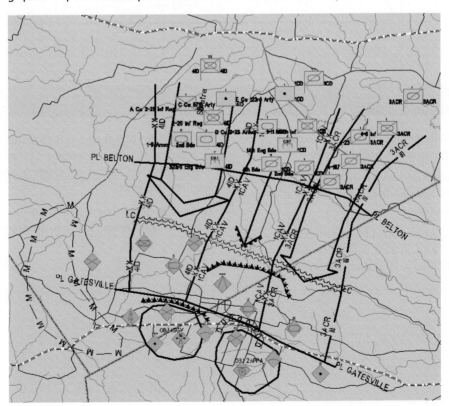

cartography and mapmaking has changed profoundly over the past few decades and will never be the same again. Nevertheless, there remains a need to understand the nature and representational characteristics of what goes into maps if they are to provide robust and defensible aids to decision making, as well as tactical and operational support tools. In cartography, there are few hard-and-fast rules to drive map composition, but a good map is often obvious once complete. Modern advances in GIS-based cartography make it easier than ever to create large numbers of maps very quickly using automated techniques once databases and map templates have been built. The creation of databases and map templates continues to be an advanced task requiring the services of trained professionals. The type of data that are used on maps is also changing—today's maps often reuse and recycle different datasets, obtained over the Internet, that are rich in detail but may be unsystematic in collection and incompatible in terms of scale. This all underpins the importance of metadata (Section 11.2.1) to evaluate datasets in terms of scale, aggregation, and representativeness prior to mapping. Collectively, these changes are driving the development of new applications founded on the emerging advances in scientific visualization that will be discussed in the next chapter.

Questions for Further Study

1. Identify the criteria that you would use to design a military mapping system for use during the Afghanistan War. How might the system subsequently be adapted for use in humanitarian relief?

2. Why are there so many classification schemes for area-based choropleth maps? What criteria would you use to select the best scheme?

3. Using Table 12.1 as a guide, create an equivalent table that identifies example maps for all the map types.

4. Using a GIS of your choice, create a map of your local area containing all seven elements listed in Section 12.3.1. What makes this a good map?

Further Reading

Brewer, C.A. 2005. *Designing Better Maps: A Guide for GIS Users*. Redlands, CA: ESRI Press.

Kraak, M-J. and Ormerling, F. 2003. *Cartography: Visualization of Spatial Data*. (2nd ed.). Harlow, UK: Pearson Education.

Robinson, A.H., Morrison, J.L., Muehrcke, P.C., Kimerling, A.J., and Guptill, S.C. 1995. *Elements of Cartography* (6th ed.). Hoboken, NJ: Wiley.

Slocum, T.A., McMaster, R.B., Kessler, F.C., and Howard, H.H. 2008. *Thematic Cartography and Geographic Visualization*. Upper Saddle River, NJ: Prentice-Hall.

Tufte, E.R. 2001. *The Visual Display of Quantitative Information*. Cheshire, CT: Graphics Press.

13 Geovisualization

This chapter builds on the cartographic design principles set out in Chapter 12 to describe a range of novel ways in which information can be presented visually to the user. Using techniques of geovisualization, GIS provides a far richer and more flexible medium for portraying attribute distributions than paper mapping. First, through techniques of spatial query, it allows users to explore, synthesize, present (communicate), and analyze the meaning of a representation. Second, it facilitates map transformation using techniques such as cartograms and dasymetric mapping. Third, geovisualization allows the user to interact with the real world from a distance, through interaction with and even immersion in artificial worlds. Together, these functions broaden the user base of GIS and have implications for the supply of volunteered geographic information (VGI) and for public participation in GIS (PPGIS).

LEARNING OBJECTIVES

After studying this chapter you will understand:

- How GIS facilitates visual communication.

- The ways in which good user interfaces can help to resolve spatial queries.

- Some of the ways in which GIS-based representations may be transformed.

- How 3-D geovisualization and virtual worlds can improve our understanding of the world.

13.1 Introduction: Uses, Users, Messages, and Media

Effective decision support through GIS requires that computer-held representations (as discussed in Chapters 3 and 6) are readily interpretable in the minds of the users who need them to make decisions. Because representations are necessarily partial and selective, they are unlikely to be accepted as objective. Critiques of GIS (Section 1.8) have suggested that even widely available digital maps and virtual Earths (such as the Google and Microsoft offerings) present a privileged' or "God's eye" view of the world because the selectivity inherent in representation may be used to reinforce the message or objectives of the interest groups that created them. This issue predates the GIS age: it is evident in the early motivations for creating maps to support military operations (Section 12.1); and the subject of geopolitics is rife with examples of the use of mapping to reinforce political propaganda. Today, Web mapping and digital spatial data infrastructures can open up politically sensitive issues and, in the era of VGI, allow them to be openly contested (see Box 13.1).

Today's data sources are much more numerous, voluminous, complex, and structurally diverse than was the case even a decade ago. Geovisualization techniques offer the prospect of rendering them intelligible to users and avoiding information overload, by weeding out distracting or extraneous detail. Other niche applications in historical GIS are concerned with assembling digital shards of evidence that may be scattered across space and time, or with providing wider spatial context to painstaking local archive studies. In such cases, geovisualization provides the context and framework for building representations, sometimes around sparse records. This technique is different from cartography and map production (see Chapter 12) in that it typically uses an interactive computer environment for data exploration, it entails the creation of multiple (including 3-D) representations of spatial data, or it allows the representation of changes over time.

Geovisualization entails multiple representations of large and complex datasets, on the fly.

Mapping makes it possible to communicate the meaning of a spatial representation of the real world to users. Historically, the paper map was the only available interface between the mapmaker and the user: it was permanent, contained a fixed array of attributes, was of predetermined and invariant scale, and rarely provided any quantitative or qualitative

---

Applications Box 13.1

The Legacy of Conflict, "Tag Wars" and Cohesive Communities

OpenStreetMap (www.openstreetmap.org and Section 19.3.2.2) is an ongoing collaborative VGI project that is seeking to create a free digital map base of the entire world. Online maps are created using GPS traces that are recorded, uploaded, and annotated by volunteers, along with aerial photography and other free sources. Most volunteers have detailed local knowledge of the areas for which they capture data, and all volunteers are empowered to edit the database as well as add to it, in order to generate coverage and to correct for errors. Local knowledge is, of course, situated in direct experience (Section 3.1), and open-source maps may be heavily contested if, as in the case of Cyprus, received wisdom is heavily conditioned by cultural affiliation.

Cyprus has been divided since 1974 when Turkey invaded the north of the Island. The south is the Greek-dominated Republic of Cyprus (which is internationally recognized), and the north is the Turkish Republic of Northern Cyprus (which is not). After 1974, places in the north of the island were renamed in Turkish, and road signs were made consistent with them. Some volunteers believe that Turkish names should be used, with Greek names relegated to the metadata; this stance is pragmatic not least because OpenStreetMap is used as a navigational aid. Yet other volunteers, mindful perhaps of the island's unhappy recent history and the illegal status of the north, have persisted in replacing Turkish names with their pre-1974 Greek counterparts. The ensuing "tag war" marked by the successive editing of Turkish and Greek names can be viewed at www.openstreetmap.org/browse/node/276379679/history, and some of the changes to the map are illustrated in Figure 13.1. Geovisualization enables dynamic mapping of the tensions that surround this disputed territory and that underpinned Greek-Cypriot rejection of the United Nations' Annan Plan to reunite the island in 2004.

▶

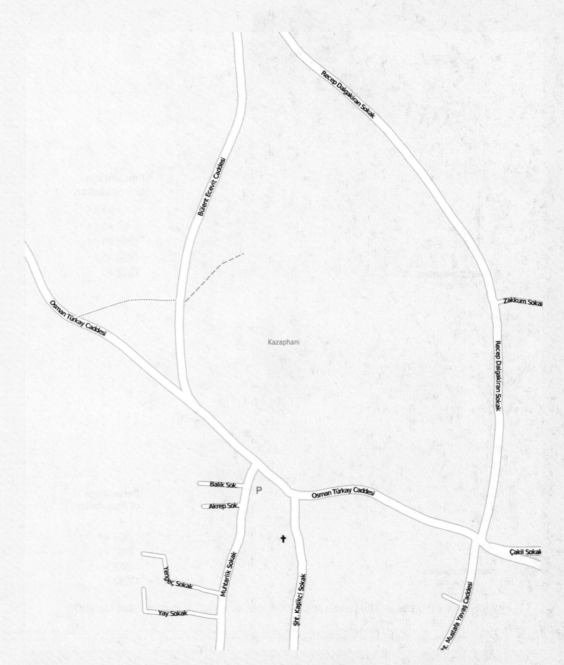

Figure 13.1 Greek and Turkish edits to OpenStreetMap in Cyprus.

The power of mapping to depict cohesion in multicultural societies presents a rosier picture in Greater London. "Onomap" is a geocomputational (Section 16.1) classification of the cultural, ethnic, and linguistic connotations of people's names that makes it possible to map the likely detailed geographies of very detailed classes at neighborhood scales (Box 6.2 and www.onomap.org). Using the classification, geographer Pablo Mateos has mapped the detailed geography of people with Greek and Turkish names, and has shown that the two groups share many residential neighborhoods in north London (Figure 13.2).

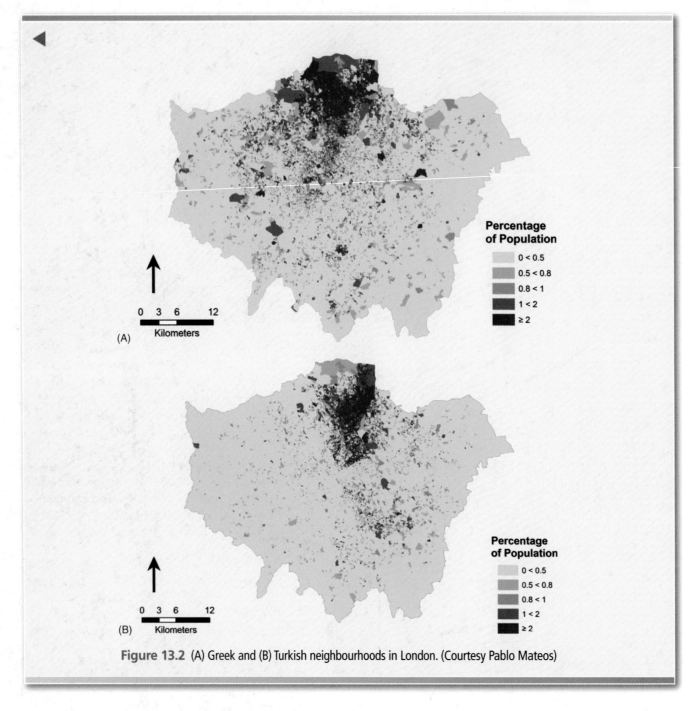

Percentage of Population

- ☐ 0 < 0.5
- ☐ 0.5 < 0.8
- ☐ 0.8 < 1
- ☐ 1 < 2
- ☐ ≥ 2

(A)

Percentage of Population

- ☐ 0 < 0.5
- ☐ 0.5 < 0.8
- ☐ 0.8 < 1
- ☐ 1 < 2
- ☐ ≥ 2

(B)

Figure 13.2 (A) Greek and (B) Turkish neighbourhoods in London. (Courtesy Pablo Mateos)

indications as to whether it was safe to use (see Section 3.7). Moreover, it was not possible to differentiate between the needs of different users. These attributes severely limit the usefulness of the paper map in today's applications environment, which can be of seemingly unfathomable complexity. Today it is often necessary to visualize data that are very detailed, continuously updated, or scattered across the Internet. These developments are reinvigorating the debate over the meaning of maps and the potential of mapping.

Some propose that geovisualization makes the selective nature of representation more transparent, whereas others have suggested the opposite (see Box 3.5). Elsewhere in this chapter we take the former view, consistent with successful use of good GIS design principles.

In technical terms, geovisualization builds on the established tenets of map production and display. Geographer Alan MacEachren characterizes geovisualization as the creation and use of visual representations to facilitate thinking, understanding, and knowledge construction about human and physical environments, at geographic scales of measurement. It is also a research-led field that integrates approaches from visualization in scientific computing (ViSC: see Section 13.4.4), cartography (Chapter 12), image analysis (Section 9.2.1), information visualization, and exploratory spatial data analysis (ESDA: Chapters 14 and 15), as well as GIS. The core motivation for this activity is to develop theories, methods, and tools for visual exploration, analysis, synthesis, and presentation of spatial data.

Geovisualization is used to explore, analyze, synthesize, and present spatial data.

As such, today's geovisualization is much more than conventional map design. It has developed into an applied area of activity that leverages geographic data resources to meet a very wide range of scientific and social needs, as well as into a research field that is developing new visual methods and tools to facilitate better user interaction with such data. In this chapter, we discuss how this is achieved through *query* and *transformation* of geographic data, and user *immersion* within them. Query and transformation are discussed in a substantially different context in Chapters 14 and 15, where they are encountered as types of spatial analysis.

 ## 13.2 Geovisualization, Spatial Query, and User Interaction

13.2.1 Overview

Fundamental to effective geovisualization is an understanding of how human cognition shapes GIS usage, how people think about space and time, and how spatial environments might be better represented using computers and digital data. The conventions of map production, presented in Section 12.3, are central to the use of mapping as a decision support tool. Many of these conventions are common to paper and digital mapping, although GIS allows far greater flexibility and customization of map design. Geovisualization develops and extends these concepts in a number of new and innovative ways in pursuance of the following objectives:

1. *Exploration*: for example, to establish whether and to what extent the general message of a dataset is sensitive to inclusion or exclusion of particular data elements.

2. *Synthesis*: to present the range, complexity, and detail of one or more datasets in ways that can be readily assimilated by users. Good geovisualization should enable the user to see the wood for the trees and may be particularly important in highlighting the time dimension.

3. *Presentation*: to communicate the overall message of a representation in a readily intelligible manner and to enable the user to understand the likely overall quality of the representation.

4. *Analysis:* to provide a medium to support a range of methods and techniques of spatial analysis (see Chapters 14 and 15).

These objectives cut across the full range of different applications tasks and are pursued by users of varying expertise that desire different degrees of interaction with their data.

Geovisualization allows users to explore, synthesize, present, and analyze their data more thoroughly than was possible hitherto.

These motivations may be considered in relation to the conceptual model of uncertainty that was presented in Figure 6.1, which is presented with additions in Figure 13.3. This revised conceptual model encourages us to think of geographic analysis not as an endpoint, but rather as the start of an iterative process of feedbacks and what-if scenario testing. It is thus appropriate to think of geovisualization as imposing a further filter (U4 in Figure 13.3) both on the results of analysis and on the conception and representation of geographic phenomena. The most straightforward way in which reformulation and evaluation of a representation of the real world can take place is through posing *spatial queries* to ask generic spatial and temporal questions such as:

- Where is . . .?

- What is at location . . .?

- What is the spatial relation between . . .?

- What is similar to . . .?

- Where has . . . occurred?

- What has changed since . . .?

- Is there a general spatial pattern, and what are the anomalies?

These questions are articulated through a graphical user interface (GUI) that is based on **W**indows, **I**cons, **M**enus and **P**ointers—the so-called WIMP interface illustrated in Figure 13.4 (see also Section 7.2). The familiar actions of pointing, clicking, and dragging windows and icons are the most common ways of interrogating a geographic database and summarizing results in map and tabular form. More recently, data and map manipulation has become possible through gestures applied to devices that have multitouch

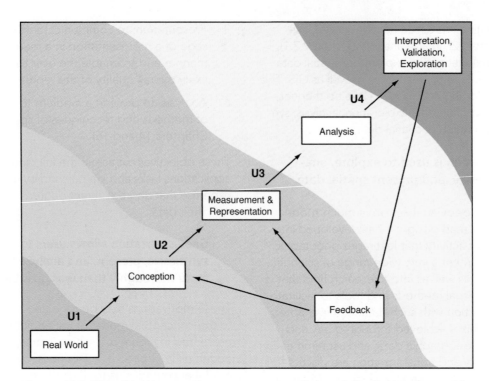

Figure 13.3 Filters U1–U4: conception, measurement, analysis, and visualization. Geographic analysis is not as an end point, but rather the start of an iterative process of feedbacks and 'what if?' scenario testing. (See also Figure 6.1.)

Figure 13.4 Geovisualization using multiple dynamically linked views in the GeoDa software. The data that the user has highlighted in blue in the table are identified in the maps using yellow hatching or coloring, while the red circles denote outliers in the analysis. The scatterplot summarises the relationship between two variables under consideration.

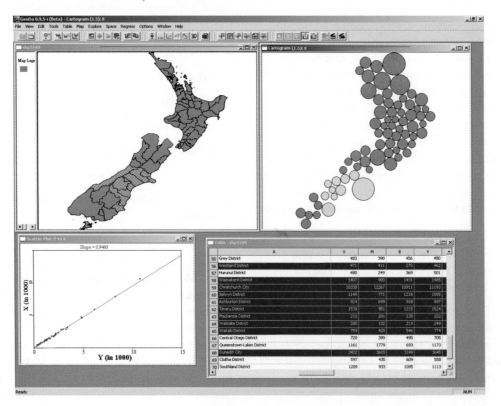

capabilities on their screens or track pads. Zooming is one such interactive gesture, accomplished by emulating a pinching motion on the screen of the hardware device. High-end applications increasingly use multiple displays and projections of maps (see Section 5.8) along with aspatial summaries such as bar charts and scatterplots. Together these make it possible to build up a picture of the spatial and other properties of a representation. The user is able to link these representations by "brushing" data in one of the multiple representations and viewing the same observations in the other views. This facilitates learning about a representation in a data-led way.

The WIMP interface allows spatial querying through pointing, clicking, and dragging windows and icons.

The WIMP interface has long been associated with desktop and notebook machines, but the last decade has seen its implementation in hand-held devices; increasing numbers of these devices are location aware because of the integration of GPS receivers. These technological developments have increased the challenges to geovisualization because a much more restricted screen size creates the need to devise ingenious solutions for effective interaction.

The advent of affordable hand-held devices has opened up a vast range of new applications, with the implications that many new users are nonexpert at interacting with computers. GIServices (see Section 1.6.3) GIS Web services (see Section 1.6.4), ranging from the general (such as Google Maps) to the very specialized, have been developed to exploit these applications. The implications for geovisualization are profound and far reaching in that a new, larger, and much more diverse user base is seeking to use pocket devices to solve an ever wider range of location-specific problems, on the fly and in real time. Figure 13.5 shows the user

Figure 13.5 (A) The user interface for the London EcoTEXT environmental information service; (B) searches based upon the multiple post (zip) codes supplied for work, home and usual recreation locations; and (C) illustrative text alerts arising from use of the service. (Courtesy Hanif Rahemtulla)

(A)

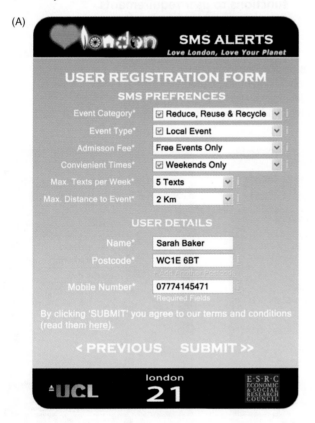

(B)

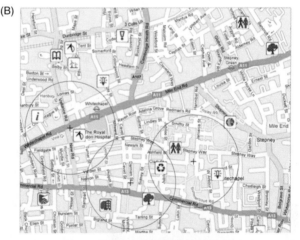

(C)

interface of a pocket device used to register for information about local environmental initiatives, along with the spatial queries that this might initiate and the kind of text alerts that might arise as a consequence.

The process of text registration raises the issue of selectivity in the supply of location-based services. Geographer Martin Raubal has investigated the ways in which the user needs depend on personal circumstances, prior experience, decision-making strategies, and basic cognitive abilities. He observes that understanding the diverse needs, experiences, and cognitive abilities of today's vast user base is crucial to fostering effective computer interaction—with the implication that hardware and software developers need to understand better the personal and situational circumstances of *all* users. Looking to the future, it is likely that the generic interfaces that have hitherto characterized many applications will be supplemented or even replaced by interfaces that can be tailored to serve group or individual needs, as illustrated in Figure 13.6. For example it is possible to envisage that a navigation service offering landmark-based wayfinding instructions might use only particular types of landmarks, or that the number and type of landmarks might

be changed according to the user's familiarity with particular physical settings.

This area of current research is termed *cognitive engineering*, and in the future it is likely that cognitively engineered GIServices will be more closely attuned to the user's ability to interpret spatiotemporal information, in order to meet specific user requirements. In practice, the software used to accomplish this may come from a wide variety of proprietary and open sources, and is not restricted to GIS packages as conventionally defined (see Section 1.5.1.2). Moreover, the advent of volunteered geographic information (VGI) and Web 2.0 GIS is increasing the range of sources that may be used to tailor GIServices; such sources are increasingly being used to devise niche maps for specialist applications. Figure 13.7 illustrates this with regard to the creation of maps for use in orienteering, created using public domain software and data. A closely related application is geocaching, which is an outdoor treasure-hunting game in which participants use a GPS receiver or other navigational techniques to hide and seek caches, or containers, as shown in Figure 13.8.

Cognitive engineering matches device functions to user requirements.

Figure 13.6 A cognitively engineered mobile hotel finder for a pocket device. The application allows users to select and weight different criteria (tourist users may be on a tighter budget than people on business), pursue different decision strategies (regular travellers may be less risk averse), and calculate the route to the optimal hotel (using different travel modes, consistent with budget and time constraints).

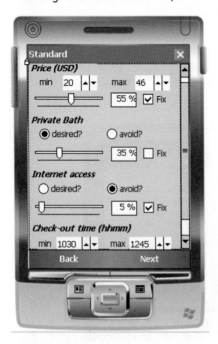

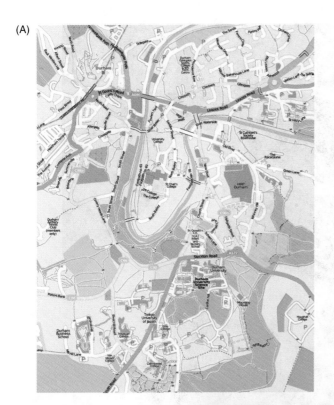

(A)

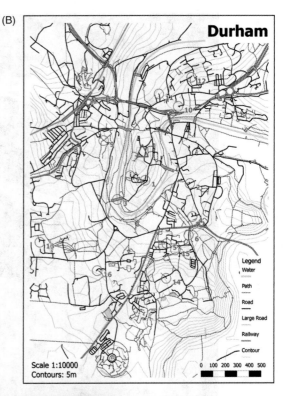

(B)

(C)

Figure 13.7 (A) The general purpose map layer base of OpenStreetMap for Durham, UK; and (B) a specialist orienteering map created using the OpenStreetMap base along with the contour data derived from Shuttle Radar Topography Mission data. These sources were integrated using the MapWindow open source GIS (www.mapwindow.org) (C) A competitor registering at an orienteering post. (Images courtesy Ollie O'Brien). Map data © OpenStreetMap and contributions, CC-BY-5A

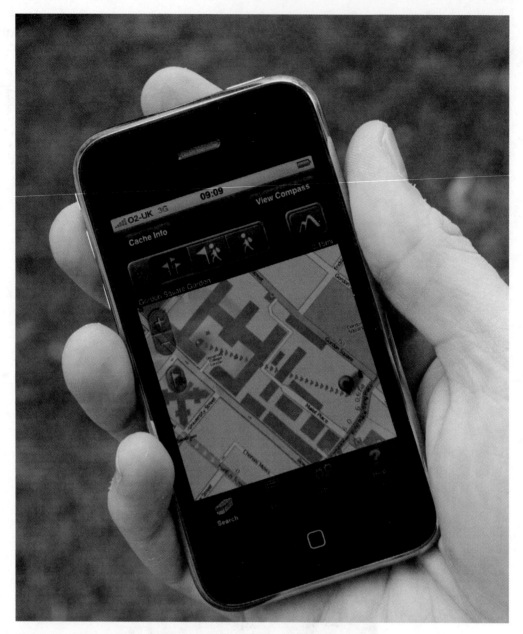

Figure 13.8 Geocaching is a global treasure hunting game where participants locate hidden containers, called geocaches, hidden outdoors and then share their experiences online. (Image courtesy Alex Singleton)

13.2.2 Spatial Query Online and the Geoweb

The spatial query functions illustrated in Figure 13.5 are central to many Internet GIS applications, and the different computer architectures set out in Chapter 11 can be used to query remote datasets using similarly remotely located software. For many users, spatial query is the end objective of a GIS application, as for example with queries about the taxable value of properties (Section 2.3.2), customer-care applications about service availability (Section 2.3.3), or Internet-site queries about real-time traffic conditions (Section

2.3.4). In other applications, spatial query is a precursor to more advanced spatial analysis. Spatial query may appear routine, but may entail complex operations, particularly when the datasets are continuously updated (refreshed) in real time—as, for example, when optimal traffic routes are updated in the light of traffic conditions or road closures (see Section 19.2).

The innovation of Web 2.0 has enabled bidirectional collaboration between Web sites (see Section 1.5.1) and has important implications for geovisualization, including the development of consolidator sites that take real-time feeds from third-party applications. Box 13.2

MapTube (www.maptube.org): A Place for Maps

MapTube is a Web site for sharing Web maps that are hosted in different locations and for overlaying them upon each other for visual comparison of data. For example, Figure 13.9 was created by overlaying a map of the London Underground network upon the UK Index of Multiple Deprivation (a UK measure of hardship) by searching for the constituent maps in the MapTube catalog, overlaying them, and adjusting the opacity of the layers to achieve the best visual balance of the two layers and background Google Maps data. OpenLayers and OpenStreetMap are also used to provide an open-source alternative to the Google map base.

The maps themselves are not stored on the MapTube site, but rather are created using feeds from other sites on the Web where the map is already published. When maps are shared, metadata about the maps and their provenance are entered by the owner and are stored by MapTube along with the link to where the map is published.

The sharing of geographically referenced data between Web sites is often hamstrung by issues of intellectual property and copyright (see Box 18.4). MapTube neatly circumvents these problems because it does not use raw data, instead using only prerendered maps that have been tiled using the public domain GMapCreator software. As such, the site offers map donors a safe way of sharing a map without giving away the raw data that were used to create it. Moreover, using this software architecture, only the owner of a map can edit its details.

MapTube was developed by Richard Milton at University College London's Centre for Advanced Spatial Analysis (CASA). The range of datasets available at www.maptube.org makes it possible to generate hypotheses ranging from the policy that is relevant (as with transport access and hardship shown in Figure 13.9) to the policy that is mildly preposterous (one can compare mortality with Big Mac prices, for example): all are presented in a vivid, interactive manner with useful slider controls and legend-placement facilities. Not-for-profit organizations can download the GMapCreator software free of charge from www.maptube.org/gmc-licence.aspx.

Figure 13.9 The geography of hardship in London and the London Underground network, viewed using Maptube.

describes one such application, MapTube, which acts as a clearinghouse for a range of Web mapping services.

13.3 Geovisualization and Interactive Transformation

13.3.1 Overview

Chapter 12 illustrated circumstances in which it was appropriate to adjust the classification intervals of maps, in order to highlight the salient characteristics of spatial distributions. We have also seen how changing the scale at which attributes are measured can help us to represent spatial phenomena, as with the scaling relations that may be used to characterize a fractal coastline (Box 4.6). More generally still, our discussion of the nature of geographic data (Chapter 4) illustrated some of the potentially difficult consequences arising out of the absence of natural units in geography—as when choropleth map counts are not standardized by any numerical or areal base measure. Standard map production and display functions in GIS allow us to standardize according to numerical base categories, yet large but unimportant (in attribute terms) zones maintain greater visual dominance in mapping than is warranted.

Such problems can be addressed by using GIS to transform the shape and extent of areal units. Our use of the term *transformation* here is in the cartographic sense of taking one vector space (the real world) and transforming it into another (the geovisualization). We consider cartographic transformation here, rather than in Chapter 12, because while the final outcome of the map transformation process may well be a paper map, this is likely to be the outcome of a series of map transformations, such as that illustrated in Figure 13.4.

GIS is a flexible medium for the cartographic transformation of maps.

Some examples of the ways in which real-world phenomena of different dimensions (Box 3.3) may or may not be transformed by GIS are shown in Table 13.1. These illustrate how the representation filter of Figure 13.3 may have the effect of transforming some objects, such as population distributions, but not others, such as agricultural fields. Further transformation may occur in order to present the information to the user in the most intelligible and parsimonious way. Thus, cartographic transformation through the U3 filter in Figure 13.3 may result from the imposition of artificial units, such as census tracts for population, or selective abstraction of data, as in the generalization of cartographic lines (Section 3.8). The standard conventions of map production and display are not always sufficient to make the user aware of the transformations that have taken place between conception and measurement—as in choropleth mapping, for example, where mapped attributes are required to take on the proportions of the zones to which they pertain.

Geovisualization techniques make it possible to manipulate the shape and form of mapped boundaries using GIS, and in many circumstances it may be appropriate to do so. Indeed, where a standard mapping projection obscures the message of the attribute distribution, map transformation becomes necessary. There is nothing untoward in doing so: one of the messages of Chapter 5 was that all

Table 13.1 Some examples of coordinate and cartographic transformations of spatial phenomena of different dimensionality (based on D Martin 1996 Geographic Information Systems: Socioeconomic Applications (Second edition). London, Routledge: 65).

Dimension	0	1	2	3
Conception of spatial phenomenon	Population distribution	Coastline	Agricultural field	Land surface
Coordinate transformation of real world arising in GIS representation/measurement				
Measurement	Imposed areal aggregation	Sequence of digitized points	Digitized polygon boundary	Arrangement of spot heights
Cartographic transformation to aid interpretation of representation				
Visualization	Cartogram or dasymetric map	Generalized line	Integral/natural area	Digital elevation model

conventional mapping of geographically extensive areas entails transformation, in order to represent the curved surface of the Earth on a flat screen or a sheet of paper. Similarly, in Section 3.8, we described how generalization procedures may be applied to line data in order to maintain their structure and character in small-scale maps. There is nothing sacrosanct about popular or conventional representations of the world, although the widely used transformations and projections do confer advantages in terms of wide user recognition, hence interpretability.

13.3.2 Cartograms

Cartograms are maps that lack planimetric correctness and distort area or distance in the interests of some specific objective. The usual objective is to reveal patterns that might not be readily apparent from a conventional map or, more generally, to promote legibility. Thus, the integrity of the spatial object, in terms of areal extent, location, contiguity, geometry, and/or topology, is made subservient to an emphasis on attribute values or particular aspects of spatial relations. One of the best known cartograms (strictly speaking a *linear cartogram*) is

the London Underground map, devised in 1933 by Harry Beck to fulfill the specific purpose of helping travelers to navigate across the network. The central-area cartogram that is widely used today is a descendant of Beck's 1933 map and provides a widely recognized representation of connectivity in London, using conventions that are well suited to the attributes of spacing, configuration, scale, and linkage of the London Underground system. The attributes of public transit systems differ between cities, as do the cultural conventions of transit users, and thus it is unsurprising that cartograms pertaining to transit systems elsewhere in the world appear quite different.

Cartograms are map transformations that distort area or distance in the interests of some specific objective.

A central tenet of Chapter 3 was that all representations are necessarily selective abstractions of reality. Cartograms depict transformed, hence artificial, realities, using particular exaggerations that are deliberately chosen. Figure 13.10 presents a regional map of the UK, along with an *equal population cartogram* of the same territory. The cartogram is a

Figure 13.10 (A) A regional map of the UK and (B) its equal population cartogram transformation. (Courtesy: Daniel Dorling and Bethan Thomas: from Dorling and Thomas 2004), page (vii)

(A)

(B)

map projection that ensures that every area is drawn approximately in proportion to its population, and thus that every individual is accorded approximately equal weight. Wherever possible, neighboring areas are kept together, and the overall shape and compass orientation of the country are kept roughly correct— although the real-world pattern of zone contiguity and topology is to some extent compromised. When attributes are mapped, the variations within cities are revealed, while the variations in the countryside are reduced in size so as not to dominate the image and divert the eye from what is happening to the majority of the population. Therefore, the transformed map of wealthy areas in the UK, shown in Figure 13.11B, depicts the considerable diversity in wealth in densely populated city areas. This phenomenon is not apparent in the usual (scaled Transverse Mercator) projection (Figure 13.11A and see Section 5.8.2).

There are semiautomated ways of producing cartograms, but representations such as those shown in Figures 13.11B and 13.12B also almost invariably entail human judgment and design. Throughout this book, one of the recurrent themes has been the value of GIS as a medium for data sharing, and this is most easily achieved if pooled

data share common coordinate systems. Yet a range of cartometric transformations is useful for accommodating the way that humans think about or experience space, as with the measurement of distance as travel time or travel cost. As such, the visual transformations inherent in cartograms can provide improved means of envisioning spatial structure, and geovisualization offers an interactive and dynamic medium for experimentation with different transformations.

13.3.3 Remodeling Spatial Distributions as Dasymetric Maps

The cartogram offers a radical means of transforming space, and hence restoring spatial balance, but sacrifices the familiar spatial framework valued by most users. It forces the user to make a stark choice between assigning each mapped element equal weight and being able to relate spatial objects to real locations on the Earth's surface. The interactive environment of geovisualization and scalable cached datasets provides a means by which the data-integrative power of GIS can be used to remodel spatial distributions and hence assign spatial attributes to meaningful, recognizable spatial objects. One

Figure 13.11 Wealthy areas of the UK, as measured by property values, property size, and levels of multiple car ownership: (A) conventional map of local authority districts and (B) its equal population cartogram transformation. (Courtesy: Daniel Dorling and Bethan Thomas: from Dorling and Thomas (2004), page 16)

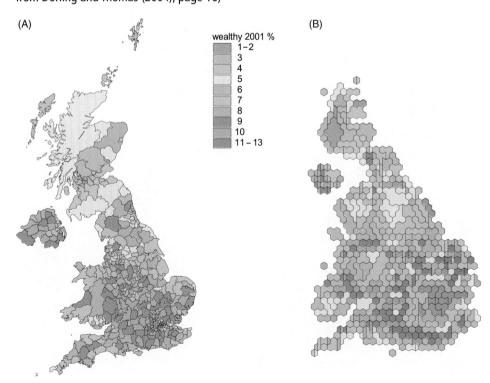

example of the way in which ancillary sources of information may be used to improve representation of a spatial distribution is known as *dasymetric mapping*. Here, the intersection of two datasets is used to suggest more precise estimates of a spatial distribution.

Figure 13.12A presents a census tract geography for which small-area population totals are known, while Figure 13.12B shows the spatial distribution of built structures in an urban area (which might be obtained from a cadaster or very-fine-resolution satellite imagery, for example). A reasonable assumption (in the absence of evidence of mixed land use or very different residential structures such as high-rise apartments and widely spaced bungalows) is that all of the built structures house resident populations at uniform density. Figure 13.12C shows how this assumption, plus an overlay of the areal extent of built structures, allows population figures to be allocated to smaller areas than census tracts, and allows calculation of indicators of residential density. The practical usefulness of dasymetric mapping is illustrated in Figure 13.13.

Figure 13.13 shows the location of an elongated census block in the City of Bristol, UK, which is characterized by both a high unemployment *rate* (Figure 13.13A) and a low absolute *incidence* of unemployment (Figure 13.13B). The resolution to this seeming paradox is that the tract houses few people, an unusually large proportion of whom are unemployed (see Section 6.4.3 for discussion of related issues of ecological fallacy). However, this hypothesis alone would not help us much in deciding whether or not the zone (or part of it) should be included in an inner-city workfare program, for example. Use of GIS to overlay fine-resolution aerial photography (Figure 13.13C) reveals the tract to be largely empty of population apart from a small extension to a large housing estate. It would thus appear sensible to assign this zone the same policy status as the zone to its west.

> **Dasymetric mapping uses the intersection of two datasets (or layers in the same dataset) to obtain more precise estimates of a spatial distribution.**

Dasymetric mapping and related techniques present a window on reality that looks more convincing than conventional choropleth mapping. However, it is important to remain aware that the visualization of reality is only as good as the assumptions that are used to create it. The information about population concentration used in Figure 13.13, for example, is defined only on the basis of common sense and is

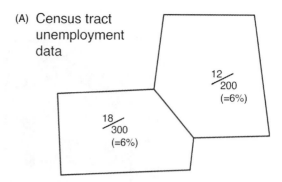

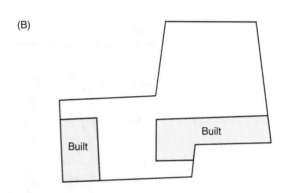

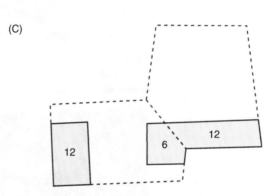

Figure 13.12 Modeling a spatial distribution in an urban area using dasymetric mapping: (A) zonal distribution of census population; (B) distribution of built structures; and (C) overlay of (A) and (B) to obtain a more contained measure of population distribution.

likely to be error prone (some built forms may be offices and shops, for example), and this will inevitably feed through into inaccuracies in visualization. Inference of land use from land cover, in particular, is an uncertain and error-prone process (see Section 6.3.2), and there is a developing literature on best practice for the classification of land use employing land cover information and classifying different (e.g., domestic versus non-domestic) land uses.

> **Inferring land use from land use is uncertain and error prone.**

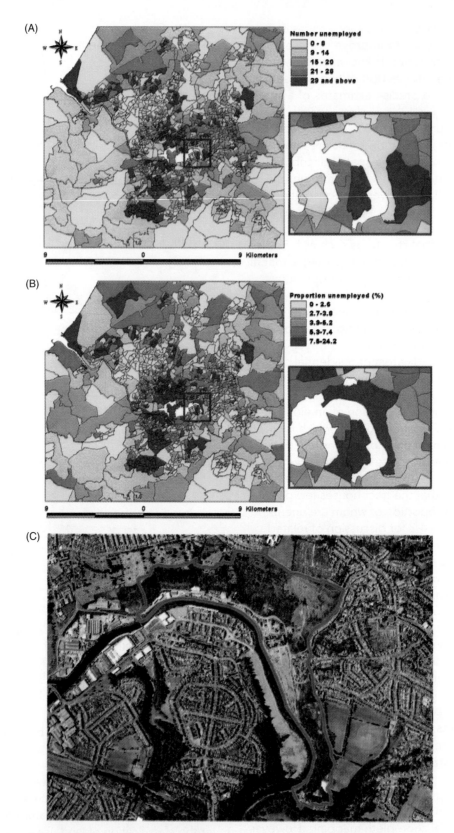

Figure 13.13 Dasymetric mapping in practice, in Bristol, UK: (A) numbers employed by census enumeration district (block); (B) proportions unemployed using the same geography; and (C) orthorectified aerial photograph of part of the area of interest. (Courtesy: Rich Harris)

13.4 Participation, Interaction, and Immersion

13.4.1 Public Participation in GIS (PPGIS)

It has been suggested that users first desire an overview of the totality of information, and then the facility to narrow the focus to information that is of immediate relevance to particular requirements. Geodemographers (see Section 2.3.3) have investigated the ways in which individuals and different groups interact with new information and communication technologies (NICTs), and have suggested a three-stage model. Initially, people use NICTs to access information; many then graduate to using them to conduct transactions (e.g., through online shopping); and user engagement culminates with people using NICTs as a gateway to participation in networked decision making. In a GIS context, this progression depends on confidence in perceiving, manipulating, and exploring spatial phenomena through geovisualization.

GIS can be used alongside a range of other computer media to facilitate the active involvement of many different groups in the discussion and management of change through planning, and may act as a bulwark against officialdom or big business. Good human—computer interaction is key to this process, and the use of GIS to foster public participation in (the use of) GIS is often termed PPGIS. The focus of PPGIS is on how people perceive, manipulate, and interact with representations of the real world as manifested in GIS: its other concerns include the way people evaluate options through multicriteria decision making (Section 16.4), and the social issues of how GIS usage remains concentrated within networks of established interests that are sometimes perceived to control the use of technology (Section 1.8). A related theme is the use of GIS to create multiple representations—capturing and maintaining the different perspectives of stakeholders rather than framing debate in the terms of a single prevailing authoritative view.

Geovisualization has a range of uses in PPGIS, including:

- Making the growing complexity of land-use planning, resource use, and community development intelligible to communities and different government departments.

- Radically transforming the planning profession through use of new tools for community design and decision making.

- Unlocking the potential of the many digital data sources that are collected, but not used, at the local level.

- Helping communities shift land-use decisions from regulatory processes to performance-based strategies, and making the community decision-making process more proactive and less reactive.

- Improving community education about local environmental and social issues.

- Improving the feed of information between public and government in emergency planning and management (see Figure 13.14).

Figure 13.14 (A) The Zaca fire began on July 4, 2007 northeast of Buellton, California, in Santa Barbara County. It consumed more than 100,000 hectares (240,000 acres), and for much of the summer threatened to displace tens of thousands of people and destroy thousands of homes in Santa Barbara and Ventura counties. It was eventually declared controlled October 29. (B) Daily briefings inside the Santa Barbara County Emergency Operations Center relied heavily on GIS data fed from members of the public as well as GIS-based maps. (Image (B) courtesy of William H. Boyer/County of Santa Barbara).

(A)

(B)

In each of these applications, geovisualization has a strong cognitive component. Users need to feel equipped and empowered to interrogate representations in order to reveal otherwise-hidden information. This requires dynamic and interactive *software* environments and the *people* skills that are key to extracting meaning from a representation. Geovisualization allows people to use software to manipulate and represent data in multiple ways, in order to create what-if scenarios or to pose questions that prompt the discovery of useful relations or patterns. This is a core remit of PPGIS, where the geovisualization environment is used to support a process of knowledge construction and acquisition that is guided by the user's knowledge of the application. PPGIS research entails usability evaluations of structured tasks, using a mixture of computer-based usability evaluation techniques and traditional qualitative research methods, in order to identify cognitive activities and user problems associated with GIS applications.

13.4.2 2.5-D and 3-D Representation

As indicated in Section 12.3.2, extruded 2.5-D representations may be used to reveal aspects of data that are not easily observable in two dimensions. This is illustrated in Figure 13.15. The map shows London's strongly monocentric structure and the heavy specialization in office facilities both in the historic City of London and London's Docklands. The use of

the extruded 2.5-D representation highlights urban density and mix-of-uses, which are relevant to issues of sustainability and local travel. Metropolitan centers such as Croydon (south of the city) are dwarfed by the main center, suggesting they are struggling to compete. Another trend is the emergence of monofunctional and car-dependent "edge city" developments around Heathrow Airport.

The third dimension can also be used to represent built form, and in recent years, online 3-D models have become available for geographically extensive parts of many cities throughout the world. The virtual globe offerings from Microsoft and Google have arisen following the wide availability of very-fine-resolution height data from airborne instruments, such as Light Detection and Ranging (LiDAR: see Section 9.2.2). Augmentation of these models is increasingly straightforward, given the wide availability of free software such as Google Sketchup that enables even the nonspecialist to create 3-D representations of individual buildings and site layouts (see Figure 13.16). These representations provide the general user with valuable context to augment online two-dimensional maps and street views, and they have been used for general applications as diverse as city marketing and online social networking. Research applications can benefit from general comparisons with 3-D built form, as with geographer Paul Torrens's work relating the built form of the city to its WiFi geography (Figure 13.17). "Off-the-shelf" 3-D representations can provide very

Figure 13.15 A 3-D representation of office and retail land use in London. The height of the bars identifies 'rateable value'—a UK indicator of the value of the property/real estate in the grid squares–that is dedicated to office and retail functions. (Courtesy Duncan Smith: data © Valuation Office Agency 2005.)

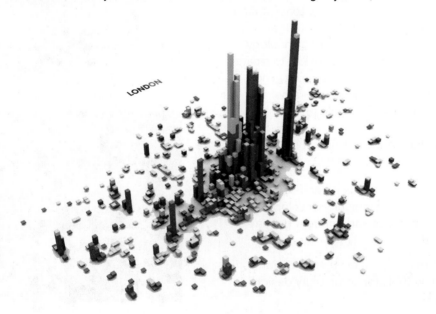

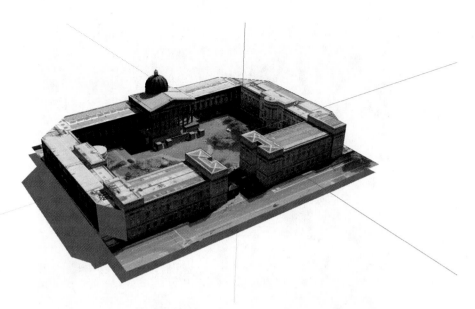

Figure 13.16 A 3-D building representation, created using Google Sketchup and suitable for inclusion in a 3-D city model.

Figure 13.17 A dense network of Wi-Fi infrastructure viewed above the built environment of Salt Lake City, Utah. The pattern of Wi-Fi transmissions suggests no centralized organization, yet does mirror urban built form. Wi-Fi activity is most prominent in the city's traditional commercial core, yet also reaches out to interstitial and peripheral parts of the city. (Areal extent of the Wi-Fi cloud on the ground is ~3 km²: 'high' and 'low' refers to the strength of Wi-Fi signals, measured in decibels relative to 1 mW.) (Courtesy Paul Torrens)

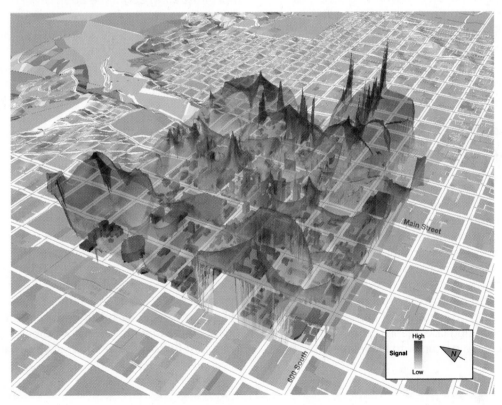

Figure 13.18 Visualizing nitrogen oxide (NOx) pollution in the Virtual London model. This pollutant, often nicknamed 'urban smog' is largely derived from vehicle emissions: red identifies higher levels, whilst blue represents lower levels. (Courtesy Steve Evans: Ordnance Survey data © Crown Copyright; NOx data courtesy of Environmental Research Group, Kings College London and the Greater London Authority).

useful contextual information, although it is important to remember that the provenance of the data that were used to create them is largely unknown, and that the motivation for their creation may be more closely attuned to the generation of advertising revenues than provision of spatial data infrastructure for specific applications (Section 19.3).

Other more detailed 3-D city models, such as Virtual London (see Figure 13.18) and Virtual Kyoto (Box 13.3), have been developed in the academic domain for purposes such as contextualizing historical cultural studies or modeling the concentration and circulation of pollutants in cities. In many respects, they crystallize both what is most exciting and what is most frustrating about the application of cutting-edge GIS techniques. They are typically assembled from diverse sources by stakeholders in local and central government, in utilities, in transportation, and increasingly from members of the public. The very detailed yet disparate nature of data holdings suitable for

Keiji Yano, Digital Humanities, and Virtual Kyoto

Keiji Yano is professor of human geography and geographic information science at Ritsumeikan University, Kyoto, where he researches geovisualization, geodemographics, and modeling of urban systems. In recent years he has led the *Kyoto Virtual Time-Space* project, which is using GIS technologies to re-create the City of Kyoto in a virtual reality space. A baseline virtual representation of present-day Kyoto was developed using laser-scanning data, and this has been used as a precursor to hindcasting (the opposite of forecasting) the ways in which the city has evolved since the Heian period (AD 794–1192: see Figure 13.19). The work has been carried out with the use of historic topographic, cadastral, and picture maps, along with archaeological data. The result is a vivid representation of changes in Kyoto's cityscape that successfully introduces a time dimension into the 3-D city model. Virtual Kyoto is part of a wider digital humanities initiative at Ritsumeikan and provides a useful framework around which various other

▶

(A)

(B)

Figure 13.19 The Gion Festival as viewed in Virtual Kyoto for (A) c. 1930 and (B) the present day. Note the stronger visual impact of the procession set against the natural landscape in the earlier image. (Source: www.rgis.lt.ritsumei.ac.jp/~keiji.yano/Gion_festival.zip)

digital archives of visual and performing arts can be organized. Related to this, Keiji maintains an active personal interest in cataloging the locations and characteristics of historic buildings in Kyoto (Figure 13.20).

Kyoto lies in an earthquake-prone part of the world, and a further use of Virtual Kyoto is in planning for disaster management. The model has been used to produce hazard maps that allow scientists and city residents to see how natural disasters might affect the city. People previously found it difficult to understand two-dimensional (digital or paper) maps, and Virtual Kyoto now allows stakeholders to share information about the changing visual landscape of Kyoto. The work is thus important in helping to build social consensus about urban planning and disaster management.

Check out the Virtual Kyoto Web site at www. geo.lt.ritsumei.ac.jp/webgis/ritscoe.html.

Figure 13.20 Keiji Yano, Geovisualization expert and civic historian (Source: www.rgis.lt.ritsumei.ac.jp/~keiji.yano/DSC00797.JPG)

inclusion in 3-D models makes networked GIS the ideal medium for assembling and integrating data and for communicating and sharing the results. Yet the very nature of such partnership activity can lead to conflicts of interest if developments in technology open up new potential uses that were not originally envisaged. The Virtual London model (see Figure 13.18), for example, potentially provide a valuable forum for planning through consultation on a range of environmental and economic development issues, yet this was not possible at the time because of issues of data ownership by Great Britain's national mapping agency (Ordnance Survey) in particular (see Box 18.4). As with the crowd-sourced OpenStreetMap project (Figures 11.8 and 13.7, and Box 18.8), however, volunteered geographic information is incrementally providing ways beyond this impasse. We address these broader management issues in Chapter 18.

These different themes share a common focus of broadening the base of users that feels confident interacting with GIS. In this context, Box 13.3 introduces Keiji Yano, who has led the development of two contrasting representations of Kyoto—one, a data-rich 3-D representation of Virtual Kyoto, and the second, a detailed historical geography of the development of the built form of the city.

It is important to consider the development of fine-scale city models in the context of the seamless scale-free mapping and street-view systems that allow users to view raster and vector information, including features such as buildings, trees, and automobiles, alongside synthetic and photorealistic global and local displays. In global visualization systems, datasets are projected onto a global TIN-based data structure (see Section 8.2.3.4). Rapid interactive rotation and panning of the globe is enabled by caching data in an efficient in-memory data structure. Multiple levels of detail, implemented as nested TINs and reduced-resolution datasets, allow fast zooming in and out. When the observer is close to the surface of the globe, these systems allow the display angle to be tilted to provide a perspective view of the Earth's surface. The user is able to roam smoothly over large volumes of global geographic information. This enables a better understanding of the distribution and abundance of globally distributed phenomena (e.g., ocean currents, atmospheric temperatures, and shipping lanes), as well as detailed examination of local features in the natural and built environment (e.g., optimum location of cell phone towers, or the impact of tree felling on the viewscape of tourist areas).

13.4.3 Hand-Held Computing and Augmented Reality

Immersion of desk- and studio-bound users in 3-D virtual reality presents one way of promoting better remote interaction with the real world, using high-power computing and very high-bandwidth computer networks. The same digital infrastructure is also being used to foster improved and more direct interaction with the real world by the development of a range of hand-held, in-vehicle, and wearable computer devices. These are discussed, along with some of the geovisualization conventions they entail, in Sections 7.6.4 and 11.3. Figure 13.21 illustrates how semi-immersive systems are being used in field computing, taking the example of a geography field course in the UK Lake District.

Figure 13.21A illustrates how students previously visualized past landscapes, by aligning a computer-generated viewshed with the real world. This approach exemplifies many of the shortcomings of traditional mapping: perfect alignment is possible from only a single point on the landscape, only a single prerendered coverage can be viewed, and there is no way of interrogating the database that was used to create the augmented representation. Use of the equipment shown in Figure 13.21B enables the user to establish the locations and directions of viewpoints, and loaded images are automatically displayed as the user approaches any of the numbered points. The hand-held device makes it possible to annotate the view and to display a range of surface and subsurface characteristics (Figure 13.21C). As discussed in Section 11.3, the current generation of augmented-reality devices remains unable to render viewsheds on the fly, for reasons of computer and bandwidth capacity. Progress continues to be made in this direction, for example, through the integration of digital compasses into mobile devices.

13.4.4 Scientific Visualization (ViSC) and Virtual Reality

Fast processing using the latest hardware, improved broadband connectivity, and more sophisticated computer graphics (including animation) have together led to a range of new computer media solutions for visualizing spatial phenomena, often described using the umbrella term *Visualization in Scientific Computing* (ViSC). ViSC provides new and more sophisticated ways of visualizing and interacting with the world than conventional mapping alone. Some of the advanced techniques are not explicitly spatial, while others are considered elsewhere in this book: these include line generalization (Section 3.8; Box 4.6), cartographic smoothing (see the discussion of interpolation in Box 4.3), and visualization of uncertainty (Section 6.4.2). ViSC entails use of a wide range of computer devices in different settings, ranging from use of hand-held computers to navigate through the real world through systems that augment user perceptions of the real world to lab-based systems that immerse users in the artificial worlds of *virtual reality*.

New computer technologies can be used to partially or wholly immerse users in artificial worlds.

These various activities pose many computational and technical challenges to computer and information

Figure 13.21 (A) Aligning a viewshed transparency to augment a view of the landscape; (B) use of computer devices to achieve the same effect for (C) a number of field sites and different surface characteristics, with the facility to annotate landscape views. (Courtesy Gary Priestnall)

Figure 13.22 An extruded 3-D map of Figure 13.16, embedded within a games engine (Courtesy Andy Hudson-Smith)

science, but together they provide new tools to help users better understand geographic data through representation, and hence improve understanding of real-world patterns and processes. In software terms, much can be achieved using the functionality of GIS packages, although the facilities available for high-end geovisualization and locationally aware user interaction are in many senses better developed in games engines (see Figure 13.22). Semi-immersive and immersive virtual reality (VR) is beginning to make the transition from the realm of computer gaming to GIS application, with new methods of spatial object manipulation and exploration using head-mounted displays and sound (see Section 11.3.1). Such applications raise the prospect of collaborative GIS applications using games software.

The advent of immersive and semi-immersive systems has important implications for participation by a broad user base since they allow:

- Users to access virtual environments and select different views of phenomena.

- Incremental changes in these perspectives to permit real-time *fly-throughs*.

- Repositioning or rearrangement of the objects that make up virtual scenes.

- Users to be represented graphically as *avatars*— digital representations of themselves.

- Engagement with avatars connected at different remote locations, in a networked virtual world.

- The development, using avatars, of new kinds of representation and modeling.

- The linkage of networked virtual worlds with *virtual reality* (VR) systems.

The developments that have fostered the emergence of 3-D visualization in GIS have also encouraged the development of online virtual worlds (e.g., Second Life: www.secondlife.com) in which online users (represented as avatars) engage, interact, and collaborate in a peer-to-peer Web 2.0 environment. Figure 13.23 illustrates how this environment might be utilized to foster collaboration in issues of urban planning and public debate in a visually collaborative environment, where manipulation of perspective and communication contributes to the interaction and sharing of ideas. Second Life provides a platform with the ability to tie information, actions, and rules to objects, opening the possibility of a true multiuser GIS.

There exists the prospect of infusing the collaborative environment of Second Life and other online spaces into virtual Earth applications (see Section 11.3.1) and creating an occupied Digital Earth. It is also likely that the kinds of advanced models discussed in Chapter 16 (e.g., agent-based models: Section 16.2) will also be assimilated into these virtual worlds in the near future. Users of Second Life already typically share 23,000

Figure 13.23 A collaborative planning application in Second Life, using elements of the Virtual London 3-D model (www.secondlife.com). (Courtesy Andy Hudson-Smith)

hours a day creating artificial representations, and the site is an excellent example of crowd-sourcing of digital geography in that the rolling fields, rivers, valleys, mountains, hamlets, and towns that occupy the ever-growing space have been created piece by piece by the millions of users. This panoply of space is editable and hence intensely collaborative.

13.5 Consolidation

Geovisualization can make a powerful contribution to decision making and can be used to simulate changes to reality. Yet, although it is governed by scientific principles, the limitations of human cognition mean that it necessarily provides a further selective filter on the reality that it seeks to represent. Mapping is about seeing the detail as well as the big picture, yet the wealth of detail that is available in today's digital environments can sometimes create information overload and threaten to overwhelm the message of the map. Geovisualization is fostering greater user interaction and participation in the use of GIS as a decision support tool.

At its worst, greater sophistication of display may transmit information overload and obscure the inherent uncertainty in most spatial data: at their best, the media of visualization allow the user to undertake a balanced appraisal of the message of a geographic representation, and to create better description, explanation, prediction, and understanding of GI. The contribution of geovisualization to PPGIS can be viewed as developing and broadening the base of users that not only sources information through GIS, but increasingly uses GIS as a medium for information exchange and participation in decision making. The old adage that seeing is believing only holds if visualization and user interaction are subservient to the underlying motivation of building representations that are fit for purpose.

1. Figure 13.24 is a cartogram redrawn from a newspaper feature on the costs of air travel from London in 1992. Using current advertisements in the press and on the Internet, create a similar cartogram of travel costs, in local currency, from the nearest international air hub to your place of study.

2. How can Web-based multimedia GIS tools be used to improve community participation in decision making?

3. Produce two (computer-generated) maps of the Israeli West Bank security fence barrier to illustrate opposing views of its effects. For example, the first might illustrate how it helps to preserve an Israeli "Lebensraum," while the second might emphasize its negative effects on Palestinian communities. In a separate short annotative commentary, describe the structure and character of the fence at a range of spatial scales.

4. Review the common *sources* of uncertainty in geographic representation and the ways in which they can be *manifested* through geovisualization.

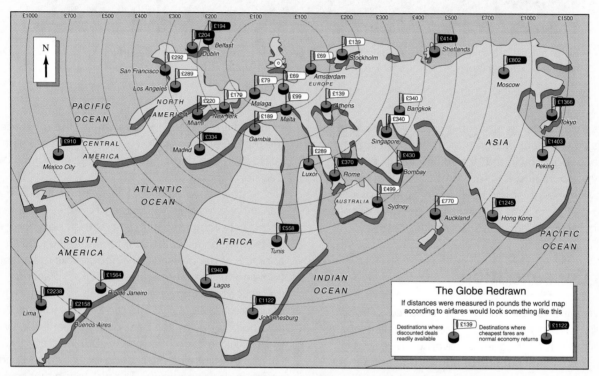

Figure 13.24 The globe redrawn in terms of 1992 travel costs from London. (Source: redrawn from newspaper article by Frank Barrett in *The Independent* on 9 February 1992, page 6)

Further Reading

Dorling, D. and Thomas, B. 2004. *People and Places: A 2001 Census Atlas of the UK*. Bristol, UK: Policy Press.

Dykes, J., MacEachren, A.M., and Kraak, M-J. (eds.) 2005. *Exploring Geovisualization*. London: Elsevier Science.

Egenhofer, M. and Kuhn, W. 2005. Interacting with GIS. In Longley, P.A., Goodchild, M.F., Maguire, D.J., and Rhind, D.W. (eds.). *Geographical Information Systems: Principles, Techniques, Management and Applications (abridged edition)*. Hoboken, NJ: Wiley.

MacEachren, A. 1995. *How Maps Work: Representation, Visualization and Design*. New York: Guilford Press.

Raubal, M. 2009. Cognitive engineering for geographic information science. *Geography Compass* 3: 1–18.

Sieber, R. 2006. Public Participation and Geographic Information Systems: A Literature Review and Framework. *Annals of the American Association of Geographers*, 96: 491–507.

Tufte, E.R. 2001. *The Visual Display of Quantitative Information* (2nd ed.). Cheshire, CT: Graphics Press.

14 Spatial Data Analysis

LEARNING OBJECTIVES

This chapter is the first in a set of three dealing with geographic analysis and modeling methods. The chapter begins with a review of the relevant terms and presents an outline of the major topics covered in the three chapters. Analysis and modeling are grounded in spatial concepts, allowing the investigator to examine the role those concepts play in understanding the world around us. This chapter also examines methods constructed around the concepts of location, distance, and area. Location provides a common key between datasets, allowing the investigator to discover relationships and correlations between properties of places. Distance defines the separation of places in geographic space and acts as an important variable in many of the processes that impact the geographic landscape. Areas define the neighborhood context of processes and events, and also capture the important concept of scale.

After studying this chapter you will understand:

- Definitions of spatial data analysis and tests to determine whether a method is spatial.

- Techniques for detecting relationships between the various properties of places and for preparing data for such tests.

- Methods to examine distance effects, in the creation of clusters, hotspots, and anomalies.

- The applications of convolution in GIS, including density estimation and the characterization of neighborhoods.

14.1 Introduction: What Is Spatial Analysis?

The techniques covered in these three chapters are generally termed *spatial* rather than *geographic* because they can be applied to data arrayed in any space, not only geographic space (see Section 1.1.1). Many of the methods might potentially be used in analysis of outer space by astronomers, or in analysis of brain scans by neuroscientists. So the term *spatial* is used consistently throughout these chapters.

Spatial analysis is in many ways the crux of GIS because it includes all of the transformations, manipulations, and methods that can be applied to geographic data to add value to them, to support decisions, and to reveal patterns and anomalies that are not immediately obvious. In other words, spatial analysis is the process by which we turn raw data into useful information, in pursuit of scientific discovery, or more effective decision making. If GIS is a method of communicating information about the Earth's surface from one person to another, then the transformations of spatial analysis are ways in which the sender tries to inform the receiver, by adding greater informative content and value, and by revealing things that the receiver might not otherwise see.

Some methods of spatial analysis were developed long before the advent of GIS, when they were carried out by hand or by the use of measuring devices like the ruler. The term *analytical cartography* is sometimes used to refer to methods of analysis that can be applied to maps to make them more useful and informative, and spatial analysis using GIS is in many ways its logical successor. But it is much more powerful because it covers not only the contents of maps but any type of geographic data.

Spatial analysis can reveal things that might otherwise be invisible—it can make what is implicit explicit.

In this chapter we will look first at some definitions and basic concepts of spatial analysis. Concepts such as location, distance, and area—the topics discussed in this chapter—have already been encountered at various points in this book, but here they serve to provide an organizing framework for the vast array of methods that fall under the heading of spatial analysis. More advanced concepts, as well as the methods used to elucidate them, are discussed in Chapter 15. Chapter 16 addresses the use of GIS to examine dynamic processes, primarily by simulation, and the use of GIS in design, when its power is directed not to understanding the world but to improving it according to specific goals and objectives.

Spatial analysis is the crux of GIS, the means of adding value to geographic data and of turning data into useful information.

Methods of spatial analysis can be very sophisticated, but they can also be very simple. A large body of methods of spatial analysis has been developed over the past century or so, and some methods are highly mathematical—so much so that it might sometimes seem that mathematical complexity is an indicator of the importance of a technique. But the human eye and brain are also very sophisticated processors of geographic data, and excellent detectors of patterns and anomalies in maps and images. So the approach taken here is to regard spatial analysis as spread out along a continuum of sophistication, ranging from the simplest types that occur very quickly and intuitively when the eye and brain look at a map to the types that require complex software and sophisticated mathematical understanding. Spatial analysis is best seen as a *collaboration* between the computer and the human, in which both play vital roles.

Effective spatial analysis requires an intelligent user, not just a powerful computer.

There is an unfortunate tendency in the GIS community to regard the making of a map using a GIS as somehow less important than the performance of a mathematically sophisticated form of spatial analysis. According to this line of thought, *real* GIS involves number crunching, and users who *just* use GIS to make maps are not serious users. But every cartographer knows that the design of a map can be very sophisticated and that maps are excellent ways of conveying geographic information and knowledge by revealing patterns and processes to us (see Chapter 12). We agree; moreover, we believe that mapmaking is potentially just as important as any other application of GIS.

Spatial analysis may be defined in many possible ways, but all definitions in one way or another express the basic idea that information on locations is essential—that analysis carried out without knowledge of locations is not spatial analysis. One fairly formal statement of this idea is the following:

> Spatial analysis is a set of methods whose results are not invariant under changes in the locations of the objects being analyzed.

The double negative in this statement follows convention in mathematics, but for our purposes we can remove it:

Spatial analysis is a set of methods whose results change when the locations of the objects being analyzed change.

On this test the calculation of an average income for a group of people is not spatial analysis because it in no way depends on the locations of the people. But the calculation of the center of the U.S. population is spatial analysis, because the results depend on knowing where all U.S. residents are located. GIS is an ideal platform for spatial analysis because its data structures accommodate the storage of object locations.

The techniques discussed in these chapters are no more than the tip of the spatial analysis iceberg. Some GISs are more sophisticated than others in the range of techniques they support, and others are strongly oriented toward certain domains of application. For a more advanced review including many techniques not mentioned here, and for a detailed analysis of the techniques supported by each package, the reader is encouraged to examine the book *Geospatial Analysis: A Comprehensive Guide to Principles, Techniques and Software Tools* by De Smith, Goodchild, and Longley, which is available online at www.spatialanalysisonline.com.

14.1.1 Examples

Spatial analysis can be used to further the aims of science, by revealing patterns that were not previously recognized and that hint at undiscovered generalities and laws. Patterns in the occurrence of a disease may hint at the mechanisms that cause the disease, and some of the most famous examples of spatial analysis are of this nature, including the work of Dr John Snow in unraveling the causes of cholera (see Box 14.1).

It is interesting to speculate on what would have happened if early epidemiologists like Snow had had access to a GIS. The rules governing research today would not have allowed Snow to remove the pump handle, except after lengthy review, because the removal constituted an experiment on human subjects. To get approval, he would have had to have shown persuasive evidence in favor of his hypothesis, and it is doubtful that the map would have been sufficient because several other hypotheses might have explained the pattern equally well. First, it is conceivable that the population of Soho was inherently at risk of cholera, perhaps by being comparatively elderly or because of poor housing conditions. The map would

have been more convincing if it had shown the *rate* of incidence relative to the population at risk. For example, if cholera was highest among the elderly, the map could have shown the number of cases in each small area of Soho as a proportion of the population over 50 in each area. Second, it is still conceivable that the hypothesis of transmission through the air between carriers could have produced the same observed pattern, if the first carrier happened to live in the center of the outbreak near the pump. Snow could have eliminated this alternative if he had been able to produce a sequence of maps, showing the locations of cases as the outbreak developed. Both of these options involve simple spatial analysis of the kind that is readily available today in GIS.

GIS provides tools that are far more powerful than the map in suggesting causes of disease.

Today the causal mechanisms of diseases like cholera, which results in short, concentrated outbreaks, have long since been worked out. Much more problematic are the causal mechanisms of diseases that are rare and not sharply concentrated in space and time. The work of Stan Openshaw at the University of Leeds, using one of his Geographical Analysis Machines, illustrates the kinds of applications that make good use of the power of GIS in a more contemporary context.

Figure 14.4 shows an application of one of Openshaw's techniques to a comparatively rare but devastating disease whose causal mechanisms remain largely a mystery—childhood leukemia. The study area is northern England, the region from the Mersey to the Tyne. The analysis begins with two datasets: one of the locations of cases of the disease and the other of the numbers of people at risk in standard census reporting zones. Openshaw's technique then generates a large number of circles, of random sizes, and places them (*throws* them) randomly over the map. The computer generates and places the circles, and then analyzes their contents, by dividing the number of cases found in the circle by the size of the population at risk. If the ratio is anomalously high, the circle is drawn. After a large number of circles have been generated, and a small proportion have been drawn, a pattern emerges. Two large concentrations, or clusters of cases, are evident in the figure. The one on the left is located around Sellafield, the location of the British Nuclear Fuels processing plant and a site of various kinds of leaks of radioactive material. The other, in the upper right, is in the Tyneside region,

Dr. John Snow and the Causes of Cholera

In the 1850s cholera was poorly understood, and massive outbreaks were a common occurrence in major industrial cities. (Even today cholera remains a significant health hazard in many parts of the world, despite progress in understanding its causes and advances in treatment.) An outbreak in London in 1854 in the Soho district was typical of the time, and the deaths it caused are mapped in Figure 14.1. The map was made by Dr. John Snow (Figure 14.2), who had conceived the hypothesis that cholera was transmitted through the drinking of polluted water rather than through the air, as was commonly believed. He noticed that the outbreak appeared to be centered on a public drinking water pump in Broad Street (Figure 14.3)—and if his hypothesis was correct, the pattern shown on the map would reflect the locations of people who drank the pump's water. There appeared to be anomalies, in the sense that deaths had occurred in households that were located closer to other sources of water, but he was able to confirm that these households also drew their water from the Broad Street pump. Snow had the handle of the pump removed, and the outbreak subsided, providing direct causal evidence in favor of his hypothesis. The full story is much more complicated than this largely apocryphal version, of course; much more information is available at www.jsi.com.

Today, Snow is widely regarded as the father of modern epidemiology.

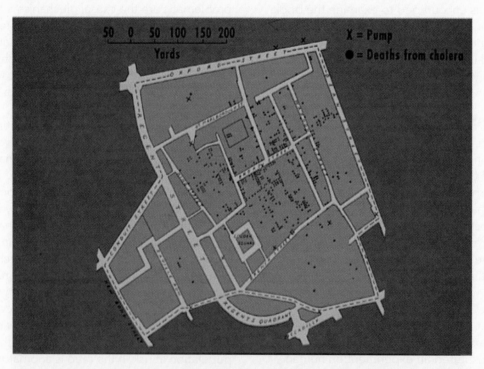

Figure 14.1 A redrafting of the map made by Dr. John Snow in 1854, showing the deaths that occurred in an outbreak of cholera in the Soho district of London. The existence of a public water pump in the center of the outbreak (the cross in Broad Street) convinced Snow that drinking water was the probable cause of the outbreak. Stronger evidence in support of this hypothesis was obtained when the water supply was cut off, and the outbreak subsided. (Courtesy: Gilbert E.W. 1958. Pioneer maps of health and disease in England. *Geographical Journal*, 124: 172–183.)

Figure 14.2 Dr. John Snow. (© TopFoto/The Image Works)

Figure 14.3 A modern replica of the pump that led Snow to the inference that drinking water transmitted cholera, located in what is now Broadwick Street in Soho, London. (Courtesy Mike Goodchild)

and Openshaw and his colleagues discuss possible local causes.

Both of these examples are instances of the use of spatial analysis for scientific discovery and decision making. Sometimes spatial analysis is used *inductively*, to examine empirical evidence in the search for patterns that might support new theories or general principles, in this case with regard to disease causation. Other uses of spatial analysis are *deductive*, focusing on the testing of known theories or principles against data. (Snow already had a theory of how cholera was transmitted, and used the map as confirmation to convince others.) Induction and deduction were also discussed in Sections 4.6 and 4.9 in the context of the representation and nature of spatial data. A third type of application is *normative*, using spatial analysis to develop or prescribe new or better designs, for the locations of new retail stores, or new

roads, or new manufacturing plant. Examples of this type appear in Section 15.4.

14.2 Analysis Based on Location

The concept of location—identifying *where* something exists or happens—is central to GIS, and the ability to compare different properties of the same place, and as a result to discover relationships and correlations and perhaps even explanations, is often presented as GIS's greatest advantage. Take Figure 14.5 as an example. It shows the age-adjusted rate of death due to cancers of the throat and lung among adult males in a 20-year period 1950–1969, compiled by county. Maps such as this immediately prompt us to ask: do I know of other properties of these counties

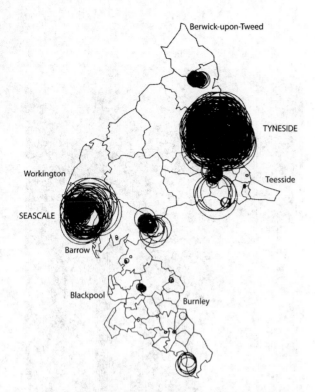

Figure 14.4 The map made by Openshaw and colleagues by applying their Geographical Analysis Machine to the incidence of childhood leukemia in northern England. A very large number of circles of random sizes is randomly placed on the map, and a circle is drawn if the number of cases it encloses substantially exceeds the number expected in that area given the size of its population at risk. (Source: Openshaw S., Charlton M., Wymer C., and Craft A. 1987. A Mark I geographical analysis machine for the automated analysis of point datasets. *International Journal of Geographical Information Systems* 1: 335–358. www.tandf.co.uk/journals)

that might explain their rates? Within seconds the mind is busy identifying counties, recalling general knowledge that might suggest cause, and checking other counties to see if they confirm suspicions. All of this depends, of course, on having a basis of knowledge in one's mind that is sufficient to identify particular counties and their general characteristics. One might note, for example, that Chicago, Detroit, Cleveland, and several other major cities are *hotspots*, but that other major cities such as Minneapolis are not—and ask why?

GIS allows us to look for explanations among the different properties of a location.

The map shows many intriguing patterns, but the pattern that is of greatest significance to the history of cancer research concerns counties distributed around the Gulf and Atlantic coasts. These were counties in-

volved in ship construction during World War II, when large amounts of asbestos were used for insulation. In the period 5 to 25 years later, the consequences of breathing asbestos fibers into the respiratory system are plain to see in the high rates of cancers in the counties containing Mobile, Alabama; Norfolk, Virginia; Jacksonville, Florida; and many other port cities.

It may seem that comparing the properties of places should be straightforward in a technology that insists on giving locations to all of the information it stores. The next subsection discusses examples of such cases. In other cases, however, comparison can be quite difficult and complex. The data models adopted in GIS and described in Chapter 8 are designed primarily to achieve efficiency in representation and to emphasize storing *information about one property for all places* over *information about all properties for one place*. We sometimes term this *horizontal* rather than *vertical* integration of data. For example, it is traditional to store all of the elevation data for a given county together, perhaps as a digital elevation model, and all of the soil data for the same county together, perhaps as a set of topologically related soil polygons. These are very efficient approaches, but they are not designed for a point-by-point comparison of elevation and soil type, or for answering questions such as "are certain soil types more likely to be found at high elevations in this county?" Subsequent subsections discuss some of the GIS techniques designed specifically for situations such as this one.

14.2.1 Analysis of Attribute Tables

In the example shown in Figure 14.5, it is quite likely that the kinds of factors responsible for high rates of cancer are already available in the attribute table of the counties, along with the cancer rates. In such cases our interest is in comparing the contents of two columns of the table, looking for possible relationships or correlations—are there counties for which cancer rates and the values of potentially causative variables are both high, or both low? Figure 14.6 shows a suitable example, where the interest lies in a possible relationship between two columns of a county attribute table. In this case the investigator suspects a pattern in the relationship between average value of house and percent black, variables that are collected and disseminated by the U.S. Bureau of the Census as county attributes. One way to examine this suspicion is to plot one variable against the other as a *scatterplot*. In Figure 14.6 median house value is

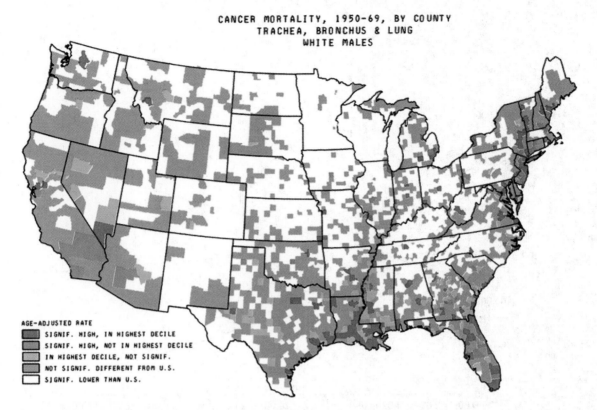

Figure 14.5 Age-adjusted rates of mortality due to cancers of the trachea, bronchus, and lung, among white males between 1950 and 1969, by county. (Source: Mason et al., *Atlas of Cancer Mortality for US Counties.* Bethesda, MD: National Cancer Institute, 1975)

shown on the vertical or *y*-axis, and percent black on the horizontal or *x*-axis. (For another example of a scatterplot see Figure 4.16.)

In a formal statistical sense, these scatterplots allow us to examine in detail the *dependence* of one variable on one or more *independent* variables. For example, we might hypothesize that the median value of houses in a county is correlated with a number of variables such as percent black, average county income, percent 65 and over, average county air pollution levels, and so forth, all stored in the attribute table associated with the properties. Formally this may be written as

$$Y = f(X_1, X_2, X_3, \ldots, X_k)$$

where *Y* is the dependent variable and X_1 through X_k are all of the possible independent variables that might affect housing value or be correlated with it. It is important to note that it is the independent variables that together predict the dependent variable, and that any hypothesized causal relationship is one way—that is, that median house value is *responsive* to average income, percent 65 and over, and so forth,

and not vice versa. For this reason the dependent variable is termed the *response* variable and the independent variables are termed *predictor* variables in some statistics textbooks.

In our case there is only a single predictor variable, and it is clear from the scatterplots that the relationship is far from perfect—that many other unidentified factors contribute to the determination of median housing value. In general this will always be true in the social and environmental sciences because it will never be possible to capture all of the factors responsible for a given outcome. So we modify the model by adding an error term ε to represent that unknown variation.

Regression analysis is the term commonly used to describe this kind of investigation, and it focuses on finding the simplest relationship indicated by the data. That simplest relationship is linear, implying that a unit increase in percent black always corresponds to a constant corresponding increase or decrease in the dependent variable. Linear relationships plot as straight lines on a scatterplot and have the equation:

$$Y = b_0 + b_1X_1 + b_2X_2 + b_3X_3 + \ldots + b_kX_k + \varepsilon$$

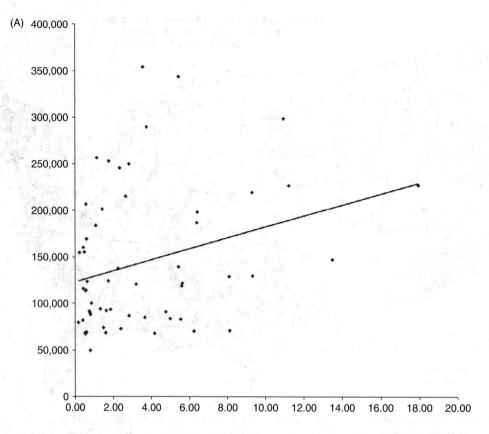

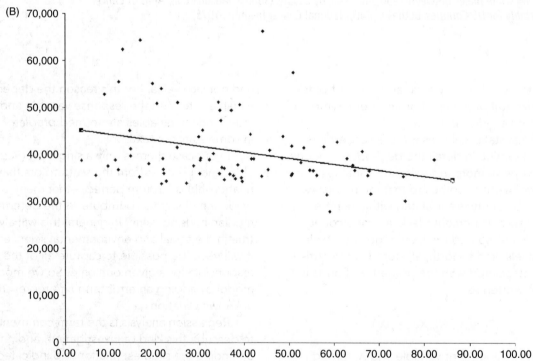

Figure 14.6 Scatterplots of median house value (*y* axis) versus percent black (*x* axis) for U.S. counties in 1990, with linear regressions: (A) California; and (B) Mississippi. (Source: U.S. Bureau of the Census)

though in the case of Figure 14.6 there is only one independent variable X_1. b_1 through b_k are termed regression *parameters* and measure the direction and strength of the influence of the independent variables X_1 through X_k on Y. b_0 is termed the *constant* or *intercept* term.

Figure 14.6A shows the data for the counties of California, and Figure 14.6B for the counties of Mississippi. Notice the difference—counties with a high percentage of blacks have *lower* housing values in Mississippi but *higher* housing values in California. In Figure 14.6 *best fit* lines have been drawn through the scatters of points. The gradient of this line is calculated as the b_1 parameter of the regression; it is positive in Figure 14.6A and negative in Figure 14.6B, indicating positive and negative trends, respectively. The value where the regression line intersects the *Y*-axis identifies the (hypothetical) median housing value when percent black is zero, and gives us the intercept value b_0. The more general multiple regression case works by extension of this principle, and each of the b parameters gauges the marginal effects of its respective X variable.

There are two lessons to be learned here. First, the relationship we have uncovered varies across space, from state to state, being an example of the general issue termed *spatial heterogeneity* that we discussed briefly in Section 4.2. Traditionally, science has concerned itself with finding patterns and relationships that exist everywhere—with *general* principles and laws that we described in Section 1.3 as *nomothetic*. In that sense what we have uncovered here is something different, a general principle (housing values are correlated with percent black) that varies in its specifics from state to state (a positive relationship in California, a negative one in Mississippi). Recently, geographers have developed a set of techniques that recognize such heterogeneity explicitly and focus not so much on what is true everywhere, but on how things vary across the geographic world. A prominent example is *geographically weighted regression*, a technique originally developed by Stewart Fotheringham, Martin Charlton, and Chris Brunsdon (Box 14.2) at the University of Newcastle-upon-Tyne. Rather than look for a single regression line, it examines how the slope and intercept vary across space in ways that may be related to other factors.

The second lesson to be learned concerns the use of counties to unveil this relationship. Counties in the United States are particularly awkward units of observation, since they vary enormously in size and population and mask variations in space that are often dramatic. The list of 58 counties of California, for example, includes one (Alpine) with a population of about 1,000 and another (Los Angeles) with a population of about 10 million. In Virginia individual cities are often their own counties and may be tiny in comparison with the size of San Bernardino County, California, the largest county in the continental United States. The lesson to be learned here is that the results of this analysis depend intimately on the

units of analysis chosen—that if we had repeated the analysis with other units, such as watersheds, our results might have been entirely different. California reverses the Mississippi trend in this case because the wealthy urban counties of California (San Francisco, Los Angeles, Alameda, etc.) are also the counties where most blacks live, whereas in Mississippi it is the rural counties with low housing values that house most blacks. But this pattern might not hold at all if we were to use watersheds rather than counties as the units of analysis. This issue is known in general as the *modifiable areal unit problem* and is also discussed in Section 6.4.3.

Earlier we described spatial analysis as a set of techniques whose results depend on the locations of the objects of study. But note that in this case the locations of the counties do not affect the results, and latitude and longitude nowhere play a role. Locations are useful for making maps of the inputs, but this analysis hardly rates the term *spatial*. In fact, many types of statistical analysis can be applied to the contents of attribute tables, without ever needing to know the locations or geometric shapes of features. The techniques discussed in the next section address similar questions, but to do so must take advantage of some of the most basic operations of a GIS.

14.2.2 Spatial Joins

In the previous section the variables needed for the analysis were all present in the same attribute table. Often, however, it is necessary to perform some basic operations first to bring the relevant variables together. Suppose, for example, that we wish to conduct a nationwide analysis of the average income of major cities, and have a GIS database containing cities as points, together with some relevant attributes of those cities: the dependent variable average income, plus some independent variables such as percent with college degrees and percent retired people. But other variables that are potentially relevant to the analysis are available only at the state level, including the state's percent unemployed. In Section 10.3 we described a *relational join* or simply a *join*—a fundamental operation in databases that is used to combine the contents of two tables using a common key. In this case the key is "state," a variable that exists in each city record to indicate the state containing the city, and in each state record as an identifier. The result of the join will be that each city record now includes the attributes of the city's containing state, allowing us to add the state-level variables to the analysis. Figure 10.2 shows a simple example of a join. One of the most powerful features of a GIS is the ability to join tables together in this way based on common geographic location. To return to the point made at the outset of this discussion of location, we have now achieved what GIS has always promised—the ability to link together information based on location.

But this example is simple because common location is represented here by a simple key indicating the state containing the city. This spatial relationship was explicitly coded in the city attribute table in this case, but in other cases it will need to be computed. The next two subsections describe circumstances in which the performance of a spatial join is a much more complex operation.

14.2.3 The Point-in-Polygon Operation

Comparing the properties of points to those of containing areas is a common operation in GIS. It occurs in many areas of social science, when an investigator is interested in the extent to which the behavior of an individual is determined by properties of the individual's neighborhood. It occurs when point-like events, such as instances of a disease, must be compared to properties of the surrounding environment. In other applications of GIS, it occurs when a company needs to determine the property on which an asset such as an oil well or a transformer lies. In its simplest form, the point-in-polygon operation determines whether a given point lies inside or outside a given polygon. In more elaborate forms there may be many polygons, and many points, and the task is to assign points to polygons. If the polygons overlap, it is possible that a given point lies in one, many, or no polygons, depending on its location. Figure 14.8 illustrates the task.

Figure 14.8 The point in polygon problem, shown in the continuous-field case (the point must by definition lie in exactly one polygon, or outside the project area). In only one instance (the orange polygon) is there an odd number of intersections between the polygon boundary and a line drawn vertically upward from the point.

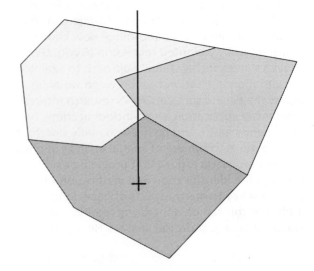

The point-in-polygon operation makes sense from both the discrete-object and the continuous-field perspectives (see Section 3.5 for a discussion of these two perspectives). From a discrete-object perspective both points and polygons are objects, and the task is simply to determine enclosure. From a continuous-field perspective, polygons representing a variable such as land ownership cannot overlap, since each polygon represents the land owned by one owner, and overlap would imply that a point is owned simultaneously by two owners. Similarly from a continuous-field perspective there can be no gaps between polygons. Consequently, the result of a point-in-polygon operation from a continuous-field perspective must assign each point to exactly one polygon.

The point in polygon operation is used to determine whether a point lies inside or outside a polygon.

Although the actual methods used by programmers to perform standard GIS operations are not normally addressed in this book, the standard approach or *algorithm* for the point-in-polygon operation is sufficiently simple and interesting to merit a short description. In essence, it consists of drawing a line from the point to infinity (see Figure 14.8), in this case parallel to the *y*-axis, and determining the number of intersections between the line and the polygon's boundary. If the number is odd, the point is inside the polygon, and if it is even, the point is outside. The algorithm must deal successfully with special cases—for example, if the point lies directly below a vertex (corner point) of the polygon. Some algorithms extend the task to include a third option, when the point lies exactly on the boundary. But others ignore this, on the grounds that it is never possible in practice to determine location with perfect accuracy, and so it is never possible to determine whether an infinitely small point lies on or off an infinitely thin boundary line.

14.2.4 Polygon Overlay

Polygon overlay is similar to the point-in-polygon operation in the sense that two sets of objects are involved, but in this case both are polygons. It exists in two forms, depending on whether a continuous-field or discrete-object perspective is taken. The development of effective algorithms for polygon overlay was one of the most significant challenges of early GIS, and the task remains one of the most complex and difficult to program.

The complexity of computing a polygon overlay was one of the greatest barriers to the development of vector GIS.

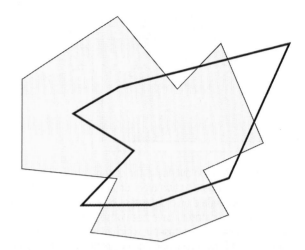

Figure 14.9 Polygon overlay, in the discrete object case. Here the overlay of two polygons produces nine distinct polygons. One has the properties of both polygons, four have the properties of the yellow shaded polygon but not the blue (bounded) polygon, and four are outside the yellow polygon but inside the blue polygon.

From the discrete-object perspective, the task is to determine whether two area objects overlap, to determine the area of overlap, and to define the area formed by the overlap as one or more new area objects (the overlay of two polygons can produce a large number of distinct area objects; see Figure 14.9). This operation is useful for determining answers to such queries as:

- How much of this proposed clearcut lies in this riparian zone?

- How much of the projected catchment area of this proposed retail store lies in the catchment of this other existing store in the same chain?

- How much of this land parcel is affected by this easement?

- What proportion of the land area of the United States lies in areas managed by the Bureau of Land Management?

From the continuous-field perspective the task is somewhat different. Figure 14.10 shows two datasets, both of which are representations of fields: one differentiates areas according to land ownership, and the other differentiates the same region according to land cover class. In the terminology of ESRI's ArcGIS, both datasets are instances of *area coverages*, or fields of nominal variables represented by nonoverlapping polygons. The methods discussed earlier in this chapter could be used to interrogate either dataset separately, but there are numerous queries

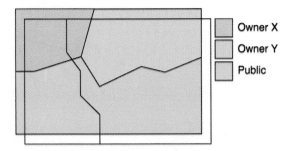

Owner X
Owner Y
Public

Figure 14.10 Polygon overlay in the continuous-field case. Here a dataset representing two types of land cover (A on the left, B on the right) is overlaid on a dataset representing three types of ownership (the two datasets have been offset for visual clarity). The result will be a single dataset in which every point is identified with one land cover type and one ownership type. It will have five polygons, since land cover A intersects with two ownership types, and land cover B intersects with three.

that require simultaneous access to both datasets. For example:

- What is the land cover class, and who is the owner of the point indicated by the user?

- What is the total area of land owned by X and with land cover class A?

- Where are the areas that lie on publicly owned land and have land cover class B?

None of these queries can be answered by interrogating one of the datasets alone—the datasets must somehow be combined so that interrogation can be directed simultaneously at both of them.

The continuous-field version of polygon overlay does this by first computing a new dataset in which the region is partitioned into smaller areas that have uniform characteristics on both variables. Each area in the new dataset will have two sets of attributes—those obtained from one of the input datasets and those obtained from the other. In effect, then, we will have performed a spatial join by creating a new table that combines both sets of attributes, though in this case we will also have created a new set of features. All of the boundaries will be retained, but they will be broken into shorter fragments by the intersections that occur between boundaries in one input dataset and boundaries in the other. Note the unusual characteristics of the new dataset shown in Figure 14.10. Unlike the two input datasets, where boundaries meet in junctions of three lines, the new map contains a new junction of four lines, formed by the new intersection discovered during the overlay process. Because the results of overlay are distinct in this way, it is almost always possible to discover whether a GIS dataset was formed by overlaying two earlier datasets.

Polygon overlay has different meanings from the continuous-field and discrete-object perspectives.

With a single dataset that combines both inputs, it is an easy matter to answer all of the queries listed above through simple interrogation, or to look for relationships between the attributes. It is also easy to reverse the overlay process. If neighboring areas that share the same land cover class are merged, for example, the result is the land ownership map, and vice versa.

Polygon overlay is a computationally complex operation, and as noted earlier much work has gone into developing algorithms that function efficiently for large datasets. One of the issues that must be tackled by a practically useful algorithm is known as the *spurious polygon* or *coastline weave* problem. It is almost inevitable that there will be instances in any practical application where the same line on the ground occurs in both datasets. This happens, for example, when a coastal region is being analyzed because the coastline is almost certain to appear in every dataset of the region. Rivers and roads often form boundaries in many different datasets—a river may function both as a land cover class boundary and as a land ownership boundary, for example. But although the same line is represented in both datasets, its representations will almost certainly not be the same. They may have been digitized from different maps, digitized using different numbers of points, subjected to different manipulations, obtained from entirely different sources (an air photograph and a topographic map, for example), and subjected to different measurement errors. When overlaid, the result is a series of small slivers. Paradoxically, the more care one takes in digitizing or processing, the worse the problem appears, as the result is simply more slivers, albeit smaller in size.

In two vector datasets of the same area, there will almost certainly be instances where lines in each dataset represent the same feature on the ground.

Table 14.1 shows an example of the consequences of slivers and how a GIS can be rapidly overwhelmed if it fails to anticipate and deal with them adequately. Today, a GIS will offer various methods for dealing with the problem, the most common of which is the specification of a *tolerance*. If two lines fall within this distance of each other, the GIS will treat them as a single line, and not create slivers (see also Section 9.3.2.2). The resulting overlay contains just one version of the line, not two. But at least one of the

Table 14.1 Numbers of polygons resulting from an overlay of five datasets, illustrating the spurious polygon problem. The datasets come from the Canada Geographic Information System discussed in Section 1.4.1, and all are representations of continuous fields. Dataset 1 is a representation of a map of soil capability for agriculture, Datasets 2 through 4 are land-use maps of the same area at different times (the probability of finding the same real boundary in more than one such map is very high), and Dataset 5 is a map of land capability for recreation. The final three columns show the numbers of polygons in overlays of three, four, and five of the input datasets.

Acres	1	2	3	4	5	1+2+5	1+2+3+5	1+2+3+4+5
0–1	0	0	0	1	2	2,640	27,566	77,346
1–5	0	165	182	131	31	2,195	7,521	7,330
5–10	5	498	515	408	10	1,421	2,108	2,201
10–25	1	784	775	688	38	1,590	2,106	2,129
25–50	4	353	373	382	61	801	853	827
50–100	9	238	249	232	64	462	462	413
100–200	12	155	152	158	72	248	208	197
200–500	21	71	83	89	92	133	105	99
500–1,000	9	32	31	33	56	39	34	34
1,000–5,000	19	25	27	21	50	27	24	22
>5,000	8	6	7	6	11	2	1	1
Totals	88	2,327	2,394	2,149	487	9,558	39,188	90,599

input lines has been moved, and if the tolerance is set too high the movement can be substantial and can lead to problems later.

14.2.5 Raster Analysis

Many of the complications addressed in the previous subsections disappear if the data are structured in raster form. For example, suppose we are interested in the agricultural productivity of land and have data in the form of a raster, each 10 m by 10 m cell giving the average annual yield of corn in the cell. We might investigate the degree to which these yield values are predictable from other properties of each cell, using rasters of fertilizer quantity applied, depth to water table, percent organic matter, and so forth, each provided for the same set of 10 m by 10 m cells.

In such cases there is no need for complicated overlay operations because all of the attributes are already available for the same set of spatial features, the cells of the raster. Overlay in raster is thus an altogether simpler operation, and this has often been cited as a good reason to adopt raster rather than vector structures. Attributes from different rasters can be readily combined for a variety of purposes, as long as the rasters consist of identically defined arrays of cells. The power of this raster-based approach to GIS is such that it belongs in Chapter 16, where it is discussed in Section 16.2.4 under the heading "Cartographic Modeling and Map Algebra."

In other cases, however, the different variables needed for an analysis may not use identical rasters. For example, it may be necessary to compare data derived from the AVHRR sensor with a cell size of 1 km by 1 km with other data derived from the MODIS sensor with a cell size of 250 m by 250 m—and the two rasters will also likely be at different orientations and may even use different map projections to flatten the Earth. (See Section 9.2.1 for a discussion of satellite remote sensing and Section 5.8 for a discussion of map projections.) In such cases it is necessary to employ some form of *resampling* to transform each dataset to a common raster. Box 14.3 shows a simple example of the use of resampling to join point data. Resampling is a form of spatial interpolation, a technique discussed at length in Section 14.3.6.

14.3 Analysis Based on Distance

The second fundamental spatial concept considered in this chapter is *distance*, the separation of places on the Earth's surface. The ability to calculate and manipulate distances underlies many forms of spatial analysis, some of the most important of which are reviewed in this section. All are based on the concept that the separation of features or events on the Earth's surface can tell us something useful, either about the mechanisms responsible for their presence or properties—in other words, to *explain* their patterns—or as input to

Comparing Attributes When Points Do Not Coincide Spatially

GIS users often encounter situations in which attributes must be compared, but for different sets of features. Figure 14.11 shows a study area with two sets of points, one being the locations where levels of ambient sound were measured using recorders mounted on telephone poles, the other the locations of interviews conducted with local residents to determine attitudes to noise. We would like to know about the relationship between sound and attitudes, but the locations and numbers of cases (the *spatial support*) are different. A simple expedient is to use spatial interpolation (Section 14.3.6), a form of intelligent guesswork that provides estimates of the values of continuous fields at locations where no measurements have been taken. In other words, it *resamples* each field at different points. Given such a method it would be possible to conduct the analysis in any of three ways:

- by interpolating the second dataset to the locations at which the first attribute was measured;

- by the reverse—interpolating the first dataset to the locations at which the second attribute was measured; or

- by interpolating both datasets to a common geometric base, such as a raster.

In the third case, note that it is possible to create a vast amount of data by using a sufficiently detailed raster. In essence, all of these options involve manufacturing information, and the results will depend to some extent on the nature and suitability of the method used to do the spatial interpolation. We normally think of a relationship that is demonstrated with a large number of observations as stronger and more convincing than one based on a small number of observations. But the ability to manufacture data upsets this standard view. The implications of this power of GIS to manufacture data are addressed in Section 15.5.

Figure 14.11 shows a possible solution using this third option. The number of grid points has been determined by the smaller number of cases—in this case, approximately 10 (12 are shown)—to minimize concerns about manufacturing information.

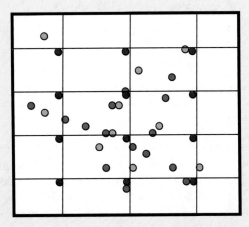

Figure 14.11 Coping with the comparison of two sets of attributes when the respective objects do not coincide. In this instance, attitudes regarding ambient noise have been obtained through a household survey of 15 residents (green dots), and are to be compared to ambient noise levels measured at 10 observation points (brown dots). The solution is to interpolate both sets of data to 12 comparison points (blue dots), using the methods discussed in Section 14.3.6.

decision-making processes. The first subsection looks at the measurement of distance and length in a GIS, as well as some of the issues involved. The second discusses the construction of buffers, together with their use in a wide range of applications. The identification of an anomalous concentration of points in space was the trigger that led to Dr. Snow's work on the causes of cholera transmission, and it is discussed in its general case as *cluster detection* in the third subsection. The concept of *spatial dependence* or *dependence at a distance*, first introduced in Chapter 4 as a fundamental property of geographic data, is given operational meaning in the fourth subsection. The fifth subsection addresses *density estimation*, based on the concept of averaging over defined distances, and the final subsection discusses the related distance-based operation of *spatial interpolation*.

14.3.1 Measuring Distance and Length

A *metric* is a rule for determining distance between points in a space. Several kinds of metrics are used in GIS, depending on the application. The simplest is the rule for determining the shortest distance between two points in a flat plane, called the Pythagorean or straight-line metric. If the two points are defined by the coordinates (x_1, y_1) and (x_2, y_2), then the distance D between them is the length of the hypotenuse of

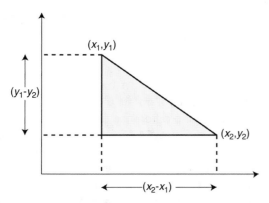

Figure 14.12 Pythagoras's Theorem and the straight-line distance between two points on a plane. The square of the length of the hypotenuse is equal to the sum of the squares of the lengths of the other two sides of the right-angled triangle.

a right-angled triangle (Figure 14.12). Pythagoras's theorem tells us that the square of this length is equal to the sum of the squares of the lengths of the other two sides. So a simple formula results:

$$D = \sqrt{(x_2 - x_1)^2 + (y_2 - y_1)^2}$$

A metric is a rule for determining distance between points in space.

The Pythagorean metric gives a simple and straightforward solution for a plane, if the coordinates x and y are comparable, as they are in any coordinate system based on a projection, such as the UTM or State Plane, or National Grid (see Chapter 5). But the metric will not work for latitude and longitude, reflecting a common source of problems in GIS—the temptation to treat latitude and longitude as if they were equivalent to plane coordinates. This issue is discussed in detail in Section 5.8.1.

For points widely separated on the curved surface of the Earth the assumption of a flat plane leads to significant distortion, and distance must be measured using the metric for a spherical Earth given in Section 5.6 and based on a great circle. For some purposes even this is not sufficiently accurate because of the nonspherical nature of the Earth, and even more complex procedures must be used to estimate distance that take nonsphericity into account. Figure 14.13 shows an example of the differences that the curved surface of the Earth makes when flying long distances.

In many applications the simple rules—the Pythagorean and great circle equations—are not sufficiently accurate estimates of actual travel distance, and we are forced to resort to summing the actual

Figure 14.13 The effects of the Earth's curvature on the measurement of distance, and the choice of shortest paths. The map shows the North Atlantic on the Mercator projection. The red line shows the track made by steering a constant course of 79 degrees from Los Angeles, and is 9807 km long. The shortest path from Los Angeles to London is actually the black line, the trace of the great circle connecting them, with a length of roughly 8800 km. This is typically the route followed by aircraft flying from London to Los Angeles. Flying in the other direction, aircraft may sometimes follow other, longer tracks, such as the red line, if by doing so they can take advantage of jet–stream winds.

lengths of travel routes. In GIS this normally means summing the lengths of links in a network representation, and many forms of GIS analysis use this approach. If a line is represented as a polyline, or a series of straight segments, then its length is simply the sum of the lengths of each segment, and each segment length can be calculated using the Pythagorean formula and the coordinates of its endpoints. But it is worth being aware of two problems with this simple approach.

First, a polyline is often only a rough version of the true object's geometry. A river, for example, never makes sudden changes of direction, and Figure 14.14 shows how smoothly curving streets have to be approximated by the sharp corners of a polyline. Because there is a general tendency for polylines to shortcut corners, *the length of a polyline tends to be shorter than the length of the object it represents*. There are some exceptions, of course—surveyed boundaries are often truly straight between corner points, and streets are often truly straight between intersections. But in general the lengths of linear objects estimated in a GIS—and this includes the lengths of the perimeters of areas represented as polygons—are often substantially shorter than their counterparts on the ground (and note the discussion of fractals in Box 4.6). Note that this is not similarly true of area estimates because shortcutting corners tends to produce both underestimates and overestimates of area, and these tend to cancel out.

A GIS will almost always underestimate the true length of a geographic line.

Second, the length of a line in a two-dimensional GIS representation will always be the length of the line's planar projection, not its true length in three dimensions, and the difference can be substantial if the line is steep (Figure 14.15). In most jurisdictions the area of a parcel of land is the area of its horizontal projection, not its true surface area. A GIS that stores the third dimension for every point is able to calculate both versions of length and area, but not a GIS that stores only the two horizontal dimensions.

14.3.2 Buffering

One of the most important operations available to the GIS user is the calculation of a *buffer* (see also Section 10.5). Given any set of objects, which may include points, lines, or areas, a buffer operation builds a new object or objects by identifying all areas that are within a certain specified distance of the original objects. Figure 14.16 shows instances of a point, a line, and an area, as well as the results of buffering. Buffers have many uses, and they are among the most popular of GIS functions:

- The owner of a land parcel has applied for planning permission to rebuild—the local planning authority could build a buffer around the parcel in

---

Figure 14.14 The polyline representations of smooth curves tend to be shorter in length, as illustrated by this street map (note how curves are replaced by straight-line segments). But estimates of area tend not to show systematic bias because the effects of overshoots and undershoots tend to cancel out to some extent.

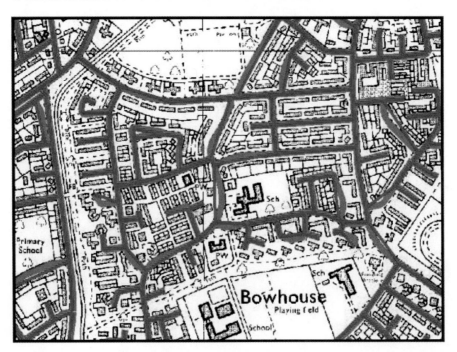

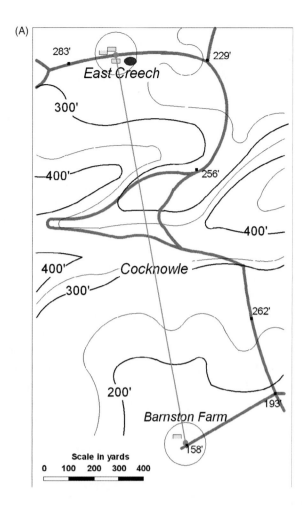

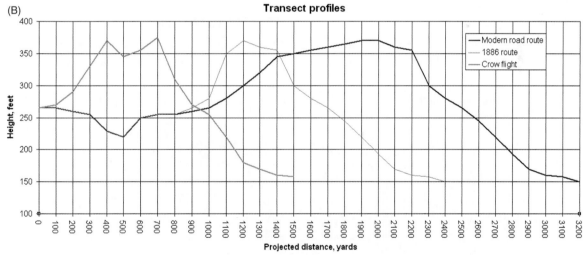

Figure 14.15 The length of a path as traveled on the Earth's surface (red line) may be substantially longer than the length of its horizontal projection as evaluated in a two-dimensional GIS. (A) shows three paths across part of Dorset in the UK. The green path is the straight route, the red path is the modern road system, and the gray path represents the route followed by the road in 1886. (B) Shows the vertical profiles of all three routes, with elevation plotted against the distance traveled horizontally in each case. 1 ft = 0.3048 m, 1 yd = 0.9144 m. (Courtesy Michael De Smith)

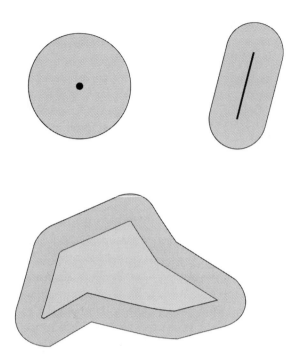

Figure 14.16 Buffers (dilations) of constant width drawn around a point, a polyline, and a polygon.

order to identify all homeowners who live within the legally mandated distance for notification of proposed redevelopments.

- A logging company wishes to clearcut an area but is required to avoid cutting in areas within 100 m of streams; the company could build buffers 100 m wide around all streams to identify these protected riparian areas.

- A retailer is considering developing a new store on a site, of a type that is able to draw consumers from up to 4 km away from its stores. The retailer could build a buffer around the site to identify the number of consumers living within 4 km of the site, in order to estimate the new store's potential sales (see Section 2.3.3.3).

- A new law is passed requiring people once convicted of sexual offenses involving young children to live more than ½ mile (0.8 km) from any school. Figure 14.17 shows the implications of such a law for an area of Los Angeles. The buffers are based on point locations; if the school grounds had been represented as polygons the buffered areas would be larger.

Buffering is possible in both raster and vector GIS. In the raster case, the result is the classification of cells according to whether they lie inside or outside the buffer, while the result in the vector case is a new set of objects (Figure 14.16). But there is an additional possibility in the raster case that makes buffering more useful in some situations. Rather than buffer according to distance, we can ask a raster GIS to *spread* outward from one or more features at rates determined by *friction*, *travel speed*, or *cost* values stored in each cell. This form of analysis is discussed in Section 15.3.2.

Buffering is one of the most useful transformations in a GIS, and is possible in both raster and vector formats.

14.3.3 Cluster Detection

One of the questions most commonly asked about distributions of features, particularly point-like features,

Figure 14.17 Buffers representing ½-mile exclusion zones around all schools in part of Los Angeles.

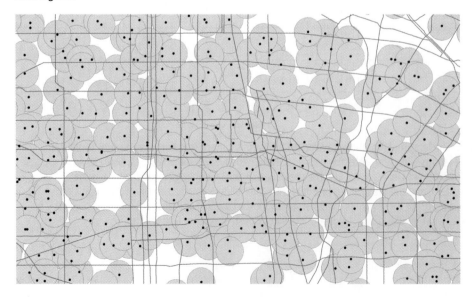

is whether they display a random pattern, in the sense that all locations are equally likely to contain a point, or whether some locations are more likely than others—and particularly, whether the presence of one point makes other points either more or less likely in its immediate neighborhood. This leads to three possibilities:

- The pattern is *random* (points are located independently, and all locations are equally likely).

- The pattern is *clustered* (some locations are more likely than others, and the presence of one point may attract others to its vicinity).

- The pattern is *dispersed* (the presence of one point may make others less likely in its vicinity).

Establishing the existence of clusters is often of great interest, since it may point to possible causal factors, as, for example, with the case of childhood leukemia studied by Stan Openshaw (Figure 14.4). Dispersed patterns are the typical result of competition for space, as each point establishes its own territory and excludes others. Thus such patterns are commonly found among organisms that exhibit territorial behavior, as well as among market towns in rural areas and among retail outlets.

Point patterns can be identified as clustered, dispersed, or random.

It is helpful to distinguish two kinds of processes responsible for point patterns. *First-order* processes involve points being located independently, but may still result in clusters because of varying point density. For example, the drinking-water hypothesis investigated by Dr. John Snow and described in Box 14.1 led to a higher density of points around the pump because of greater access. Similarly, the density of organisms of a particular species may vary over an area because of varying suitability of habitat. *Second-order* processes involve interaction between points, leading to clusters when the interactions are attractive in nature and to dispersion when they are competitive or repulsive. In the cholera case, the contagion hypothesis rejected by Snow is a second-order process and results in clustering even in situations when all other density-controlling factors are perfectly uniform. Unfortunately, as argued in Section 14.1, Snow's evidence did not allow him to resolve with complete confidence between first-order and second-order processes, and in general it is not possible to determine whether a given clustered point pattern was created by varying density factors or by interactions. On the other hand, dispersed patterns can only be created by second-order processes.

Clustering can be produced by two distinct mechanisms, identified as first-order and second-order.

Many tests are available for clusters, and some excellent books have been written on the subject. Only one test will be discussed in this section, to illustrate the method. This is the K function, and unlike many such statistics it provides an analysis of clustering and dispersion over a range of scales. In the interests of brevity the technical details will be omitted, but they can be found in the texts listed at the end of this chapter. They include procedures for dealing with the effects of the study area boundary, which is an important distorting factor for many pattern statistics.

$K(d)$ is defined as the expected number of points within a distance d of an arbitrarily chosen point, divided by the density of points per unit area. When the pattern is random, this number is πd^2, so the normal practice is to plot the function:

$$\hat{L}(d) = \sqrt{K(d)/\pi}$$

since $\hat{L}(d)$ will equal d for all d in a random pattern, and a plot of $\hat{L}(d)$ against d will be a straight line with a slope of 1. Clustering at certain distances is indicated by departures of $\hat{L}$ above the line, and dispersion by departures below the line.

Figures 14.18 and 14.19 show a simple example, used to discover how trees are spaced relative to each other in a forest. The locations of trees are shown in Figure 14.18. In Figure 14.19A locations are analyzed in relation to Tree A. At very short distances

Figure 14.18 Point pattern of individual tree locations. A, B, and C identify the individual trees analyzed in Figure 14.19. (Source: Getis A. and Franklin J. 1987. Second-order neighborhood analysis of mapped point patterns. *Ecology* 68(3): 473–477).

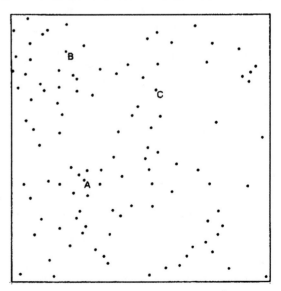

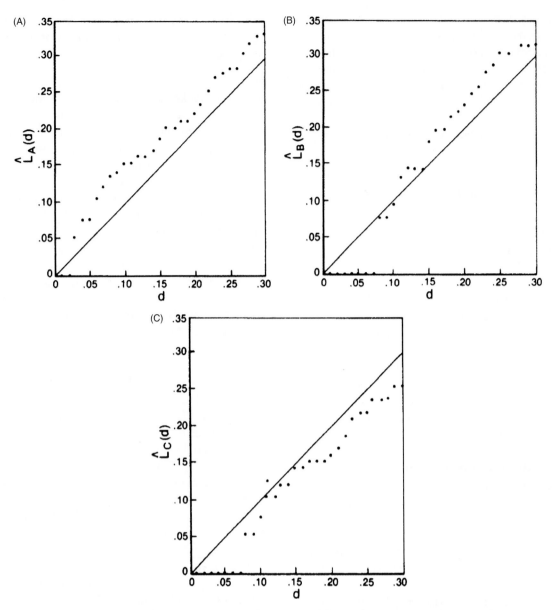

Figure 14.19 Analysis of the local distribution of trees around three reference trees in Figure 14.18 (see text for discussion). (Source: Getis A. and Franklin J. 1987. Second-order neighborhood analysis of mapped point patterns. *Ecology* 68(3): 473–477).

there are fewer trees than would occur in a random pattern, but for most of the range up to a distance of 30% of the width (and height) of the study area there are *more* trees than would be expected, showing a degree of clustering. Tree B (Figure 14.19B) has no nearby neighbors, but shows a degree of clustering at longer distances. Tree C (Figure 14.19C) shows fewer trees than expected in a random pattern over most distances, and it is evident in Figure 14.18 that C is in a comparatively widely spaced area of forest.

14.3.4 Dependence at a Distance

The concept of distance decay—that interactions and similarities decline over space in ways that are often systematic—was introduced in Section 4.5 as a fundamental property of geographic data. It is so fundamental in fact that it is often identified somewhat informally as Tobler's First Law of Geography, first discussed in Chapter 4. This subsection examines techniques for measuring spatial dependence effects—in other words, the ways in which the characteristics of one location are correlated with characteristics of other nearby locations. Underlying this are concepts similar to those examined in the previous subsection on cluster detection—the potential for *hotspots* of anomalously high (or low) values of some property.

One way to look for such effects is to think of the features as fixed and their attributes as displaying interesting or anomalous patterns. Are attribute values randomly distributed over the features, or do extreme values tend to cluster: high values surrounded by high values, and low values surrounded by low values? In such investigations, the processes that determined the locations of features, and were the major concern in the previous section, tend to be ignored and may have nothing to do with the processes that created the pattern of their attributes. For example, the concern might be with some attribute of counties—their average house value, or percent married—and hypotheses about the processes that lead to counties having different values on these indicators. The processes that led to the locations and shapes of counties, on the other hand, which were political processes operating perhaps 100 years ago, would be of no interest. The usefulness, or otherwise, of different areal units was discussed in Section 6.1.

The Moran statistic described in Chapter 4 is designed precisely for this purpose, to indicate the general properties of the pattern of attributes. It distinguishes between positively autocorrelated patterns, in which high values tend to be surrounded by high values and low values by low values; random patterns, in which neighboring values are independent of each other; and dispersed patterns, in which high values tend to be surrounded by low and vice versa. Chapter 4 described various ways of defining the weights needed to calculate the Moran statistic, and also showed how it is possible to use various measures of separation or distance as a basis for the weights. A common expedient, described in Box 4.4, is to use a simple binary indicator of whether or not two areas are adjacent as a surrogate for the distance between them.

The Moran statistic looks for patterns among the attributes assigned to features.

In recent years there has been much interest in going beyond these global measures of spatial dependence to identify dependences locally. Is it possible, for example, to identify individual hotspots, areas where high values are surrounded by high values, and coldspots where low values are surrounded by low? Is it possible to identify anomalies, where high values are surrounded by low or vice versa? Local versions of the Moran statistic are among this group, and along with several others now form a useful resource that is easily implemented in GIS.

Figure 14.20 shows an example, using the Local Moran statistic to differentiate states according to their roles in the pattern of one variable, the median value of housing. The weights have been defined by adjacency, such that pairs of states that share a common

boundary are given a weight of 1 and all other pairs a weight of zero.

14.3.5 Density Estimation

One of the strongest arguments for spatial analysis is that it is capable of addressing *context*, of asking to what extent events or properties at some location are related to or determined by the location's surroundings. Does living in a polluted neighborhood make a person more likely to suffer from diseases such as asthma; does living near a liquor store make binge drinking more likely among young people; can obesity be blamed in part on a lack of nearby parks or exercise facilities? Buffering is one approach, defining a neighborhood as contained within a circle of appropriate radius around the person's location. The liquor store question, for example, could be addressed by building buffers around each individual, and through a point-in-polygon operation determining the number of liquor stores within a defined distance. But this would imply that any liquor store within the buffer was equally important regardless of their distance, and every store outside the buffer was irrelevant. Instead some kind of attenuating effect of distance, such as the options shown in Figure 4.7, would seem to be more appropriate and realistic.

The general term for this kind of technique is *convolution*. A weighting function is chosen, based on a suitable distance decay function, and applied to the nearby features. So the net impact P of a set of liquor stores on a person at a given location **x** might be represented as the sum:

$$P(\mathbf{x}) = \sum_i w_i z_i$$

where i denotes a liquor store, z_i is a measure of the liquor store's relative size or importance, and w_i is a weight that declines with distance according to a distance decay function. Functions such as P defined in this way are often termed *potential* functions and have many uses in spatial analysis. They can be used to measure the potential impact of different population centers on a recreational facility, or the potential expenditures of the surrounding populations at a proposed retail facility. In all such cases the intent is to capture influence *at a distance*.

In effect, P measures the density of features in the neighborhood of **x**. The most obvious example is the estimation of population density, and that example is used in this discussion. But it could be equally well applied to the density of different kinds of diseases, or animals, or any other set of well-defined points. Consider a collection of point objects, such as those

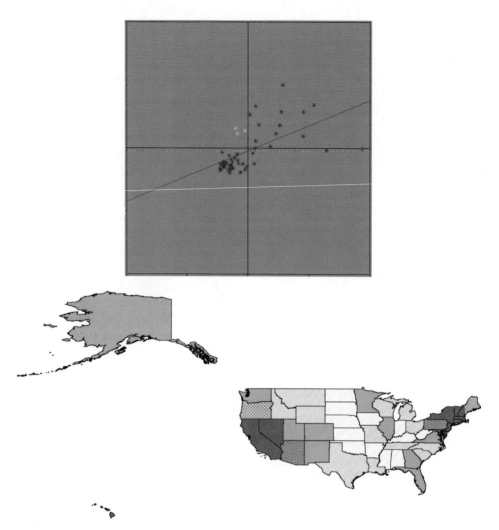

Figure 14.20 The Local Moran statistic, applied using the GeoDa software (available via geodacenter.asu.edu) to describe local aspects of the pattern of housing value among U.S. states. In the map window, the states are colored according to median value, with the darker shades corresponding to more expensive housing. In the scatterplot window, the median value appears on the x-axis, while on the y-axis is the weighted average of neighboring states. The three points colored yellow are instances where a state of below-average housing value is surrounded by states of above-average value. The windows are linked, and the three points are identified as Oregon, Arizona, and Pennsylvania. The global Moran statistic is +0.4011, indicating a general tendency for clustering of similar values.

shown in Figure 14.21. The surface shown in the figure is an example of a *kernel function*, the central idea in density estimation. Any kernel function has an associated length measure, and in the case of the function shown, which is a Gaussian distribution, the length measure is a parameter of the distribution. We can generate Gaussian distributions with any value of this parameter, and they become flatter and wider as the value increases. In density estimation, each point is replaced by its kernel function, and the various kernel functions are added to obtain an aggregate surface, or continuous field of density. If one thinks of each kernel as a pile of sand, then each pile has the same total weight of one unit. The total weight of all

piles of sand is equal to the number of points, and the total weight of sand within a given area, such as the area shown in the figure, is an estimate of the total population in that area. Mathematically, if the population density is represented by a field $\rho(x,y)$, then the total population within area A is the integral of the field function over that area, that is:

$$P = \int_A \rho \, dA$$

A variety of kernel functions are used in density estimation, but the form shown in Figure 14.21B is perhaps the most common. This is the traditional

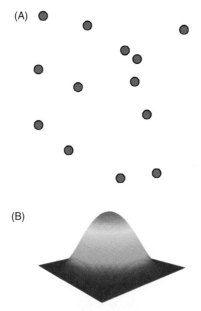

(A)

(B)

Figure 14.21 (A) A collection of point objects, and (B) a kernel function. The kernel's shape depends on a distance parameter—increasing the value of the parameter results in a broader and lower kernel, and reducing it results in a narrower and sharper kernel. When each point is replaced by a kernel and the kernels are added, the result is a density surface whose smoothness depends on the value of the distance parameter.

bell curve or Gaussian distribution of statistics, and is encountered elsewhere in this book in connection with errors in the measurement of position in two dimensions (Section 6.3.2.2). By adjusting the width of the bell, it is possible to produce a range of density surfaces of different amounts of smoothness. Figure 14.22 contrasts two density estimations from the same data, one using a comparatively narrow bell to produce a complex surface and the other using a broader bell to produce a smoother surface.

An important lesson to extract from this discussion concerns the importance of scale. Density is an abstraction, created by taking discrete objects and convolving their distribution with a kernel function. The result depends explicitly on the width of the kernel, which we recognize as a length measure. Change the measure and the resulting density surface changes, as Figure 14.22 shows. Density estimation is just one example of a common phenomenon in geographic data—the importance of scale in the definition of many data types, and hence the importance of knowing what that scale is explicitly when dealing with geographic data.

> **Many geographic data types have a scale built into their definition; it is important therefore to know what that scale is as precisely as possible.**

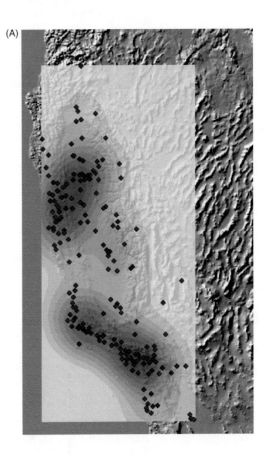

(A)

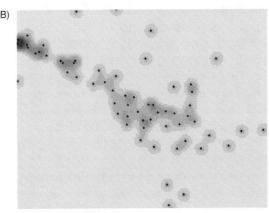

(B)

Figure 14.22 Density estimation using two different distance parameters in the respective kernel functions. In (A), the surface shows the density of ozone-monitoring stations in California, using a kernel radius of 150 km. In (B), zoomed to an area of Southern California, a kernel radius of 16 km is too small for this dataset, as it leaves each kernel isolated from its neighbors.

14.3.6 Spatial Interpolation

Spatial interpolation is a pervasive operation in GIS. Although it is often used explicitly in analysis, it is also used implicitly in various operations such as the preparation of a contour map display, where spatial interpolation is invoked without the user's direct involvement. Spatial interpolation is a process of intelligent guesswork in which the investigator (and the GIS) attempt to make a reasonable estimate of the value of a continuous field at places where the field has not actually been measured. All of the methods use distance, based on the belief that the value at a location is more similar to the values measured at nearby sample points than to the values at distant sample points, a direct use of Tobler's First Law (discussed throughout Chapter 4). Spatial interpolation is an operation that makes sense only from the continuous-field perspective. The principles of spatial interpolation are discussed in Section 4.5; here the emphasis is on practical applications of the technique and on commonly used implementations of the principles.

Spatial interpolation finds applications in many areas:

- In estimating rainfall, temperature, and other attributes at places that are not weather stations, and where no direct measurements of these variables are available.

- In estimating the elevation of the surface between the measured locations of a DEM.

- In *resampling* rasters (Section 14.2.5), the operation that must take place whenever raster data must be transformed to another grid.

- In contouring, when it is necessary to guess where to place contours between measured locations.

In all of these instances spatial interpolation calls for intelligent guesswork, and the one principle that underlies all spatial interpolation is the Tobler Law (Section 3.1): "nearby things are more related than distant things." In other words, the best guess as to the value of a field at some point is the value measured at the closest observation points—the rainfall *here* is likely to be more similar to the rainfall recorded at the nearest weather stations than to the rainfall recorded at more distant weather stations. A corollary of this same principle is that in the absence of better information, it is reasonable to assume that any continuous field exhibits relatively smooth variation; fields tend to vary slowly and to exhibit strong positive spatial autocorrelation, a property of geographic data discussed in Section 4.6.

Spatial interpolation is the GIS version of intelligent guesswork.

In this section three methods of spatial interpolation are discussed, all distance-based: Thiessen polygons; inverse-distance weighting (IDW), which is the simplest commonly used method; and Kriging, a popular statistical method that is grounded in the theory of regionalized variables and falls within the field of *geostatistics*. These methods are discussed at greater length in de Smith, Goodchild, and Longley's *Geospatial Analysis* reader and at www. spatialanalysisonline.com.

14.3.6.1 Thiessen Polygons

Thiessen polygons were suggested by Thiessen as a way of interpolating rainfall estimates from a few rain gauges to obtain estimates at other locations where rainfall had not been measured. The method is very simple: to estimate rainfall at any point, take the rainfall measured at the closest gauge. This leads to a map in which rainfall is constant within polygons surrounding each gauge and changes sharply as polygon boundaries are crossed. Although many GIS users associate polygons defined in this way with Thiessen, they are also known as Voronoi and Dirichlet polygons. They have many other uses besides spatial interpolation:

- Thiessen polygons can be used to estimate the trade areas of each of a set of retail stores or shopping centers.

- They are used internally in the GIS as a means of speeding up certain geometric operations, such as search for a point's nearest neighbor.

- They are the basis of some of the more powerful methods for generalizing vector databases.

As a method of spatial interpolation they leave something to be desired, however, because the sharp change in interpolated values at polygon boundaries is often implausible.

Figure 14.23 shows a typical set of Thiessen polygons. If each pair of points that share a Thiessen polygon boundary is connected, the result is the network of irregular triangles. These are named after Delaunay and are frequently used as the basis for the triangles of a TIN representation of terrain (Chapter 8).

Figure 14.23 Thiessen polygons drawn around each station in part of the Southern California ozone-monitoring network. Note how the polygons, which enclose the area closest to each point, in theory extend off the map to infinity and so must be truncated by the GIS at the edge of the map.

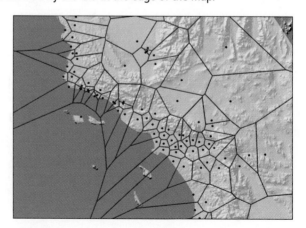

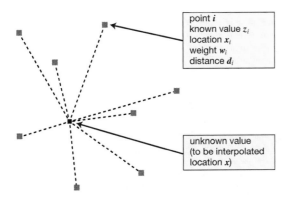

point i
known value z_i
location x_i
weight w_i
distance d_i

unknown value
(to be interpolated
location x)

Figure 14.24 Notation used in the equations defining spatial interpolation.

14.3.6.2 Inverse-Distance Weighting

IDW is the workhorse of spatial interpolation, the method that is most often used by GIS analysts. It employs the Tobler Law by estimating unknown measurements as weighted averages over the known measurements at nearby points, giving the greatest weight to the nearest points.

More specifically, denote the point of interest as $\mathbf{x}$, and the points where measurements were taken as $\mathbf{x}_i$, where i runs from 1 to n, if there are n data points. Denote the unknown value as $z(\mathbf{x})$ and the known measurements as z_i. Give each of these points a weight w_i, which will be evaluated based on the distance d_i from $\mathbf{x}_i$ to $\mathbf{x}$. Figure 14.24 explains this notation with a diagram. Then the weighted average computed at $\mathbf{x}$ is:

$$z(\mathbf{x}) = \sum_i w_i z_i / \sum_i w_i$$

In other words, the interpolated value is an average over the observed values, weighted by the w's. Notice the similarity between this equation and that used to define potential $P(\mathbf{x})$ in the previous subsection. The only difference is the presence here of a denominator, reflecting the nature of IDW as an averaging process rather than a summation.

There are various ways of defining the weights, but the option most often employed is to compute them as the inverse squares of distances—in other words (compare the options discussed in Section 4.5):

$$w_i = 1/d_i^2$$

This means that the weight given to a point drops by a factor of 4 when the distance to the point doubles (or by a factor of 9 when the distance triples). In addition, most software gives the user the option of ignoring altogether points that are further than some specified distance away, or of limiting the average to a specified number of nearest points, or of averaging over the closest points in each of a number of direc-

tion sectors. But if these values are not specified the software will assign default values to them.

> **IDW provides a simple way of guessing the values of a continuous field at locations where no measurement is available.**

IDW achieves the desired objective of creating a smooth surface whose value at any point is more like the values at nearby points than the values at distant points. If it is used to determine z at a location where z has already been measured, it will return the measured value because the weight assigned to a point at zero distance is infinite. For this reason IDW is described as an *exact* method of interpolation because its interpolated results honor the data points exactly. (An *approximate* method is allowed to deviate from the measured values in the interests of greater smoothness, a property that is often useful if deviations are interpreted as indicating possible errors of measurement, or local deviations that are to be separated from the general trend of the surface.)

But because IDW is an average it suffers from certain specific characteristics that are generally undesirable. A weighted average that uses weights that are never negative must always return a value that is between the limits of the measured values. No point on the interpolated surface can have an interpolated z that is more than the largest measured z, or less than the smallest measured z. Imagine an elevation surface with some peaks and pits, but suppose that the peaks and pits have not actually been measured, but are merely indicated by the values of the measured points. Figure 14.25 shows a cross section of such a surface. Instead of interpolating peaks and pits as one might expect, IDW produces the kind of result shown in the figure—small pits where there should be peaks, and small peaks where there should be pits. This behavior is often obvious in GIS output that has been generated using IDW. A related problem concerns extrapolation: if a trend is indicated by the data, as shown in Figure 14.25, IDW will inappropriately indicate a regression to the mean outside the area of the data points.

> **IDW interpolation may produce counterintuitive results in areas of peaks and pits, and outside the area covered by the data points.**

In short, the results of IDW are not always what one would want. There are many better methods of spatial interpolation that address the problems that were just identified, but IDW's ease of programming and its conceptual simplicity make it among the most popular. Users should simply beware, and take care

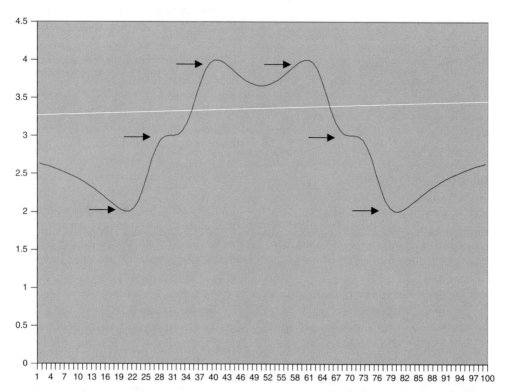

Figure 14.25 Potentially undesirable characteristics of IDW interpolation. Data points located at 20, 30, 40, 60, 70, and 80 have measured values of 2, 3, 4, 4, 3, and 2, respectively. The interpolated profile shows a pit between the two highest values, and regression to the overall mean value of 3 outside the area covered by the data.

to examine the results of interpolation to ensure that they make good sense.

14.3.6.3 Kriging

Of all of the common methods of spatial interpolation it is Kriging that makes the most convincing claim to be grounded in good theoretical principles. The basic idea is to discover something about the general properties of the surface, as revealed by the measured values, and then to apply these properties in estimating the missing parts of the surface. Smoothness is the most important property (note the inherent conflict between this and the properties of fractals, Section 4.7), and it is operationalized in Kriging in a statistically meaningful way. There are many forms of Kriging, and the overview provided here is very brief. Further reading is identified at the end of the chapter.

> **There are many forms of Kriging, but all are firmly grounded in theory.**

Suppose we take a point **x** as a reference and start comparing the values of the field there with the values at other locations at increasing distances from the reference point. If the field is smooth (if the Tobler Law is true, i.e., if there is positive spatial autocorrelation), the values nearby will not be very different—$z(\mathbf{x})$ will not be very different from $z(\mathbf{x}_i)$. To measure the amount, we take the difference and square it, since the sign of the difference is not important: $(z(\mathbf{x}) - z(\mathbf{x}_i))^2$. We could do this with any pair of points in the area.

As distance increases, this measure will likely increase also, and in general a monotonic (consistent) increase in squared difference with distance is observed for most geographic fields. (Note that z must be measured on a scale that is at least interval, though *indicator Kriging* has been developed to deal with the analysis of nominal fields.) In Figure 14.26, each point represents one pair of values drawn from the total set of data points at which measurements have been taken. The vertical axis represents one-half of the squared difference (one-half is taken for mathematical reasons), and the graph is known as the *semivariogram* (or *variogram* for short—the difference of a factor of two is often overlooked in practice, though it is important

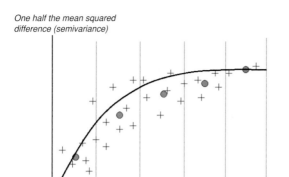

One half the mean squared difference (semivariance)

distance

Figure 14.26 A semivariogram. Each cross represents a pair of points. The solid circles are obtained by averaging within the ranges or *buckets* of the distance axis. The solid line is the best fit to these five points, using one of a small number of standard mathematical functions.

mathematically). To express its contents in summary form, the distance axis is divided into a number of ranges or *bins*, as shown, and points within each range are averaged to define the heavy points shown in the figure.

This semivariogram has been drawn without regard to the *directions* between points in a pair. As such, it is said to be an *isotropic* variogram. Sometimes there is sharp variation in the behavior in different directions, and *anisotropic* semivariograms are created for different ranges of direction (e.g., for pairs in each 90 degree sector).

An anisotropic variogram asks how spatial dependence changes in different directions.

Note how the points of this typical variogram show a steady increase in squared difference up to a certain limit and how that increase then slackens off and virtually ceases. Again, this pattern is widely observed for fields, and it indicates that difference in value tends to increase up to a certain limit, but then to increase no further. In effect, there is a distance beyond which there are no more geographic surprises. This distance is known as the *range*, and the value of difference at this distance as the *sill*.

Note also what happens at the other, lower end of the distance range. As distance shrinks, corresponding to pairs of points that are closer and closer together, the semivariance falls, but there is a suggestion that it never quite falls to zero, even at zero distance. In other words, if two points were sampled a vanishingly small distance apart, they would give different values. This is known as the *nugget* of the semivariogram. A nonzero nugget occurs when there is substantial error in the measuring instrument, such

that measurements taken a very small distance apart would be different due to error, or when there is some other source of local noise that prevents the surface from being truly smooth. Accurate estimation of a nugget depends on whether there are pairs of data points sufficiently close together. In practice, the sample points may have been located at some time in the past, outside the user's control, or may have been spread out to capture the overall variation in the surface, so it is often difficult to make a good estimate of the nugget.

The nugget can be interpreted as the variation among repeated measurements at the same point.

To make estimates using Kriging, we need to reduce the semivariogram to a mathematical function, so that semivariance can be evaluated at any distance, not just at the midpoints of buckets as shown in Figure 14.26. In practice, this means selecting one from a set of standard functional forms and fitting that form to the observed data points to get the best possible fit. This is shown in the figure. The user of a Kriging function in a GIS will have control over the selection of distance ranges and functional forms, and whether a nugget is allowed.

Finally, the fitted semivariogram is used to estimate the values of the field at points of interest. As with IDW, the estimate is obtained as a weighted combination of neighboring values, but the estimate is designed to be the best possible given the evidence of the semivariogram. In general, nearby values are given greater weight, but unlike IDW direction is also important—a point can be *shielded* from influence if it lies behind another point, since the latter's greater proximity suggests greater importance in determining the estimated value, whereas relative direction is unimportant in an IDW estimate. The process of maximizing the quality of the estimate is carried out mathematically, using the precise measures available in the semivariogram.

Kriging responds both to the proximity of sample points and to their directions.

Unlike IDW, Kriging has a solid theoretical foundation, but it also includes a number of options (e.g., the choice of the mathematical function for the semivariogram) that require attention from the user. In that sense it is definitely not a *black box* that can be executed blindly and automatically (Section 2.3.5.5), but instead forces the user to

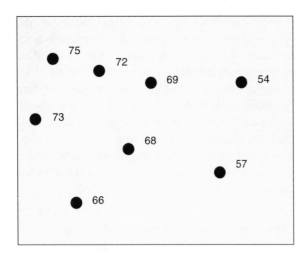

Figure 14.27 A dataset with two possible interpretations: first, a continuous field of atmospheric temperature measured at eight irregularly spaced sample points, and second, eight discrete objects representing cities, with associated populations in thousands. Spatial interpolation makes sense only for the former, and density estimation only for the latter.

become directly involved in the estimation process. For that reason GIS software designers will likely continue to offer several different methods, depending on whether the user wants something that is quick, despite its obvious faults, or better but more demanding of the user.

14.3.6.4 A Final Word of Caution

Spatial interpolation and density estimation are in many ways logical twins: both begin with points and end with surfaces. Moreover we already noted the similarity in the equations for density estimation and IDW. But conceptually the two approaches could not be more different because spatial interpolation seeks to estimate the missing parts of a continuous field from samples of the field taken at data points, while density estimation creates a continuous field from discrete objects. The values interpolated by spatial interpolation have the same measurement scale as the input values, but in density estimation the result is simply a count per unit area.

Figure 14.27 illustrates this difference. The dataset can be interpreted in two sharply different ways. In the first, it is interpreted as sample measurements from a continuous field, and in the second as a collection of discrete objects. In the discrete-object view there is nothing between the objects but empty space—no missing field to be filled in through spatial interpolation. It would make no sense at all to apply spatial interpolation to a collection of discrete objects and no sense at all to apply density estimation to samples of a field.

Density estimation makes sense only from the discrete-object perspective, and spatial interpolation only from the field perspective.

 Conclusion

This chapter has discussed some basic methods of spatial analysis based on two concepts: location and distance. Chapter 15 continues with techniques based on more advanced concepts, and Chapter 16 examines spatial modeling. Several general issues have been raised throughout the discussion: issues of scale and resolution, and accuracy and uncertainty, which are discussed in greater detail in Chapter 6.

Questions for Further Study

1. Did Dr. John Snow actually make his inference strictly from looking at his map? What information can you find on the Web on this issue? (try www.jsi.com)

2. You are given a map showing the home locations of the customers of an insurance agent, and are asked to construct a map showing the agent's market area. Would spatial interpolation or density estimation be more appropriate, and why?

3. What is conditional simulation, and how does it differ from Kriging? Under what circumstances might it be useful?

4. What are the most important characteristics of the three methods of spatial interpolation discussed in this chapter? Using a test dataset of your own choosing, compute and describe the major features of the surfaces interpolated by each.

Further Reading

Bailey, T.C. and Gatrell, A.C. 1995. *Interactive Spatial Data Analysis.* Harlow, UK: Longman Scientific and Technical.

Burrough, P.A. and McDonnell, R.A. 1998. *Principles of Geographical Information Systems.* New York: Oxford University Press.

De Smith, M.J., Goodchild, M.F., and Longley, P.A. 2009. *Geospatial Analysis: A Comprehensive Guide to Principles, Techniques and Software Tools.* Third Edition. Winchelsea: Winchelsea Press. www. spatialanalysisonline.com.

Fotheringham, A.S., Brunsdon, C., and Charlton, M. 2002. *Geographically Weighted Regression: The Analysis of Spatially Varying Relationships.* Hoboken, NJ: Wiley.

Isaaks, E.H. and Srivastava, R.M. 1989. *Applied Geostatistics.* New York: Oxford University Press.

O'Sullivan, D. and Unwin, D.J. 2010. *Geographic Information Analysis* (Second Edition). Hoboken, NJ: Wiley.

Silverman, B.W. 1986. *Density Estimation for Statistics and Data Analysis.* New York: Chapman and Hall.

15 Spatial Analysis and Inference

The second of two chapters on spatial analysis focuses on five areas: analyses that address concepts of area and centrality, analyses of surfaces, analyses that are oriented to design, and statistical inference.

The chapter begins with area-based techniques, including measures of area and shape. Techniques based on the concept of centrality seek to find central or representative points for geographic distributions. Surface techniques include measures of slope, aspect, and visibility, along with the delineation of watersheds and channels. Design is concerned with choosing locations on the Earth's surface for various kinds of activities and with the positioning of points, lines, and areas and associated attributes to achieve defined objectives. Statistical inference concerns the extent to which it is possible to reason from limited samples or study areas to conclusions about larger populations or areas, but is often complicated by the nature of geographic data.

LEARNING OBJECTIVES

After studying this chapter you will understand:

- Methods for measuring properties of areas.

- Measures that can be used to capture the centrality of geographic phenomena.

- Techniques for analyzing surfaces and for determining their hydrologic properties.

- Techniques for the support of spatial decisions and the design of landscapes according to specific objectives.

- Methods for generalizing from samples, and the problems of applying methods of statistical inference to geographic data.

15.1 The Purpose of Area-Based Analyses

One of the ways in which humans simplify geography and address the Earth's infinite complexity is by ascribing characteristics to entire areas rather than to individual points. *Regional* geography relies heavily on this process to make parsimonious descriptions of the geographic world, a practice discussed earlier in Chapter 3 in the context of representation, and in Chapter 4 as one aspect of the nature of geographic data. At a technical level, this results in the creation not of points but of polygons (Section 8.2). In Chapter 6 we discussed some of the issues of uncertainty that arise as a result of this process, when areas are not truly homogeneous or when their boundaries are not precisely known.

15.1.1 Measurement of Area

Many types of interrogation ask for measurements—we might want to know the total area of a parcel of land, or the distance between two points, or the length of a stretch of road—and in principle all of these measurements are obtainable by simple calculations inside a GIS. Comparable measurements by hand from maps can be very tedious and error-prone. In fact, it was the ability of the computer to make accurate evaluations of area quickly that led the Canadian government to fund the development of the world's first GIS, the Canada Geographic Information System, in the mid-1960s (see the brief history of GIS in Section 1.4), despite the primitive state and high costs of computing at that time. Evaluation of area by hand is a messy and soul-destroying business. The *dot-counting* method uses transparent sheets on which randomly located dots have been printed—an area on the map is estimated by counting the number of dots falling within it. In the *planimeter* method a mechanical device is used to trace the area's boundary, and the required measure accumulates on a dial on the machine.

> **Humans have never devised good manual tools for making measurements from maps, particularly measurements of area.**

By comparison, measurement of the area of a digitally represented polygon is trivial and totally reliable. The common procedure or algorithm calculates and sums the areas of a series of trapezia, formed by dropping perpendiculars to the *x*-axis, as shown

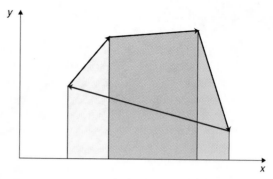

Figure 15.1 The algorithm for calculation of the area of a polygon given the coordinates of the polygon's vertices. The polygon consists of the three black arrows, plus the blue arrow forming the fourth side. Trapezia are dropped from each edge to the *x*-axis, and their areas are calculated as (difference in *x*) times (average of *y*). The trapezia for the first three edges, shown in yellow, orange, and blue, are summed. When the fourth trapezium is formed from the blue arrow, its area is negative because its start point has a larger *x* than its endpoint. When this area is subtracted from the total, the result is the correct area of the polygon.

in Figure 15.1. By making a simple change to the algorithm, it is also possible to use it to compute a polygon's centroid (see Section 15.2.1 for a discussion of centroids; note that the term *centroid* is often used in GIS whenever a polygon is collapsed to a point, whether or not the point is technically the centroid of the polygon). It is advisable to be cautious, however, to ensure that the coordinate system is appropriate. For example, if a polygon's vertices are coded in latitude and longitude, the result of computing area may be a measurement in "square degrees." But except at the equator the length of a degree of longitude varies with latitude and is always less than the length of a degree of latitude. On the other hand, measurement of area using projected coordinates, such as UTM or State Plane (Sections 5.8.2 and 5.8.3), will give a result in square meters or square feet, respectively. Note, however, that neither of these projected coordinate systems has equal-area properties, so the result will be somewhat distorted. Studies that require accurate measures of area should use only equal-area projections.

15.1.2 Measurement of Shape

GIS are also used to characterize the *shapes* of areas. In many countries the system of political representation is based on the concept of districts or constituencies, which are used to define who will vote for each place in the legislature (Box 15.1). In the United States and the UK, and in many other countries that derived their system of representation from the UK, one

Shape and the 12th Congressional District of North Carolina

In 1992, following the release of population data from the 1990 Census, new boundaries were proposed for the voting districts of North Carolina (Figure 15.2). For the first time race was used as an explicit criterion, and districts were drawn that as far as possible grouped minorities (notably blacks) into districts in which they were in the majority. The intent was to avoid the historic tendency for minorities to be thinly spread in all districts, and thus to be unable to return their own representative to Congress. Blacks were in a majority in the new 12th District, but in order to achieve this the district had to be drawn in a highly contorted shape.

The new district, and the criteria used in the redistricting, were appealed to the U.S. Supreme Court. In striking down (rejecting) the new districting scheme, Chief Justice William Rehnquist wrote for the 5–4 majority that "A generalized assertion of past discrimination in a particular industry or region is not adequate because it provides no guidance for a legislative body to determine the precise scope of the injury it seeks to remedy. Accordingly, an effort to alleviate the effects of societal discrimination is not a compelling interest."

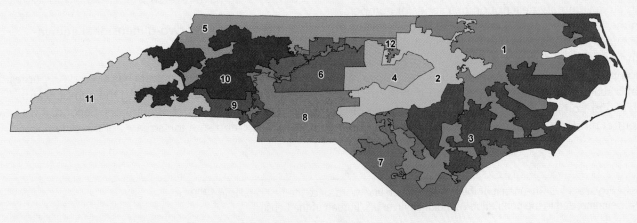

Figure 15.2 The boundaries of the 12th Congressional District of North Carolina drawn in 1992 show a very contorted shape and were appealed to the U.S. Supreme Court. (Source: Copyright Durham Herald Company, Inc., www.herald-sun.com)

place is reserved in the legislature for each district. It is expected that districts will be compact in shape, and the manipulation of a district's shape to achieve certain overt or covert objectives is termed gerrymandering, after an early governor of Massachusetts, Elbridge Gerry (the shape of one of the state's districts was thought to resemble a salamander, with the implication that it had been manipulated to achieve a certain outcome in the voting). The construction of voting districts is an example of the principles of aggregation and zone design discussed in Section 6.2.1.

Anomalous shape is the primary means of detecting gerrymanders of political districts.

Geometric shape was the aspect that alerted Gerry's political opponents to the manipulation of districts, and today shape is measured whenever GIS is used to aid in the drawing of political district boundaries, as must occur by law in the United States after every decennial census. An easy way to define shape is by comparing the perimeter length of an area to its area measure. Normally the square root of area is used, to ensure that the numerator and denominator are both measured in the same units. A common measure of shape or compactness is:

$$S = P/3.54\sqrt{A}$$

where P is the perimeter length and A is the area. The factor 3.54 (twice the square root of π) ensures that the most compact shape, a circle, returns a shape of 1.0, and the most distended and contorted shapes return much higher values.

15.2 Centrality

The topics of generalization and abstraction were discussed earlier in Section 3.8 as ways of reducing the complexity of data. This section reviews a related topic, that of numerical summaries. If we want to describe the nature of summer weather in an area we cite the *average* or *mean*, knowing that there is substantial variation around this value, but that it nevertheless gives a reasonable *expectation* about what the weather will be like on any given day. The mean (the more formal term) is one of a number of measures of *central tendency*, all of which attempt to create a summary description of a series of numbers in the form of a single number. Another is the *median*, the value such that one-half of the numbers are larger and one-half are smaller. Although the mean can be computed only for numbers measured on interval or ratio scales, the median can be computed for ordinal data. For nominal data the appropriate measure of central tendency is the *mode*, or the most common

value. For definitions of nominal, ordinal, interval, and ratio see Box 3.3. Special methods must be used to measure central tendency for cyclic data; they are discussed in texts on directional data, for example, by Mardia and Jupp.

15.2.1 Centers

The spatial equivalent of the mean would be some kind of center, calculated to summarize the positions of a number of points. Early in U.S. history the Bureau of the Census adopted a practice of calculating a representative center for the U.S. population. As agricultural settlement advanced across the West in the Nineteenth Century, the repositioning of the center every 10 years captured the popular imagination. Today, the movement west has slowed and shifted more to the south (Figure 15.3) and by the next census may even have reversed.

Centers are the two-dimensional equivalent of the mean.

The mean of a set of numbers has several properties. First, it is calculated by summing the numbers and dividing by the number of numbers. Second, if we take any value d and sum the squares of the differences

Figure 15.3 The march of the U.S. population westward since the first census in 1790, as summarized in the population centroid. (Source: U.S. Bureau of the Census)

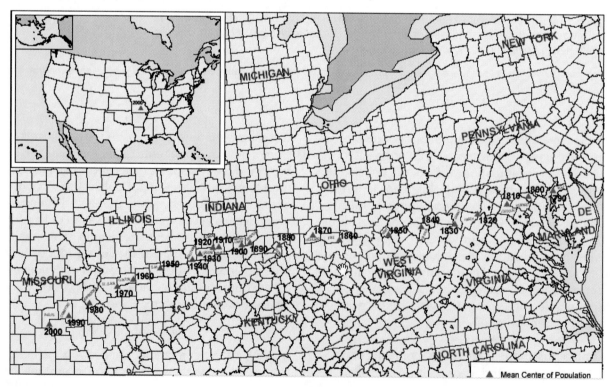

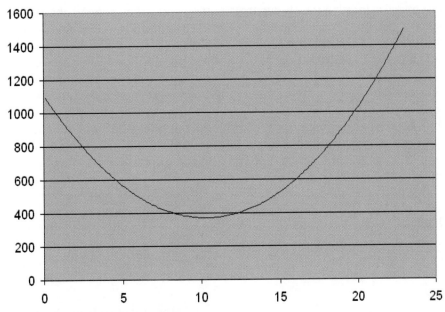

Figure 15.4 Seven points are distributed along a line, at coordinates 1, 3, 5, 11, 12, 18, and 22. The curve shows the sum of distances squared from these points, and how it is minimized at the mean [(1+3+5+11+12+18+22)/7 = 10.3].

between the numbers and *d*, then when *d* is set equal to the mean this sum is minimized (Figure 15.4). Third, the mean is the point about which the set of numbers would balance if we made a physical model such as the one shown in Figure 15.5 and suspended it.

These properties extend easily into two dimensions. Figure 15.6 shows a set of points on a flat plane, each one located at a point (x_i, y_i) and with weight w_i. The *centroid* or *mean center* is found by taking the weighted average of the x and y coordinates:

$$\bar{x} = \sum_i w_i x_i / \sum_i w_i$$

$$\bar{y} = \sum_i w_i y_i / \sum_i w_i$$

It also is the point that minimizes the sum of squared distances, and it is the balance point.

Just like the mean, the centroid is a useful summary of a distribution of points. Although any single centroid may not be very interesting, a comparison of centroids for different sets of points or for different times can provide useful insights.

The centroid is the most convenient way of summarizing the locations of a set of points.

The property of minimizing functions of distance (the square of distance in the case of the centroid) makes centers useful for different reasons. Of particular interest is the location that minimizes the sum of distances, rather than the sum of squared distances, since

Figure 15.6 The centroid or mean center replicates the balance-point property in two dimensions—the point about which the two-dimensional pattern would balance if it were transferred to a weightless, rigid plane and suspended.

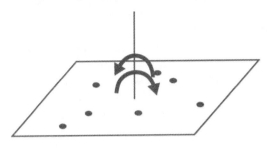

Figure 15.5 The mean is also the balance point, the point about which the distribution would balance if it were modeled as a set of equal weights on a weightless, rigid rod.

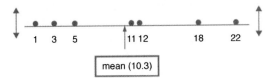

this could be the most effective location for any service that is intended to serve a dispersed population. The point that minimizes total straight-line distance is known as the *point of minimum aggregate travel* or MAT and is discussed in detail in Section 15.4, which is devoted to methods of design.

All of the methods described in this and the following section are based on plane geometry, and simple *x,y* coordinate systems. If the curvature of the Earth's surface is taken into account, the centroid of a set of points must be calculated in three dimensions and will always lie under the surface. More useful perhaps are versions of the MAT and centroid that minimize distance over the curved surface (the minimum total distance and the minimum total of squared distances, respectively, using great circle distances: see Sections 5.7 and 14.3.1).

15.2.2 Dispersion

Central tendency is the obvious choice if a set of measurements must be summarized in a single value, but what if there is the opportunity for a second summary value? Here the measure of choice for measurements with interval or ratio properties is the *standard deviation*, or the square root of the mean squared difference from the mean:

$$s = \sqrt{\sum_i (x_i - \bar{x})^2 / n}$$

where *n* is the number of observations, *s* is the standard deviation, x_i refers to the *i*th observation, and $\bar{x}$ is the mean of the observations. In weighted form the equation becomes:

$$s = \sqrt{\sum_i w_i(x_i - \bar{x})^2 / \sum_i w_i}$$

where w_i is the weight given to the *i*th observation. The *variance*, or the square of the standard deviation (the mean squared difference from the mean), is often encountered, but it is not as convenient a measure for descriptive purposes. Standard deviation and variance are considered more appropriate measures of dispersion than the *range* (the difference between the highest and lowest numbers) because as averages they are less sensitive to the specific values of the extremes.

The standard deviation has also been encountered in Section 6.3.2.2 in a different guise, as the root mean squared error (RMSE), a measure of dispersion of observations about a true value. Just as in that instance, the Gaussian distribution provides a basis for generalizing about the contents of a sample of numbers, using the mean and standard deviation as the parameters of a simple bell curve. If data follow a Gaussian distribution, then approximately 68% of values lie within one standard deviation of the mean and approximately 5% of values lie outside two standard deviations.

These ideas convert very easily to the two-dimensional case. A simple measure of dispersion in two dimensions is the *mean distance from the centroid*. In some applications it may be desirable to give greater weight to more distant points. For example, if a school is being located, then students living at distant locations are comparatively disadvantaged. They can be given greater weight if each distance is squared, such that a student twice as far away receives four times the weight. This property is minimized by locating the school at the centroid.

Mean distance from the centroid is a useful summary of dispersion.

Measures of dispersion can be found in many areas of GIS. The breadth of the kernel function of density estimation (Section 14.3.5) can be thought of as a measure of how broadly a pile of sand associated with each point is dispersed. RMSE is a measure of the dispersion inherent in positional errors (Section 6.3.2.2).

15.3 Analysis of Surfaces

Continuous fields of elevation provide the basis for many types of analysis in GIS. More generally, any field formed by measurements of an interval or ratio variable, such as air temperature, rainfall, or soil pH, can also be conceptualized as a surface and analyzed using the same set of tools, though the results may make little sense in some cases. This section is devoted first to simple measurement of surface slope and aspect. It then introduces techniques for determining paths over surfaces, watersheds and channels, and finally intervisibility.

The various ways of representing continuous fields were first discussed in Section 3.5.2 and later in greater technical detail in Section 8.2. The techniques discussed in this section begin with a raster representation, termed a digital elevation model (DEM) in the case of terrain representation. In most cases the value recorded will be the elevation at the center of each raster cell, though in some cases it may be the mean elevation over the cell; it is always important to check a dataset's documentation before using it.

The digital elevation model is the most useful representation of terrain in a GIS.

15.3.1 Slope and Aspect

Knowing the exact elevation of a point above sea level is important for some applications, including prediction of the effects of global warming and rising sea levels on coastal cities. For many applications, however, the value of a DEM lies in its ability to produce derivative measures through transformation, specifically measures of slope and aspect, both of which are also conceptualized as fields. Imagine taking a large sheet of plywood and laying it on the Earth's surface so that it touches at the point of interest. The magnitude of steepest tilt of the sheet defines the *slope* at that point, and the direction of steepest tilt defines the *aspect* (Box 15.2).

This sounds straightforward, but it is complicated by a number of issues. First, what if the plywood fails to sit firmly on the surface, but instead pivots, because the point of interest happens to be a peak, or a ridge? In mathematical terms, we say that the surface at this point *lacks a well-defined tangent*, or that the surface at this point is *not differentiable*, meaning that it fails to obey the normal rules of continuous mathematical functions and differential calculus. The surface of the Earth has numerous instances of sharp breaks of slope, rocky outcrops, cliffs, canyons, and deep gullies that defy this simple mathematical approach to slope, and this is one of the issues that led Benoît Mandelbrot to develop his theory of fractals (see Section 4.7).

A simple and satisfactory alternative is to take the view that slope must be measured at a particular resolution. To measure slope at a 30-m resolution, for example, we evaluate elevation at points 30 m apart and compute slope by comparing them (equivalent in concept to using a plywood sheet 30 m across). The value this gives is specific to the 30-m spacing, and a different spacing (or different-sized sheet of plywood) would have given a different result. In other words, *slope is a function of resolution* or scale, and it makes no sense to talk about slope without at the same time

---

Technical Box (15.2)

Calculation of Slope Based on the Elevations of a Point and Its Eight Neighbors

There are many ways of estimating the slope in each cell of a DEM, depending on the equations that are used. Described here is the most common, though by no means the only approach (refer to Figure 15.7 for point numbering):

$$b = (z_3 + 2z_6 + z_9 - z_1 - 2z_4 - z_7)/8D$$
$$c = (z_1 + 2z_2 + z_3 - z_7 - 2z_8 - z_9)/8D$$

where b and c are tan(slope) in the x and y directions, respectively, D is the grid point spacing, and z_i denotes elevation at the ith point, as shown in Figure 15.7. These equations give the four diagonal neighbors of Point 5 only half the weight of the other four neighbors in determining slope at Point 5.

$$\tan(slope) = \sqrt{b^2 + c^2}$$

where *slope* is the angle of slope in the steepest direction.

$$\tan(aspect) = b/c$$

where *aspect* is the angle between the y-axis and the direction of steepest slope, measured clockwise.

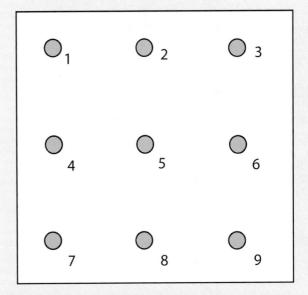

Figure 15.7 Calculation of the slope at Point 5 based on the elevation of it and its eight neighbors.

Since *aspect* varies from 0 to 360, an additional test is necessary that adds 180 to *aspect* if c is positive.

talking about a specific resolution or level of detail. This is convenient because slope is easily computed in this way from a DEM with the appropriate resolution.

The spatial resolution used to calculate slope and aspect should always be specified.

A second issue is the existence of several alternative *measures* of slope, and it is important to know which one is used in a particular software package and application. Slope can be measured as an *angle*, varying from 0 to 90 degrees as the surface ranges from horizontal to vertical. But it can also be measured as a percentage or ratio, defined as *rise over run*, and unfortunately there are two different ways of defining run. Figure 15.8 shows the two options, depending on whether run means the horizontal distance covered between two points, or the diagonal distance (the *adjacent* or the *hypoteneuse* of the right-angled triangle respectively). In the first case (opposite over adjacent), slope as a ratio is equal to the tangent of the angle of slope and ranges from zero (horizontal) through 1 (45 degrees) to infinity (vertical). In the second case (opposite over hypotenuse), slope as a ratio is equal to the sine of the angle of slope and ranges from zero (horizontal) through 0.707 (45 degrees) to 1 (vertical). To avoid confusion we will use the term *slope* only to refer to the measurement in degrees and will call the other options tan(slope) and sin(slope), respectively.

When a GIS calculates slope and aspect from a DEM, it does so by estimating slope at each of the data points of the DEM, by comparing the elevation at that point to the elevations of surrounding points. But the number of surrounding points used in the calculation varies, as do the weights given to each of the surrounding points in the calculation. Box 15.2 shows this idea in practice, using one of the most common methods, which employs eight surrounding points and gives them different weights depending on how far away they are.

15.3.2 Modeling Travel on a Surface

One way to interpret a buffer drawn around a point (Figure 14.16) is that it represents the distance that could be traveled from the point in a given time, assuming constant travel speed. But if travel speed were not uniform, and instead were represented by a continuous field of *friction*, then the buffer would be modified to expand more in some directions than in others. This function is often termed a *spread*, and like many techniques discussed in this section it is an example of a function that is far easier to execute on a raster representation. The total friction associated with a route is calculated by summing the friction values of the cells along the route, and since there are many possible routes from an origin to a destination it is necessary to select that route which minimizes total friction. But first one must select a *move set*, the set of possible moves that can be made between cells along the route. Figure 15.9 shows the two most commonly used move sets. When a diagonal move is made, it is necessary to multiply the associated friction by 1.414 ($\sqrt{2}$) to account for the move's greater length.

Figure 15.10 shows an example DEM, in this case with 30 m point spacing and covering an area roughly corresponding to Orange County, California. Archaeologists are often interested in the impact of

---

Figure 15.8 Three alternative definitions of slope. To avoid ambiguity we use the angle, which varies between 0 and 90 degrees.

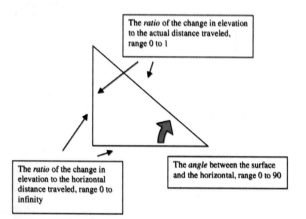

The *ratio* of the change in elevation to the actual distance traveled, range 0 to 1

The *ratio* of the change in elevation to the horizontal distance traveled, range 0 to infinity

The *angle* between the surface and the horizontal, range 0 to 90

Figure 15.9 The rook's-case (left) and queen's-case (right) move sets, defining the possible moves from one cell to another in solving the problem of optimum routing across a friction surface.

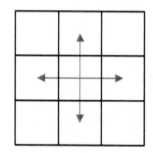
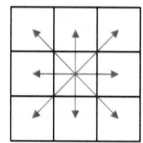

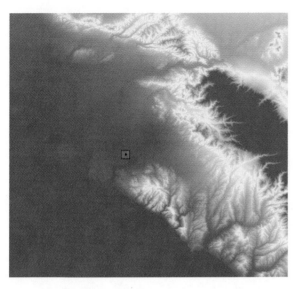

Figure 15.10 A digital elevation model of an area of Southern California, including most of Orange County. High ground is shown in blue. The Pacific Ocean covers the lower left, and the red symbol is located at Santa Ana (John Wayne) Airport.

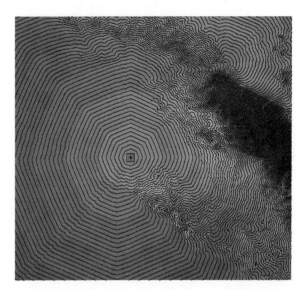

Figure 15.11 Contours of equal travel cost from an origin. Friction or travel "cost" is determined by a combination of ground elevation and ground slope, reflecting the conditions faced by pre-Columbian populations in this part of California. Note the effect of the Santa Ana River gorge in providing easy access to the Inland Empire (upper right).

terrain on travel, since it may help to explain ancient settlement patterns and communication paths. If we assume that travelers will avoid steep slopes and high terrain, then a possible measure of friction would combine slope s and elevation e in an expression such as $s + e/100$ where s is measured in degrees and e in feet. Applying this to the Orange County DEM, Figure 15.11 shows the effects of traveling away from a location near the current Santa Ana airport. Contours show points of equal total friction ("cost") from the start point. Notice how the gorge formed by the Santa Ana River provides an easy route across the high, steep ground and into what is now the Inland Empire.

15.3.3 Computing Watersheds and Channels

A DEM provides an easy basis for predicting how water will flow over a surface, and therefore of many useful hydrologic properties. Consider the DEM shown in Figure 15.12. We assume that water can flow from a cell to any of its eight neighboring cells, down the direction of steepest slope. Since only eight such directions are possible in this raster representation, we assume that water will flow to the lowest of the eight neighbors, provided at least one neighbor is lower. If no neighbor is lower, water is instead assumed to pond, perhaps forming a lake, until it rises high enough to overflow. Figure 15.12A

shows the outflow directions predicted for each cell and its accumulation in the one cell that has no lower neighbor. Since this cell is at the edge of the area, we assume that it will spill over the boundary, forming the area's outflow.

A *watershed* is defined as the area upstream of a point—in other words the area that drains through that point. Once we know the flow directions, it is easy to identify a point and the associated upstream area that forms that point's watershed. Note that any point on the map has an associated watershed and that watersheds therefore may overlap.

This solution represents what a hydrologist would term *overland* flow. When flow accumulates sufficiently, it begins to erode its bed and form a channel. If we could establish an appropriate threshold, expressed in terms of the number of upstream cells that drain through a given cell, then we could map a network of channels. Figure 15.12B shows an example using a threshold of four cells.

In reality some landscapes have closed depressions that fill with water to form lakes. Other landscapes, particularly those developed on soluble rocks such as limestone or gypsum, contain closed depressions that drain underground. But a DEM generated by any of the conventional means will also likely contain elevation errors, some of which will appear as artificial closed depressions. So a GIS will commonly include a

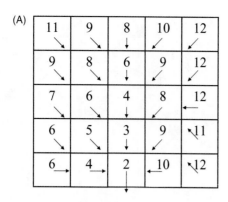

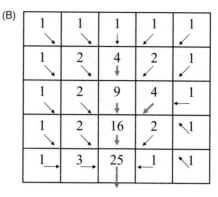

Figure 15.12 Hydrologic analysis of a sample DEM: (A) the DEM and inferred flow directions using the rook's case move set; (B) accumulated flows in each cell and eroded channels based on a threshold flow of 4 units.

routine to "fill" any such "closed depressions," allowing them to overflow and add to the general surface runoff. This filling step will need to be conducted before any useful hydrologic analysis of a landscape can be made.

Figure 15.13 shows the result of applying a filling step and then computing the drainage network for the Orange County DEM, using a channel-forming threshold of 10,000 cells or 9 sq km. The analysis correctly predicts a large river flowing into the top

right of the area shown, through the Santa Ana River gorge. However, rather than flowing south to reach the Pacific below the middle center of the figure as the real river does, the predicted channel flows several kilometers further west before turning south, entering the Pacific near Seal Beach. Interestingly, this is the historic course of the river, before several severe floods in the Nineteenth Century. The current concrete channel, clearly visible in the figure, is delimited by levées, which are

Figure 15.13 Analysis of the Orange County DEM predicts the river channels shown in red in this Google Earth mashup. The Santa Ana River appears to flow out of the gorge shown in the upper right, and then far to the west before emptying into the Pacific near Seal Beach. In reality, it turns south and empties near Newport Beach in the bottom center. See text for explanation.

not large enough to affect the DEM given its 30 m point spacing.

15.3.4 Computing Visibility

One of the most powerful forms of surface analysis centers on the ability to compute intervisibility: can Point A on the surface be seen from Point B? This often takes the form of determining a point's *viewshed*—the area of surface that can be seen by an observer located at the point whose eye is elevated some specified height above the surface. Viewsheds have been used to plan the locations of observation points, transmitters, and mobile-phone towers. They have been used to analyze the locations adopted for prehistoric burial mounds, to test the hypothesis that people sought to be buried at conspicuous points where their remains could dominate the landscape. An interesting GIS analysis by Anne Knowles (Box 15.3) showed that a significant factor in the outcome of the Battle of Gettysburg, a turning point in the American Civil War, was the visibility of the battlefield achieved by the two commanders from their respective headquarters.

Anne Knowles: Illuminating History Through GIS

Anne Kelly Knowles (Figure 15.14) discovered geography while editing thematic maps for a U.S. history textbook, becoming convinced that placing history in its geographic context made the past remarkably vivid. Ever since, Knowles has used geographic methods, particularly mapping and GIS, to unearth the geographic stories embedded in historical sources. Her first study used migration data contained in immigrant obituaries to trace patterns of Welsh settlement in Nineteenth-Century America. Next, she built a historical GIS of nearly 1,000 iron works to determine why it took decades for American iron manufacturers to catch up with their British rivals. With undergraduates at Middlebury College, she is now using GIS to explore the geographies of the Holocaust during World War II, including the

Figure 15.14 Anne Kelly Knowles meeting with Middlebury College research assistants Alexander Yule, Jack Cuneo, and Chester Harvey. The map, by Harvey and Stephanie Ellis, depicts a death march from Auschwitz. (Provided by Anne Kelly Knowles, Middlebury College, Photo taken by Angela Evancie)

development of the Nazi camp system in relation to German manufacturing and the routes and experiences of prisoners evacuated from Auschwitz near the end of the war.

Historical sources are rarely complete or consistent, which makes them challenging for GIS analysis. At the same time, historical GIS is opening new windows on the past by exploiting previously untapped information contained in manuscripts and historical maps. By extracting elevation contours from an 1874 map of the battlefield of Gettysburg, Knowles and Middlebury students were able to create a more historically accurate terrain model than present-day data could provide. Viewshed analysis then made it possible to approximate what General Robert E. Lee could—and could not—see from known viewpoints during the battle (Figure 15.15). To estimate lines of sight and zones of relative safety and danger at Auschwitz, Knowles and her team will run visibility analysis on a digital rendering of the camp that they are building from architectural plans and photographs. In both cases, reconstructing historical environments provides valuable new evidence for exploring the past.

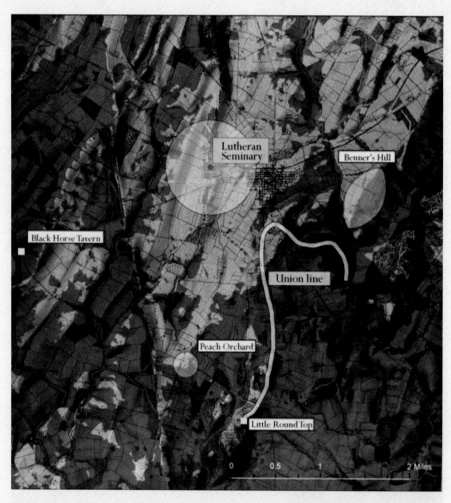

Figure 15.15 This viewshed image, draped over a georectified 1874 map, estimates which parts of the Gettysburg battlefield were probably visible from Lee's viewpoint high in the cupola of the Lutheran Seminary on July 2, 1863. It suggests he saw much more than historians previously thought (tinted white), though little of the hotly contested ground around Little Round Top. (Image created by Caitrin Abshere, Charlie Wirene, and Anne Kelly Knowles; 1874 map courtesy the National Archives and Records Administration.)

Figure 15.16 The area visible from a 200-ft tower located on the hills near Newport Beach, California.

Figure 15.16 shows the viewshed of an observation point located on the low hills near Newport Beach, again using the Orange County DEM. In this case the eye has been elevated 200 ft above the surface, enough to command a view of the flatter areas in the northwest, but still not enough to view areas to the north and east of the higher hills in the area.

15.4 Design

In this section we look at the analysis of spatial data not for the purpose of discovering anomalies, or testing hypotheses about process, as in previous sections, but with the objective of creating improved designs. These objectives might include the minimization of travel distance or the costs of construction of some new development, or the maximization of someone's profit. The three principles of retailing are often said to be *location*, *location*, and *location*, and over the years many GIS applications have been directed at applications that involve, in one way or another, the search for optimum designs. The methods for finding centers described in Section 15.2.1 were shown to have useful design-oriented properties, and the modeling of travel across a surface discussed in Section 15.3.2 can also be interpreted as a design problem, for routing power lines, highways, or tanks. This section includes discussion of a wider selection of these so-called *normative* methods, or methods developed for application to the solution of practical problems of design.

Normative methods apply well-defined objectives to the design of systems.

Design methods are often implemented as components of systems built to support decision making—so-called *spatial-decision support systems*, or SDSS. Complex decisions are often contentious, with many *stakeholders* interested in the outcome and in arguing for one position or another. SDSS are specially adapted GIS that can be used during the decision-making process to provide instant feedback on the implications of various proposals and the evaluation of what-if scenarios. SDSS typically have special user interfaces that present only those functions relevant to the application.

The methods discussed in this section fall into several categories. The next section discusses methods for the optimum location of points and extends the method introduced earlier for the MAT. The second section discusses routing on a network, and its manifestation in the *traveling-salesperson problem* (TSP). The methods also divide between those that are designed to locate points and routes on networks and those designed to locate points and routes in continuous space without respect to the existence of roads or other transportation links.

15.4.1 Point Location

The MAT problem is an instance of location in continuous space and finds the location that minimizes total distance with respect to a number of points. The analogous problem on a network would involve finding that location on the network that minimizes total distance to a number of points, also located on the network, using routes that are constrained to the network. Figure 15.17 shows the contrast between continuous and network views, and Chapter 7 discusses data models for networks.

A very useful theorem first proved by Louis Hakimi reduces the complexity of many location problems on networks. Figure 15.17B shows a typical basis for network location. The links of the network come together in *nodes*. The weighted points are also located on the network and also form nodes. For example, the task might be to find the location that minimizes total distance to a distribution of customers, with all customers aggregated into these weighted points. The weights in this case would be counts of customers. The Hakimi theorem proves that for this problem of minimizing distance the only locations that have to be considered are the nodes; it is impossible for the optimum location to be anywhere else. It is easy

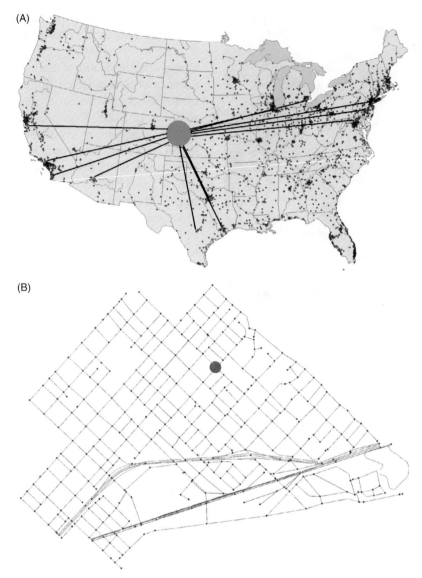

Figure 15.17 Search for the best locations for a central facility to serve dispersed customers. In (A) the problem is solved in continuous space, with straight-line travel, for a warehouse to serve the 12 largest U.S. cities. In continuous space there is an infinite number of possible locations for the site. In (B) a similar problem is solved at the scale of a city neighborhood on a network, where Hakimi's theorem states that only junctions (nodes) in the network and places where there is weight need to be considered, making the problem much simpler, but where travel must follow the street network.

to see why this should be so. Think of a trial point located in the middle of a link, away from any node. Think of moving it slightly in one direction along the link. This moves it toward some weights and away from others, but every unit of movement results in the same increase or decrease in total weighted distance. In other words, the total distance traveled to the location is a linear function of the location along the link. Since the function is linear, it cannot have a minimum midlink, so the minimum must occur at a node.

Optimum location problems can be solved in either discrete or continuous space, depending largely on scale.

The MAT problem on a network is known as the *1-median* problem, and the *p-median* problem seeks optimum locations for any number *p* of central facilities such that the sum of the distances between each weight and the *nearest* facility is minimized. A typical practical application of this problem is in the location of central public facilities, such as libraries, schools, or agency offices, when the objective is to locate for maximum total accessibility.

Many problems of this nature have been defined for different applications and implemented in GIS. While the median problems seek to minimize total distance, the *coverage* problems seek to minimize the *furthest* distance traveled, on the grounds that dealing with the worst case of accessibility is often more attractive than dealing with average accessibility. For example, it may make more sense to a city fire department to locate so that a response is possible to *every* property in less than five minutes, than to worry about minimizing the *average* response time. Coverage problems find applications in the location of emergency facilities, such as fire stations (Figure 15.18), where it is desirable that every possible emergency be covered within a fixed number of minutes of response time, or when the objective is to minimize the worst-case response time, to the furthest possible point.

Figure 15.18 GIS can be used to find locations for fire stations that result in better response times to emergencies. (PhotoDisc/ Getty Images)

All of these problems are referred to as *location-allocation* problems because they involve two types of decisions: where to *locate* and how to *allocate* demand for service to the central facilities. A typical location-allocation problem might involve the selection of sites for supermarkets. In some cases the allocation of demand to sites is controlled by the designer, as it is in the case of school districts when students have no choice of schools. In other cases allocation is a matter of choice, and good designs depend on the ability to predict how consumers will choose among the available options. Models that make such predictions are known as *spatial interaction models*, and their use is an important area of GIS application in market research.

Location-allocation involves two types of decisions: where to locate and how to allocate demand for a service.

15.4.2 Routing Problems

Point-location problems are concerned with the design of fixed locations. Another area of optimization is in routing and scheduling, or decisions about the optimum tracks followed by vehicles. A commonly encountered example is in the routing of delivery vehicles (Box 15.4). These examples show a base location, a depot that serves as the origin and final destination of delivery vehicles; and a series of stops that need to be made. There may be restrictions on the times at which stops must be made. For example, a vehicle delivering home appliances may be required to visit certain houses at certain times, when the residents are home. Vehicle routing and scheduling solutions are used by parcel delivery companies, school buses, on-demand public transport vehicles, and many other applications.

Underlying all routing problems is the concept of the *shortest path*—the path through the network between a defined origin and destination that minimizes distance, or some other measure based on distance, such as travel time. Attributes associated with the network's links, such as length, travel speed, restrictions on travel direction, and level of congestion are often taken into account. Many people are now familiar with the routine solution of the shortest path problem by Web sites such as MapQuest.com (Section 1.5.1.1), which solve many millions of such problems per day for travelers, and by similar technology used by in-vehicle navigation systems. They use standard algorithms developed decades ago, long before

Routing Service Technicians for Sears

Sears manages one of the largest home-appliance repair businesses in the world, with six distinct geographic regions that include 50 independent districts. More than 10,000 technicians throughout the United States complete approximately 11 million in-home service orders each year. Decisions on dividing the daily orders among its technician teams, and on routing each team optimally, used to be made by dispatchers who relied on their own intuition and personal knowledge. But human intuition can be misleading and is lost when dispatchers are sick or retire.

Several years ago Sears teamed with ESRI Inc. (Redlands, CA) to build the Computer-Aided Routing System (CARS) and the Capacity Area Management System (CAMS). CAMS manages the planned capacity of available service technicians assigned to geographic work areas, and CARS provides daily street-level geocoding and optimized routing for the mobile service technicians (Figure 15.19). The mobile Sears Smart Toolbox application provides service technicians with repair information for products, such as schematic diagrams. It also contains a GIS module for mobile mapping and routing, which gives in-vehicle navigation capabilities to assist in finding service locations and minimizing travel time. TeleAtlas ('s-Hertogenbosch, the Netherlands) provides the accurate street data that is critical for supporting geocoding and routing.

The system has had a major impact on Sears' business, saving the business tens of millions of dollars annually.

Figure 15.19 Screenshot of the system used by drivers for Sears to schedule and navigate a day's workload.

the advent of GIS. The path that is strictly shortest is often not suitable because it involves too many turns or uses too many narrow streets, and algorithms will often be programmed to find longer routes that use faster highways, particularly freeways. Routes in Los Angeles, for example, can often be caricatured as (1) shortest route from origin to nearest freeway, (2) follow freeway network, and (3) shortest route from nearest freeway to destination, even though this route may be far from the shortest. The latest generation of route finders is also beginning to accommodate time of day, time of the week, and real-time information on traffic congestion.

The simplest routing problem with multiple destinations is the so-called traveling-salesperson problem, or TSP. In this problem there are a number of places that must be visited in a tour from the depot, and the distances between pairs of places are known. The problem is to select the best tour out of all possible orderings, in order to minimize the total distance traveled. In other words, the optimum is to be selected out of the available tours. If there are *n* places to be visited, including the depot, then there are (*n*-1)! possible tours (the symbol ! indicates the product of the integers from 1 up to and including the number, known as the number's *factorial*). Since it is irrelevant, however, whether any given tour is conducted in a forward or backward direction, the effective number of options is (*n*-1)!/2. Unfortunately, this number grows very rapidly with the number of places to be visited, as Table 15.1 shows.

A GIS can be very effective at solving routing problems because it is able to examine vast numbers of possible solutions quickly.

The TSP is an instance of a problem that becomes quickly unsolvable for large *n*. Instead, designers adopt procedures known as *heuristics*, which are algorithms designed to work quickly and to come close to providing the best answer, while not guaranteeing that the best answer will be found. One not very good heuristic for the TSP is to proceed always to the closest unvisited destination (a so-called *greedy*

approach), and finally to return to the start. Many spatial optimization problems, including location-allocation and routing problems, are solved today by the use of sophisticated heuristics.

15.5 Hypothesis Testing

This last section reviews a major area of statistics—the testing of hypotheses and the drawing of inferences—and its relationship to GIS and spatial analysis. Much work in statistics is *inferential*; that is, it uses information obtained from samples to make general conclusions about a larger population, on the assumption that the sample came from that population. The concept of inference was introduced in Section 4.4 as a way of reasoning about the properties of a larger group from the properties of a sample. At that point several problems associated with inference from geographic data were raised. This section revisits and elaborates on that topic, and discusses the particularly thorny issue of spatial hypothesis testing.

For example, suppose we were to take a random and independent sample of 1,000 people and ask them how they might vote in the next election. By *random and independent*, we mean that every person of voting age in the general population has an equal chance of being chosen and that the choice of one person does not make the choice of any others—parents, neighbors—more or less likely. Suppose that 45% of the sample said they would support Barack Obama. Statistical theory then allows us to give 45% as the best estimate of the proportion who would vote for Obama among the *general* population, and it also allows us to state a *margin of error*, or an estimate of how much the true proportion among the population will differ from the proportion among the sample. A suitable expression of margin of error is given by the 95% confidence limits, or the range within which the true value is expected to lie 19 times out of 20. In other words, if we took 20 different samples, all of size 1000, there would be a scatter of outcomes, and 19 out of 20 of them would lie within these 95% confidence limits. In this case a simple analysis using the *binomial distribution* shows that the 95% confidence limits are 3%; in other words, 19 times out of 20 the true proportion lies between 42 and 48%.

This example illustrates the *confidence limits* approach to inference, in which the effects of sampling are expressed in the form of uncertainty about the properties of the population. An alternative that is

---

Table 15.1 The number of possible tours in a traveling-salesperson problem.

Number of places to visit	Number of possible tours
3	1
4	3
5	12
6	60
7	360
8	2520
9	20160
10	181440

commonly used in scientific reasoning is the *hypothesis-testing* approach. In this case our objective is to test some general statement about the population— for example, that 50% will support Obama in the next election (and 50% will support the other candidate—in other words, there is no real preference in the electorate). We take a sample, and we then ask whether the evidence from the sample supports the general statement. Because there is uncertainty associated with any sample, unless it includes the entire population, the answer is never absolutely certain. In this example and using our confidence limits approach, we know that if 45% were found to support Obama in the sample, and if the margin of error was 3%, it is highly unlikely that the true proportion in the population is as high as 50%. Alternatively, we could state the 50% proportion in the population as a *null hypothesis* (we use the term *null* to reflect the absence of something, in this case a clear choice), and determine how frequently a sample of 1,000 from such a population would yield a proportion as low as 45%. Again the answer is very small; in fact, the probability is 0.0008. But it is not zero, and its value represents the chance of making an error of inference—of rejecting the hypothesis when in fact it is true.

Methods of inference reason from information about a sample to more general information about a larger population.

These two concepts—confidence limits and inferential tests—are the basis for statistical testing and form the core of introductory statistics texts. There is no point in reproducing those introductions here, and the reader is simply referred to them for discussions of the standard tests—F, t, χ^2, and so on. The focus here is on the problems associated with using these approaches with geographic data, in a GIS context. Several problems were already discussed in Section 4.6 as violations of the assumptions of inference from regression. The next section reviews the inferential tests associated with one popular descriptive statistic for spatial data, the Moran index of spatial dependence that was discussed in Section 4.6. The following section discusses the general issues and points to ways of resolving them.

15.5.1 Hypothesis Tests on Geographic Data

Although inferential tests are standard practice in much of science, they are very problematic for geographic data. The reasons have to do with fundamental properties of geographic data, many of which were introduced in Chapter 4, and others have been encountered at various stages in this book.

First, many inferential tests propose the existence of a population, from which the sample has been obtained by some well-defined process. We saw in Section 4.4 how difficult it is to think of a geographic dataset as a sample of all of the datasets that might have been. It is equally difficult to think of a dataset as a sample of some larger area of the Earth's surface, for two major reasons.

First, the samples in standard statistical inference are obtained independently (Section 4.4). But a geographic dataset is often *all there is* in a given area—it *is* the population. Perhaps we could regard a dataset as a sample of a larger area. But in this case the sample would not have been obtained randomly; instead, it would have been obtained by systematically selecting all cases within the area of interest. Moreover, the samples would not have been independent. Because of spatial dependence, which we have understood to be a pervasive property of geographic data, it is very likely that there will be similarities between neighboring observations.

A GIS project often analyzes all the data there is about a given area, rather than a sample.

Figure 15.20 shows a typical instance of geographic sampling. In this case the objective is to explore the relationship between topographic elevation and vegetation cover class, based on a DEM and a map of vegetation, in the area north of Santa Barbara. We might suspect, for example, that certain types of vegetation are encountered only at higher altitudes. One way to do this would be by randomly sampling at a set of points, recording the elevation and vegetation cover class at each, and then examining the results in a routine statistical analysis. In Figure 15.20, roughly 50 sample points are displayed. But why 50? If we increased the number to 500, our statistical test would have more data, and consequently more power to detect and evaluate any relationship. But why stop at 500? Here as in Box 14.3 the geographic case appears to create the potential for endless proliferation of data. Tobler's First Law (Sections 3.1 and 14.3.4) tells us, however, that after a while additional data values will not be real, since they could have been predicted from previously sampled values. We cannot have it both ways—if we believe in spatial interpolation (Section 14.3.6), we cannot at the same time believe in independence of geographic samples, despite the fact that this is a basic assumption of statistical tests. In effect, a geographic area can only yield a limited

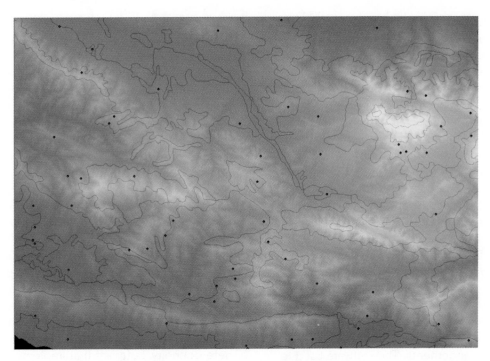

Figure 15.20 A randomly placed sample of points used to examine the relationship between vegetation cover class (delimited by the boundaries shown) and elevation (whiter areas are higher), in an area north of Santa Barbara.

number of truly independent samples, from points spaced sufficiently far apart. Beyond that number, the new samples are not truly independent and in reality add nothing to the power of a test.

Finally, the issue of spatial *heterogeneity* (Section 4.2) also gets in the way of inferential testing. The Earth's surface is highly variable, and there is no such thing as an average place on it. The characteristics observed on one map sheet are likely substantially different from those on other map sheets, even when the map sheets are neighbors. So the census tracts of a city are certainly not acceptable as a random and an independent sample of all census tracts, even the census tracts of an entire nation. They are not independent, and they are not random. Consequently, it is very risky to try to infer the properties of *all* census tracts from the properties of all of the tracts in any *one* area. The concept of sampling, which is the basis for statistical inference, does not transfer easily to the spatial context.

The Earth's surface is very heterogeneous, making it difficult to take samples that are truly representative of any large region.

Before using inferential tests on geographic data, therefore, it is advisable to ask two fundamental questions:

- Can I conceive of a larger *population* that I want to make inferences about?

- Are my data acceptable as a *random* and an *independent* sample of that population?

If the answer to either of these questions is *no*, then inferential tests are not appropriate.

Given these arguments, what options are available? One strategy that is sometimes used is to discard data until the proposition of independence becomes acceptable—until the remaining data points are so far apart that they can be regarded as essentially independent. But no scientist is happy throwing away good data.

Another approach is to abandon inference entirely. In this case the results obtained from the data are descriptive of the study area, and no attempt is made to generalize. This approach, which uses local statistics to observe the *differences* in the results of analysis over space, represents an interesting compromise between the nomothetic and idiographic positions outlined in Section 1.3.1. Generalization is very tempting, but the heterogeneous nature of the Earth's surface makes it very difficult. If generalization is required, then it can be accomplished by appropriate experimental design—by replicating the study in a sufficient number of distinct areas to warrant confidence in a generalization.

Another, more successful approach exploits the special nature of spatial analysis and its concern with detecting pattern. Consider the example in Figure 14.20, where the Moran index was computed at +0.4011, an indication that high values tend to be surrounded by high values, and low values by low values—a positive spatial autocorrelation. It is reasonable to ask whether such a value of the Moran index could have arisen by chance, since even a random arrangement of a limited number of values typically will not give the theoretical value of 0 corresponding to no spatial dependence (actually the theoretical value is very slightly negative). In this case 51 values are involved, arranged over the 51 features on the map (the 50 states plus the District of Columbia). If the values were arranged randomly, how far would the resulting values of the Moran index differ from 0? Would they differ by as much as 0.4011? Intuition is not good at providing answers.

In such cases a simple test can be run by simulating random arrangements. The software used to prepare this illustration includes the ability to make such simulations, and Figure 15.21 shows the results of simulating 999 random rearrangements. It is clear that the actual value is well outside the range of what is possible for random arrangements, leading to the conclusion that the apparent spatial dependence is real.

How does this *randomization* test fit within the normal framework of statistics? The null hypothesis being evaluated is that the distribution of values over the 51 features is random, each feature receiving a value that is independent of neighboring values. The population is the set of all possible arrangements, of which the data represent a sample of one. The test then involves comparing the test statistic—the Moran Index—for the actual pattern against the distribution of values produced by the null hypothesis. So the test is a perfect example of standard hypothesis testing, adapted to the special nature of spatial data and the common objective of discovering pattern.

Randomization tests are uniquely adapted to testing hypotheses about spatial pattern.

Finally, a large amount of research has been devoted to devising versions of inferential tests that cope effectively with spatial dependence and spatial heterogeneity. Software that implements these tests is now widely available, and interested readers are urged to consult the appropriate sources. GeoDa is an excellent comprehensive software environment for such tests (available via geodacenter.asu.edu), and many of these methods are available as extensions of standard GIS packages.

Figure 15.21 Randomization test of the Moran Index computed in Figure 14.20. The histogram shows the results of computing the index for 999 rearrangements of the 51 values on the map. The yellow line on the right shows the actual value, which is very unlikely to occur in a random arrangement, reinforcing the conclusion that there is positive spatial dependence in the data.

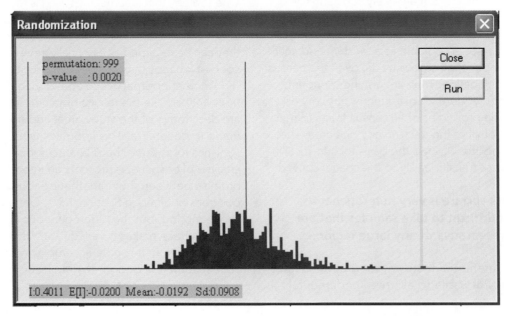

15.6 Conclusion

This chapter has covered the conceptual basis of many of the more sophisticated techniques of spatial analysis that are available in GIS. The last section in particular raised some fundamental issues associated with applying methods and theories that were developed for nonspatial data to the spatial case. Spatial analysis is clearly not a simple and straightforward extension of nonspatial analysis, but instead raises many distinct problems, as well as some exciting opportunities. The two chapters on spatial analysis have only scratched the surface of this large and rapidly expanding field.

Questions for Further Study

1. Parks and other conservation areas have geometric shapes that can be measured by comparing park perimeter length to park area, using the methods reviewed in this chapter. Discuss the implications of shape for park management, in the context of (a) wildlife ecology and (b) neighborhood security.

2. What exactly are *multicriteria* methods? Examine one or more of the methods in the Eastman chapter referenced in Further Reading, summarizing the issues associated with (a) measuring variables to support multiple criteria, (b) mixing variables that have been measured on different scales (e.g., dollars and distances), and (c) finding solutions to problems involving multiple criteria.

3. Besides being the basis for useful summary measures, fractals also provide interesting ways of simulating geographic phenomena and patterns. Browse the Web for sites that offer fractal simulation software, or investigate one of many commercially available packages. What other uses of fractals in GIS can you imagine?

4. Every point on the Earth's surface has an antipodal point—the point that would be reached by drilling an imaginary hole straight through the Earth's center. Britain, for example, is approximately antipodal to New Zealand. If one-third of the Earth's surface is land, you might expect that one-third of all of the land area would be antipodal to points that are also on land, but a quick look at an atlas will show that the proportion is actually far less than that. In fact, the only substantial areas of land that have antipodal land are in South America (and their antipodal points in China). How is spatial dependence relevant here, and why does it suggest that the Earth is not so surprising after all?

Further Reading

De Smith, M.J., Goodchild, M.F., and Longley, P.A. 2009. *Geospatial Analysis: A Comprehensive Guide to Principles, Techniques and Software Tools* (3rd ed.). Winchelsea, UK: Winchelsea Press. www.spatialanalysisonline.com. See particularly Section 6 on Surface and Field Analysis and Section 7 on Network and Location Analysis.

Eastman, J.R. 2005. Multicriteria methods. In Longley, P.A., Goodchild, M.F., Maguire, D.J., Rhind, D.W. (eds.). *Geographical Information Systems: Principles, Techniques, Management and Applications* (abridged ed.). Hoboken, NJ: Wiley, pp. 225–234.

Ghosh, A. and Rushton, G. (ed.) 1987. *Spatial Analysis and Location-Allocation Models*. New York: Van Nostrand Reinhold.

Mardia, K.V. and Jup, P.E. 2000. *Directional Statistics*. New York: Wiley.

O'Sullivan, D. and Unwin, D.J. 2010. *Geographic Information Analysis* (2nd Edition). Hoboken, NJ: Wiley.

16 Spatial Modeling with GIS

Models are used in many different ways, ranging from simulations of how the world works to evaluations of planning scenarios to the creation of indicators. In all of these cases the GIS is used to carry out a series of transformations or analyses of geographic space, either at one point in time or at a number of intervals. The chapter begins with the necessary definitions and presents a taxonomy of models, with examples. It addresses the difference between analysis, the subject of Chapters 14 and 15, and this chapter's focus upon modeling. The alternative software environments for modeling are reviewed, along with capabilities for cataloging and sharing models, which are developing rapidly. The chapter ends with a look into the future of modeling and associated GIS developments.

LEARNING OBJECTIVES

After studying this chapter you will:

- Know what modeling means in the context of GIS.

- Be familiar with the important types of models and their applications.

- Be familiar with the software environments in which modeling takes place.

- Understand the needs of modeling and how these needs are being addressed by current trends in GIS software.

16.1 Introduction

This chapter identifies many of the distinct types of models supported by GIS and gives examples of their applications. After *system* and *object*, *model* is probably one of the most overworked terms in the English language, especially in the language of GIS, with many distinct meanings in the context of GIS and even of this book. So first it is important to address the meaning of the term as it is used in this chapter.

Model is one of the most overworked terms in the English language.

A clear distinction needs to be made between the *data models* discussed in Chapters 3 and 8 and the spatial models that are the subject of this chapter. A data model is a template for data, a framework into which specific details of relevant aspects of the Earth's surface can be fitted, and a set of assumptions about the nature of data. For example, the raster data model forces all knowledge to be expressed as properties of the cells of a regular grid laid on the Earth. A data model is in essence a statement about form or about how the world *looks*, limiting the options open to the data model's user to those allowed by its template. Models in this chapter are expressions of how the world is believed to *work*; in other words they are expressions of process (see Section 1.3.1 on how these both relate to the science of problem solving). They may include dynamic simulation models of natural processes such as erosion, tectonic uplift, the migration of elephants, or the movement of ocean currents. They may include models of social processes, such as residential segregation or the movements of cars on a congested highway. They may include processes designed by humans to search for optimum alternatives, for example, in finding locations for a new retail store. Finally, they may include simple calculations of indicators or predictors, such as happens when layers of geographic information are combined into measures of groundwater vulnerability or social deprivation.

The common element in all of these examples is the manipulation of geographic information in multiple stages. In some cases these stages will perform a simple transformation or analysis of inputs to create an output, and in other cases the stages will loop to simulate the development of the modeled system through time, in a series of iterations. Only when all the loops or iterations are complete will there be a final output. There will be intermediate outputs along the way, and it is often desirable to save some or all of these, in case the model needs to be rerun or if parts of the model need to be changed.

All of the models discussed in this chapter are *digital* or *computational* models, meaning that the operations occur in a computer and are expressed in a language of 0s and 1s. In Chapter 3 representation was seen as a matter of expressing geographic form in 0s and 1s; this chapter looks for ways of expressing geographic *process* in 0s and 1s. The term *geocomputation* is often used to describe the application of computational models to geographic problems.

All of the models discussed in this chapter are also *spatial* models (and deal of course with *geographic* space). There are two key requirements of such a model:

1. There is variation across the space being manipulated by the model (an essential requirement of all GIS applications, of course).

2. The results of modeling change when the locations of objects change—location *matters* (this is also a key requirement of spatial analysis as defined in Section 14.1).

Models do not have to be digital, and it is worth spending a few moments considering the other type, known as *analog*. The term was defined briefly in Section 3.7 as describing a representation, such as a paper map, that is a scaled physical replica of reality. Analog models can be very efficient, and they are widely used to test engineering construction projects and proposed airplanes. They have two major disadvantages relative to digital models, however: they can be expensive to construct and operate, and unlike digital models they are virtually impossible to copy, store, or share. Box 16.1 describes the analog experiments conducted in World War II in support of a famous bombing mission.

An analog model is a scaled physical representation of some aspect of reality.

The level of detail of any analog model is measured by its *representative fraction* (Section 3.7, Box 4.2), the ratio of distance on the model to distance in the real world. Like digital data, computational models do not have a well-defined representative fraction; instead, level of detail is measured as *spatial resolution*, defined as the shortest distance over which change is recorded. *Temporal resolution* is also important, being defined as the shortest time over which change is recorded, and corresponding in the case of many dynamic models to the time interval between iterations.

Modeling the Effects of Bombing on a Dam

In World War II the British conceived a scheme for attacking three dams upstream of the strategically important Ruhr valley, the center of the German iron and steel industry. Dr. Barnes Wallis envisioned a roughly spherical bomb that would bounce on the water upstream of the dam, settling next to the dam wall before exploding. Tests were conducted on scaled models (Figure 16.1) to prove the concept. This approach assumes that all aspects of the real problem scale equally, so that conclusions reached using the scale model are also true in reality. And because this assumption may be shaky, it is advisable to test the conclusions of the model later with full-scale prototypes. The successful attacks by the Royal Air Force's 617 Squadron were immortalized in a 1951 book (*The Dam Busters* by Paul Brickhill) and in a 1954 movie (*The Dam Busters* starring Richard Todd as Wing Commander Guy Gibson).

With modern computer power, much of the experimental work could be done quickly and cheaply using computational models instead of scaled analog models. It would be necessary first to build a digital representation of the system (the dam, the lake, and the bomb), and then to build a digital model of the processes that occur when a bomb is dropped at a shallow angle onto a lake surface, bounces on the surface, and then explodes under water in close proximity to a concrete dam wall.

About Us Plan a Visit   What's On News Collections Exhibitions Education Shop

RAF MUSEUM

You are here: RAF Museum Home / Hendon / Exhibitions / Dam Busters /

Model Dam Experiments

One of the models built across the Bricket Wood Stream in early 1941.
B689

While researching the means to attack Nazi industry, Barnes Wallis was assisted by the Building Research Laboratory.

A team led by Dr Norman Davey built, then blew up, several large model dams to determine the best possible point of detonation to cause maximum destruction.

Unfortunately these tests weren't to everyone's convenience as an Air Ministry press release relates 'allotment holders...were bewildered and annoyed... when a mysterious and sudden onrush of water swept down...and inundated their plots'.

Figure 16.1 An analog model built in 1941 to validate the "bouncing bomb" ideas of Dr. Barnes Wallis, in preparation for the successful Royal Air Force assault on German dams in 1943. (Reproduced by permission of Royal Air Force Museum)

Spatial and temporal resolution are critical factors in models. They define what is left out of the model, in the form of variation that occurs over distances or times that are less than the appropriate resolution. They also therefore define one major source of uncertainty in the model's outcomes. Uncertainty in this context can be best defined through a comparison between the model's outcomes and the outcomes of the real processes that the model seeks to emulate. Any model leaves its user uncertain to some degree about what the real world will do; a measure of uncertainty attempts to give that degree of uncertainty an explicit magnitude. Uncertainty has been discussed in the context of geographic data in Chapter 6; its meaning and treatment in the context of spatial modeling are discussed later in this chapter.

Any model leaves its user uncertain to some degree about what the real world will do.

Spatial and temporal resolution also determine the cost of acquiring the data, because in general it is more costly to collect a fine-resolution representation than a coarse-resolution one. More observations have to be made, and more effort is consumed in making them. They also determine the cost of running the model, since execution time expands as more data have to be processed and as more iterations have to be made. One benchmark is of critical importance for many dynamic models: they must run faster than the processes they seek to simulate, if the results are to be useful for planning. One expects this to be true of computational models, but in practice the amount of computing can be so large that the model simulation slows to unacceptable speed.

Spatial resolution is a major factor in the cost both of acquiring data for modeling, and of actually running the model.

16.1.1 Why Model?

Models are built for a number of reasons. First, a model might be built to support a decision or design process in which the user wishes to find a solution to a spatial problem in support of a decision, perhaps a solution that optimizes some objective. This concept was discussed in Section 15.4. Often the decision or design process will involve multiple criteria, an issue discussed in Section 16.4.

Second, a model might be built to allow the user to experiment on a replica of the world rather than on the real thing. This is a particularly useful approach when the costs of experimenting with the real thing are prohibitive, or when unacceptable impacts would result, or when results can be obtained much faster with a model. Medical students now routinely learn anatomy and the basics of surgery by working with digital representations of the human body rather than with expensive and hard-to-get cadavers. Humanity is currently conducting an unprecedented experiment on the global atmosphere by pumping vast amounts of CO_2 into it. How much better it would have been if we could have run the experiment on a digital replica and understood the consequences of CO_2 emissions before the experiment began.

Experiments embody the notion of *what-if scenarios*, or policy alternatives that can be plugged into a model in order to evaluate their outcomes. The ability to examine such options quickly and effectively is one of the major reasons for modeling (Figure 16.2).

Models allow planners to experiment with what-if scenarios.

Finally, models allow the user to examine dynamic outcomes by viewing the modeled system as it evolves and responds to inputs. As discussed in the next section, such dynamic visualizations are far more compelling and convincing than descriptions of outcomes or statistical summaries when shown to stakeholders and the public. Scenarios evaluated with dynamic models are thus a very effective way of motivating and supporting debates over policies and decisions.

16.1.2 To Analyze or to Model?

Section 2.3.4.2 used the example of emergency evacuation to illustrate GIS applications in the areas of transportation and logistics. The example of evacuating neighborhoods in Santa Barbara, California, illustrated a fundamental difference of approach that explains the relationship between Chapters 14 and 15 on the one hand, and Chapter 16 on the other. Tom Cova's work, illustrated in Figure 2.16, consisted of an analysis of streets and population, with the aim of mapping difficulty of evacuation, measured as the number of people evacuated per lane of road. This analysis was inherently static, capturing what planners might need to know in a single map display. On the other hand, Paul Torrens, in simulating the actual process of growth in the American Mid-West, discussed in Section 2.3.5.1, illustrated the use of a GIS to replicate a dynamic *process*. These simulations allow the researcher to examine what-if scenarios by varying a range of conditions, including zoning controls and new highways. Simulations like these can galvanize a community into action far more effectively than static analysis.

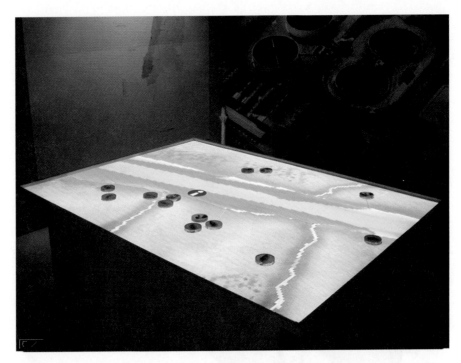

Figure 16.2 A "live" table constructed by Antonio Câmara and his group as a tool for examining alternative planning scenarios. The image on the table is projected from a computer, and users are able to "move" objects around on the image representing sources of pollution by interacting directly with the display. Each new position results in the calculation of a hydrologic model of the impacts of the sources and the display of the resulting pattern of impacts on the table. (Reproduced by permission of Antonio Camara/Ydreams)

Models can be used for dynamic simulation, providing decision makers with dramatic visualizations of alternative futures.

In summary, analysis as described in Chapters 14 and 15 is characterized by

- A static approach, at one point in time.

- The search for patterns or anomalies, leading to new ideas and hypotheses.

- Manipulation of data to reveal what would otherwise be invisible.

By contrast, modeling, as we define it in this chapter, is characterized by

- Multiple stages, perhaps representing different points in time.

- Implementing ideas and hypotheses about the behavior of the real world.

- Experimenting with policy options and scenarios.

16.2 Types of Models

16.2.1 Static Models and Indicators

A static model represents a single point in time and typically combines multiple inputs into a single output. There are no time steps and no loops in a static model, but the results are often of great value as predictors or indicators. For example, the Universal Soil Loss Equation (USLE, first discussed in Section 1.3.1) falls into this category. It predicts soil loss at a point, based on five input variables, by evaluating the equation:

$$A = R \times K \times LS \times C \times P$$

where A denotes the predicted erosion rate, R is the Rainfall and Runoff Factor, K is the Soil Erodibility Factor, LS is the Slope Length Gradient Factor, C is the Crop/Vegetation and Management Factor, and P is the Support Practice Factor. Full definitions of each of these variables and their measurement or estimation can be found in descriptions of the USLE (see, for example, www.co.dane.wi.us/landconservation/uslepg. htm).

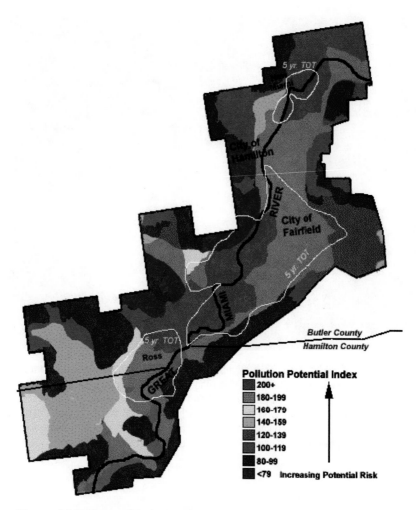

Pollution Potential Index

- ■ 200+
- ■ 180-199
- □ 160-179
- ■ 140-159
- ■ 120-139
- ■ 100-119
- ■ 80-99
- ■ <79 Increasing Potential Risk

Figure 16.3 The results of using the DRASTIC groundwater vulnerability model in an area of Ohio. The model combines GIS layers representing factors important in determining groundwater vulnerability and displays the results as a map of vulnerability ratings. (Reproduced by permission of Hamilton to New Baltimore Grand Water Consortium Pollution Potential (DRASTIC) map reproduced by permission of Ohio Department of Natural Resources)

A static model represents a system at a single point in time.

The USLE passes the first test of a spatial model in that many if not all of its inputs will vary spatially when applied to a given area. But it does not pass the second test, since moving the points at which A is evaluated will not affect the results. Why, then, use a GIS to evaluate the USLE? There are four good reasons: (1) because some of the inputs, particularly LS, require a GIS for their calculation from readily available data, such as digital elevation models; (2) because the inputs and outputs are best expressed, visualized, and used in map form rather than as tables of point observations; (3) because the inputs and outputs of the USLE are often integrated with other types

of data, for further analysis that may require a GIS; and (4) because the data management capabilities of a GIS, and its ability to interface with other systems, may be the best immediately available. Nevertheless, it is possible to evaluate the USLE in a simple spreadsheet application such as Excel; the Web site cited in the previous paragraph includes a downloadable Excel macro for this purpose.

Models that combine a variety of inputs to produce an output are widely used, particularly in environmental modeling. The DRASTIC model calculates an index of groundwater vulnerability from input layers, by applying appropriate weights to the inputs (Figure 16.3). Box 16.2 describes another application, the calculation of a groundwater protection model from several inputs in a karst environment

Building a Groundwater Protection Model in a Karst Environment

Rhonda Pfaff (ESRI staff) and Alan Glennon (a doctoral candidate at the University of California, Santa Barbara) describe a simple but elegant application of modeling to the determination of groundwater vulnerability in Kentucky's Mammoth Cave watershed. Mammoth Cave is protected as a national park, containing extensive and unique environments, but is subject to potentially damaging runoff from areas in the watershed outside park boundaries and therefore not subject to the same levels of environmental protection. Figure 16.4 shows a graphic rendering of the model, in ESRI's ModelBuilder. Each operation is shown as a rectangle, and each dataset as an ellipse. Reading from top left, the model first clips the slope layer to the extent of the watershed, then selects slopes greater than or equal to 5 degrees. A land-use

layer is analyzed to select fields used for growing crops, and these are then combined with the steep-slopes layer to identify crop fields on steep slopes. A dataset of streams is buffered to 300 m, and finally this is combined to form a layer identifying all areas that are crop fields, on steep slopes, within 300 m of a stream. Such areas are particularly likely to experience soil erosion and to generate runoff contaminated with agricultural chemicals, which will then impact the downstream cave environment with its endangered populations of sightless fish. Figure 16.5 shows the resulting map.

A detailed description of this application is available at www.esri.com/news/arcuser/0704/files/modelbuilder.pdf, and the datasets are available at www.esri.com/news/arcuser/0704/summer2004.html.

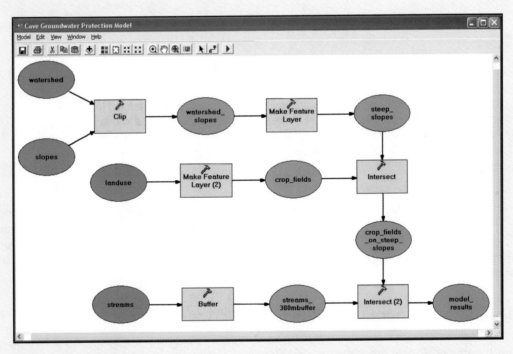

Figure 16.4 Graphic representation of the groundwater protection model developed by Rhonda Pfaff and Alan Glennon for analysis of groundwater vulnerability in the Mammoth Cave watershed, Kentucky.

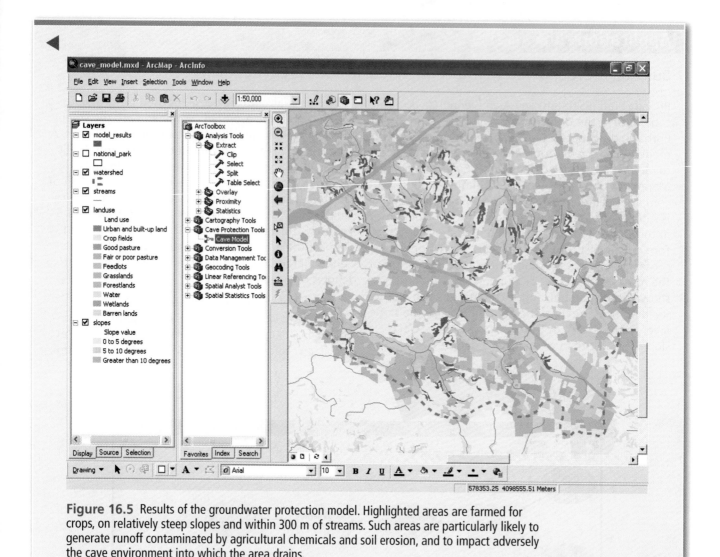

Figure 16.5 Results of the groundwater protection model. Highlighted areas are farmed for crops, on relatively steep slopes and within 300 m of streams. Such areas are particularly likely to generate runoff contaminated by agricultural chemicals and soil erosion, and to impact adversely the cave environment into which the area drains.

(an area underlain by potentially soluble limestone, and therefore having substantial and rapid ground-water flow through cave passages). It uses ESRI's ModelBuilder software, which is described later in Section 16.3.1.

16.2.2 Individual and Aggregate Models

The simulation models used by transportation planners to examine traffic patterns work at the individual level by attempting to forecast the behavior of each driver and vehicle in the study area. By contrast, it would clearly be impossible to model the behavior of every molecule in the Mammoth Cave watershed (Box 16.2). Instead, any modeling of groundwater movement must be done at an aggregate level by predicting the movement of water as a continuous fluid. In general, models of physical systems are forced to adopt aggregate approaches because of the enormous number of

individual objects involved, whereas it is much more feasible to model individuals in human systems, or in studies of animal behavior. Even when modeling the movement of water as a continuous fluid, it is still necessary to break the continuum into discrete pieces, as it is in the representation of continuous fields (see Figure 3.7). Some models adopt a raster approach and are commonly called *cellular* models (see Section 16.2.3). Other models break the world into irregular pieces or polygons, as in the case of the groundwater protection model described in Box 16.2.

> **Aggregate models are used when it is impossible to model the behavior of every individual element in a system.**

Models of individuals are often termed *agent-based* models (ABM) or *autonomous agent* models, implying the existence of discrete agents with defined decision-making behaviors. Each agent might represent an individual, a corporation, or a government agency. With the massive computing power now available on the desktop, together with techniques of object-oriented programming, it is comparatively easy to build and execute ABMs, even when the number of agents is itself massive. Such models have been used to analyze and simulate many types of animal behavior, as well as the behavior of pedestrians in streets (Box 16.3), shoppers in stores, and drivers on congested roads. They have also been used to model the behavior of decision makers, not through their movements in space but through the decisions they make regarding spatial features. ABMs have been constructed to model the decisions made over land use in rural areas by formulating the rules that govern individual decisions by landowners in response to varying market conditions, policies, and regulations. Of key interest in such models is the impact of these factors on the fragmentation of the landscape, with its implications for wildlife habitat.

For example, Dan Brown (Box 16.4) of the University of Michigan has spent many years developing models of land-use transition, including models of such transitions in southeast Michigan. Figure 16.6 shows another example, the modeling of land-use transition in the Amazon Basin. These models combine knowledge of existing patterns of land use with information about the interests and

Figure 16.6 Simulation of land-cover transition in part of the Amazon Basin. (A) Predictions of a model based on eight years of transitions of individual cells starting with the observed pattern in 1997, using rules that include proximity to roads, changing agricultural conditions, and so on. (B) Observed pattern after eight years of transitions. (Courtesy: Gilberto Câmara, Director, INPE, the Brazilian remote sensing agency)

(A)
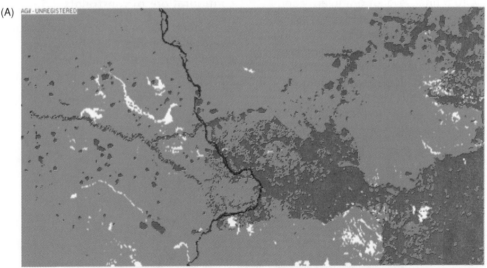

(B)
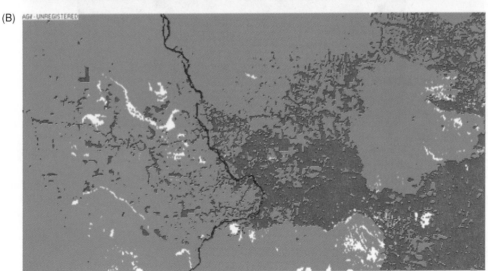

Agent-based Models of Movement in Crowded Spaces

Working at the University College London Centre for Advanced Spatial Analysis, geographer and planner Michael Batty (see Box 6.5) uses GIS to simulate the disasters and emergencies that can occur when large-scale events generate congestion and panic in crowds (Figure 16.7). The events involve large concentrations of people in small spaces that can arise because of accidents, terrorist attacks, or simply the build-up of congestion through the convergence of large numbers of people into spaces with too little capacity. He has investigated scenarios for a number of major events, such as the movement of very large numbers of people to Mecca to celebrate the Hajj (a holy event in the Muslim calendar; Figure 16.7) and the Notting Hill Carnival (Europe's largest street festival, held annually in West Central London; Figure 16.8). His work uses agent-based models that take GIScience to a finer level of granularity and incorporate temporal processes as well as spatial structure. Fundamental to agent-based modeling of such situations is the need to understand how individuals in crowds interact with each other and with the geometry of the local environment.

Batty and his colleagues have been concerned with modeling the behavior of individual human agents under a range of crowding scenarios. This modeling style is called agent-based because it depends on simulating the aggregate effect of the behaviors of individual objects or persons, in an inductive (bottom-up: Section 4.6) way. The modeling is based on averages—on modeling an aggregate outcome (such as the onset of panic)—although the particular behavior of each agent is a combination of responses that are based on how routine behavior is modified by the presence of other agents and by the particular environment in which the agent finds himself or herself. It is possible to simulate a diverse range of outcomes, each based on different styles of interaction between individual agents and different geometrical configurations of constraints.

Figure 16.8 shows one of Batty's simulations of the interactions between two crowds. Here a parade is moving around a street intersection—the central portion of the movement (walkers in white) is the

Figure 16.7 Massive crowds congregate in Mecca during the annual Hajj. On February 1, 2004, 244 pilgrims were trampled when panic stampeded the crowd. This was a unique but unfortunately not a freak occurrence: 50 pilgrims were killed in 2002, 35 in 2001, 107 in 1998, and 1425 were killed in a pedestrian tunnel in 1990. (© Kazuyoshi Nomachi/© Corbis)

Figure 16.8 Simulation of the movement of individuals during a parade. Parade walkers are in white, watchers in red. The watchers (A) build up pressure on restraining barriers and crowd control personnel, and (B) break through into the parade. (Reproduced by permission of Michael Batty)

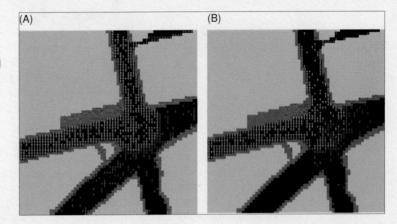

(A) (B)

parade and the walkers around this in gray/red are the watchers. This model can be used to simulate the build-up of pressure through random motion, which then generates a breakthrough of the watchers into the parade, an event that often leads to disasters of the kind experienced in festivals, rock concerts, and football (soccer) matches as well as ritual situations like the Hajj.

Biographical Box (16.4)

Dan Brown, Spatial Modeler

Dan Brown (Figure 16.9) is professor in the School of Natural Resources and Environment at the University of Michigan. He teaches courses on GIS, spatial analysis, and spatial modeling, all in the context of environmental science and management. His research has aimed at understanding human–environment interactions through a focus on land-use and land-cover changes, through modeling these changes, and through spatial analysis and remote sensing for mapping landscape patterns. He chairs the Land Use Steering Group, under the auspices of the U.S. Climate Change Science Program, and has served as a member of the NASA Land Cover and Land Use Change Science Team, on a variety of panels for the National Research Council, NASA, and the National Science Foundation, and on the Editorial Boards for *Computers, Environment and Urban Systems, Landscape Ecology,* and the *Journal of Land Use Science.*

He writes:

> While in college I decided I wanted to work on solving environmental challenges. I took a GIS class and was sold on its power in combining and analyzing interactions between multiple types of information to understand environmental processes and human impacts on them. Beginning in graduate school I became interested in how we can use GIS to address science questions, like why do certain species occur where they do? Or what are the effects of different land-use patterns on environmental processes in different regions? This has led to an interest in modeling. GIS helps us model in two ways: we can develop models that describe the patterns we see, and we can develop models

Figure 16.9 Dan Brown. (UM Photo Services, Scott R. Galvin)

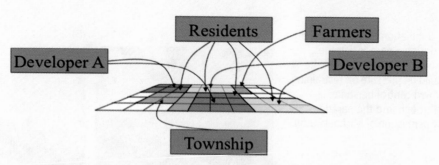

Figure 16.10 Brown's agent-based models represent the actions of multiple actors, including farmers who sell land, townships that buy properties to preserve and set zoning policies, developers who use different subdivision designs, and residents who are looking for a place to live that meets their needs. The models are being used to understand the ecological consequences of land-use change and to evaluate policy interventions.

that describe the processes that we think might be causing those patterns. One of the most exciting challenges is how to combine descriptions of patterns in GIS with descriptions of processes. Lately, I've been working with agent-based models as a way of modeling land-use change (Figure 16.10) because they enable us to represent the attributions, decisions, and actions of multiple actors affecting spatial landscapes.

behaviors of decision makers to project changes in the human and physical landscape into the future. While the accuracy of the predictions (Section 16.5) cannot be guaranteed, models such as these are nevertheless very useful in helping planners, decision makers, and citizens to examine the future impacts of alternative decisions and policy frameworks.

16.2.3 Cellular Models

Cellular models represent the surface of the Earth as a raster. Each cell in the fixed raster has a number of possible states, which change through time as a result of the application of transition rules. Typically, the rules are defined over each cell's neighborhood and determine the outcome of each stage in the simulation based on the cell's state, the states of its neighbors, and the values of cell attributes. The study of cellular models was first popularized by the work of John Conway, professor of mathematics at Princeton, who studied the properties of a model he called the Game of Life (Box 16.5). This has been used by geographer Keith Clarke to model urban growth.

Cellular models represent the surface of the Earth as a raster, each cell having a number of states that are changed at each iteration by the execution of rules.

16.2.4 Cartographic Modeling and Map Algebra

In essence, modeling consists of combining many stages of transformation and manipulation into a single whole, for a single purpose. In the example in Box 16.2, the number of stages was quite small—only six—whereas in the Clarke urban growth model several stages are executed many times, as the model iterates through its annual steps.

The individual stages can consist of a vast number of options, encompassing all of the basic transformations of which GIS is capable. In Section 14.1.2 it was argued that GIS manipulations could be organized into six categories, depending on the broad conceptual objectives of each manipulation. But the six are certainly not the only way of classifying and organizing the numerous methods of spatial analysis into a simple scheme. Perhaps the most successful is the one developed by Dana Tomlin and known as *cartographic modeling* or *map algebra*. It classifies all GIS transformations of rasters into four basic classes, and it is used in several raster-centric GISs as the basis for their analysis languages:

- *Local* operations examine rasters cell by cell, examining the value in a cell in one layer and perhaps comparing it with the values in the same cell in other layers.

- *Focal* operations compare the value in each cell with the values in its neighboring cells—most often eight neighbors.

The Game of Life

The game is played on a raster. Each cell has two states, *live* and *dead*, and there are no additional cell attributes (no variables to differentiate the space), as there would be in GIS applications. Each cell has eight neighbors (the Queen's case, Figure 15.9). There are three transition rules at each time-step in the game:

1. A dead cell with exactly three live neighbors becomes a live cell.

2. A live cell with two or three live neighbors stays alive.

3. In all other cases a cell dies or remains dead.

With these simple rules it is possible to produce an amazing array of patterns and movements, and some surprisingly simple and elegant patterns emerge out of the chaos. (In the field of agent-based modeling these unexpected and simple patterns are known as *emergent properties*.) The page www.math.com/students/wonders/life/life.html includes extensive details on the game, examples of particularly interesting patterns, and executable Java code. Figure 16.11 shows three stages in an execution of the Game of Life.

Several interesting applications of cellular methods have been identified, and particularly

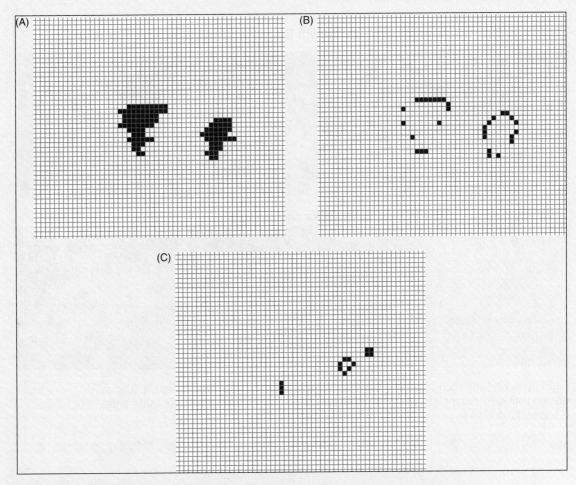

Figure 16.11 Three stages in an execution of the Game of Life: (A) the starting configuration, (B) the pattern after one time-step, and (C) the pattern after 14 time-steps. At this point all features in the pattern remain stable.

outstanding are the efforts to apply them to urban growth simulation. The likelihood of a parcel of land developing depends on many factors, including its slope, access to transportation routes, status in zoning or conservation plans, but above all its proximity to other development. These models express this last factor as a simple modification of the rules of the Game of Life—the more developed the state of neighboring cells, the more likely a cell is to make the transition from undeveloped to developed. Figure 16.12 shows an illustration from one such model, that developed by Keith Clarke and his co-workers at the University of California, Santa Barbara, to predict growth patterns in the Santa Barbara area through 2040. The model iterates on an annual basis, and the effects of neighboring states in promoting infilling are clearly evident. The inputs for the model—the transportation network, the topography, and other factors—are readily prepared in GIS, and the GIS is used to output the results of the simulation.

One of the most important issues in such modeling is calibration and validation—how do we know the rules are right, and how do we confirm the accuracy of the results? Clarke's group has calibrated the model by *brute force*, that is, by running the model forward from some point in the past under different sets of rules, and comparing the results to the actual development history from then to the present. This method is extremely time-consuming, since the model has to be run under vast numbers of combinations of rules, but it provides at least some degree of confidence in the results. The issue of accuracy is addressed in more detail, and with reference to modeling in general, later in this chapter.

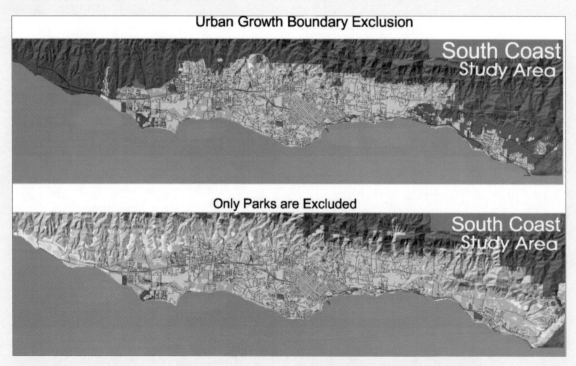

Figure 16.12 Simulation of future urban growth patterns in Santa Barbara, California. (Upper) Growth limited by current urban growth boundary. (Lower) growth limited only by existing parks. (Courtesy Keith Clarke)

- *Global* operations produce results that are true of the entire layer, such as its mean value.

- *Zonal* operations compute results for blocks of contiguous cells that share the same value, such as the calculation of shape for contiguous areas of the same land use, and attach their results to all of the cells in each contiguous block.

With this simple schema, it is possible to express any model, such as Clarke's urban growth simulation, as a series of processing functions, and to compile a sequence of such functions into a *script* in a well-defined language, allowing the sequence to be executed repeatedly with a simple command. The only constraint is that the model inputs and outputs be in raster form.

Map algebra provides a simple language in which to express a model as a script.

A more elaborate map algebra has been devised and implemented in the PCRaster package, developed for spatial modeling at the University of Utrecht (pcraster.geog.uu.nl). In this language, a symbol refers to an entire map layer, so the command $A = B + C$ takes the values in each cell of layers B and C, adds them, and stores the result as layer A.

 ## 16.3 Technology for Modeling

16.3.1 Operationalizing Models in GIS

Many of the ideas needed to implement models at a practical level have already been introduced. In this section they are organized more coherently in a review of the technical basis for modeling.

Models can be defined as sequences of operations, and we have already seen how such sequences can be expressed either as graphic flowcharts or as scripts. One of the oldest graphic platforms for modeling is Stella, now very widely distributed and popular as a means of conceptualizing and implementing models. Although it has good links to GIS, its approach is conceptually nonspatial, so it will not be described in detail here. Instead, the focus will be on similar environments that are directly integrated with GIS. Of these, the first may have been ERDAS's Imagine software (www.gis.leica-geosystems.com/Products/Imagine), which allows the user to build complex modeling sequences from primitive operations, with a focus on the manipulations needed in image processing and remote sensing. ESRI introduced a graphic interface to modeling in ArcView 3.x Spatial Modeler and enhanced it in ArcGIS 9.0 ModelBuilder (see Figure 16.4).

Any model can be expressed as a script, or visually as a flowchart.

In these interfaces datasets are typically represented as ellipses, operations as rectangles, and the sequence of the model as arrows. The user is able to modify and control the operation sequence by interacting directly with the graphic display.

Fully equivalent, but less visual and therefore more demanding on the user, are *scripts*, which express models as sequences of commands, allowing the user to execute an entire sequence by simply invoking a script. Initial efforts to introduce scripting to GIS were cumbersome, requiring the user to learn a product-specific language such as ESRI's AML (Arc Macro Language) or Avenue. Today, it is common for GIS model scripts to be written in such industry-standard languages as Visual Basic, Perl, JScript, or Python. Scripts can be used to execute GIS operations, request input from the user, display results, or invoke other software. With interoperability standards such as Microsoft's COM and .NET (Chapter 7), it is possible to call operations from any compliant package from the same script. For example, a script might combine GIS operations with calls to Excel for simple tabular functions, or even calls to another compliant GIS.

With suitable standards it is possible for a script to call operations from several distinct software packages.

16.3.2 Model Coupling

The previous section described the implementation of models as direct extensions of an underlying GIS, through either graphic model-building or scripts. This approach makes two assumptions: first, that all of the operations needed by the model are available in the GIS (or in another package that can be called by the model), and second, that the GIS provides sufficient performance to handle the execution of the model. In practice, a GIS will often fail to provide adequate performance, especially with very large datasets and large numbers of iterations, because it has been designed as a general-purpose software system, rather than specifically optimized for modeling. Instead, the user is forced to resort to specialized code, written in a lower-level language such as C, C++, C#, or Java. Clarke's model, for example, is programmed in C, and the GIS is used only for preparing input and visualizing output. Other models may be spatial only in the sense of having geographically differentiated inputs and outputs, as discussed in the case of the USLE in Section 16.2.1, making the use of a GIS optional rather than essential.

In reality, therefore, much spatial modeling is done by *coupling* GIS with other software. A model is said to be *loosely coupled* to a GIS when it is run as a separate piece of software, and data are exchanged to and from the GIS in the form of files. Since many GIS formats are proprietary, it is common for the exchange files to be in openly published interchange format, and to rely on translators or direct-read technology at either end of the transfer. In extreme cases it may be necessary to write a special program to convert the files during transfer, when no common format is available. Clarke's is an example of a model that is loosely coupled to GIS. A model is said to be *closely coupled* to a GIS when both the model and the GIS read and write the same files, obviating any need for translation. When the model is executed as a GIS script or through a graphic GIS interface, it is said to be *embedded*. Section 7.3.3 describes the process of GIS customization in more detail.

16.3.3 Cataloging and Sharing Models

In the digital computer everything—the program, the data, the metadata—must be expressed eventually in bits. The value of each bit or byte (Box 3.1) depends on its meaning, and clearly some bits are more valuable than others. A bit in Microsoft's operating systems Windows XP or Vista is clearly extremely valuable, even though there are hundreds of millions of them, since the operating system is used by hundreds of millions of computers worldwide. On the other hand, a bit in a remote-sensing image that happens to have been taken on a cloudy day has almost no value at all, except perhaps to someone studying clouds. In general, one would expect bits of programs to be more valuable than bits of data, especially when those programs are usable by large numbers of people.

This line of argument suggests that the GIS scripts, models, and other representations of process are potentially valuable to many users, and well worth sharing. But almost all of the investment society has made in the sharing of digital information has been devoted to data and text. In Section 11.2 we saw how geolibraries, data warehouses, and metadata standards have been devised and implemented to facilitate sharing of geographic data, and parallel efforts by libraries, publishers, and builders of search engines have made it easy nowadays to share text via the Web. But very little effort to date has gone into making it possible to share *process objects*, digital representations of the process of GIS use.

A process object, such as a script, captures the process of GIS use in digital form.

Notable exceptions include ESRI's ArcScripts, a library of GIS scripts contributed by users and maintained by the vendor as a service to its customers. The library is easily searched for scripts to perform specific types of analysis and modeling. At time of writing it included 4957 scripts, ranging from one to implement the StormWater Management Model (SWMM) to a set of basic tools for crime analysis and mapping. ArcGIS 9.x includes the ability to save, catalog, and share scripts as GIS functions.

The term *GIService* is often used to describe a GIS function being offered by a server, for use by any user connected to the Internet (see Chapters 7 and 11). In essence, GIServices offer an alternative to locally installed GIS, allowing the user to send a request to a remote site instead of to his or her GIS. GIServices are particularly effective in the following circumstances:

- When it would be too complex, costly, or time-consuming to install and run the service locally.

- When the service relies on up-to-date data that would be too difficult or costly for the user to constantly update.

- When the person offering the service wishes to maintain control over his or her intellectual property. This would be important, for example, when the function is based on a substantial amount of knowledge and expertise.

To date, several standard GIS functions are available as GIServices, satisfying one or more of the above requirements. There are several instances on the Web of *gazetteer services*, which allow a user to send a place-name and to receive in return one or more coordinates of places that match the place-name. Several companies offer *geocoding services*, returning coordinates in response to a street address (try www.geocode.com). ESRI's ArcGIS Online (www.arcgisonline.com) provides a searchable catalog to GIServices. Such services are important components of the *service-oriented architecture* concept (Section 11.1), in which a user's needs are satisfied by chaining together sequences of such services.

16.4 Multicriteria Methods

The model developed by Pfaff and Glennon and depicted in Figure 16.4 rates vulnerability to runoff based on three factors—cropland, slope, and distance from stream—but treats all three as simple binary measures and produces a simple binary result. Land is vulnerable if slope is greater than 5%, land use is cropping, and distance from stream is less than 300 m. In reality, of course, 5% is not a precise cutoff between nonvulnerable and vulnerable slopes, and neither is 300 m a precise cutoff between vulnerable distances and nonvulnerable distances. The issues surrounding rules such as these, and their fuzzy alternatives, were discussed at length in Section 6.2.

A more general and powerful conceptual framework for models like this would be constructed as follows. A number of factors influence vulnerability, denoted by x_1 through x_n. The impact of each factor on vulnerability is determined by a transformation of the factor $f(x)$. For example, the factor *distance* would be transformed so that its impact *decreases* with increasing distance (as in Section 4.5), whereas the impact of *slope* would be *increasing*. Then the combined impact of all of the factors is obtained by weighting and adding them, each factor i having a weight w_i:

$$I = \sum_{i=1}^{n} w_i f(x_i)$$

In this framework both the functions f and the weights w need to be determined. In the example the f for slope was resolved by a simple step function, but it seems more likely that the function should be continuously decreasing, as shown in Figure 16.13. The U-shaped function also shown in the figure would be appropriate in cases where the impact of a factor declines in both directions from some peak value (for example, smoke from a tall smokestack has its greatest impact at some distance downwind).

Many decisions depend on identifying relevant factors and adding their appropriately weighted values.

Figure 16.13 Three possible impact functions: (red) the step function used to assess slope in Figure 16.4; (blue) a decreasing linear function; and (black) a function showing impact rising to a maximum and then decreasing.

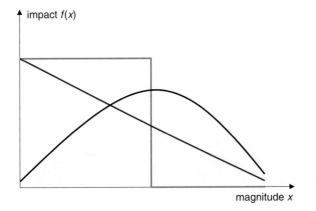

This approach provides a good conceptual framework both for the indicator models typified by Box 16.2 and for many models of design processes. In both cases it is possible that multiple views might exist about appropriate functions and weights, particularly when modeling a decision over an important development with impact on the environment. Different *stakeholders* in the design process can be anticipated to have different views about what is important, how that importance should be measured, and how the various important factors should be combined. Such processes are termed *multicriteria decision making* or *MCDM*, and are commonly encountered whenever decisions are controversial. Some stakeholders may feel that environmental factors deserve high weight, others that cost factors are the most important, and still others that impact on the social life of communities is all-important.

An important maxim of MCDM is that it is better for stakeholders to argue in principle about the merits of different factors and how their impacts should be measured than to argue in practice about alternative decisions. For example, arguing about whether slope is a more important factor than distance, and about how each should be measured, is better than arguing about the relative merits of Solution A, which might color half of John Smith's field red, over Solution B, which might color all of David Taylor's field red. Ideally, all of the controversy should be over once the factors, functions, and weights are decided, and the solution they produce should be acceptable to all, since all accepted the inputs. Would it were so!

Each stakeholder in a decision may have his or her own assessment of the importance of each relevant factor.

Although many GIS have implemented various approaches to MCDM, Clark University's Idrisi offers the most extensive functionality as well as detailed tutorials and examples (www.clarklabs.org). Developed as nonprofit software for GIS and image processing, Idrisi has many tens of thousands of users worldwide. One example of Idrisi's support for MCDM is the Analytical Hierarchy Process (AHP) devised by Thomas Saaty, which focuses on capturing each stakeholder's view of the appropriate weights to give to each impact factor. The impact of each factor is first expressed as a function, choosing from options such as those shown in Figure 16.13. Then each stakeholder is asked to compare each pair of factors (with n factors there are $n(n-1)/2$ pairs) and to assess their relative importance in ratio form. A matrix is created for each stakeholder,

Table 16.1 An example of the weights assigned to three factors by one stakeholder. For example, the entry "7" in Row 1 Column 2 (and the 1/7 in Row 2 Column 1) indicates that the stakeholder felt that Factor 1 (slope) is seven times as important as Factor 2 (land use).

	Slope	Land use	Distance from stream
Slope		7	2
Land use	1/7		1/3
Distance from stream	1/2	3	

as in the example shown in Table 16.1. The matrices are then combined and analyzed, and a single set of weights extracted that represent a consensus view (Figure 16.14). These weights would then be inserted as parameters in the spatial model, to produce a final result. The mathematical details of the method can be found in books or tutorials on the AHP.

16.5 Accuracy and Validity: Testing the Model

Models are complex structures, and their outputs are often forecasts of the future. How, then, can a model be tested, and how does one know whether its results can be trusted? Unfortunately, there is an inherent tendency to trust the results of computer models because they appear in numerical form (and numbers carry innate authority) and because they come from computers (which also appear authoritative). Scientists normally test their results against reality, but in the case of forecasts reality is not yet available, and by the time it is there is likely little interest in testing. So

Figure 16.14 Screenshot of an AHP application using Idrisi (www.clarklabs.org). The five layers in the upper left part of the screen represent five factors important to the decision. In the lower left the image shows the table of relative weights compiled by one stakeholder. All of the weights' matrices are combined and analyzed to obtain the consensus weights shown in the lower right, together with measures to evaluate consistency among the stakeholders. (Courtesy Clark Labs)

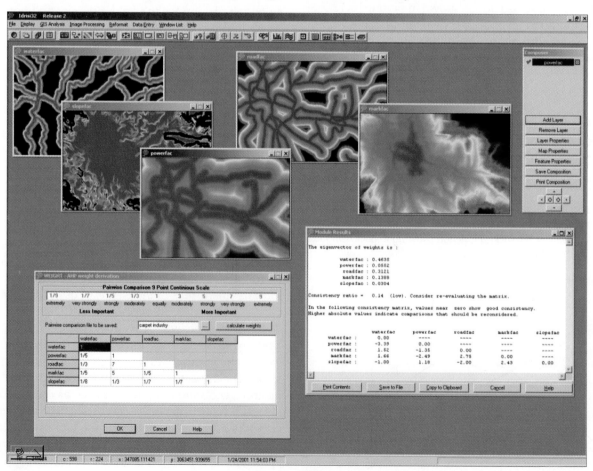

modelers must resort to other methods to verify and validate their predictions.

Results from computers tend to carry innate authority.

Models can often be tested by comparison with past history by running the model not into the future, but forward in time from some previous point. But these are often the data used to *calibrate* the model, to determine its parameters and rules, so the same data are not available for testing. Instead, many modelers resort to *cross-validation*, a process in which a subset of data are used for calibration and the remainder for validating results. Cross-validation can be done by separating the data into two time periods or into two areas, using one for calibration and one for validation. Both are potentially dangerous if the process being modeled is also changing through time or across space (it is *nonstationary* in a statistical sense), but forecasting is dangerous in these circumstances as well.

Models of real-world processes can be validated by experiment by proving that each component in the model correctly reflects reality. For example, the Game of Life would be an accurate model if some real process actually behaved according to its rules; and Clarke's urban growth model could be tested in the same way by examining the rules that account for each past land-use transition. In reality, it is unlikely that real-world processes will be found to behave as simply as model rules, and it is also unlikely that the model will capture every real-world process impacting the system.

If models are no better than approximations to reality, then are they of any value? Certainly, human society is so complex that no model will ever fit perfectly. As Ernest Rutherford, the experimental physicist and Nobel Laureate is said to have once remarked, probably in frustration with social scientist colleagues, "The only result that can possibly be obtained in the social sciences is, some do, and some don't." Neither will a model of a physical system ever perfectly replicate reality. Instead, the outputs of models must always be taken advisedly, bearing in mind several important arguments:

- A model may reflect behavior under ideal circumstances and therefore provide a norm against which to compare reality. For example, many economic models assume a perfectly informed, rational decision maker. Although humans rarely behave this way, it is still useful to know what would happen if they did, as a basis for comparison.

- A model should not be measured by how closely its results match reality, but by how much it reduces uncertainty about the future. If a model can narrow the options, then it is useful. It follows that any forecast should also be accompanied by a realistic measure of uncertainty (see Section 6.5).

- A model is a mechanism for assembling knowledge from a range of sources and presenting conclusions based on that knowledge in readily used form. It is often not so much a way of discovering how the world works, as a way of presenting existing knowledge in a form helpful to decision makers.

- Just as the British politician Denis Healey once remarked that capitalism was the worst possible economic system "apart from all of the others," so modeling often offers the only robust, transparent analytical framework that is likely to garner any respect among decision makers with competing objectives and interests.

Any model forecast should be accompanied by a realistic measure of uncertainty.

Several forms of uncertainty are associated with models, and it is important to distinguish between them. First, models are subject to the uncertainty present in their inputs. Uncertainty *propagation* was discussed in Section 6.4.2. Here it refers to the impacts of uncertainty in the inputs of a model on uncertainty in the outputs. In some cases propagation may be such that an error or uncertainty in an input produces a proportionate error in an output. In other cases a very small error in an input may produce a massive change in output; and in other cases outputs can be relatively insensitive to errors in inputs. It is important to know which of these cases holds in a given instance, and the normal approach is through repeated numerical simulation (often called Monte Carlo simulation because it mimics the random processes of that city's famous gambling casino), adding random distortions to inputs and observing their impacts on outputs. In some limited instances it may even be possible to determine the effects of propagation mathematically.

Second, models are subject to uncertainty over their parameters. A model builder will do the best possible job of calibrating the model, but inevitably there will be uncertainty about the correct values, and no model will fit the data available for calibration perfectly. It is important, therefore, that model users conduct some form of *sensitivity analysis*, examining each parameter in turn to see how much influence it has on the results. This can be done by raising and lowering each parameter value, for example by +10% and –10%, rerunning the model, and comparing the results. If changing a parameter by 10% produces a less-than-10% change in results, the model can be said to be relatively insensitive to that parameter. This allows the modeler to focus attention on those parameters that produce the greatest impact on results, making every effort to ensure that their values are correct.

Sensitivity analysis tests a model's response to changes in its parameters and assumptions.

Third, uncertainty is introduced because a model is run over a limited geographic area or *extent*. In reality, there is always more of the world outside the extent, except in a few cases of truly global models. The outside world both impacts the area modeled and is impacted by it. Changes in land-use practices within the extent will influence areas downstream through water pollution and downwind through atmospheric pollution; in turn they will be impacted within the extent by what happens in other areas upstream or upwind. National strategies will be influenced by in-migration, whether legal or illegal. In all such cases the effect is to make the results of any modeling uncertain.

Fourth, and most importantly, models are subject to uncertainty because of the *labeling* of their results. Consider the model of Box 16.2, which computes an indicator of the need for groundwater protection. In truth, the areas identified by the model have three characteristics: slope greater than 5%, used for crops, and less than 300 m from a stream. Described this way, there is little uncertainty about the results, though there will be some uncertainty due to inaccuracies in the data. But once the areas selected are described as *vulnerable*, and in need of management to ensure that groundwater is *protected*, a substantial leap of logic has occurred. Whether or not that leap is valid will depend on the reputation of the modeler as an expert in groundwater hydrology and biological conservation; on the reputation of the organization sponsoring the work; and on the background science that led to the choice of parameters (5% and 300 m). In essence, this third type of uncertainty, which arises whenever labels are attached to results that may or may not correctly reflect their meaning, is related to how results are described, in other words to their metadata, rather than to any innate characteristic of the data themselves.

16.6 Conclusion

Modeling was defined at the outset of this chapter as a process involving multiple stages, often in emulation of some real physical process. Modeling is often dynamic, and current interest in modeling is stretching the capabilities of GIS software, most of which was designed for the comparatively leisurely process of analysis, rather than the intensive and rapid iterations of a dynamic model. In many ways, then, modeling represents the cutting edge of GIS, and the next few years are likely to see very rapid growth both in GIS users' interest in modeling and in GIS vendors' interest in software development. The results are certain to be interesting.

Questions for Further Study

1. Write down the set of rules that you would use to implement a simple version of the Clarke urban growth model, using the following layers: transportation (cell contains road or not), protected (cell is not available for development), already developed, and slope.

2. Review the steps you would take and the arguments you would use to justify the validity of a GIS-based model.

3. Select a domain of environmental or social science, such as species habitat prediction or residential segregation. Search the Web and your library for published models of processes in this domain, and summarize their conceptual structure, technical implementation, and application history.

4. Compare the relative advantages of the different types of dynamic models discussed in this chapter when applied to a selected area of environmental or social science.

Further Reading

De Smith, M.J., Goodchild, M.F., and Longley, P.A. 2009. *Geospatial Analysis: A Comprehensive Guide to Principles, Techniques and Software Tools* (3rd ed.). Leicester: Troubador and www.spatialanalysisonline.com.

Heuvelink, G.B.M. 1998. *Error Propagation in Environmental Modeling with GIS*. London: Taylor and Francis.

Maguire, D., Batty, M., and Goodchild, M.F. (eds.). 2005. *GIS, Spatial Analysis, and Modeling*. Redlands, CA: ESRI Press.

Saaty, T.L. 1980. *The Analytical Hierarchy Process: Planning, Priority Setting, Resource Allocation*. New York: McGraw-Hill.

Tomlin, C.D. 1990. *Geographic Information Systems and Cartographic Modeling*. Englewood Cliffs, NJ: Prentice Hall.

⑤ MANAGEMENT AND POLICY

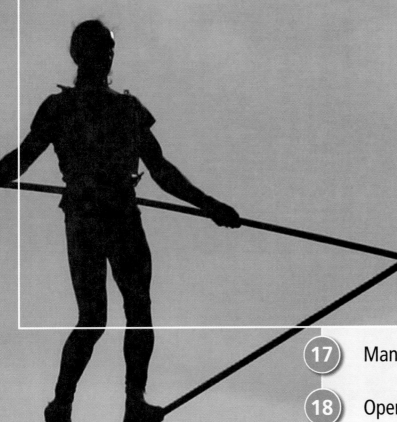

(Photo by Chung/Sun-Jun/Getty Images, Inc.) **425**

17 Managing GIS

Much of the material in earlier parts of this book assumes that you have a GIS and can use it effectively to meet your organization's goals. This chapter addresses the essential "front end": How to choose, implement, and manage operational GIS. Later chapters deal with how GIS contributes to the business of an organization. Procurement and operational management involves four key stages: the analysis of needs, the formal specification, the evaluation of alternatives, and the implementation of the chosen system. In particular, implementing GIS requires consideration of issues such as return on investment, planning, support, communication, resource management, and funding. Successful ongoing management of an operational GIS has five key dimensions: support for customers, operations, data management, application development, and project management.

LEARNING OBJECTIVES

After studying this chapter, you will understand:

- Return on investment concepts.

- How to go about choosing a GIS to meet your needs.

- Key GIS implementation issues.

- How to manage an operational GIS effectively with limited resources and ambitious goals.

- Why GIS projects fail—some pitfalls to avoid and some useful tips about how to succeed.

- The roles of staff members in a GIS project.

- Where to go for more detailed advice.

17.1 Introduction

This chapter examines the practical aspects of obtaining and managing an operational GIS. It is deliberately embedded in the part of the book that focuses on high-level management concepts because we believe that success comes from combining strategy and implementation. It is the role of management in GIS projects to ensure that operations are carried out effectively and efficiently, and that a healthy, sustainable GIS can be maintained—one that meets the organization's strategic objectives.

Obtaining and running a GIS seems at first sight to be a routine and an apparently mechanical process. It is certainly not rocket science, but neither is it simple and without tried and tested principles and best practices. The consequences of failure can be catastrophic, both for an organization and for careers. Success involves constant sharing of experience and knowledge with other people, keeping good records, and making numerous judgments where the answer is not preordained.

17.2 The Case for GIS: ROI

The most fundamental question an organization can ask about GIS is: do we need one? This question can be posed in many ways, and it covers quite a number of issues. The key strategic questions that senior executives will have are likely to be:

- What value will an investment in GIS have for our organization?

- When will the benefits of a GIS be delivered?

- Who will be the recipients of the benefits?

- What is the level of investment needed, both initially and in an ongoing operational basis?

- Who is going to deliver these benefits, and what resources are required—both internally and externally—to realize the expected benefits?

- What is the proven financial case—that is, does the investment in GIS provide the financial or other value to make it worthwhile?

For many years GIS projects have been initiated based largely on qualitative, value-added reasoning; for example, "if we implement this technology, then we will be able to perform these additional services." However, in recent years greater pressure for financial accountability in both the public and private sectors, combined with a better understanding of difficulties associated with implementing enterprise IT systems

and processes, and a realization that proper project accountability is a key aspect of good management, have all led to wider adoption of return on investment methods in project planning and evaluation. To be successful, GIS strategies must be aligned with business strategies, and GIS processes must reflect business processes. GIS that exist in a vacuum and are disconnected from an organization's business processes may be more of a business concern than no GIS at all.

Although there are many ways to describe the success or failure of a GIS project, the most widely used is return on investment (ROI). ROI studies use a combination of qualitative and quantitative measures to assess the utility that an organization will obtain from an investment. Although both qualitative and quantitative measures are often used together, much greater importance is almost always placed on fact-based, quantitative measures given their greater objectivity and persuasiveness as far as senior executives are concerned.

ROI uses a combination of qualitative and quantitative measures to assess the utility that an organization will obtain from an investment.

Maguire, Kouyoumjian, and Smith have developed an ROI methodology suitable for use in GIS projects. This comprises a sequence of 10 interrelated steps designed to be performed by a GIS professional supported by a small project team. Shown in outline form in Figure 17.1, the methodology begins with a series of planning and investigation activities that lay the groundwork for subsequent steps. Step 1, preparing for the ROI project, requires a review of an organization's mission statement(s) and an understanding of its past and present landscape of GIS. Step 2 comprises a series of interviews with key stakeholders to elicit, with guidance, how GIS can contribute to an organization's mission, collecting information concerning the high-level business issues and challenges that it faces. These insights are then organized into a series of business opportunities, which are prioritized in Step 3. The next group of steps is concerned with GIS program definition. The information gathered in earlier stages is used to define a program of GIS projects in Step 4 and dictates how these projects will be governed and managed in Step 5. The next series of steps form the core of the methodology and are concerned with business analysis. In Step 6, the defined projects are broken down into constituent parts and the resource costs are determined, from which estimated benefits

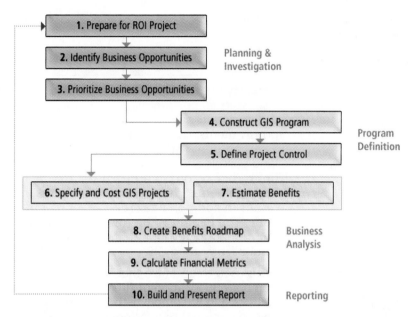

Figure 17.1 Overview of GIS ROI methodology. (Reproduced from Maguire et al. 2008)

will be detailed in Step 7. In Step 8, a benefits road map is created that shows when an organization will realize the benefits. In Step 9, ROI and other relevant financial metrics are calculated in order to demonstrate quantitatively the value of GIS to an organization. In Step 10, a compelling report is created by aggregating the information and research completed previously. If the GIS ROI project is successful, this report will concisely show how GIS can contribute value to an organization, including its cost, benefits, time to implement, resources required, governance, and the return on investment an organization will realize.

In most IT projects—and GIS projects are no exception—it is much easier to measure costs than benefits. Costs can simply be determined by making an inventory of all the goods and services required (hardware, software, data, consultants, etc.). The term *benefit* usually means any type of material value obtained from a GIS project. It is useful to distinguish between tangible and intangible benefits. Tangible benefits, also sometimes referred to as hard or economic benefits, are those that can be precisely defined and to which we can assign a specific monetary value. Examples of tangible benefits include cost avoidance—such as, increasing the number of valve inspections a water utility engineer can complete per route—and therefore decreasing inspection time; increased revenue from additional property taxes as a result of using a GIS to manage land records; and the value of time saved by creating reports more quickly using a GIS.

Intangible benefits, also sometimes referred to as soft savings or institutional benefits, are those to which we cannot assign a monetary value. Examples of intangible GIS benefits include increased morale of employees due to improved information systems, improved citizen satisfaction with government as a result of readily available online access to maps and data, and better customer relationships through more efficient information management. Although it is important to include both types of benefits in a GIS business justification case, depending on the organization, executive decision makers are usually much more easily persuaded by tangible, measurable benefits than by those that are intangible.

Table 17.1 describes some examples from the government and utility sectors of the main types of tangible and intangible benefits that have been used in the past to justify GIS projects. Since the general principles are common across all these areas, it should easily be possible to extrapolate from the details of government and utility benefits to other sectors and industries. These lists are by no means exhaustive, and there is a degree of overlap between several categories (for example, revenue growth, cost reduction, and cost avoidance), but they can be a guide that will help focus the process of researching GIS program benefits. The tables represent facts-based benefits at a high level, without considering details, and can be used as a starting point to express the values of the benefits of a GIS. Table 17.1 begins with very tangible benefits and ends with very intangible benefits.

Table 17.1 Examples of tangible and intangible benefit types for public and private organizations. Both tangible and intangible benefits are ordered strongest to weakest.

	Benefit Type	Government	Utility
STRONG ↓ ↓ ↓ ↓ ↓ **Tangible** ↓ ↓ ↓ ↓ WEAK	Revenue Growth—how can a GIS generate revenue (strictly speaking profit) for the organization?	Property taxes account for a substantial portion of the income for many local governments. GIS is used to assess accurately the size of land parcels and keep an up-to-date record of property improvements. This typically results in additional tax revenue. The benefit is the total additional tax revenue that results from using GIS.	There is great demand for accurate, detailed utility data for emergency management, construction coordination, and other purposes. Utilities sell network data products to other utilities and governments in the same geography for such purposes. The total benefit is the sum of all income.
	Revenue Protection and Assurance—how can GIS save the organization money?	GIS is used to optimize the number and location of fire stations and to comply with key government regulations about service response times. By using GIS to analyze fire station locations, road networks, and the distribution of population, the location of existing or new stations can be optimized. Avoiding the cost of building a new fire station, or saving money by combining existing stations, saves governments millions of dollars a year. The benefit is the amount of money saved by optimizing the location of the fire stations.	Costly reactive and emergency repairs to sewer networks can be eliminated by creating a planned cycle of inspections and maintenance work orders. This enables network infrastructure to be properly maintained. The total benefit is the sum of the average costs of an emergency repair compared with those of planned repairs.
	Health and Safety—how can GIS save the lives (or reduce injury) of employees or citizens? Although some might take the view that lives are invaluable, it is commonplace to ascribe a monetary value to loss of life in, for example, the insurance industry.	The most important role of governments is to protect the lives of citizens. Police forces are usually tasked with monitoring the security of major public events. GIS is increasingly regarded as a key component of emergency operations centers where it is used to store, analyze, and visualize data about events. Data about, for example, suspicious packages can be used to help support decision making about the evacuation of surrounding areas. The benefit of GIS is based on an estimate of the monetary value of lives saved as a result of using the GIS.	Integrated AM/FM/GIS electric network databases and work-order management systems are being used to create map products for use in Call-Before-You-Dig systems that reduce the likelihood of electrocution from hitting electrical conductors. Before-and-after comparison of lives saved multiplied by an agreed value of a life can lead to large benefits.
	Cost Reduction—this is different from Cost Avoidance because here we are assuming that this is an activity that an	Local government planning departments are reducing the cost of creating land-use plans and zoning maps by	Multiple entry of "as-built" work orders by several departments is reduced by centralizing data entry in

Table 17.1 (*continued*)

	Benefit Type	Government	Utility
STRONG ↓ ↓ ↓ ↓ **Tangible** ↓ ↓ ↓ ↓ **WEAK**	organization has to perform, and the objective becomes how to perform the activity with minimal net expenditure.	building databases and map templates that can be reused many times. The benefit is the difference in the cost of manual and GIS-based plan and map creation.	a single department. The total benefit is the cost of the as-built work-order entry multiplied by the reduction in the number of work orders entered.
	Cost Avoidance—rather than reducing costs, it is sometimes possible to avoid them altogether.	Local government departments issue permits for many things such as public events and roadside dumpsters (skips). By using GIS, a government department can automate the process of finding the location, issuing the permit, and tracking its status. The benefit is the reduction in the cost of issuing and tracking each permit, multiplied by the number of permits issued. There are often additional benefits of reduced time to obtain a permit and improved tracking.	Forecasting the demand for gas by integrating geodemographic data with a model of the gas network will avoid costly overbuilding of distribution capacity. The benefit is the reduced cost of construction and operation of the unwanted gas facilities.
	Increase Efficiency and Productivity—how can the organization do more with less resource?	Fire is a major hazard to forests in many drier parts of the world. Firefighters need access in their fire trucks to maps of structures to better fight fires. Field-based GIS are used to map structures more efficiently. These data are then transferred to a GIS database, and map books are produced automatically, thus greatly improving the productivity of the surveyors and cartographers (not to mention the firefighters). The benefit of increased efficiency and productivity can be quantified by comparing manual with GIS-based operations. The hours saved can be converted into dollars based on the hourly rates of workers.	Automating map production improves the efficiency and productivity of staff, freeing them up to perform other tasks and in some cases avoiding hiring new staff. The benefit is equal to the number of person-hours that can be assigned to other tasks, multiplied by their hourly rate.
	Save Time—if a process is carried out using a GIS, how much faster can it be completed?	Searches are typically performed whenever a property is bought/sold to determine ownership rights and encumbrances (rights of way, mineral rights, etc.). This requires a search to be sent to many departments. Historically this has been a manual,	The amount of time it takes to inspect outside plant will be reduced by providing better routing between inspection sites and field-based data-entry tools. The benefit is the additional number of inspections per inspector per day multiplied by the cost of an inspection.

(*continued*)

Table 17.1 (*continued*)

	Benefit Type	Government	Utility
STRONG ↓ ↓ ↓ ↓ **Tangible** ↓ ↓ ↓ **WEAK**		paper-based process taking several weeks. By implementing this in a GIS, the process can be managed electronically, performed in parallel, and responses automatically produced as a report. The time taken to complete the process can be reduced by several days or weeks. The benefit is the monetary value of the time saved multiplied by the number of searches performed. There are additional benefits of improved service to citizens.	There are additional benefits of reduced vehicle use (less congestion, pollution, and gasoline consumption).
	Increase Regulatory Compliance—if an organization has to comply with mandatory regulations, can it be done cheaper and faster with GIS?	Various e-government initiatives around the world mandate that federal/central government agencies expand the use of the Internet for delivering services in a citizen-centric, results-oriented, market-based way. GIS is a cost effective way to achieve this goal because it is a commercial off-the-shelf system that can be relatively easily implemented across multiple departments. The objective of government is generally to minimize the cost of compliance. This is a cost (a negative benefit), and the benefit derives from minimizing the costs of compliance and avoiding fines for noncompliance.	Telecom utilities are bound by law to comply with several exacting operating regulations. GIS can be used to maintain a database of the status of resources and to produce timely reports that summarize compliance levels. Given that this is a cost (a negative benefit) to the utility, the benefit derives from minimizing the costs of compliance and avoiding fines for noncompliance.
	Improve Effectiveness—if a process is carried out using a GIS, how much more effective will it be?	GIS is used to fight crime using hotspot analysis and thematic mapping techniques to illustrate geographic patterns. The results from this work are being used to catch criminals and as the basis for crime prevention campaigns (e.g., increase police presence on the streets and advise members of the public to lock garages). The effectiveness of the GIS is quantified by assessing the reduction in crime and assigning a monetary value to it.	GIS outage-management systems improve the speed and accuracy with which the location of outages (loss of service) can be identified and repair crews can be dispatched. The benefit is the value of total reduction in time for repairing outages, which in turn translates into electricity, water, or gas consumed.

Table 17.1 (*continued*)

	Benefit Type	Government	Utility
STRONG ↓ **Tangible** ↓ **WEAK**	Add New Capability—what new activities or services does GIS make possible, and what value do they add to the organization?	Infection by West Nile virus has resulted in a number of deaths in the past few years. GIS is a key component of the virus-spread monitoring process because it allows data to be analyzed and visualized geographically. This is a new capability that has allowed medical staff to understand and mitigate the spread of the virus. The benefit is the value of the GIS in terms of reduced medical costs for treating infected carriers.	GIS allows utilities to create map products that show the status of construction, to facilitate coordination of participating contractors. The benefit is the value of the improved level of coordination.
STRONG ↓ ↓ **Intangible** ↓ ↓ **WEAK**	Improve Service and Excellence Image—in what ways will use of a GIS cast the organization in a better light (e.g., forward looking, better organized, or more responsive)?	A number of government departments have front-desk clerks that respond to requests for information from citizens (e.g., status of a request for planning consent, rezoning request, or water service connection). GIS can be used to speed up responses to requests, and in some cases access can be provided in public facilities such as libraries or on the Web. The benefit is the value of improved organizational image.	By creating a publicly accessible Web site that shows planned electrical network maintenance and outages, an electrical utility can improve its image. The benefit is determined by the value of a page view, multiplied by the total number of page views.
	Enhanced Citizen/Customer Satisfaction—can GIS be used to enhance the satisfaction levels of citizens/customers?	Providing access to accurate and timely information is not only a requirement for governments, but it is also desirable because more information generally leads to higher levels of satisfaction. Changes in the satisfaction of citizens can be measured by questionnaires.	Utilities can enhance their customer care capability by providing access to accurate and up-to-date outside-plant network information. This enables customers to find out about themselves and their services.
	Improve Staff Well-being—can GIS be used to enhance the well-being and morale of employees?	Staff are generally happier if they can see that an organization is progressive and is improving effectiveness and efficiency, as well as spending public money wisely. GIS is a tool that can contribute to all these and therefore to worker satisfaction and well-being.	Automating boring and repetitive data-entry tasks can increase staff retention. There is a tangible benefit in reducing recruitment costs and an intangible value in having happy and contented staff.

Ross Smith, GIS Consultant

Ross Smith's (Figure 17.2) interest in GIS started during his undergraduate studies at Wilfrid Laurier University in Ontario, Canada, while working in the Computer Science Department as an internal IT consultant. Upon graduating, Ross went to work for the Ministry of Natural Resources where he developed GIS applications and digitized massive amounts of data—in those days on VAX/VMS and Tektronix machines.

Recognizing the potential of the technology, Ross pursued his M.Sc. in GIS at the University of Edinburgh, Scotland, and then embarked on a UK-based consulting career. After several years in the UK, Ross moved to New Zealand where he worked independently for major utilities, telecommunications providers, and GIS vendors. He then went to Poland where he helped launch and lead an offshore network broadband design center that was heavily reliant on GIS technology. While based in Poland, Ross also became involved in work for the United States Air Force in Europe, working in Italy and Germany on the Air Force bases and running multiple GIS projects that ranged from antiterror/force protection to land-use management. Ross returned to North America in 2003 and joined PA Consulting Group, a global management consulting firm, where he leads the Denver-based IT consulting team, as well as the global geospatial consulting team.

Ross is known for structuring and delivering large complex technology programs, and he is recognized as

Figure 17.2 Ross Smith, GIS Consultant. (Courtesy Ross Smith)

an industry leader in geospatial strategy development and return on investment models. His primary focus is on the telecommunication, utilities, and defense sectors and on IT-enabled business transformation. Ross is a co–author of a GIS business case and strategy book (*The Business Benefits of GIS: An ROI Approach*); and he is also the author of numerous articles on strategic planning, return on investment modeling, and organizational design. Ross comments:

It has become obvious to me that GIS is not just about technology, although this is clearly important; it is also critical that business and organizational strategies are given due consideration if a project is to be successful and sustainable in the long term.

Examining the generic benefits in Table 17.1 shows that these textual descriptions are imprecise, and they often leave the reader wishing that the benefit had been expressed in monetary terms rather than, for example, in time saved or processes improved. In some cases, it is possible to undertake further analysis that will allow an increase in the usefulness of weakly tangible or intangible benefits, so that they can be turned into strongly tangible benefits. The general point to make here is that the value that can be assigned to a benefit is proportional to the amount of work that is put in to extract that value. In choosing benefits, it is important to avoid generalized statements such as "GIS will reduce costs" or "With GIS we can improve operational performance." The best benefits are specific, measurable, and relevant to an organization. Maguire, Kouyoumjian and Smith's ROI methodology shows how to create a number of

specific measures of business success, though these cannot be discussed here due to space limitations. Box 17.1 highlights the work of Ross Smith, a GIS consultant who has used ROI analysis in many client engagements.

17.3 The Process of Developing a Sustainable GIS

GIS projects are similar to many other large IT projects in that they can be broken down into four major life-cycle phases (Figure 17.3). For our simplified purposes, these are:

- Business planning (strategic analysis and requirements gathering)

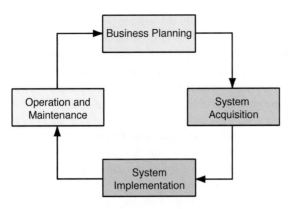

Figure 17.3 GIS project life-cycle stages.

- System acquisition (choosing and purchasing a system)
- System implementation (assembling all the various components and creating a functional solution)
- Operation and maintenance (keeping a system running)

These phases are iterative. Over a decade or more, several iterations may occur, often using different generations of GIS technology and methodologies. Variations on this model include prototyping and rapid application development—but space does not permit much discussion of them here.

GIS projects comprise four major life-cycle phases: business planning; system acquisition; system implementation; and operation and maintenance.

Roger Tomlinson is regarded as the father of GIS because in the spring of 1962 he embarked on the first GIS project, which was to build a land-use inventory of Canada. He has also developed a methodology for obtaining a GIS that is likely to fulfill user needs (Table 17.2). This high-level approach is very practical and is designed to ensure that a resulting GIS will match user expectations.

Table 17.2 The Tomlinson methodology for getting a GIS that meets user needs (Source: Adapted from R. Tomlinson 2007).

Stage	Action	Commentary
1	Consider the strategic purpose	This is the guiding light. The system that gets implemented must be aligned with the purpose of the organization as a whole.
2	Plan for the planning	Since the GIS planning process will take time and resources, you will need to get approval and commitment at the front end from senior managers in the organization.
3	Conduct a technology seminar	Think of the technology seminar as a sort of town-hall meeting between the GIS planning team, the various staff, and other stakeholders in the organization.
4	Describe the information products	Know what you want to get out of it.
5	Define the system scope	Scoping the system means defining the actual data, hardware, software, and timing.
6	Create a data design	The data landscape has changed dramatically with the advent of the Internet and the proliferation of commercial datasets. Developing a systematic procedure for safely navigating this landscape is critical.
7	Choose a logical data model	The new generation of object-oriented data models is ushering in a host of new GIS capabilities and should be considered for all new implementations. Yet the relational model is still prevalent, and the savvy GIS user will be conversant in both.
8	Determine system requirements	Getting the system requirements right at the outset (and providing the capacity for their evolution) is critical to successful GIS planning.
9	Return on investment, migration, and risk analysis	The most critical aspect of doing an ROI analysis is the commitment to include all the costs that will be involved. Too often managers gloss over the real costs, only to regret it later.
10	Make an implementation plan	The implementation plan should illuminate the road to GIS success.

17.3.1 Choosing a GIS

Drew Clarke has proposed a general model of how to specify, evaluate, and choose a GIS, variations of which organizations have used over the past 25 years or so. The model (Figure 17.4) is based on 14 steps grouped into four stages: analysis of requirements; specification of requirements; evaluation of alternatives; and implementation of the system. Such a process is both time-consuming and expensive. It is only appropriate for large GIS implementations (contracts over $250,000 in initial value), where it is particularly important to have investment and risk appraisals. We describe the model here so that those involved with smaller systems can judiciously select those elements relevant to them. On the basis of painful experience, however, we urge the use of formalized approaches to evaluating the need for any subsequent acquisition of a system. It is amazing how small projects, carried out quickly because they are small, evolve into big and costly ones!

Choosing a GIS involves four stages: analysis of requirements; specification of requirements; evaluation of alternatives; and implementation of system.

For organizations undertaking acquisition for the first time, huge benefits can be accrued through partnering with other organizations that are more advanced, especially if they are in the same field (see Chapter 19). This is often possible in the public sector, for example, where local governments have similar tasks to meet. But a surprising number of private-sector organizations are also prepared to share their experiences and documents.

Stage 1: Analysis of Requirements The first stage in choosing a GIS is an iterative process for identifying and refining user requirements, and for determining the business case for acquisition. The deliverable for each step is a report that should be discussed with users and the management team. It is important to keep records of the discussions and share them with those involved so that there can be no argument at a later stage about what was agreed! The results of each report help determine successive stages.

Step 1: **Definition of objectives**
This is often a major decision for any organization. The rational process of choosing a GIS begins with and spins out of the development of the organization's strategic plan—and an outline decision that GIS can play a role in the implementation of this plan. Strategic and tactical objectives must be stated in a form that is understandable to managers. The outcome from Step 1 is a document that managers and users can endorse as a plan to proceed with

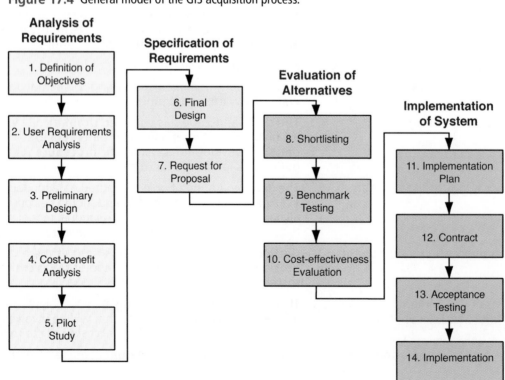

Figure 17.4 General model of the GIS acquisition process.

the acquisition; that is, the relevant managers believe there is sufficient promise to proceed to the next step and commit the initial funding required.

Step 2: User requirements analysis

The analysis will determine how the GIS is designed and evaluated. Analysis should focus on what information is presently being used, who is using it, and how the source is being collected, stored, and maintained. This is a map of existing processes (which may possibly be improved as well as being replicated by the GIS) or of processes newly designed after a business reengineering exercise. The necessary information can be obtained through interviews, documentation, reviews, and workshops. The report for this phase should be in the form of workflows, lists of information sources, and current operation costs. The clear definition of likely or possible change (e.g., future applications—see Figure 17.5), new information products (e.g., maps and reports—Figure 17.6), or different utilization of functions and new data requirements is essential to successful GIS implementation.

Step 3: Preliminary design

This stage of the design is based on results from Step 2. The results will be used for subsequent cost-benefit analysis (Step 4) and will enable specification of the pilot study. The four key tasks are: develop preliminary database specifications; create preliminary functional specifications; design preliminary system models; and survey the market for potential systems. Database specifications involve estimating the amount and type of data (see Chapter 9). Many consultants maintain checklists, and vendors frequently publish descriptions of their systems on their Web sites. The choice of system model involves decisions about raster and vector data models, and a survey should be undertaken to assess the capabilities of commercial off-the-shelf (COTS) systems. This might involve a formal Request for Information (RFI) to a wide range of vendors. A balance needs to be struck between creating a document so open that the vendor has problems identifying what needs are paramount and a document that is so prescriptive and closed that no flexibility or innovation is possible.

APPLICATION

Display zoning map information for a user-defined area.

FUNCTIONS USED IN THE APPLICATION:

Review and prepare zoning changes.

DESCRIPTION OF APPLICATION:

This application uses zoning and related parcel-based data from the database to display existing information related to zoning for a specific area that is defined by the user. The application must be available interactively at a workstation when the user invokes a request and identifies the subject land parcel. The application will define a search area based upon the search distance defined and input by the user, and will display all required data for the area within the specified distance from the outer boundary of the subject parcel.

DATA INPUTS:

User defined: Parcel identifier
              Search distance

Database: Zoning boundaries
              Zoning dimensions
              Zoning codes
              Parcel boundaries
              Parcel dimensions
              Parcel numbers
              Street names
              Addresses

PRODUCTS OUTPUT:

1. Zoning map screen display with subject parcel highlighted, search area boundaries, search distance, all zoning data, parcel data, street names and addresses.

2. Hard copy map of the above.

Figure 17.5 Sample application definition form.

Whether to buy or to build a GIS used to be a major decision (Chapter 7). This occurred especially at "green field" sites—where no GIS technology has hitherto been used—and at sites where a GIS has already been implemented but was in need of modernization. But the situation is now quite different: use of general-purpose COTS solutions is the norm. COTS GIS have ongoing programs of enhancement and maintenance and can normally be used for multiple projects. Typically, they are better documented, and more people in the job market have experience of them. As a consequence, risk arising from loss of key personnel is reduced.

There has been a major move in GIS away from building proprietary GIS toward buying COTS solutions.

Step 4: Cost-benefit analysis

Purchase and implementation of a GIS is a nontrivial exercise because of the expense involved in both money and staff resources (typically management time). It is quite common

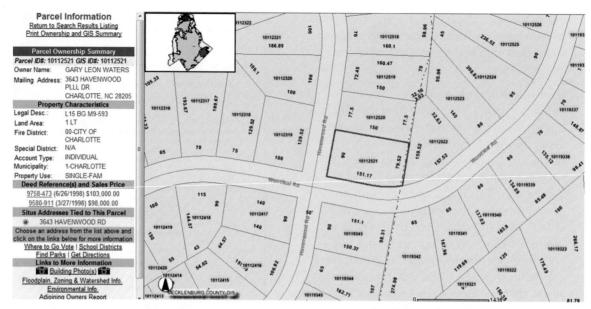

Figure 17.6 A report produced by a local government GIS. (Source: Mecklenburg County, North Carolina GIS; see polaris.mecklenburgcountync.gov/website/redesign/viewer.htm)

for organizations to undertake a cost-benefit analysis to justify the effort and expense, and to compare it against the alternative of continuing with the current data, processes, and products—the status quo. Cost-benefit cases are normally presented as a spreadsheet, along with a report that summarizes the main findings and suggests whether the project should be continued or halted. Senior managers then need to assess the merits of this project in comparison with any others that are competing for their resources. Cost-benefit analysis can be considered a simplified form of return-on-investment calculation (see Section 17.2).

Step 5: Pilot study

A pilot study is a miniature version of the full GIS implementation that aims to test several facets of the project. The primary objective is to test a possible or likely system design before finalizing the system specification and committing significant resources. Secondary objectives are to develop the understanding and confidence of users and sponsors, to test samples of data if a data capture project is part of the implementation, and to provide a test bed for application development.

A pilot is a miniature version of a full GIS implementation designed to test as many aspects of the final system as possible.

It is normal to use existing hardware or to lease hardware similar to that which is expected to be used in the full implementation. A reasonable cross sec-tion of all the main types of data, applications, and product deliverables should be used during the pilot. But the temptation must be resisted to try to build the whole system at this stage, regardless of how easy the "techies" may claim it to be! Users should be prepared to discard *everything* after the pilot if the selected technology or application style does not live up to expectations.

The outcome of a pilot study is a document containing an evaluation of the technology and approach adopted, an assessment of the cost-benefit case, and details of the project risks and impacts. Risk analysis is an important activity, even at this early stage. Assessing what can go wrong can help avoid potentially expensive disasters in the future. The risk analysis should focus on the actual acquisition processes as well as on implementation and operation.

Stage 2: Specification of Requirements The second stage is concerned with developing a formal specification that can be used in the structured process of soliciting and evaluating proposals for the system.

Step 6: Final design

This creates the final design specifications for inclusion in a Request for Proposals (RFP:

also called an Invitation to Tender, or ITT) to vendors. Key activities include finalizing the database design, defining the functional and performance specifications, and creating a list of possible constraints. From these, requirements are classified as mandatory, desirable, or optional. The deliverable is the final design document. This document should provide a clear description of essential requirements—without being so prescriptive that innovation is stifled, costs escalate, or insufficient vendors feel able to respond.

Step 7: Request for proposals

The RFP document combines the final design document with the contractual requirements of the organization. These documents will vary from organization to organization but are likely to include legal details of copyright of the design and documentation, intellectual property ownership, payment schedules, procurement timetable, and other draft terms and conditions. Once the RFP is released to vendors by official advertisement and/or personal letter, a minimum period of several weeks is required for vendors to evaluate and respond. For complex systems, it is usual to hold an open meeting to discuss technical and business issues.

Stage 3: Evaluation of Alternatives

Step 8: Short-listing

In situations where several vendors are expected to reply, it is customary to have a short-listing process. Submitted proposals must first be evaluated, usually using a weighted scoring system, and the list of potential suppliers needs to be narrowed down to between two and four. Good practice is to have the scoring system done by several individuals acting independently and to compare the results. This whole process allows both the prospective purchaser and supplier organizations to allocate their resources in a focused way. Short-listed vendors are then invited to attend a benchmark-setting meeting.

Step 9: Benchmarking

The primary purpose of a benchmark is to evaluate the proposal, people, and technology of each selected vendor. Each one is expected to create a prototype of the final system that will be used to perform representative tests. The prospective purchaser scores the results of these tests. Scores are

also assigned for the original vendor proposal and the vendor presentations about their company. Together, these scores form the basis of the final system selection. Unfortunately, benchmarks are often conducted in a rather secretive and confrontational way, with vendors expected to guess the relative priorities (and the weighting of the scores) of the prospective purchaser. While it is essential to follow a fair and transparent process, maintain a good audit trail, and remain completely impartial: a more open cooperative approach usually produces a better evaluation of vendors and their proposals. If vendors know which functions have the greatest value to customers, they can tune their systems appropriately.

Step 10: Cost-effectiveness evaluation

Next, surviving proposals are evaluated for their cost-effectiveness. This is again more complex than it might seem. For example, GIS software systems vary quite widely in the type of hardware they use; some need additional database management system (DBMS) licenses, customization costs will vary, and maintenance will often be calculated in different ways. The goal of this stage is to normalize all the proposals to a common format for comparative purposes. The weighting used for different parts must be chosen carefully because this can have a significant impact on the final selection. Good practice involves debate within the user community—for they should have a strong say—on the weighting to be used and some sensitivity testing to check whether very different answers would have been obtained if the weights were slightly different. The deliverable from this stage is a ranking of vendors' offerings.

Stage 4: Implementation of System

The final stage is planning the implementation, contracting with the selected vendor, testing the delivered system, and actually using the GIS.

Step 11: Implementation plan

A structured, appropriately paced implementation plan is an essential ingredient of a successful GIS implementation. The plan commences with identification of priorities, definition of an implementation schedule, and creation of a resource budget and management plan. Typical activities that need to be included in the schedule are installation and acceptance testing, staff training, data collection, and customization. Implementation should be coordinated with both users and suppliers.

Step 12: Contract

An award is subject to final contractual negotiation to agree on general and specific terms and conditions, what elements of the vendor proposal will be delivered, when they will be delivered, and at what price. General conditions include contract period, payment schedule, responsibilities of the parties, insurance, warranty, indemnity, arbitration, and provision of penalties and contract termination arrangements.

Step 13: Acceptance testing

This step is to ensure that the delivered GIS matches the specification agreed in the contract. Part of the payment should be withheld until this step is successfully completed. Activities include installation plus tests of functionality, performance, and reliability. A system seldom passes all tests the first time, and so provision should be made to repeat aspects of the testing.

Step 14: Implementation

This is the final step at the end of what can be a long road. The entire GIS acquisition period can stretch over many months or even longer. Activities include training users and support staff, data collection, system maintenance, and performance monitoring. Customers may also need to be "educated" as well! Once the system is successfully in operation, it may be appropriate to publicize its success for enhancement of the brand image, or political purposes.

17.3.1.1 Discussion of the Classical Acquisition Model

The general model outlined earlier has been widely employed as the primary mechanism for large GIS procurements in public organizations. It is rare, however, that one size fits all, and although the model has many advantages, it also has some significant shortcomings:

- The process is expensive and time-consuming for both suppliers and vendors. A supplier can spend as much as 20% of the contract value on winning the business, and a purchaser can spend a similar amount in staff time, external consultancy fees, and equipment rental. This ultimately makes sys-

tems more expensive—though competition does drive down cost.

- Because it takes a long time and because GIS is a fast-developing field, proposals can become technologically obsolete within several months.

- The short-listing process requires multiple vendors, which can end up lowering the minimum technical selection threshold in order to ensure enough bidders are available.

- In practice, the evaluation process often focuses undue attention on price rather than on the long-term organizational and technical merits of the different solutions.

- This type of procurement can be highly adversarial. As a result, it can lay the foundations for an uncomfortable implementation partnership (see Chapter 19), and it often does not lead to full development of the best solution. Every implementation is a trade-off between functionality, time, price, and risk. A full and frank discussion between purchaser and vendor on this subject can generate major long-term benefits.

- Many organizations have little idea about what they *really* need. Furthermore, it is very difficult to specify precisely in any contract what a system must perform. As users learn more, their aspirations also rise, resulting in "feature creep" (the addition of more capabilities) often without any acceptance of an increase in budget. On the other hand, some vendors take a minimalist view of the capabilities of the system featured in their proposal and make all modifications during implementation and maintenance through chargeable change orders. All this makes the entire system acquisition cost far higher than was originally anticipated; the personal consequences for the budget holders concerned can be unfortunate.

- Increasingly, most organizations already have some type of GIS; the classical model works best in a "green field" situation.

As a result of these problems, this type of acquisition model is not used in small or even some larger procurements, especially where the facilities can be augmented rather than totally replaced. A less complex and formal selection method is *prototyping*. Here a vendor or pair of vendors is selected early on using a smaller version of the evaluation process outlined in the preceding part of this chapter. The vendor(s) is/are then funded to build a prototype in collaboration with the user organization. This fosters a close partnership to exploit the technical capabilities of systems and developers, and it helps to maintain system flexibility in the light of changing requirements and technology. This approach works best for those procurements—sometimes even some large ones—where there is some uncertainty about the most appropriate technical solution and where the organizations involved are mature, able to control the process, and not subject to draconian procurement rules.

Prototyping is a useful alternative to classical, linear system acquisition exercises. It is especially useful for smaller procurements where the best approach and outcome is more uncertain.

17.3.2 Implementing a GIS

This section provides a checklist of important management issues to consider when implementing a GIS.

17.3.2.1 Plan Effectively
Good planning is essential through the full life cycle of all GIS projects. Both strategic planning and operational planning are important to the success of a project. Strategic planning involves reviewing overall organizational goals and setting specific GIS objectives. Operational planning is more concerned with the day-to-day management of resources. Several general project management productivity tools are available that can be used in GIS projects. Figure 17.7 shows one diagramming tool called a Gantt chart. Several other implementation techniques and tools are summarized in Table 17.3.

17.3.2.2 Obtain Support
If a GIS project is to prosper, it is essential to garner support from *all* key stakeholders. This can entail several activities, including establishing executive (director-level) leadership support; developing a public relations strategy by, for example, exhibiting key information products or distributing free maps; holding an open house to explain the work on the GIS team; and participating in GIS seminars and workshops, locally and sometimes nationally.

17.3.2.3 Communicate with Users
Involving users from the very earliest stages of a project will lead to a better system design and will help with user acceptance. Seminars, newsletters, and frequent updates about the status of a project are good ways to educate and involve users. Setting expectations about capabilities, throughput, and turnaround at reasonable levels is crucial to avoid any later misunderstandings with users and managers.

Figure 17.7 Gantt chart of a basic GIS project. This chart shows task resource requirements over time, with task dependencies. This example presents a straightforward chart, with a small number of tasks. (Reproduced by permission of Washington State Department of Transportation)

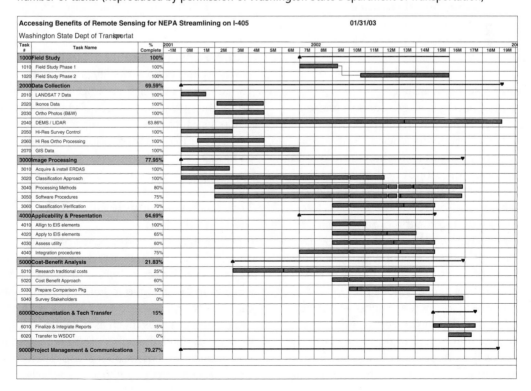

Table 17.3 GIS implementation tools and techniques (After Heywood et al. 2006, with additions).

Technique	Purpose
SWOT analysis	This is a management technique used to establish strengths, weaknesses, opportunities, and threats (hence SWOT) in a GIS implementation. The output is a list and narrative.
ROI analysis	This methodology is used to assess the value added by comparing outputs with inputs (see Section 17.2).
Rich picture analysis	Major participants are asked to create a schematic/picture showing their understanding of a problem using agreed conventions. These are then discussed as part of a consensus-forming process.
Demonstration systems	Many vendors and GIS project teams create prototype demonstrations to stimulate interest and educate users/funding agencies.
Interviews and data audits	These aim to define problems and determine current data holdings. The output is a report and recommendations.
Organization charts, system flowcharts, and decision trees	These are all examples of flowcharts that show the movement of information, the systems used, and how decisions are currently reached.
Data flow diagrams and dictionaries	These are charts that track the flow of information and computerized lists of data in an organization.
Project management tools	Gantt charts (see Figure 17.7) and PERT (program evaluation and review technique) are tools for managing time and resources.
Object-model diagrams	These show objects to be modeled in a GIS and the relationships between them (see, for example, Figure 8.18).

17.3.2.4 Anticipate and Avoid Obstacles

These obstacles may involve staffing, hardware, software, databases, organization/procedures, time frame, and funding. Be prepared!

17.3.2.5 Avoid False Economies

Money saved by not paying staff a reasonable (market value) wage or by insufficient training is often manifested in reduced staff efficiencies. Furthermore, poorly paid or poorly trained staff often leave through frustration. This situation cannot be prevented by contractual means and must be tackled by paying market-rate salaries and/or building a team culture where staff enjoy working for the organization.

Cutting back on hardware and software costs by, for example, obtaining less powerful systems or canceling maintenance contracts may save money in the short term but will likely cause serious problems in the future when workloads increase and the systems get older. Failing to account for depreciation and replacement costs, that is, by failing to amortize the GIS investment, will store up trouble ahead. The amortization period will vary greatly—hardware may be depreciated to zero value after, say, four years while buildings may be amortized over 30 years.

17.3.2.6 Ensure Database Quality and Security

Investing in database quality is essential at all stages from design onward. Catastrophic results may ensue if any of the updates or (especially) the database itself is lost in a system crash or corrupted by hacking, and the like. This requires not only good precautions but also contingency (disaster recovery and business continuity) plans and periodic serious trials of them.

17.3.2.7 Accommodate GIS within the Organization

Building a system to replicate old and inefficient ones is not a good idea; nor is it wise to go to the other extreme and expect the whole organization's ways of working to be changed to fit better with what the GIS can do! Too much change at any one time can destroy organizations just as much as too little change can ossify them. In general, the GIS must be managed in a way that fits with the organizational aspirations and culture if it is to be a success. All this is especially a problem because GIS projects often blaze the trail in terms of introducing new technology, interdepartmental resource sharing, and generating new sources of income.

17.3.2.8 Avoid Unreasonable Time Frames and Expectations

Inexperienced managers often underestimate the time it takes to implement GIS. Good tools, risk analysis, and time allocated for contingencies are important methods of mitigating potential problems. The best guide to how long a project will take is experience in other similar projects—though the differences between the organizations, staffing, tasks, and so on, need to be taken into account.

17.3.2.9 Funding

Securing ongoing, stable funding is a major task of a GIS manager. Substantial GIS projects will require core funding from one or more of the stakeholders. None of these will commit to the project without a business case and risk analysis (Section 17.4.3). Additional funding for special projects, and from information and service sales, is likely to be less certain. In many GIS projects the operational budget will often change significantly over time as the system matures. The three main components are staff, goods and services, and capital investments. A commonly experienced distribution of costs between these three elements is shown in Table 17.4.

17.3.2.10 Prevent Meltdown

Avoiding the cessation of GIS activities is the ultimate responsibility of the GIS manager. According to Roger Tomlinson, some of the main reasons for the failure of GIS projects are as follows:

- Lack of executive-level commitment
- Inadequate oversight of key participants
- Inexperienced managers
- Unsupportive organizational structure

Table 17.4 Percentage distribution of GIS operational budget elements over three time periods (after Sugarbaker 2005).

Budget item	Year 1–2	Year 3–6	Year 6–12
Staff and benefits	30	46	51
Goods and services	26	30	27
Equipment and software	44	24	22
Total	100	100	100

- Political pressures, especially where these change rapidly
- Inability to demonstrate benefits
- Unrealistic deadlines
- Poor planning
- Lack of core funding

17.3.3 Managing a Sustainable, Operational GIS

Larry Sugarbaker has characterized the many operational management issues throughout the life cycle of a GIS project as customer support; effective operations; data management; and application development and support. Success in any one—or even all—of these areas does not guarantee project success, but they certainly help to produce a healthy project. Each is now considered in turn.

> **Success in operational management of GIS requires customer support, effective operations, data management, and support for applications development.**

Applications Box 17.2 describes a very successful and well-managed GIS in South Korea.

17.3.3.1 Customer Support

In progressive organizations *all* users of a system and its products are referred to as customers. A critical function of an operational GIS is a customer support service. This could be a physical desk with support staff, or, increasingly, it is a networked electronic mail and telephone service. Since this is likely to be the main interaction with GIS support staff, it is essential that the support service creates a good impression and delivers the type of service users need. The unit will typically perform key tasks, including technical support and problem logging plus supplying requests for data, maps, training, and other products. Performing these tasks will require both GIS analyst-level and administrative skills. It is imperative that *all* customer interaction is logged and that procedures are put into place to handle requests and complaints in an organized and structured fashion. This is both to provide an effective service and to correct systemic problems.

Customer support is not always seen as the most glamorous of GIS activities. However, a GIS manager who recognizes the importance of this function and delivers an efficient and effective service will be rewarded with happy customers. Happy customers remain customers. Effective staff management includes finding staff with the right interests and aspirations,

Managing Land Information in Korea through GIS

In South Korea, a complex and rapidly changing society, local government authorities administer the public land through assessment of land prices, management of land transactions, land-use planning and management, and civil services. In many cases, more than one department of a local government authority produces and manages the same or similar land and property information; this has led to discrepancies in the information held across local government. With the large number of public land administration responsibilities and the control of each given to the local authorities, many problems arose in the past. This led to the decision to develop a GIS-based method for sharing the information produced or required for administering land in the public and private sectors (Figure 17.8B). The Korean Land Management

(A)

(B)

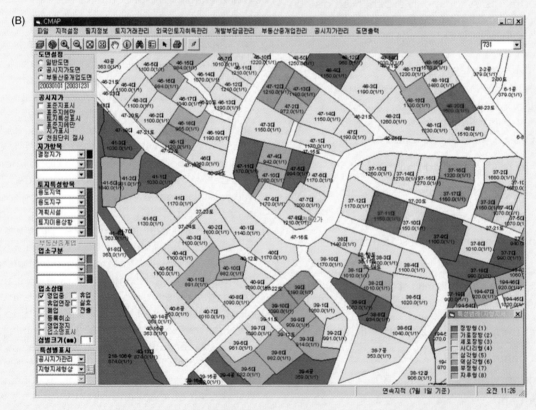

Figure 17.8 (A) Seoul by night. (B) Land information map for part of Seoul. (Reproduced by permission of Corbis) (© David Ball/© Corbis)

Information System (LMIS) was established in 1998. The purpose of this GIS is to provide land information, increase productivity in public land administration, and support the operation of the land planning policies of the Korean Ministry of Construction and Transportation (MOCT). The LMIS database includes many spatial data such as topographic, cadastral, and land-use district maps.

Hyunrai Kim, vice director of the Land Management Division of Seoul Metropolitan City, summarizes the advantages of this system thus: "By means of the Internet-based Land Information Service System, citizens can get land information easily at home. They don't have to visit the office, which may be located far from their homes." The system has also resulted in time and cost savings. With the development of the Korean Land Price Management System,

it is also possible to compute land prices directly and produce maps of variations in land price. Initially, the focus was mainly on the administrative aspects of data management and system development; however, attention then turned to the expansion and development of a decision support system using various data analyses. It is intended that the Land Legal Information Service System will also be able to inform land users of regulations on land use. In essence, LMIS is becoming a crucial element of e-government. This case study highlights the role that GIS can play beyond the obvious one of information management, analysis, and dissemination. It highlights the value of GIS in enabling organizational integration and the reality of generating benefits through improved staff productivity.

rotating GIS analysts through posts, and setting the right (high) level of expectation in the performance of all staff. Managers can learn much by taking a turn in the hot seat of a customer support role!

17.3.3.2 Operations Support

Operations support includes system administration, maintenance, security, backups, technology acquisitions, and many other support functions. In small projects, everyone is charged with some aspects of system administration and operations support. But as projects grow beyond five or more staff, it is worthwhile designating someone specifically to fulfill what becomes a core, even crucial, role. As projects become larger, this grows into a full-time function. System administration is a highly technical and mission-critical task requiring a dedicated, properly trained, and paid person.

Perhaps more than in any other role, clear written descriptions are required for this function to ensure that a high level of service is maintained. For example, large, expensive databases will require a well-organized security and backup plan to ensure that they are *never* lost or corrupted. Part of this plan should be a disaster recovery strategy. What would happen, for example, if there were a fire in the building housing the database server or some other major problem?

17.3.3.3 Data Management Support

The concept that geographic data are an important part of an organization's critical infrastructure is becoming widely accepted. Large, multiuser geographic databases use DBMS software to allocate

resources, control access, and ensure long-term usability (see Chapter 10). DBMS can be sophisticated and complicated, requiring skilled system administrators for this critical function.

A database administrator (DBA) is a person responsible for ensuring that all data meet all of the standards of accuracy, integrity, and compatibility required by the organization. A DBA will also typically be tasked with planning future data resource requirements—derived from continuing interaction with current and potential customers—and the technology necessary to store and manage them. Similar comments to those outlined above for system administrators also apply to this position.

17.3.3.4 Application Development and Support

Although a considerable amount of application development is usual at the onset of a project, it is also likely that there will be an ongoing requirement for this type of work. Sources of application development work include improvements/enhancements to existing applications, as well as new users and new project areas starting to adopt GIS.

Software development tools and methodologies are constantly in a state of flux, and GIS managers must invest appropriately in training and new software tools. The choice of which language to use for GIS application development is often a difficult one. Consistent with the general movement away from proprietary GIS languages, wherever possible GIS managers should try to use mainstream, open languages that are likely to have a long lifetime (see Chapter 7).

Ideally, application developers should be assigned full-time to a project and should become permanent members of the GIS group to ensure continuity (but often this does not occur).

17.4 Sustaining a GIS—The People and Their Competences

Throughout this chapter we have sometimes highlighted and sometimes hinted at the key role of staff as assets in all organizations. If they do not function well—both individually and as a team—nothing of merit will be achieved.

17.4.1 GIS Staff and the Teams Involved

Several different staff will carry out the operational functions of a GIS. The exact number of staff and their precise roles will vary from project to project. The same staff member may carry out several roles (e.g., it is quite common for administration and application development to be performed by a GIS technical person), and several staff members may be required for the same task (e.g., there may be many digitizing technicians and application developers). Figure 17.9 shows a generalized view of the main staff roles in medium to large GIS projects.

All significant GIS projects will be overseen by a management board built up of a senior sponsor (usually a director or vice-president), members of the user community, and the GIS manager. It is also useful to have one or more independent members to offer disinterested advice. Although this group may seem intimidating and restrictive to some, used in the right way it can be a superb source of funding, advice, support, and encouragement.

Typically, day-to-day GIS work involves three key groups of people: the GIS team itself; the GIS users; and external consultants. The GIS team comprises the dedicated GIS staff at the heart of the project, with the GIS manager designated as the team leader. This individual needs to be skilled in project and staff management and have sufficient understanding of GIS technology and the organization's business to handle the liaisons involved. Larger projects will have specialist staff experienced in project management, system administration, and application development.

GIS users are the customers of the system. There are two main types of user (other than the leaders of organizations who may rely on GIS indirectly to provide information on which they base key decisions). These are professional users and clerical staff/technicians. Professional users include engineers, planners, scientists, conservationists, social workers, and technologists who utilize output from GIS for their professional work. Such users

Figure 17.9 The GIS staff roles in a medium to large size GIS project.

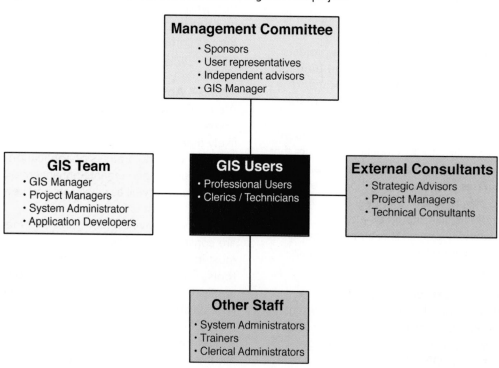

are typically well-educated in their specific field, but may lack advanced computer skills and knowledge of the GIS. They are usually able to learn how to use the system themselves and can tolerate changes to the service.

Clerical and technical users are frequently employed as part of the wider GIS project initiative to perform tasks such as data collection, map creation, routing, and service call response. Typically, the members of this group have limited training and skills for solving ad hoc problems. They need robust, reliable support. They may also include staff and stakeholders in other departments or projects that assist the GIS project on either a full- or part-time basis—for example, system administrators, clerical assistants, or software engineers provided from a common resource pool, or managers of other databases or systems with which the GIS must interface. Finally, many GIS projects utilize the services of external consultants. They could be strategic advisors, project managers, or technical consultants able to supplement the available staffing. Although these consultants may appear expensive at first sight, they are often well-trained and highly focused. They can be a valuable addition to a project, especially if internal knowledge and/or resources are limited and for benchmarking against approaches elsewhere. But the in-house team must not rely too heavily on consultants lest, when they go, all key knowledge and high-level experience goes with them.

The key groups involved in GIS are the management board; the GIS team (headed by a GIS manager); the users; external consultants; and various customers.

17.4.2 Project Managers

A GIS project will almost certainly have several subprojects or project stages and hence require a structured approach to project management. The GIS manager may take on this role personally, although in large projects it is customary to have one or more specialist project managers. The role of the project manager is to establish user requirements, to participate in system design, and to ensure that projects are completed on time, within budget, and according to an agreed-upon quality plan. Good project managers are rare creatures and must be nurtured for the good of the organization. One of their characteristics is that, once one project is completed, they like to move on to another, so retaining them is only possible in an enterprising environment. Transferring their expertise and knowledge into the heads and files of others is a priority before they leave a project (see also Section 1.2).

17.4.3 Coping with Uncertainty

As we have seen, geographic information varies hugely in its characteristics, and rarely is the available information ideal for the task in hand. Staff need a clear understanding of the concepts and implications of uncertainty (see Chapter 6) and the related concepts of accuracy, error, and sensitivity. Understanding of business risk arising from GIS use and of how GIS can help reduce organizational risk is also essential. This section focuses on the practical aspects relevant to managers of operational GIS—and hence on the skills, attitudes, and other competences they need to bring to their work.

Organizations must determine how much uncertainty they can tolerate before information is deemed useless. This can be difficult because it is application-specific. An error of 10 m in the location of a building is irrelevant for business geodemographic analysis, but it could be critical for a water utility maintenance application that requires digging holes to locate underground pipes. Some errors in GIS can be reduced but sometimes at a considerable cost. It is common experience that trying to remove the last 10% of error typically costs 90% of the overall sum. As we concluded in Section 6.5, uncertainty in GIS-based representations is almost always something that we have to live with to a greater or lesser extent. The key issue here is identifying the amount of uncertainty that can be tolerated for a given application, and what can be done at least partially to eliminate it or ameliorate its consequences. Some of this, at least, can only be done by judgment informed by past experience.

A conceptual framework for considering uncertainty was developed in Section 6.1. This discussion also introduced the notion of measurement error. Some practical examples of errors in operational GIS are as follows:

- Referential errors in the identity of objects (e.g., a street address could be wrong, resulting in incorrect property identification during an electric network trace).

- Topological errors (e.g., a highway network could have missing segments or unconnected links, resulting in erroneous routing of service or delivery vehicles).

- Relative positioning errors (e.g., a gas station incorrectly located on the wrong side of a divided highway or dual-carriageway road could have major implications for transportation models).

- Absolute errors in the real location of objects in the real world (e.g., tests for whether factories are within a smoke control zone or floodplain could provide erroneous results if the locations are incorrect). This could lead to litigation.

- Attribute errors (e.g., incorrectly entering land-use codes would give errors in agricultural production returns to government agencies).

Managing errors requires use of quality assurance techniques to identify them and assess their magnitude. A key task is determining the error tolerance that is acceptable for each data layer, information product, and application. It follows that both data creators and users must make analyses of possible errors and their likely effects, based on a form of ROI analysis (Section 17.2). And sitting in the midst of all this is the GIS manager who must know enough about uncertainty to ask the right questions; if the system provides what subsequently turns out to be nonsense, the manager is likely to be the first to be blamed! In no sense is the good and long-lasting GIS manager simply someone who ensures that the "wheels go round."

17.5 Conclusions

Any management function in GIS (and indeed elsewhere) is mostly about motivating, organizing or steering, enhancing skills, and monitoring the work of other people.

Managing a GIS project *is* different from using GIS in decision making. Normally, managing a GIS project requires good GIS expertise and first-class project management skills. In contrast, those involved at different levels of the organization's management chain need some awareness of GIS, its capabilities, and its limitations— scientific and practical—alongside their substantial leadership skills. But our experience is that the division is not clear-cut. GIS project managers cannot succeed unless they understand the objectives of the organization, the business drivers, and the culture in which they operate, as well as something of how to value, exploit, and protect their assets (Chapter 18). Moreover, decision makers can make good decisions only if they understand more of the scientific and technological background than they may wish to do: running or relying on a good GIS service involves much more than the networking of a few PCs running one piece of software.

So, good management of GIS requires excellent people, technical and business skills, and the capacity to ensure mutual respect and team working between the users and the experts.

Questions for Further Study

1. How can we assess the potential value of a proposed GIS?

2. Prepare a new sample GIS application definition form using Figure 17.5 as a guide.

3. List 10 tasks critical to a GIS project that a GIS manager must perform and the roles of the main members of a GIS project team.

4. Why might a GIS project fail? Draw on information from various chapters in this book.

Further Reading

Douglas, B. 2008. *Achieving Management Success with GIS.* Hoboken, NJ: Wiley.

Heywood, I., Cornelius, S., and Carver, S. 2006. *An Introduction to Geographical Information Systems* (3rd ed.). Harlow: Pearson Education Ltd.

Maguire, D.J., Kouyoumjian, V., and Smith, R. 2008. *The Business Benefits of GIS: An ROI Approach.* Redlands, CA: ESRI Press.

Sugarbaker, L.J. 2005. Managing an Operational GIS. In Longley, P.A., Goodchild, M.F., Maguire, D.J., and Rhind, D.W. (eds.) *Geographical Information Systems: Principles, Techniques, Management and Applications* (abridged ed.). Hoboken, NJ: Wiley.

Tomlinson, R. 2007. *Thinking about GIS: Geographic Information System Planning for Managers* (3rd ed.). Redlands, CA: ESRI Press.

18 Operating Safely with GIS

In earlier chapters we covered the fundamentals of how to use GIS tools by understanding the underlying science. In Chapter 17 we set out how to select and manage the tools. In this chapter we focus on how best to use this sophisticated science and technology suite to make safe decisions. By this we mean decisions that will enable you to seize opportunities without losing public trust or incurring excessive costs, while not normally being challenged by lawyers or citizens. This necessitates some understanding of the role nation-states play in setting the ground rules and the extent to which they are active participants and regulators in GI markets. Naturally, there is a geography to this contextual information; the world is far from homogeneous.

LEARNING OBJECTIVES

After studying this chapter, you will understand:

- The nature of trade-offs in GIS-based decision making.

- The common and different drivers for organizations and employees using GIS.

- The characteristics of information in general and GI in particular.

- Some important elements of the operating environment for safe GIS-based decision making—the law, trading in GI and the role of governments, protecting or undermining privacy, the ethics of behavior, and public trust.

- How operating environments can vary through time and in different jurisdictions.

- Where to go for more detailed information and advice.

18.1 Introduction

Most readers of this book live in a capitalist society, with all of its advantages and disadvantages (see Chapter 20). Our tangible wealth and physical well-being come increasingly from the exploitation of our knowledge, skills, and available information. Much of that information is geographic in nature and tells us where things are happening, where there is high attraction to live and work, which is a generally safe area, how much is the local cost of living, and so on.

But even if the hidden hand of capitalism shapes how societies evolve and prosper, most of us also live in a managed economy. Managers exist to make decisions and implement them successfully. Increasingly, there is a demand to ensure that decision making in such managed economies relies on good evidence and that evidence is available for widespread scrutiny. Each and every organization— businesses, governments, voluntary organizations, and everything else (even universities)—has managers, and everyone at some stage is going to be a manager in some function. Given that much of the information we use is geographic, it follows that GIS and GI are central to many management functions and much decision making.

Geography shapes decisions and the consequences.

Good decision making is difficult and entails taking account of many environmental or contextual factors as well as simpler operational ones. Box 18.1 describes some of the general issues that anyone involved in GIS use for decision making soon discovers.

Geography is increasingly being recognized as a central concern for all organizations, and this book contains many examples of why this is so. But in the real world, rarely is geography the *only* issue with which managers have to grapple. And rarely do managers have the luxury of being solely in charge of anything or have all the information needed to make truly excellent decisions. The reality is about operating in a situation where knowledge of the options and the outcomes of different decisions is incomplete (see Chapter 6).

18.2 GIS and Decision Making

It is easy (but misleading) to consider all decisions as isolated activities. In reality, changing one thing influences another. *Any* policy or management action taken has consequences, some of which are unforeseen. Except for those situations where there is not a zero-sum game, a trade-off almost always exists between the benefits accrued by someone from this policy or action and the disbenefits accrued by another (see Box 18.2). Sometimes these benefits or disbenefits provide advantages only to a small number of people such as the staff of a business selling a service. Sometimes the benefits apply to the whole of society, though at some cost to something else that society holds dear. And finally, the balance between private gain and public benefit that is acceptable differs in different societies: there is a geography of trust, privacy, openness, and commercial exploitation.

Everything is interconnected: normally, some gain and some lose in decision making.

Applications Box 18.1

Some GIS Facts of Life

- People cause more problems than technology: they are, in effect, complex adaptive systems embedded within adaptive societal systems!

- Things change increasingly rapidly—both the technology and the user expectations.

- Complexity is normal, but some of it is self-imposed—and all organizations tend to believe they are unique and complicated.

- Uncertainty is always with us.

- Interdependence is inescapable. Creation and use of a GIS often have far wider ramifications than originally imagined and hence impact on many more people, at least indirectly.

- There are considerable differences in national or even regional cultures. These can have significant impacts on the use of GIS in different countries.

The Trade-off between Geographical Detail and Privacy

It is now technologically possible to produce crime maps on a street-by-street basis, showing individual crime locations. Figure 18.1A shows one such example, portraying individual homicides by weapon type for part of central Manhattan. In 2008, the UK government made the publication of crime maps mandatory by all English police authorities. The results vary by police force, but a typical example is shown in Figure 18.1B. These maps are much less detailed than the U.S. example, the crimes being summarized by wards (areas of about 10,000 people). This use of areas to summarize crimes was not done for technical reasons but ostensibly because the authorities were persuaded that too much geographical detail would undermine the housing market and contravene the personal privacy of those who had already suffered loss. As such, the decision benefits one group of individuals at a cost to others.

(A)

(B)

Figure 18.1 (A) Individual homicides between 2003 and 2009 by weapon type in central Manhattan, courtesy of the *New York Times*. (B) Total notified offenses for part of Central London in April 2009, as produced by the Metropolitan Police Authority. The results are produced at the electoral ward level rather than for individual streets, even though data are geocoded to that level or finer, Fig 18.1B, (Source: Ordnance Survey. Crown copyright)

18.3 Organizational Context

Some highly successful people—such as Jimmy Wales, the co-founder of Wikipedia—have personal objectives that they translate into organizational objectives. The founders of Google set the organization's "do no evil" mission. But most individuals are hired hands, and their work-related objectives are nominally at least those of the organization.

It is also important to recognize that we see some breakdown of previously discrete sectors: commercial considerations impinge more widely, and there is some convergence between the activities and operations of commerce and industry, government, the not-for-profit sector, and academia. Increasingly, everyone is concerned with explicit goals (many of which share similar traits) and with knowledge management. Moreover, in the GIS world at least, we also see significant overlap in functions. For instance, both government and the private sector are important producers of geographic information. The not-for-profit sector increasingly acts as an agent for or supplement to government and often operates in a very business-like fashion. The result is that previous distinctions are becoming blurred, and movement of staff from one sector to another is becoming more common.

To an extent, organizational objectives are selected from a common set. Every organization now has to listen and respond to customers, clients, or other stakeholders. Every organization also has to pay attention to citizens whose power sometimes can be mobilized successfully against even the largest corporations. Every organization has to plan strategically and deliver more for less input, meeting (sometimes public) targets (e.g., profitability). Everyone is expected to be innovative and to deliver successful new products or services much more frequently than in the past. Everyone has to act and be seen to be acting within the laws, regulatory frameworks, and some conventions. Finally, everyone has to be concerned with risk minimization, knowledge management, and protection of the organization's reputation and assets.

For these reasons, we use *business* as a single term to describe the corporate activities in all four sectors identified above—commerce and industry, government, not for profit and academia. Accordingly, we think of the GIS world as being driven by organizational and individual objectives, using scientific understanding and raw material (data, information, evidence, knowledge, or wisdom: see Section 1.2), tools (geographic information systems software), and human capital (skills, insight, attitudes, and experience) to achieve them. Along the way, the managers have to adhere to the rules of the game as set out in laws, preserve reputations, behave ethically, and create new products or services (e.g., other information or knowledge). Failure to adhere to these rules can have disastrous organizational and individual consequences—hence the title of this chapter.

The GIS world is driven by organizational and personal objectives, scientific understanding, and raw materials (GIS, skills, experience, and GI).

The emphasis placed on these objectives differs significantly between organizations in different sectors, and this translates into the parameters defining the decision-making space within which an individual operates (see Table 18.1). Thus commonality between the different sectors should not be exaggerated. For example, in publicly owned commercial organizations in the United States, pleasing the stockholders every quarter is a commonplace requirement. Takeovers and mergers are much more common in the commercial than the government sector—as in the purchase of Navteq, the Chicago-based firm that among other things provides digital maps for car guidance, by Nokia, the Finnish mobile telecommunications business, in July 2008 for $8.1 billion. But it is still realistic to regard all good organizations as operating in a business-like fashion.

18.4 Geographic Information

GIS and the science underlying such systems have been described throughout this book. Here we focus on geographic information (GI)—the fuel required for many decisions, with GIS as the engine. The characteristics of information and GI in particular shape the opportunities and pitfalls that GIS users face.

18.4.1 The Characteristics of Information

Information, as seen from a management perspective, has a number of unusual characteristics as a commodity. In particular, it does not wear out through use, though it may diminish in value as time passes.

It is usually argued that information is in general a public good. A pure public good has very specific characteristics:

Table 18.1 Some business drivers and typical responses. Note that some of the responses are common to different drivers and some drivers are common to different types of organization.

Sector	Selected Business Drivers	Possible Response	GIS Example
Private	Create bottom-line profit and return part of it to shareholders. Build tangible and intangible assets of firm. Build brand awareness.	Get first mover advantage; create or buy best possible products, hire best (and most aggressive?) staff; take over competitors or promising start-ups to obtain new assets; invest as much as needed in good time; ensure effective marketing and "awareness raising" by any means possible; reduce internal cost base.	Purchase and exploitation of Map-Guide software by Autodesk and subsequent development—one of the earliest Web mapping tools. Engagement of ESRI in collaboration with educational sector since about 1980, leading to 80%+ penetration of that market and most students then becoming ESRI software-literate. Purchase of GDT Technologies by Tele-Atlas to obtain comprehensive, consolidated U.S. database and to remove major competitor.
	Control risk.	Set up risk management procedures, arrange partnerships of different skills with other firms, establish secret cartel with competitors (illegal)/gain *de facto* monopoly. Establish tracking of technology and of the legal and political environment. Avoid damage to the organization's brand image—a key business asset.	Typically, GIS firms will partner with other information and communication technology organizations (often as the junior partner) to build, install, or operate major Information Technology (IT) systems. Many GIS software suppliers have partners who develop core software, build related value-added software, or act as system integrators, resellers, or consultants. Data creation and service companies often establish partnerships with like bodies in different countries to create pan-national seamless coverage. Avoid partnerships with fly by night operations. Know what is coming via network of industry, government, and academic contacts.
	Get more from existing assets.	Sweat assets, e.g., find new markets, which can be met from existing data resources, reorganized if necessary.	Target marketing: use data on existing customers to identify like-minded consumers and then target them using geodemographic information systems (see Box 18.9).
	Create new business.	Anticipate future trends and developments and secure them.	Go to GIS conferences and monitor developments, for example, via competitors' staff advertisements. Buy start-ups with good ideas (e.g., ESRI and the Spatial Database Engine or SDE). Network with others in GI industry and adjacent ones and in academia to anticipate new opportunities.

(continued)

Table 18.1 (*continued*)

Sector	Selected Business Drivers	Possible Response	GIS Example
Government	Seek to meet the policies and promises of elected representatives or justify actions to politicians so as to get funding from taxes.	Identify why and to what extent proposed actions will impact on policy priorities of government and lobby as necessary for tax appropriations. Obtain political champions for proposed actions: ensure they become heroes if these succeed.	Attempts to create national GIS/GI strategies and NSDIs (see Chapter 19). Impact of President Clinton's Executive Order 12906. Force the pace of progress on interoperability to meet the needs of Homeland Security, minimize environmental hazards such as flooding. Bring data together from different departments of state to demonstrate how government funding is distributed geographically in relation to need or agreed political imperative (e.g., additional funding per caput in Northern Ireland).
	Provide good Value for Money (VfM) to taxpayer.	Demonstrate effectiveness (meeting specification, on time and within budget) and efficiency (via benchmarking against other organizations).	Constant reviews of VfM of British government bodies (including Ordnance Survey, Meteorological Office) over 15+ years, including comparison with private-sector providers of services. National Performance Reviews of government (including GIS users) in the United States and the impact of the e-government and other initiatives.
	Respond to citizens' needs for information or for enhanced services via the Web.	Identify these; set up/encourage delivery infrastructure; set laws that make some information availability mandatory.	Setting up the U.S. national GI Clearing House under Executive Order 12906, and equivalent developments elsewhere (see Chapter 19). Subsequent development of hundreds of GI portals worldwide. Seek to create international level playing field for information trading, for example, through European Union INSPIRE initiative (see Section 19.3.3). Hold competitions to obtain new ideas from innovators (such as www. showusabetterway.co.uk, in which over half the entries used GI).
	Control risk.	Avoid congressional/parliamentary exposure for projects going wrong and media pillorying.	Get buy-in from superiors at every stage. Adhere to risk management strategies and processes. Do numerous pilot studies in several areas.
	Act equitably and with propriety at all times.	Ensure that all citizens and organizations (clients, customers, suppliers) are treated identically and that government processes are transparent, publicized, and followed strictly.	Treat all requests for information equally (unless a pricing policy permits dealing with high value transactions as a priority). Put all suitable material on Web but also ensure material is available in other forms for citizens without access to the Internet.

- The marginal cost of providing an additional unit is close to zero. Thus, in effect, copying a small amount of GI adds nothing to the total cost of production (see Section 1.2); where large datasets are concerned, however, computing and storage costs may still be significant.

- Use by one individual does not reduce availability to others (termed nonrivalry). This characteristic is summarized in this famous Thomas Jefferson quotation: "He who receives an idea from me, receives instruction himself without lessening mine; as he who lights his taper at mine, receives light without darkening me."

- Individuals cannot be excluded from using the good or service (termed nonexcludability).

A pure public good is a special form of externality. Such externalities arise where production or consumption of a good (e.g., information) by one agent imposes costs on or delivers benefits to other producers or consumers. In this case, provision of a pure public good for one group of users will benefit other potential users because it is not possible to exclude them. Pollution of water or air and pollution by ambient noise are classic examples of externalities arising from external costs, that is, disbenefits (Figure 18.2), whereas refuse collection, education, and public health are classic examples of external benefits.

In practice, information is an optional public good, in that—unlike defense—it is possible to opt to take it or not; you may not feel the need to avail yourself of freely available U.S. Geological Survey data, for example! Moreover, the accessibility and cost of the systems to permit use of information influence whether, in practice, it is a public good in any practical sense. Since only a small if growing fraction of the world's population has ready access to GIS tools, scarcely can it be argued that geographic information in digital form is yet a universal public good in the traditional sense. To be pedantic, it may also be best to define information as a quasi-public good since it may be nonrival (see second bullet point above), but its consumption can in certain circumstances be excluded and controlled. The business cases and vast investments of a number of major commercial GI purveyors such as GeoEye and DigitalGlobe are based on this proposition. If everything they produced could be copied for free and redistributed at will by anyone, their business would be challenged (but different business models are used by other players such as Google, as discussed in Section 18.5.2).

Figure 18.2 The geography of a negative externality: patterns of noise pollution in decibels in central London. (Source: Department for the Environment and Rural Affairs. Crown copyright)

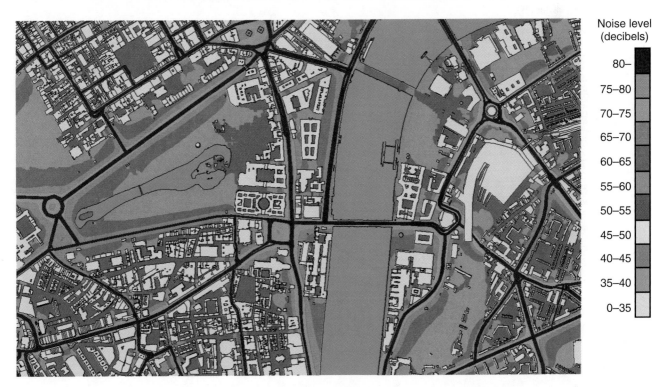

Noise level (decibels)

80–	
75–80	
70–75	
65–70	
60–65	
55–60	
50–55	
45–50	
40–45	
35–40	
0–35	

Thus the pecuniary value of information may well depend on restricting its availability while its social value may be enhanced by precisely the opposite approach—another trade-off. Finally, a set of information is also often an "experience good" that consumers find hard to value unless they have used it before.

Information and physical goods have very different characteristics.

However expensive it is to collect or create in the first instance and even to update, information in digital form can generally be copied and distributed via the Internet at near-zero marginal cost. Its widespread availability produces many positive (and some negative) externalities that arise far beyond the original creator:

- Ensuring consistency in the collection of information creates *producer externalities* by reducing the costs of creating and using data and hence potentially broadening the range of applications.

- Providing users access to the same information produces *network externalities*. It is, for example, desirable that all the emergency services use the same geographic framework data.

- Promoting the efficiency of decision making generates *consumer externalities*. One example is where access to consistent information allows pressure groups to be more effective in influencing government policy and in monitoring activities in regard to pollution.

18.4.2 Additional Characteristics of GI

GI shares all these characteristics but has others:

- Comparing one dataset for an area with another often permits the infilling of missing data, thereby cutting the cost of data collection in some cases.

- Some indication of the quality of individual GI datasets can be assessed through overlay on other, supposedly correlated but independently collected, data sets to check their consistency.

- It provides added value, almost for nothing. The number of combinations arising from overlays rises very rapidly as the number of input datasets rises (see Figure 18.3). Hence, when datasets are linked together, many more applications can be tackled and new products and services created than when they are held separately.

Figure 18.3 The number of pairs of variables that can be selected from 2 to 20 variables (= overlays) for the same area, assuming order is not important and repetition is not allowed. Note that the total number of combinations is much larger (i.e., includes all groups of 2, 3, 4 or more variables).

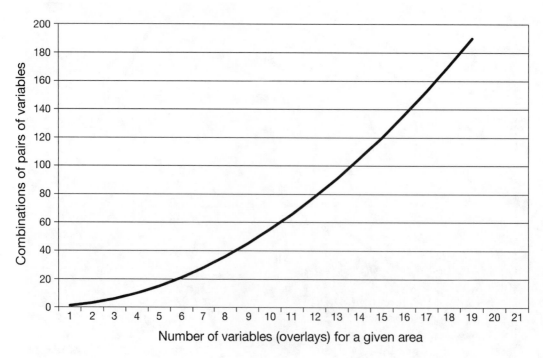

- It is difficult to quantify the quality of some types of GI, for example, area classification data like soil type, and this has ramifications when combining data by overlay and in modeling.

18.5 GIS, GI, and Key Management Issues

We have selected five different areas in which safe GIS use is impacted by management or where GIS can contribute to ameliorating management's difficulties. These are the law (especially intellectual property rights, or IPR); trading in GI and the local roles of government; protecting or undermining privacy; the ethics of behavior; and public trust.

18.5.1 The Law

The use of GIS can support implementation of the law. It underpins the efforts of many law enforcement and criminal justice staff, for example, to plan effectively for emergency responses, create and test risk mitigation strategies, analyze historical events, and project future ones. It sometimes provides crucial evidence: in a famous murder case in the UK, for example, the accused were acquitted partly on the evidence that mobile phones associated with them were used 2.4 miles away within 3 minutes of the report of the murder. But the law also influences the use of GIS: it touches everything.

Beware! The law touches everything.

During a career in GIS, we may have to deal with many manifestations of the law, including copyright and other intellectual property rights (IPR), data protection laws, public access issues enabled, for example, through Freedom of Information Acts (FOIA), and legal liability issues. But since laws of various sorts have several roles—to regulate and incentivize the behavior of citizens and to help resolve disputes and protect the individual citizen—almost all aspects of the operations of organizations and individuals are steered or constrained by them. Accordingly, in this chapter we concentrate on two aspects of the law and GIS: protecting innovation and liability.

One further complication in areas such as GIS is that the law is always doomed to trail behind the development of new technology; laws only get enacted after (sometimes long after) a technology appears. Finally, there is also a geography of the law.

The legal framework varies from country to country. For example, the Swedish tradition of open records on land ownership and many personal records dates back to the Seventeenth Century, but much less open frameworks exist even in other countries of Europe. The creation, maintenance, and dissemination of "official" (government-produced) geographic information is strongly influenced by national laws and practice.

18.5.1.1 Fostering Innovation and Protecting Exploitation

Innovation is the key to progress. Two types of innovation can be identified: where it is created internal to an organization and its exploitation is protected by law or where its creation occurred through an open innovation process where the innovation results may not be protected (though there may still be some form of licensing—see Section 18.5.1.2). Some of the most important open technological innovations of the last 400 years—notably Harrison's chronometer (Box 18.3) and the Global Positioning System—have resolved problems of geographical description. But central to most commercial activity is protecting the fruits of innovation (at least for a time) from competition, which simply seeks to clone a product, a service, or a process invented by others. Thus the investment of time and creativity that goes into imagining, creating, and refining a new algorithm, software package, or database is protected from cloning without permission by intellectual property rights agreed on by most countries of the world. There are various categories under which IPR is protected (see Box 18.4), and the type of protection varies between countries. IPR is in constant flux as a result of changes in the law and challenges facilitated by new technologies: if in doubt, it is important to seek up-to-date professional advice.

Protecting innovation underpins much commerce—and governments.

Protection of IPR is a major factor for most GIS businesses. Any Web search will provide examples of legal conflicts. For example, on June 3, 2009, Facet Technology Corp. and the Dutch navigation device maker TomTom, N.V., reached a settlement after a 16-month legal battle. The case was settled through a patent agreement under which TomTom acquired a limited license to Facet's patents on tools that facilitate capturing road sign and other data needed to underpin Advanced Driver Assistance Systems.

The Longitude Problem

On October 22, 1707, a total of 2000 British sailors and troops died when their ships ran on to rocks far to the west of where they thought they were. This and similar disasters forced the British Parliament in 1714 to offer a prize of £20,000 (worth about $4 million in 2008 prices, based on a long-run series for the retail price index) for a reliable solution to defining longitude to within an accuracy of half a degree. Initially, it was thought that the Lunar Distance Method was the best approach. That was based on the idea that, if an accurate catalog of the positions of the stars was created and the position of the moon was then precisely measured relative to the stars, tables of the moon's position would give the time at the reference point. But the longitude problem was eventually solved by a working-class carpenter with little formal education. John Harrison (Figure 18.4A) built a series of highly accurate clocks that coped with the motion of ships and changes in temperature and humidity. Longitude was computed from the difference between local time (with noon being when the sun was at its zenith) and the time at a reference point such as Greenwich, as shown by the chronometer: a 1-hour difference indicated a 15-degree difference in longitude. After much altercation with the judging committee, numerous demonstrations that his chronometers (see Figure 18.4B) could keep very accurate time on long voyages, and petitioning of the king, Harrison eventually received the prize in 1773 after 40 years of innovation.

(A)

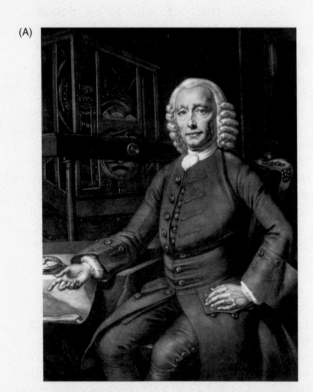

(B)

Figure 18.4 (A) John Harrison, the winner of the Eighteenth Century open innovation prize (worth $4 million at today's prices) to find a reliable method of measuring longitude. (B) The Harrison H4 chronometer with which he won the prize. (Source: Photo A: Courtesy Royal Astronomical Society/Photo Researchers, Inc; Photo B: National Maritime Museum www.nmm. ac.uk/harrison#longitude and Sobel D (1995) *Longitude*.)

In some industries, however—notably in the services sector where innovation often occurs in close association with a particular customer—IPR is not protected, and the "first to market" advantage is exploited in products with a short shelf life. Alternatively, developers may choose a nontraditional business model and seek the widest take-up of their products, with few revenue benefits obtained directly from users (e.g., Google or open-source software).

18.5.1.2 Owning and Exploiting GI

How does IPR fit into the world of GIS and geographic information? The answer is that it is complex, especially when seen from a global rather than a national perspective, for at least three reasons:

1. National or regional IPR laws vary in some elements of substance and in many details.

2. The formal interaction between the private and public sector also varies between countries.

3. Global databases are being built in some domains (e.g., for satellite navigation purposes) while many other geographical databases are national in scope, content, ontology, and reference systems.

The simplest way to summarize the situation is to answer some commonly asked questions:

Can geographic data, information, evidence, and knowledge be regarded as property? The answer to this question is normally yes, at least under certain legal systems. The situation is most obviously true in regard to data or information—under either copyright or database protection. Who owns the data is sometimes difficult to define unequivocally, notably in aggregated personal data. An example is an individual's medical records, which sometimes are subsequently geocoded to provide data for research purposes. Place-name labels may also be property (see Box 18.5)

Can geographic information always be legally protected? There is little argument about copyright where it is manifestly based on great originality and creativity, for example, in relation

---

Applications Box (18.4)

Intellectual Property Rights

Intellectual property rights (IPR) are the rights given to persons over their intellectual creations, and they usually give the creator an exclusive right to the use of his or her creation for a certain period of time. The length of time establishes a balance between the individual's rights to benefit from the inventions and the benefits to wider society. There are two main types of IPR:

Copyright and Rights Related to Copyright
The rights of authors of literary and artistic works (such as books and other writings, musical compositions, paintings, sculpture, computer programs, and films) are protected by copyright, for a minimum period of 50 years after the death of the author. The main social purpose of protecting copyright and related rights is to encourage and reward creative work.

Industrial Property
Industrial intellectual property can be divided into two main areas:

- The protection of distinctive signs, in particular trademarks (which distinguish the goods or services of one undertaking from those of other undertakings) and geographical indicators (see Box 18.5). This protection aims to stimulate and ensure fair competition and to protect consumers by enabling them to make informed choices between various goods and services. It may last indefinitely, provided the sign in question continues to be distinctive.

- Other types of industrial property. These rights are protected primarily to stimulate innovation, design, and creation of technology. In this category fall inventions (protected by patents), industrial designs, and trade secrets. The social purpose of these rights is to provide protection for the results of investment in developing new technology, thus giving the incentive and means to finance research and development activities. The protection is usually given for a finite term (typically 20 years in the case of patents).

Source: www.wto.org

Geographical Indicators

Geographical Indicators are place-name labels applied to a product; these indicators denote that the product was—and can only be—created in a particular (usually limited) geographical area. Examples include Champagne, Tequila, Parma Ham, and Roquefort or Feta cheese. Articles 22 and 28 of the World Trade Organization's Agreement on Trade Related Aspects of Intellectual Property Rights (TRIPS) argue that a product's quality, reputation, or other characteristics can be determined by its origin. All products are covered by Article 22, which defines a standard level of protection. This article states that geographical indications have to be protected in order to avoid misleading the public and to prevent unfair competition. Article 23 provides a higher or enhanced level of protection for geographical indications for wines and spirits.

to a painting. Rarely, however, is the situation as clear-cut, especially in the GIS world where some information is widely held to be in the public domain, some is regarded as facts, and some underpins either the commercial viability or the public rationale of government organizations.

Location can be the determining factor in awarding some IPR.

In the United States, the Supreme Court's ruling in the famous *Feist* case of 1991 has been widely taken to mean that factual information gathered by "the sweat of one's brow"—as opposed to original, creative activities—is not protected by copyright law. Names and addresses are regarded widely as geographic facts. Nonetheless, several jurisdictions in the United States and elsewhere have found ways to protect compilations of facts, provided that these compilations demonstrate creativity and originality. The real argument is about which compilations are sufficiently original to merit copyright protection. A number of post-*Feist* instances have occurred in which the courts have recognized maps not as facts but as creative works, involving originality in selection and arrangement of information and the reconciliation of conflicting alternatives. Thus some uncertainty still exists in U.S. law about what GI can be protected.

In Europe, different arrangements pertain—both copyright and database protection exists. For copyright protection to apply to a database, it must have originality in the selection or arrangement of the contents; and for a database right to apply, the database must be the result of substantial investment. It is, of course, entirely possible that a database will satisfy both of these requirements so that both copyright and database rights may apply. A database right is in many ways very similar to copyright so that, for example, there is no need for registration: it is an automatic right like copyright, and it begins as soon as the material that can be protected exists in a recorded form. Such a right can apply to both paper and electronic databases. It lasts 15 years from making, but, if published during this time, then the term is 15 years from publication. Protection may be extended if the database is refreshed substantially within the period. As stressed earlier, geographic differences in the law have potential geographic consequences for commercial GI activity. Protection equivalent to the database right will not necessarily exist in the rest of the world, although all members of the World Trade Organization (WTO) do have an obligation to provide copyright protection for some databases.

Can information collected directly by machine, such as a satellite sensor, be legally protected? Given the need for originality, it might be assumed that automated sensing should not be protectable by copyright because it contains no originality or creativity. Clearly, this interpretation of copyright law is not the view taken by the major players, such as GeoEye, since they include copyright claims for their material and demand contractual acceptance of this proposition before selling products to firms or individuals. Typically, as in the case of GeoEye, this contractual requirement runs as follows:

All content, graphics, audio, video, software, and other materials available via the Site, and the compilation of such materials (including, for example, postings and collections of links to other Internet resources, and the descriptions of those resources that are contained within or made available via the Site) (collectively, the

"Materials" or "Content"), are the property of GeoEye and/or its suppliers and licensors, and are protected by U.S. copyright and other intellectual property laws and international treaties.

From the point of view of enforcing copyright, it helps that there are at present only a few original suppliers of satellite imagery. The cost of getting into the satellite-sensing business is high; thus any republication or resale of an image could be tracked back to the source with relative ease. In addition, there is some evidence that some of the key customers (notably the military) are interested primarily in near-real-time results and hence require frequent updates. In any event, imagery up to 15 years old would be protected under the EC's Copyright Directive, at least within the European Union area itself.

How can tacit geographic and process knowledge—such as that held in the heads of employees and gained by experience—be legally protected? "Know-how" can migrate very rapidly. Suppose that the lead designer of a new software system is recruited by a competitor software vendor. Although no code is actually transported, the essence of the software and the lessons learned in creating the first version would greatly facilitate the creation of a competitor. This situation has happened in GIS/GI organizations from the earliest times. The only formal way to protect tacit knowledge is to write some appropriate obligations into the contract of all members of staff—and be prepared to sue the individual if he or she abrogates that agreement. Human rights legislation may complicate winning such action. Keeping your staff busy and happy is a better solution!

How can you prove theft of your data or information? It is now quite common to take other parties to court for alleged theft of your information. One of the authors previously led an organization that successfully litigated against the Automobile Association in the UK for illegal copying, amending, and reproducing (in millions of copies) portions of Ordnance Survey (OS) maps. This infringement occurred despite the terms, conditions of use, and charges for such mapping being published and well-known: the OS received over $30 million in back payment.

The legal process in such circumstances is an adversarial one. Therefore, it is crucial to have good evidence to substantiate one's case and to have good advice on the laws as they exist and as they have been previously interpreted by courts. How then do you prove that some GI is really yours and that any differences that actually exist do not simply demonstrate different provenances—especially since digital technology makes it easy to disguise the look and feel of your GI and reproduce products to very different specifications, perhaps generalized?

The solution is to be both proactive and reactive. In the first instance, the data may be watermarked in both obvious and nonvisible ways. Although watermarks can be removed by unscrupulous individuals, they serve as a warning. Scattering a series of groups of small numbers of colored pixels randomly throughout a raster dataset ("salting") can be effective. Proof acceptable to the courts requires a good audit trail within one's own organization to demonstrate that these particular pixels were established by management action for that purpose. In addition to watermarking, fingerprinting can also be used. At least one major commercial mapping organization has used occasional fictitious roads to its road maps for this purpose—the U.K. based Geographers' A–Z Map Company included the non-existent Lye Close in its Bristol maps to trap unwary copyright infringers. In areas subject to temporal change (such as due to tides or through vegetation change), it is very unlikely that any two surveys or uses of different aerial photographs could ever have produced the same result in detail. Such evidence should, in the right hands, be enough to demonstrate a good case, though rather different techniques may be called for in regard to imagery-based datasets of slow-changing areas.

Who owns information derived by adding new material to source information produced by another party? Suppose you find that you have obtained a dataset that is useful but that, by the addition of an extra element or by overlaying data, becomes immensely more valuable or useful. So who owns and can exploit the results in derivative works? The answer is both you and the originator of the first dataset. Without his or her input, yours would not have been possible. This is true even if the original data are only implicit—if, for instance, you trace off only a few key features and dump all the rest of the original. Moreover, you may be in deep trouble if you combine your data with the original data and omit to ask the originator for permission to do so. Where moral rights exist (e.g., in

European countries), the author has the right to integrity (i.e., to object to any distortion or modification of his or her work which might be prejudicial to his or her honor). He or she also has the right to attribution as the author, to decide when a work will be published, and to withdraw a work from publication under defined circumstances.

Almost all GI is licensed somehow. Read the license before using the GI!

Do I need a license? In most cases, users of GI explicitly or implicitly have accepted the terms of a license for the information they are accessing and reusing. Even where the licensing is not explicit and use of the information is free, it still exists (e.g., under terms and conditions that prevent editing of the data and then misrepresenting the work of the originator or in regard to "creative commons"; creativecommons.org). In some cases, licensing is an onerous process and can necessitate signing up to hundreds of pages of contractual details limiting the extent and nature of reuse and even the charging mechanism. In some other cases, licensing can be done online, such as via the UK government's Click Use licenses (www.opsi.gov.uk).

It is always wise to read the terms of any licensing agreement with great care before simply clicking on an "I agree" box. The late 2008 version of the license to use Google Maps, for instance, included the following statements, which provide Google with a license to use your data if it had been published using Google software:

1. You retain copyright and any other rights you already hold in Content which you submit, post or display on or through, the Services. By submitting, posting or displaying the content you give Google a perpetual; irrevocable, worldwide, royalty-free, and non-exclusive license to reproduce, adapt, modify, translate, publish, publicly perform, publicly display and distribute any Content which you submit, post or display on or through, the Services. This license is for the sole purpose of enabling Google to display, distribute and promote the Services and may be revoked for certain Services as defined in the Additional Terms of those Services.

2. You agree that this license includes a right for Google to make such Content available to other companies, organizations or individuals with whom Google has relationships for the provision of syndicated services, and to use such Content in connection with the provision of those services.

3. You understand that Google, in performing the required technical steps to provide the Services to its users, may (a) transmit or distribute your Content over various public networks and in various media; and (b) make such changes to your Content as are necessary to conform and adapt that Content to the technical requirements of connecting networks, devices, services or media. You agree that this licence shall permit Google to take these actions.

4. You confirm and warrant to Google that you have all the rights, power and authority necessary to grant the above license.

18.5.1.3 GIS, GI, and Legal Liability

Liability is a creation of the law to support a range of important social goals, such as avoiding injurious behavior, encouraging the fulfillment of obligations established by contracts, and distributing losses to those responsible for them. It is a huge and complicated issue. Apparently, Alaska is the only U.S. State that has specific provisions for limiting liability for GIS. In many cases liability in data, products, and services related to GIS will be determined by resort to contract law and warranty issues. This assumes that gross negligence has not occurred; if it has occurred, the situation can be far worse and individuals as well as corporations can be liable for damages. Other liability burdens may also arise under legislation relating to specific substantive topics such as intellectual property rights, privacy rights, antitrust laws (or noncompetition principles in a European context), and open records laws.

Clearly, this is a serious issue for all GIS practitioners. Minimizing losses for users of geographic software and data products and reducing liability exposure for creators and distributors of such products are achieved primarily through performing competent work and keeping all parties informed of their obligations. If you have a contract to provide software, data, services, or consultancy to a client, that contract should make the limits of your responsibility clear, though disclaimers rarely count for much. And you should adhere to well-recognized professional codes of conduct (see Section 18.5.4).

18.5.2 Trading in GI and the Role of Government

That GI is now a necessary part of delivering services, especially satellite navigation and online mapping, is self-evident. Google, Microsoft, and Nokia have all bought their way into accessing or owning GI as part of their normal business; IBM, Microsoft, and Oracle have created spatial data-handling software as part of their main database engines. This builds on the availability of tools provided by ESRI and other vendors to create and manage the GI and substantial investment in the data collection process by commercial organizations. In addition, many value-added resellers provide niche applications, datasets, or services built on top of the core offerings from multinational giants. So the existence, accuracy, quality, availability, accessibility, and price of GI underpins the whole industry, even if the end use of the information is not charged for directly (e.g., under business models that provide services supported by advertising).

GI is the rocket fuel and GIS the engine.

18.5.2.1 The Role of Governments

One characteristic distinguishes the GIS/GI world from many others: governments have been and are major collectors (and sometimes providers) of certain types of GI and could provide much more information that is presently locked up inside public-sector organizations. The origins of this often lie in defense needs, and in some parts of the world security concerns still constrain access to detailed GI. In general, however, the remit of government data providers varies considerably across the world. We now describe briefly these differences and their significance for information trading, both for individual users and for commercial organizations. We use the geographic framework (i.e., the basic topography of the country) as the worked example since it usually forms a template to which most other GI is fitted.

What constitutes the framework varies between different countries. In the United States, the Federal Geographic Data Committee has defined the information content of the U.S. framework as consisting of

- geodetic control (control points, datums, etc.)
- elevation data
- hydrography

- public land cadastral information
- digital orthoimagery
- transportation
- the geography of governmental/administrative units

Similar definitions, including place-names, would in practice be found in many countries. Most typically, the framework is manifested in user terms through topographic mapping. Normally, such mapping will contain all this information but partly in implicit form (e.g., geodetic control). It has long been produced by national mapping agencies; as arms of government, these have provided continuity of provision over decades or even centuries in many countries.

Commercial imperatives in selling framework data are readily understood. But a number of governments are also seeking to recover some of the heavy up-front costs of information (especially GI) collection through charging for its use. Some of this has been brought about through governments worldwide reviewing their roles, responsibilities, and taxation policies and reforming their public services as a consequence. In some cases this has led to privatization of government functions or at least outsourcing. The imperative for some reviews has been financial, with a search for greater efficiency and wiping out of duplication (e.g., through the U.S. e-government and EU INSPIRE initiatives). In other reviews, the imperative was ideological—that the state should do less and the private sector should do more.

18.5.2.2 Free Our Data?

The reform of government has led to two approaches in regard to GI. The first has been the outsourcing of production: thus large parts of the imagery previously collected by many national mapping organizations are now produced under contract by the private sector. Another approach has been simultaneously to force economies within government and get the user to pay for the bulk of GI. In some European countries, for instance, land records and the cadaster are paid for totally through user fees; since many people do not own land or property (around 30% in the UK), it is deemed unfair for them to contribute to the costs via general taxation. It is also worth stressing that such charging is nothing new in principle: historically, almost all government organizations have imposed *some* charge for data, including federal government bodies in the United States. The only issue is for what elements users should be charged, for example, the whole product or only the costs of documentation and distribution.

The development of the Internet and the World Wide Web have dramatically changed this equation.

In the last few years a new element has begun to change the information policies of various governments. This is exemplified by President Obama's first Executive Order (http://www.whitehouse.gov/the_press_office/Transparency_and_Open_Government/). This demanded that all federal bodies take steps to make available information they hold—and elicit what additional information would be helpful—to citizens. The driver for this in the US was to increase transparency and openness of government and hence improve public trust in government.

Some governments act commercially in GI.

The advantages and disadvantages of the "charging for everything" and "free access" approaches are summarized in Box 18.6, though these approaches are subject to much argument. In Britain, information policy has fluctuated over two centuries (Box 18.7), but the ongoing costs of running the national mapping organization have been met for more than a decade totally from revenues raised from users. Perhaps as a consequence of the users' power, the main elements of the highly detailed database are guaranteed never to be more than 6 months out-of-date. That said, the policy also has some real downsides, as suggested in Box 18.6, and has recently been substantially modified (Box 18.7).

Applications Box (18.6)

The Advantages and Disadvantages of the "User Pays" and "Free Access" Philosophies (*and Some Contrary Views*)

Arguments for Cost Recovery from Users

- Charging, which reflects the cost of collecting, checking, and packaging data, actually measures real need, reduces market distortions, and forces organizations to establish their real priorities (*but not all organizations have equivalent purchasing power, e.g., utility companies can pay more than charities; differential pricing also invites legal challenge in some jurisdictions*).

- Users exert more pressure where they are paying for data, and, as a consequence, data quality is usually higher and the products are more fit for purpose.

- Charging minimizes the problem of subsidy of users at the expense of taxpayers; there is a ratio of about 30,000 to 1 of taxpayers to direct GIS users globally (*but some users are acting on behalf of the whole populace, such as local governments*).

- Empirical evidence shows that governments are more prepared to part-fund data collection where users are prepared to contribute meaningful parts of the cost. Hence full data coverage and update are achieved more rapidly: the average age of the traditional U.S. basic scale government-produced mapping is over 30 times greater than that in Britain (*but once the principle is conceded, government usually seeks to raise the proportion recovered inexorably*).

- Charging minimizes frivolous or trivial requests, each consuming part of fixed resources. (*This rationale is now much less relevant given the user-driven access to the Internet in many countries.*)

- Charging enables a government to reduce taxes.

Arguments for Dissemination of Data at Zero or Copying Cost

- The data are already paid for; hence any new charge is a second charge on the taxpayer for the same goods (*but taxpayers are not the same as users—see above*).

- The cost of collecting revenue may be large in relation to the total gains, so the exercise is inefficient (*not true in the private sector—so why should it be so in the public sector?*).

- Maximum value to the citizenry comes from widespread use of the data through innovation leading to intangible benefits or through taxes paid by the private-sector added-value organizations (*a contention that is very difficult to prove one way or the other because of the many factors involved, but recent changes of policy to free or low-cost access to GI in countries like Austria and Spain have resulted in much wider GI use*).

- The citizen should have unfettered access to any information held by his or her government, other than that which is sensitive for security or environmental protection reasons (e.g., the location of nests of endangered species).

Mutations in Information Policy

Following the policies of the British government, the Ordnance Survey (OS) first took action to prevent free riding or illegal use of the government's maps at least as early as 1817. The OS expressly warned anyone infringing their copyright that this would lead to legal action (see Figure 18.5). In the 1930s world of mapping, public policy was on a "charge only for ink and paper" basis; in the 1960s it changed to charging users a fraction of the total costs of operations. In 1999, the move to a Trading Fund status effectively led to the OS having to become profitable and thus meet full costs and interest on capital.

In 2008, the UK government was seeking to nurture the use of the Web for social, economic, and educational purposes. A government-commissioned report *The Power of Information* led to some ministerial support of the merits of freely available information, along the lines of the U.S. federal model. In addition, pressure from users, the media, and commercial organizations forced a government review of the cost recovery policy so far as it applied to government organizations operating under Trading Fund status. This review included an international comparison of GI trading models. The results of this review were announced in the national budget publication in April 2009. It reaffirmed that charging would continue as far as the Trading Funds are concerned. However, some changes to the Ordnance Survey business model to affect wider and easier use of OS information were mandated.

Soon after this, the UK Prime Minister appointed Sir Tim Berners-Lee, the originator of the World Wide Web, to foster the availability and re-use of UK public sector information. Berners-Lee convinced the Prime Minister that opening out PSI of all kinds not only had transparency benefits—as recognized in the United States—but would also foster innovative uses of it by UK firms, leading in turn to better government performance, better engagement of citizens and tax benefits from successful firms. To widespread surprise, the Prime Minister then announced that the basis on which Ordnance Survey operated would change and this was re-affirmed after a public consultation on the best way ahead. As a result, a selection of OS data was made available on a free to use (for whatever purposes) basis on April 1, 2010, some five days before the national elections which removed the Prime Minister from power. The most detailed OS data—MasterMap (see Figure 18.7)—was still however subject to licensing and charging for commercial purposes though new contracts enabling their use across the whole of government were also announced. The whole episode illustrates the multiple political, financial, technological, and quality issues intertwined in setting GI policy, and the volatile impact of engagement at the highest political levels with information policy.

TRIGONOMETRICAL SURVEY OF GREAT BRITAIN

It having been represented to the Master-General and Principal Officers of His Majesty's Ordnance, that certain mapsellers and others have, through inadvertence or otherwise, copied, reduced, or incorporated into other works and published, parts of the "Trigonometrical Survey of Great Britain," a work executed under the immediate orders of the said Master-General and Board, the said Master-General and Board have thought proper to direct, that public notice be given to all mapsellers and others, cautioning them against copying, reducing, or incorporating into other works and publishing, all or any part of the said "Trigonometrical Survey," or of the Ordnance maps which may have been or may be engraven therefrom.

"Every offender after this notice given, will be proceeded against according to the provisions of the Act of Parliament made for the protection of property of this kind."

By order of the Board,
R.H.Crew, Secretary

Office of Ordnance, 24th February 1817

Figure 18.5 The earliest known warning against infringing GI copyright: the warning printed by the Board of Ordnance in the *London Gazette* of March 1, 1817.

18.5.2.3 Competition versus Collaboration

The principle that any GI held by government should be used to the benefit of the population is widely accepted. The big issue is how the benefit is measured. In some countries, the state plays a very small role apart from releasing data; in others, it forges some relationship with the private sector. In a small number of countries, the state is the only provider and severely constrains the private sector's use of the GI.

Despite the apparent synergy, the relationship between the state as a provider of GI and the private sector as a value-adder sometimes breaks down completely. One example of where this has happened is shown in Figure 18.6. This figure shows the distributions of two different sets of meteorological stations— a long-standing one run by the official German meteorological service and the other run by a consortium of private-sector bodies. The latter was set up when the consortium's members could not agree to acceptable terms with the government body for supply of the official data.

Other examples of technology-facilitated competition between state and private-sector products are shown in Figures 18.7 and 18.8. These figures show two

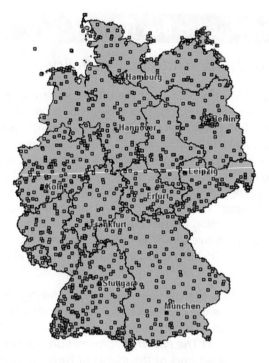

Figure 18.6 Two meteorological station networks in Germany. Blue dots: weather stations from the DWD (Germany's National Meteorological Service). Red dots: weather stations from a private meteorological information provider. (Source: www.wetterstationen.meteomedia.de)

Figure 18.7 A detailed topographic template produced by the Ordnance Survey for part of Britain, showing various layers in OS MasterMap built up from the topography, addresses, integrated transport links, and an orthophoto layer. (Source: Ordnance Survey. Crown copyright)

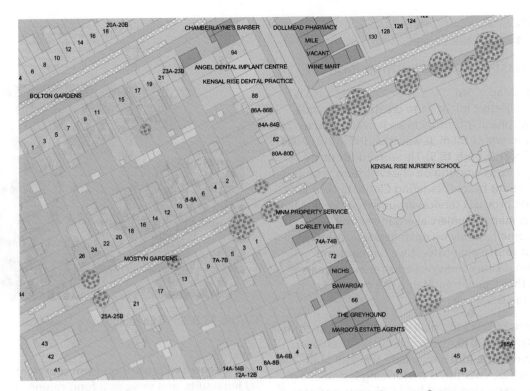

Figure 18.8 A detailed "topographic template" produced by the GeoInformation® Group, free of Ordnance Survey copyright and licensing constraints. The data set includes a base map. Overlays available include tree canopies and power lines, addresses, heights of buildings, a 5-m resolution digital terrain model, and an orthophoto layer. (Courtesy The GeoInformation® Group)

different sets of detailed topographic information for parts of the UK produced by the Ordnance Survey and by a modest-sized commercial enterprise that has taken care to assemble their information free of OS intellectual property rights. In addition, the OpenStreetMap project (www.osmfoundation.org) has harnessed voluntary efforts to produce a set of mapping for parts of many countries in the world without drawing on official sources (see Box 18.8 and Section 19.4). It is clear both that the advent of new GIS technologies lowers the entry price to GI markets and that the role of government as a producer of GI is under increasing challenge.

New GIS technologies lower the entry price to GI markets.

18.5.3 GIS and Privacy

Our personal right to privacy has become ever more sharply undermined since the bombing of the Twin Towers in New York in 2001, the subsequent global increase in surveillance to minimize the chances of similar events, and the development of new technologies for capturing detailed records of the landscape

and monitoring where individuals are within it (see Figure 18.10). Yet paradoxically, many people happily make available personal information on social networking sites and to commercial retailers but express great concern about state bodies holding equivalent information.

Google's introduction of car-based imaging to produce Street View in 2008 led to many concerns worldwide about the impact on privacy. The Greek data protection agency, for example, banned Street View's expansion in the country until it obtained acceptable undertakings on how long the images would be kept on Google's database and what measures the firm would take to make people aware of privacy rights. Reactions at the local level were varied. In the UK a spate of burglaries in a Buckinghamshire village had already put residents on the alert for any suspicious vehicles. So when the Google Street View car trundled toward Broughton with a 360-degree camera on its roof, villagers sprang into action! Forming a human chain to stop it, they harangued the driver about the "invasion of privacy," adding that the images that Google planned to put online could be used by burglars. Beginning in the summer of 2009, Google deployed tricycles mounted with its 360-degree Street

Steve Coast—OpenStreetMap Pioneer

OpenStreetMap (OSM www.openstreetmap.org) is an open, collective initiative to create and provide free geographical information such as street maps to anyone who wants them. It was founded in July 2004 by Steve Coast (Figure 18.9). He worked for many technology firms working on "heavy-lifting" computing applications before founding CloudMade in 2008. CloudMade provides services around OpenStreetMap and allows many commercial organizations, both large and small, to rely on open geodata. A long-term open source advocate, Steve has used and spoken about Linux and Wikipedia for many years and feels that OpenStreetMap is the natural extension into the GIS space.

Versions of OSM are now available in most countries. By January 2010, the project's maps had over 200,000 users. The initial map data were all built from scratch by volunteers performing systematic ground surveys using hand-held GPS units and cameras, notebooks, or voice recorders. The data were then entered into the OpenStreetMap database from a PC. More recently, the availability of aerial photography and other data sources from commercial (e.g., Yahoo) and government sources has greatly increased the speed of this work and has allowed land-use data to be collected more accurately.

In April 2006, the OpenStreetMap Foundation was established to act as a custodian for the servers and services necessary to host OpenStreetMap, pro-

Figure 18.9 Steve Coast. (Courtesy Steve Coast)

vide a degree of protection from copyright and liability legal suits, and serve as a fund-raising vehicle. The Foundation is an international nonprofit organization supporting but not controlling the project and is owned by its members: the very nature of the whole project is shown by the fact that anyone can set up servers to host the OSM data. The OSM information was initially made available under the Creative Commons Attribution-Share Alike 2.0 license. However, legal investigation work and community consultation is underway to re-license the project under the Open Database License from Open Data Commons, which is believed to be more suitable for a map dataset.

View cameras (Figure 18.11) to map areas of Britain that were inaccessible by its fleet of Street View cars. Following the forging of a partnership between Google and the official UK tourism office, some of the sites visited will be chosen by popular vote.

18.5.3.1 Geoslavery

The widespread availability of GI as in Street View and as handled in GIS is manifestly a factor in the loss of anonymity. Perhaps the most extreme viewpoint is that of Jerry Dobson and Peter Fisher who have coined the term *geoslavery*. By this term, they mean the constant monitoring of an individual's position via mobile (or cell) phones and a hypothetical extension that would reward or cause pain if the "slave" deviated from a path chosen by the "master".

Privacy is a complex concept, varying between individuals and at different moments in time.

While some people would abhor any broadcasting of their location, others—such as many users of Google—welcome the ability to tell friends where they are in town if they are looking for a social gathering (even if this then provides Google with the opportunity to tailor advertisements geographically for them). Parents concerned about the safety of their children might well contemplate fitting them with a GPS receiver and transmitter to know where they are at all times. A hill walker or mountaineer might well be happy for others to know where he is in the event of an accident. In these examples, an element of privacy is being traded off to ensure

reason
Free Minds and Free Markets

$3.50 U.S. $4.50 Canada June 2004

NICK GILLESPIE...
They Know Where You Are!
The unsung benefits of a database nation
(see next page)

Figure 18.10 Customized cover of an issue of *Reason* magazine, with an air photo of the home of each one of the 40,000 individual postal subscribers shown on the copy they received. This was achieved by merging their subscription address through a geocoded address matching process with suitably georeferenced aerial photography and generating customized, clipped digital images for the digital printing process. It heralds hyper-individualized publications but also highlights concerns about privacy. (Source: Reason Magazine http://reason.com/june-2004/samples.shtml)

enhanced safety. Thus privacy is a complex concept, varying between individuals and at different moments in time.

18.5.3.2 Privacy versus the Case for Individual Data

In many cases, the resolution of GI pertaining to an area of land is not normally a matter of contention over privacy. Central to privacy concerns, however, are data or information about human individuals or small firms. Much information in contemporary society about individual people, their actions, wealth, and relationships and about individual businesses is collected through administrative systems. Some information is collected through statistical surveys; even the population census is a sample survey of sorts since rarely is *everyone* counted. Detailed personal data exist in all governments and in many commercial organizations.

Such detailed information is often aggregated to produce information used in GIS. But the GIS

Figure 18.11 The Google Street View trike with cameras used in areas inaccessible to the car-mounted cameras. (Courtesy WENN.com/NewsCom)

also plays an important role in linking the individual data together—sometimes using an explicit and a unique common key (such as a social security number) but at other times through the approximation of a geographic address or other locational identifier. In particular, bringing together multiple sets of information about human individuals is highly contentious. The extent of the technological capabilities (notably GIS), statutory constraints on its use, and the culture of those empowered to use it determine the threat to privacy. Thus organizations such as the U.S. Bureau of Census and other U.S. federal government departments have strong protocols for which data may be linked together and for specified purposes. Such precautions do not always work: in Britain the tax authorities lost a CD containing 25 million records of individual taxpayers in 2007. But such losses are virtually unknown among organizations set up specifically as data collectors and aggregators (such as national statistical institutes).

Most socioeconomic data are collected for individuals, and so privacy matters.

As usual, there is a trade-off here. The potential loss of privacy through misuse or loss of personal data has to be traded against the considerable benefits to be gained from use of individual data. Potential benefits include the following:

- Reducing the burden on people or businesses to fill in multiple questionnaires, with data collected for one purpose being spun off from databases originally collected for another.

- Knowing where everyone is and hence the location of potential victims in the event of natural hazards occurring.

- Ensuring that wherever an individual travels, his or her detailed health records could be available to any doctor in the event of a sudden illness or accident.

- Allocating resources (e.g., social security payments) on a fair basis related to the individual's characteristics.

- Reducing the incidence of fraud by comparing living standards and so on of each individual with the norm for people or businesses of their type, often by merging multiple administrative datasets together (e.g., tax records and social security benefits).

- Tracking the life histories of individuals to study correlations between, for example exposure to environmental hazards and subsequent illnesses.

- Profiling people on the basis of their background, personal characteristics, or contacts as being predisposed toward acts of crime or terrorism or for credit-rating purposes.

- Studying inequality in society through analysis of life experiences by people in small, specifically defined groups (e.g., communities or ethnic groups), from which new policies and actions might flow.

These benefits could be of major value to individuals, businesses, and society. Such benefits are not realizable from aggregate data (e.g., for geographical areas). While it is possible to hypothesize the characteristics of human individuals by microsimulation techniques such that in total they exactly match the aggregate data, this is no more than a scenario (see Section 16.1.1). Data about *real* individuals are crucial to achieving and justifying these benefits, and a geographical (or other) code of some kind attached to each record is needed to produce the aggregate data from the individual's details.

In many—though not all—of the examples listed in this section, the name of the individual is irrelevant. Where data are linked together for research purposes, then anonymized, and kept in a secure environment, perhaps under license, there is little to fear. Whatever the reality, however, the public remains suspicious and needs to be reassured that steps have been taken to ensure that no disclosure is made of data pertaining solely to them. Statisticians have evolved various models for disclosure control. The most common

model is to aggregate individual data together into groups (often those living in a geographical area) before publication, with each group containing a minimum number of people. But this constrains what an end user outside government statisticians can do with the data. Geographers have devised more sophisticated systems of masks that trade off some of the risks of disclosure against greater utility of the data. In some countries official statisticians have set up research laboratories in which outsiders can access individual data under very strictly controlled conditions—the best solution.

Until recently, most of these benefits arose from government collecting, holding, and analyzing the raw data. But the situation has changed dramatically. A good example of how business exploits individual data provided freely by customers is presented in Box 18.9. Still greater benefits could be realized if public- and private-sector information could be merged. The ability to merge customer preferences (e.g., for food, alcohol, or cigarettes) or spending patterns with health information could provide many benefits to the treatment of disease. But such cross-sector exchanges of data are fraught with difficulties given legislation, commercial confidentiality, the partial nonrepresentative coverage of much commercial data, and the likely reaction of citizens and customers.

In conclusion, the combination of GIS and personal data brings many benefits for decision makers. Yet it is obvious that these benefits are bought at a price, at least for some people. The downside is potentially that:

- An individual's deeply valued privacy is compromised.

- Fear of misuse of the data may undermine trust in the organization collecting it (this is particularly acute for government, where providing information is often mandatory, as compared to commerce where it is normally voluntary through loyalty cards, etc.).

- Errors in data linkage (which sometimes need to be done on a fuzzy matching basis) could lead to incorrect judgments and policies.

18.5.4 GIS Ethics and Decision Making

Technology in general and GIS technology in particular is neither good nor evil, and it certainly cannot be held responsible for the sins of society. But technology can empower those who choose to engage in either good or bad behavior.

Exploiting Customer Data

Created by its eponymous founders in 1989, dunnhumby (www.dunnhumby.com) is now largely owned by Tesco, the world's third largest retailer. The firm also provides similar services to other major retailers in 20 countries (e.g., Macy's). Globalization of retailing is ensuring that many cutting-edge practices are spreading widely, and dunnhumby's business is expanding accordingly.

The main service dunnhumby provides is based on the data mining of information from sales, loyalty card use, and other data to enable retailers to understand customers and their needs, wants, and preferences. The scale of the mining, analysis, and exploitation is huge: Tesco, for example, has 13 million regular customers, and the geographical and other patterns of their purchasing are created from their shopping records and responsiveness to tactical marketing initiatives (see Section 2.3.2.1). Some 25,000 Tesco products are categorized individually to help build up a "Lifestyle DNA profile"

of each customer. On the basis of such analyses, each individual shopper is grouped together, and vouchers are printed for discounts on goods they have purchased in the past or for goods that other shoppers with similar characteristics have also purchased.

This results in a mailing to all 13 million Tesco Clubcard customers at least four times a year, with a summary of their rewards and vouchers tailored to en-courage them to return and try new goods. Some 7 million different variations of product offerings are made in each mailing. Customer take-up is between 20% and 50%, in contrast to the norm of about 2% in most direct marketing. The ranges of goods in store are also adjusted geographically in response to the habits of those who shop there. And the characteristics of new stores are planned on the basis of knowledge of people living nearby (including use of Census of Population and other externally provided data: see Section 2.3.3).

Ethics matter in operating GIS safely.

Ethics refers to principles of human conduct, or morals, as well as to the systematic study of such human values. An act is considered to be ethical if it agrees with approved moral behavior or norms in a specific society. GIS technology provides ample scope for unethical behavior, notably to produce results that knowingly benefit one individual or group rather than another or society as a whole. This behavior is facilitated by the apparent simplicity and believability of map outputs—despite the high complexity of the assumptions made in combining data, the various algorithms that might be employed, and well-known ways of misleading the eye by use of map scale, color, map projection, and symbolism (see Chapter 12). In that regard, using GIS for decision making is a little different than the challenges faced by professions in many fields.

So how do we know who to trust? How do we know, for instance, that the person who has offered to forecast environmental risk or provide disaster management at a microlevel is appropriate for the job? The simplest way to find out is to review candidates'

educational attainment and track record and to use only registered professionals who have gone through an extensive formal qualifications process and who can be disbarred for unethical conduct. However, GIS is a field where few directly relevant professional bodies exist and many GIS practitioners operate on an ad hoc basis. So the best we can often do is to create a code of conduct, ensure that practicing GIS experts sign up to that code, and provide severe penalties (e.g., to reputation) if they do not adhere to it. Such codes can never be perfect because it is impossible to legislate for every possible eventuality. The essence of a true professional, however, is that he or she abides by the spirit of the code even when it does not proscribe a risky action.

The most well-known code of ethics in the GIS field is that formulated by the Urban and Regional Information Systems Association in 2003, drawing on many preexisting professional codes. It is shown in abbreviated form in Box 18.10. The code has been adopted by the GIS Certification Institute, which has now formally granted GIS Professional status to over 3,000 applicants who have met the Institute's standards.

The URISA GIS Code of Ethics

This ethics code embraces four main sets of obligations:

Obligations to society:
  Do the best work possible.
  Contribute to the community to the extent
    possible, feasible, and advisable.
  Speak out about issues.

Obligations to employers and funders:
  Deliver quality work.
  Have a professional relationship.
  Be honest in representations.

Obligations to colleagues and the profession:
  Respect the work of others.
  Contribute to the discipline to the extent possible.

Obligations to individuals in society:
  Respect privacy.
  Respect individuals.

Each of these categories is subdivided into a total of 35 ways of behaving. The code is relentlessly positive—it enjoins GIS practitioners to ethical behavior as a way of working for the good of all, rather than under threat of penalty (de-barring professionals from working in the United States is in any case complicated by employment law). The enforcement mechanism is minimal. But it is nevertheless a good thing, and customers of GIS-based services should expect providers to adhere to the code.

18.5.5 Public Trust

Public trust is essential to the effective running of civil society and to maintaining the success of a business. It is manifested in multiple ways: in governments of different levels (federal, state, and local), in politicians, in science and scientists (e.g., in nuclear power or stem cell research), in banking and other businesses, in professional bodies, and in datasets. High levels of trust minimize time-wasting debates and facilitate the implementation of policies. But too high a level of trust is also bad: mindless acceptance of what governments propose is bad for democracy. The question is: how much trust is appropriate?

The GIS community seems to enjoy high public trust—but it could evaporate.

In GIS, we seem to enjoy a high level of public trust: the apparent simplicity of the mapping message disguises the complex algorithms used and the professional judgments made in producing them. Satisfactory experience of embedded GIS functionality in telephones, satellite navigation, and Web sites providing where is factual-type information is now almost ubiquitous and produces a spillover effect on other GIS functionality. The common GIS practice of drawing together data from multiple sources can be seen as an advantage in suggesting we do not rely solely on one view of the world. And the stunning capabilities of GIS technology driven by skilled experts dazzle the onlooker into acceptance and trust (see Section 13.1).

Yet similar trust is not shared everywhere or in all areas of science. Figure 18.12, for example, shows how levels of public trust in official statistics vary across the countries of the European Union. Experts in the area and peer review paint a rather different picture: many of the countries (Britain, France, Italy, and Germany) that are low in the ranking are large and relatively heterogeneous, with controversial politics, but also have national statistical organizations that are highly ranked by their peers; yet the public takes a different view, and this matters. Those at the top of the public trust rankings (the Netherlands and the Scandinavian countries) are much smaller, mostly more homogeneous, and more consensual in their politics. Overall, it seems clear that spillover from what people think of their government into products (like statistics) produced by it is also occurring. This is particularly true where the statistics are used to buttress government claims of progress; economic and health statistics are typically trusted less than transport ones, presumably because the public thinks governments have little to gain by manipulating transport statistics!

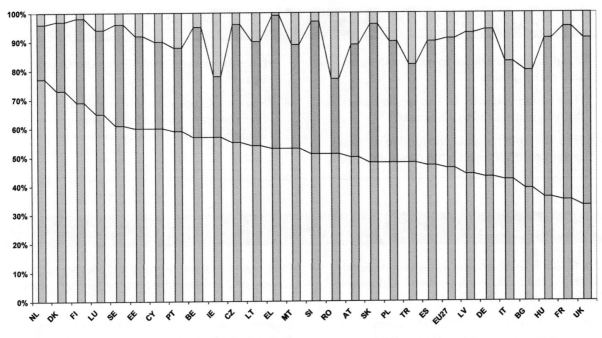

Figure 18.12 Public trust in official statistics, as shown by the European Commission Eurobarometer 67 (Spring 2007) survey across 27 EU countries plus Turkey. Blue indicates the percentage of respondents tending to trust official statistics, red shows the proportion that tend not to trust them, and gray are the don't knows. The countries are arranged from those with highest "tend to trust" ratings (left) to lowest (right). 'DK' in the caption denotes 'Don't Know'.

Such lack of trust does not in general occur with producers of reference mapping (such as government national mapping agencies) or in regard to commercial providers of software or data. Yet we know that the quality of data available varies considerably across the globe and that many of the data series are as difficult to collect as good quality statistics. Soil maps, for instance, are based upon similar field sampling principles to the assembly of population statistics (see Section 4.4). Poor sampling can fail to identify discontinuities in soil profiles, just as it can impede understanding of discontinuities in population statistics. Thus there is some danger that the hidden complexity of GIS could be revealed and lead to a diminution of public trust: that would be extremely dangerous for the continuing success of GIS. We minimize the risk of loss of trust by following a published code of practice (see Section 18.5.4), by being transparent about the inputs and processes underlying any decisions we make, and by running sensitivity analyses as part of any what-if analysis (see Section 16.5).

18.6 Conclusions

It should be obvious that exploiting a GIS successfully involves much more than an ability to select options from a menu. As argued earlier in the book, a real understanding of the science behind the GIS operations is essential, as is an understanding of efficient management practices. But an understanding of the societal context in which you are operating is also essential. To operate without concern for the legal framework in the particular jurisdiction is potentially ruinous. Lacking an understanding of the complexities of public- and private-sector interactions in that jurisdiction could lead to great expense and loss of opportunities. Performing without due regard for ethics and without an awareness of losing public trust could be suicidal. Awareness of your operating environment is thus crucial to making sound and effective decisions using GIS and GI.

Questions for Further Study

1. What factors are liable to complicate safe decision making using GIS?

2. How would you price GI and services based on it if you were running a business?

3. Find out from national and local (e.g., state) government Web sites what constraints you will have to operate within if you use GI produced by them.

4. How would you go about reassuring the public that they can retain trust in what you are doing and that their privacy will be maintained?

Further Reading

Cho, G. 2005. *Geographical Information Systems: Mastering the Legal Issues* (2nd ed.). Chichester, WileyBlackwell.

Dobson, J., and Fisher, P. 2003. Geoslavery. *IEEE Technology and Society Magazine,* Spring, 47–51.

Dodgson, M., Gann, D., and Salter, A. 2008. *The Management of Technological Innovation.* New York: Oxford University Press.

Obermeyer, N.J. and Pinto, J.K. 2008. *Managing Geographic Information Systems* (2nd ed.). New York: Guilford Press.

Shapiro, C. and Varian, H.R. 1999. *Information Rules: A Strategic Guide to the Network Economy.* Cambridge, MA: Harvard University Press.

Wise, S. and Craglia, M. (eds.) 2008. *GIS and Evidence-Based Policy Making.* London: Taylor and Francis.

19 GIS Partnerships

No single organization or individual can provide all the skills, tools, or knowledge required to carry out significant GIS-based projects. In this chapter we examine how partnerships between government bodies, between private-sector ones, or even between mixtures of them have helped to achieve desired ends. We consider what drives such partnering—commercial advantage, access to better tools or data, prestige, political factors, risk minimization, altruism or sheer necessity. We particularly focus on the concept of the spatial data infrastructure (SDI) and how it has evolved in different continents, countries, local governments, and communities. There are important lessons to be learned from SDIs that have been a success, even if success is hard to measure. Finally, we review the coming together of volunteers as a form of partnership to create geographical information, and we identify the advantages and disadvantages this partnership has in comparison with GI from official sources.

LEARNING OBJECTIVES

After studying this chapter, you will understand:

- The necessity for GIS partnerships and their potential benefits.

- The nature of the big idea—spatial data infrastructures (SDIs).

- The history of and differences in SDIs at global, continental, national, and local levels.

- The evolution in SDIs away from a technical solution to a technically supported social and institutional network.

- Other forms of partnership in GIS, notably commercial ones and combinations of individual volunteers exploiting technology to achieve critical mass.

19.1 Introduction

We live in a competitive world. But there are many occasions when the GIS user finds it best to collaborate or partner with other institutions or people. In principle, partnerships can bring missing skills, know-how, technology, finance, and better branding to the enterprise. Because the need for partnerships applies at the personal, institutional, local area, national, and global levels, there is great scope for different approaches. The form of the partnerships can range from the enforced (e.g., through legislation) and the highly formal based on contract to the informal where participation is entirely voluntary. Some partnerships result from short-term events that span national boundaries (such as the Chernobyl nuclear reactor accident in 1986 or the Asian tsunami of 2004). Others arise from long-term relationships like membership of NATO—which has had a profound effect on the standardization of GIS within the military in those participating nations.

Partnerships can add vital staff skills, technology, marketing know-how, and brand image. They can also lead to cost-and risk-sharing.

Nothing is uncaused: organizations or governments getting together to act in partnerships rarely happen without some forcing function. But the drivers behind partnerships arise from a legion of different causes, though rarely is there a conscious decision to do something to enhance the use of GIS. Almost invariably, increased usage arises because of another, larger decision. Financial difficulties or the need for scale to conquer markets figure large in commercial partnerships (which sometimes turn into mergers—almost all mergers are in fact takeovers). National emergencies such as 9/11 in 2001, wild bushfires (see Section 13.4), or the global financial crisis galvanize governments into action, with executive leadership or legislation providing the formal legitimation for change. The U.S. actions in creating the Department of Homeland Security after 9/11 and giving it leadership across government for protecting the country against terrorism, allied to the enhanced funding for the military, have raised the profile of GIS activities considerably in the United States. Moreover, technological change such as Web 2.0 makes it possible to achieve previously impossible aims such as the UK government's transformational government agenda, which was designed to enhance government services, cut costs, and personalize services to the citizens.

Partnerships are easier if someone is in overall charge (e.g., Homeland Security); this facilitates setting clear project milestones, monitoring of progress and enforcement action when things go wrong. Creation and maintenance of spatial data infrastructures in most countries (see Section 19.3) usually involves one lead government department. In practice, however, the power to make things happen is limited where no one is really in overall charge since action is heavily reliant on consensus on virtually all matters. However, it can be done, as in the development of the Web and open-source products.

19.2 Commercial Partnerships

Until the early 2000s, it was common for free-standing data suppliers such as GDT (United States), Navteq (United States) and TeleAtlas (Netherlands) to supply GI to various players, who then supplied technology and services to other businesses or the public. Since then, massive consolidation of the industry has taken place, and GI has become even bigger business. Nokia, the Finnish mobile telecommunications business, purchased Navteq, the Chicago-based firm providing digital maps for car guidance and other applications, in July 2008 for $8.1 billion. TomTom, the Dutch satellite navigation organization, purchased TeleAtlas, its primary map supplier (also based in the Netherlands), in late 2007 for $2.5 billion (and earlier TeleAtlas had taken over the U.S. GDT business). Microsoft and Google have also bought new technology from other firms. On a smaller and less visible scale, ESRI has often purchased start-up firms with smart technology, which it has then embedded in its own GIS.

Despite these acquisitions, Google, Microsoft, ESRI, and other GIS companies still operate many partnerships; for example, they provide services using data in part at least originating from other organizations. Indeed, this is the core of the Google business model, where data are acquired free or through license (see Section 18.5.1.2), processed, and then served to Google Earth and Google Maps users. A good example is Google Oceans, with data provided by around 50 other organizations, mostly in the public sector.

There are also hybrid forms of partnership. One of the most successful has been the Open Geospatial Consortium (OGC: www.opengeospatial.org), an innovative public-private partnership that operates as a voluntary consensus standards organization with nearly 400 member organizations worldwide. Its aim is to serve as a global forum for the collaboration of developers and users of spatial data products and services, and to advance the development of international standards for geospatial interoperability. From this, it aims to provide free and openly available standards to the market, tangible value to its members, and measurable benefits to users. The trigger to OGC action is identification of an interoperability problem. Working through its committees, working groups, and test beds, OGC operates a member-approved and collaborative process to define, document, and implement open specifications that solve such interoperability problems. The result has been the embedding of many OGC standards in commercial software and considerable benefits to users in terms of interoperability.

19.3 Spatial Data Infrastructures

Whatever the practical merits and scale of private-sector partnerships, the story of GI/GIS partnerships has been dominated by a single conceptual model for over 15 years: the spatial data infrastructure (SDI). Since the concept of an SDI was first widely promulgated (at least with that terminology) in the United States in 1994 (see Section 19.3.1), it has spread globally and vertically within countries. Over 150 SDIs have been described in the literature (Figure 19.1), many of them national in intent. Commercial software vendors like ESRI and Intergraph have created tools to assemble them. Many books and reports have been written on SDIs. The meaning of the term differs in different countries, though in essence it normally describes a widely available GIS search and mapping engine and additional institutional and legal elements. But Craglia and Campagna's insightful analysis of SDIs concluded in 2009 that

- Best practice has migrated from SDIs being led by data providers to being driven by users.

- The greatest scope for adding value is probably at a local or regional, rather than at the global or national, level.

- The most effective SDIs are essentially social networks of people and organizations, in which technology and data play a supportive role. The technology is cheap, the data are expensive, and social relations are invaluable.

This all represents a substantial change of emphasis since the earliest formal NSDI, that of the United States. We now examine this evolution and its ramifications for GIS users.

Figure 19.1 Some examples of SDIs and portals to them that were extant at the time of writing.

19.3.1 How It All Began

Although some pan-government agreements on sharing data and formal frameworks preceded it, the real origins of SDIs were in the early 1990s and involved the Mapping Science Committee (MSC) of the United States National Research Council, together with the U.S. Geological Survey. From the outset, this was seen as a national endeavor: an NSDI was viewed as a comprehensive and coordinated environment for the production, management, dissemination, and use of geospatial data, involving the totality of the relevant policies, technology, institutions, data, and individuals. A 1994 MSC report urged specifically the use of partnerships in creating the NSDI. The U.S. Strategy for NSDI document was initially published by the Federal Geographic Data Committee (FGDC) in 1994 and revised in 1997 and 2004. Box 19.1 summarizes FGDC's 1994 view of the problem NSDI was intended to fix.

A vitally important lever for realizing the vision was the Clinton administration's wish in early 1993 to reinvent the federal government. Staff of the Federal Geographic Data Committee (FGDC) seized an opportunity and outlined the concept of the NSDI as a means to foster better intergovernmental relations, to empower state and local governments in the development of geospatial data sets, and to improve the performance of the federal government. In April 1994, President Clinton signed Executive Order 12906—Co-ordinating Geographic Data Acquisition and Access: The National Spatial Data Infrastructure. This order directed that federal agencies, coordinated by FGDC, carry out specified tasks to implement the NSDI; it created an environment within which new partnerships were not only encouraged but required. This was all embedded further through the contents of the Office of Management and Budget's (OMB) Circular A-16 and the US E-Government Act of 2002.

A substantial impetus for the NSDI and local SDIs was given by a number of emergencies in the United States such as 9/11 in 2001 (Figure 19.2) and the destruction wrought by Hurricane Katrina in New Orleans in 2005 (Figure 19.3 and see Box 1.1). The 9/11 experience in which the New York City Emergency Operations Center was located in a destroyed building, and data and mapping support had to be provided initially by staff from Hunter College was particularly telling. (Subsequently a 50-strong GIS team worked on the creation and production of mapping for over two months in partnership with New York City staff.) Among the lessons learned was the need to duplicate data and metadata, to have wide-

---

Applications Box 19.1

The Problem to which NSDI was the Solution

In the United States, geographic data collection is a multibillion-dollar business. In many cases, however, data are duplicated. For a given piece of geography, such as a state or a watershed, there may be many organizations and individuals collecting the same data. Networked telecommunications technologies, in theory, permit data to be shared, but sharing data is difficult. Data created for one application may not be easily translated into another application. The problems are not just technical—institutions are not accustomed to working together. The best data may be collected on the local level, but they are unavailable to state and federal government planners. State governments and federal agencies may not be willing to share data with one another or with local governments. If sharing data among organizations were easier, millions could be saved annually, and governments and businesses could become more efficient and effective.

Public access to data is also a concern. Many government agencies have public-access man-dates. Private companies and some state and local governments consider public access as a way to generate a revenue stream or to recover the costs of data collection. Although geographic data have been successfully provided to the public through the Internet, current approaches suffer from invisibility. In an ocean of unrelated and poorly organized digital flotsam, the occasional site offering valuable geographic data to the public cannot easily be found.

Once found, digital data may be incomplete or incompatible, but the user may not know this because many datasets are poorly documented. The lack of metadata or information on the who, what, when, where, why, and how of databases inhibits one's ability to find and use data, and consequently, makes data sharing among organizations harder. If finding and sharing geographic data were easier and more widespread, the economic benefits to the nation could be enormous.

Figure 19.2 Ground Zero site a few days after 9/11; the New York City Emergency Operations Center had been located in the World Trade Center complex and was destroyed. (Courtesy James Tourtellotte/MAI/Landove LLC)

area access to them, and to have better arrangements in place to circumvent bureaucracy in emergencies. A fully functioning SDI would have provided all of this.

The story of the United States' NSDI is far from being the only one worldwide; important SDI developments occurred even earlier in Australia and some other countries. Moreover, many other developments have occurred at state and local levels in the United States. But the U.S.NSDI was a pioneer and serves to illustrate the need for partnerships of many different types. In a country with over 80,000 different public-sector governing bodies (e.g., counties, school boards)—perhaps the most complex geographical structure anywhere in the world, with different jurisdictions overlapping in space—no one person or body can be in overall charge. And the nature of the partnerships inevitably varies with circumstance: persuasion to participate can take the form of military force, financial incentives, executive direction, vested self-interest, or altruism.

From its original concept as a national enterprise, the SDI has spread upward to the global scale and downward to the local one. We now examine some examples of each.

SDIs are not concrete things but visions of how to make better use of scarce resources.

Figure 19.3 The depth of flooding in New Orleans as a consequence of Hurricane Katrina on August 31, 2005. (Courtesy NOAA)

19.3.2 SDI Partnerships at the Global Level

We can think of global-level partnerships as being of two types: those based on executive mandate and finances to carry out particular tasks, and those that are in some sense voluntary. Thus the United Nations' various organizations, such as the Food and Agriculture Organization (FAO), the World Meteorological Organization (WMO), or the World Health Organization (WHO), fall into the first category and the Global Spatial Data Infrastructure Association (GSDI) and OpenStreetMap (see Box 18.8) into the second.

19.3.2.1 Global Organizations and Their Partnerships

Organizations such as FAO and WHO have numerous programs to tackle existing problems of food shortage or disease reduction. Many of these programs involve use of GIS, normally in partnership with national or more local bodies on the ground. Such global organizations are also necessarily involved in the development of ontologies (see Chapter 3) and much else of relevance to GIS users. One such example arises from the need for effective management of fish stocks shared by nation-states to achieve long-term sustainable fisheries. Quite aside from the obvious political factors, this can be complicated because the data made available by different agencies are not always interoperable, and harmonization of the multiple datasets is necessary using a common ontology. Without such a preliminary stage, the apparently simple task of identifying the nature, extent and scale of the problem is likely to fail. Thus the rather straightforward mapping in the FAO database of the global distribution of fish species is based on many detailed classification decisions. Built on top of this are many examples of large-scale, GIS-based predictive modeling of the natural occurrence of marine species of vital importance for enhancing human diet and conserving fish stocks (see, for example, www.aquamaps.org).

An even more striking case of partnerships at the global level is provided by the WMO, which has a membership of 188 nation-states and territories. WMO sees partnership as working with international agencies, other organizations, academia, the media, and the private sector to improve the range and quality of critical environmental information and services. The need for partnership is obvious: no single government or agency has the necessary resources to address on its own all the challenges arising from the complexity of the Earth system and the interconnections of weather, water, climate, and related environmental processes. WMO therefore sets out to facilitate worldwide cooperation in establishing networks of meteorological stations and relevant hydrological and other geophysical observations and to promote the free and unrestricted exchange of data and information, products, and services in real or near-real time (Figure 19.4). As a

Figure 19.4 The distribution of 3,325 ARGO drifters (buoys) in the world's oceans at 0605 UTC on March 22, 2009—the data from these buoys are updated in almost real time from satellite observations. Each drifter collects data on temperature and salinity to 2000 m depth, and their movements indicate the direction and strength of ocean currents—important information for predicting weather. The ARGO project is a partnership of 44 countries. Viewing of the ARGO data is facilitated by a GIS viewer.

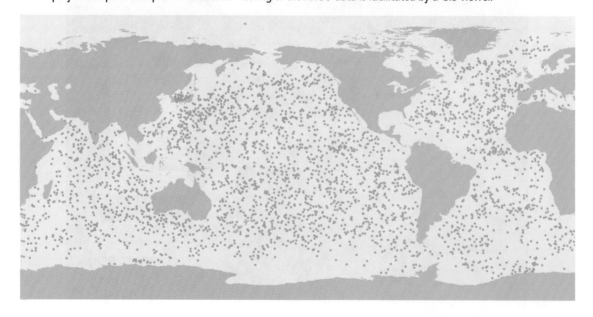

consequence, the global availability of meteorological data is now part of the taken-for-granted world.

19.3.2.2 Global Voluntary Partnerships

The Global Spatial Data Infrastructure Association (GSDI) was founded in 1989. Its aims are not to create a global SDI as such—an impossible task—but rather to

- Serve as a point of contact and effective voice for those in the global community involved in developing, implementing, and advancing spatial data infrastructure concepts.

- Foster spatial data infrastructures that support sustainable social, economic, and environmental systems integrated from local to global scales.

- Promote the informed and responsible use of geographic information and spatial technologies for the benefit of society.

The GSDI has been very active as a communication and awareness-raising device, with regional chapters and newsletters plus well-attended conferences.

Another global voluntary partnership is the International Society for Digital Earth (www.digitalearth-isde.org) largely fostered by the Chinese Academy of Sciences and the Ministry of Sciences and Technology of China but involving scientists from many countries. Formally launched in 2006 (Figure 19.5), the Society exists to promote international cooperation on the Digital Earth vision. In particular, it seeks to enhance Digital Earth technology to play a key role in global economic and social sustainable development, environmental protection, disaster mitigation, natural resources conservation, and improvement of human living standards. Aside from running international conferences, it has set up the *International Journal of Digital Earth* (www.tandf.co.uk/journals/TJDE).

A different kind of voluntary organization is represented by the International Steering Committee for Global Map, which was launched in response to AGENDA 21 formulated at the United Nations Conference on Environment and Development held in 1992. The conference recognized that geographic information is critical for understanding the current status of the global environment and monitoring changes in it. AGENDA 21 called for an action program to address this need. That same year Japan's Ministry of Land, Infrastructure and Transport began to gather support for the Global Mapping concept. To be manifested as a 1 km-resolution database capable of being mapped at a 1: 1 million scale, this was designed to be produced through international cooperation. Composed largely of national mapping organizations, by early 2009 ISCGM had completed the first version of its Global Map, which contained land cover and the proportion of tree cover plus the topography for 73 national/regional areas, and was making these available in formats that operate with open-source or free software. Figure 19.6 shows the status of the topography coverage at that time.

Figure 19.5 The launch of the International Society for Digital Earth in Beijing in May 2006 showing members of the Executive Committee. Professor Chen Shupeng, the father of GIS in China, is on the extreme left of the front row (see Dedication). (Provided by Secretary General Secretary of ISDE)

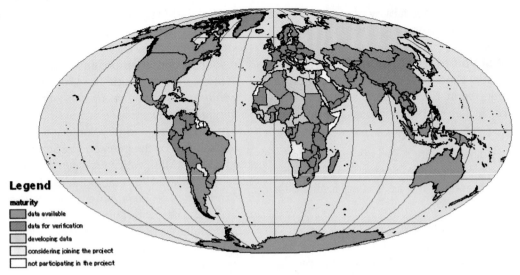

Figure 19.6 The status of the national partnerships committed to the Global Map in January 2009.

Legend

maturity

- data available
- data for verification
- developing data
- considering joining the project
- not participating in the project

Finally, the lowering of entry costs to data collection brought about by new technologies has enabled truly global partnerships of individuals to appear and create useful GI. Perhaps the best example is OpenStreetMap, a collaborative exercise designed to produce mapping for ubiquitous use under license but free of copyright restrictions (see Box 18.8). Contributors to OpenStreetMap take hand-held GPS devices with them on journeys, or go out specially to record GPS tracks. They record street names, village names, and other features using notebooks, digital cameras, and voice recorders. The contributors then upload those GPS logs showing where they traveled and trace out the roads on OpenStreetMap's collaborative database. Using their notes, they add the street names and information such as the type of road or path and the connections between roads. Those data are then processed to produce detailed street-level maps, which can be published freely on sites such as Wikipedia, used to create hand-held or in-car navigation devices, or printed and copied without restriction.

The extent and detail of the coverage of these maps differ from place to place but are already good in some countries and expanding in many others. Figure 19.7A shows a Scottish example with unique features: the classification of features included differs from traditional maps produced by national mapping agencies by including public toilets, post boxes, and named pubs (compare with Figure 18.7). Figure 19.7B shows part of an OpenStreetMap of Pyongyang in North Korea. This is remarkable for in certain countries (like North Korea), the use of GPS by individuals to map the topography is still regarded as contrary to the interests of state security, and heavy punishment can ensue.

Given the constraints under which OpenStreetMap collaborators sometimes operate and the gargantuan ambition of the project, an extraordinary amount has already been achieved over a short period. Inevitably, the longer term challenge for such collaborations is not only to extend consistent map coverage to the whole country (and world) but also to maintain the mapping up-to-date in a consistent and predictable way: doing things the first time is more fun than updating someone else's efforts!

19.3.3 SDI Partnerships at the Multicountry Level

There are many multinational partnership arrangements in GIS, especially transborder and bilateral ones. Here we concentrate on two such examples—the GIS and GI developments in the European Union and in the Asia-Pacific region.

19.3.3.1 The European Dimension

The European Union (EU) comprises 27 countries and some 500 million people. The European Commission (EC), the public service of the EU, has been considering a variety of GIS and GI issues since the 1980s. Progress was often slow in the early years, complicated by the decentralized, multicultural, and multilingual nature of decision making among the constituent countries and the relatively low priority accorded to GI matters in EC work. A serious attempt to define a GI policy framework for Europe through the GI2000 initiative failed. However, some real achievements have been made in GIS research via the coordination and funding provided by the EC's Joint Research Centre in Ispra, Italy.

Figure 19.7 A section of OpenStreetMap for (A) south Edinburgh, showing many paths and stairways, named buildings (e.g., the Institute of Geography), and other elements not shown on official mapping. This may be compared with the contents of official mapping (Figure 18.7) and a commercial competitor (Figure 18.8) in the same country. Figure 19.7B shows an OpenStreetMap for Pyongyang in North Korea. (Map data © OpenStreetMap and contributors, CC-BY-SA)

The status of SDIs in Europe as of 2007 is summarized in Figure 19.8. This summary was produced by an EC-funded study and summarizes the then situation in 32 countries (27 EU member states, one Candidate Country, and 4 European Free Trade Area countries). Each country is calibrated on 32 indicators structured around seven main components: organizational issues, legal framework and funding, reference data and core thematic data, metadata, access and other services, standards, and thematic environmental data. Details of each indicator are given in sdi.jrc.ec.europa.eu/ws/ Advanced_Regional_SDIs/arsdi_report.pdf.

Though the results are capable of different interpretations, they suggest that most European countries were seen as having a fairly well coordinated SDI approach at the national level and also one or more of the SDI components at an operational level (the first two columnar sections from the left). Data availability (related to the INSPIRE themes), metadata, network services, and standards are also claimed to be quite well developed, particularly in the Europe 15 countries while the Europe 10 countries (those countries that entered the Union in 2004) are rapidly catching up. Against these positive developments, it is clear from Figure 19.8 that the area of legal issues

and funding showed much more variable progress. The results also show that discovery and view (network) services and, to a certain extent, download services were relatively well developed in 2007, but much progress is still to be made on the provision of services (columns 29–30). Based on these findings, the study divided the 32 countries into those with first-generation SDIs (largely driven by data providers) and the second-generation ones where users play a much larger role in stimulating developments.

Part of all this development of SDIs grew up independently within each member state, but the EU itself also influenced some developments. The most productive of these mechanisms prior to 2010 was the incorporation of GI and GIS within broader EU programs. Examples include the eEurope and eContent initiatives to foster European industry, extensive use of GIS within the European Environment Agency, and the use of new location-related technologies and content by citizens.

But the most effective way to change policies and practice across the entire European Union is to get agreement on a Directive. An EU Directive introduced by the European Commission and approved by the European Parliament and the Council of Ministers must be implemented in national law within a defined

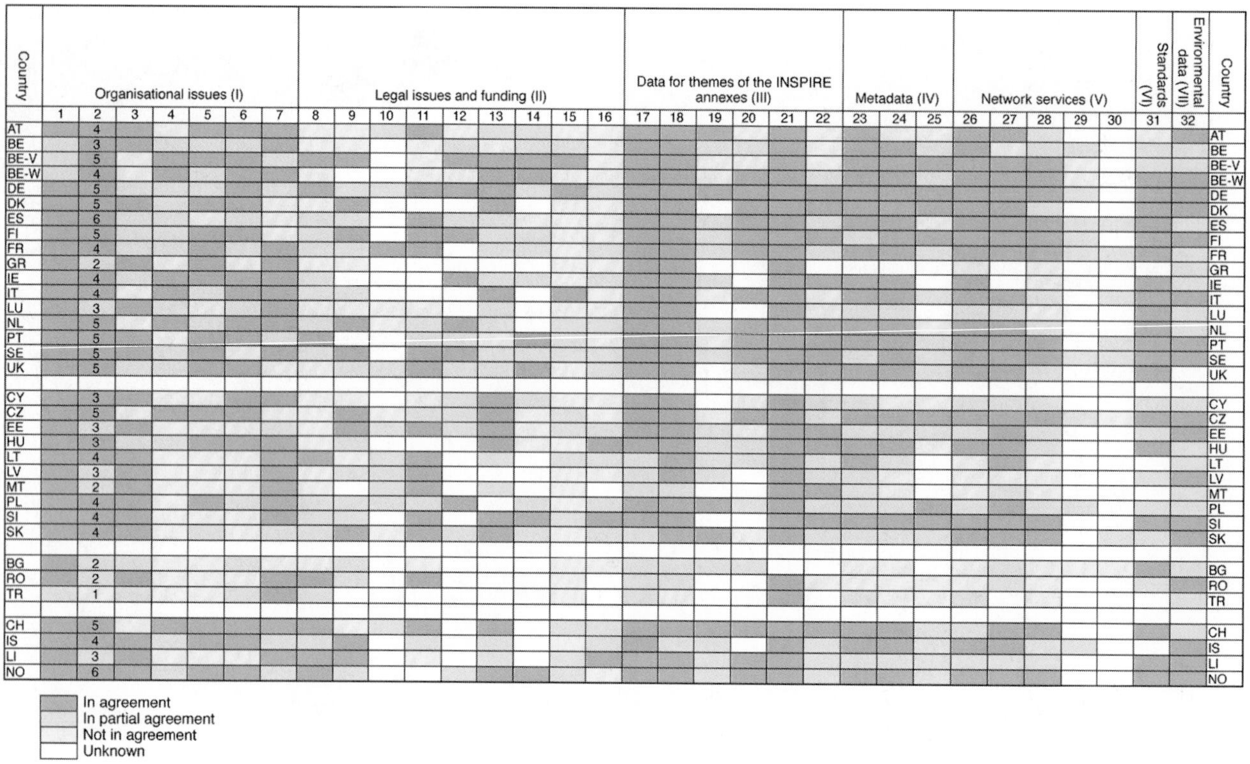

Figure 19.8 Tabular summary of the state of SDI development in 32 countries in "greater Europe" in 2007. (Source: EC study summarized in sdi.jrc.ec.europa.eu/ws/Advanced_Regional_SDIs/arsdi_report.pdf)

Legend:
- In agreement
- In partial agreement
- Not in agreement
- Unknown

period (normally two years). One example of a Directive relevant to GI is that designed to foster commercial and other reuse of information created and held by public-sector bodies across the 27 countries of the EU: much public-sector information (PSI) of this type is geographic. Although this Directive has weaknesses (e.g., it does not mandate that all countries must make *all* PSI available for reuse), it is now embedded in the laws of all EU states and the EU has taken legal action against those states which, in the Commission's view, have not implemented the Directive appropriately.

INSPIRE Certainly the most important EU Directive of the Council and the European Parliament relevant to GI thus far is INSPIRE (see Figure 19.9)—Directive 2007/2/EC. In many respects it and the original Clinton Executive Order 12906 are the most significant GI mandates yet formulated worldwide. The Directive is designed to establish a legal framework for the establishment and operation of an INfrastructure for SPatial InfoRmation in Europe (hence INSPIRE). Based on much work by the Directorate of the Environment, Eurostat (Europe's statistical agency), and the Joint Research Center, this Directive is designed to meet both community and local needs. The value of what is proposed is anticipated to be greatest to policy-makers such as legislators and public

authorities, but citizens, academics, and business are also expected to benefit.

The purpose of such an infrastructure is to support the formulation, implementation, monitoring, and evaluation of community environmental policies, as well as to overcome major barriers which the European Commission and GI industry saw as still affecting the availability and accessibility of pertinent data. These barriers include:

- Inconsistencies in spatial data collection: European spatial data are often missing or incomplete, or, alternately, the same data are collected twice by different organizations.

- Lack of documentation: description of available spatial data is often incomplete.

- Incompatible spatial datasets: hence they often cannot be combined safely with other spatial datasets.

- Incompatible geographic information initiatives: the infrastructures to find, access, and use spatial data often either do not exist or are highly inconsistent across the European Community.

- Barriers to data sharing: cultural, institutional, financial, and legal barriers prevent or delay the sharing of existing spatial data.

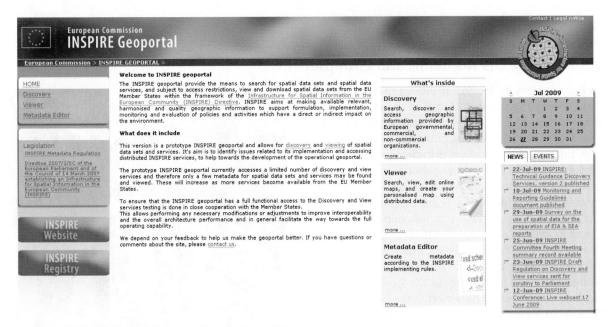

Figure 19.9 The INSPIRE European Union Directive Web site.

The key elements of the INSPIRE Directive designed to overcome these barriers include:

- Mandating the creation of metadata to describe existing information resources so that they can be more easily found and accessed.

- Harmonizing key spatial data themes needed to support environmental policies in the EU.

- Forming agreements on network services and technologies to allow discovery, viewing, downloading of information resources, and access to related services.

- Making policy agreements on sharing and access, including licensing and charging.

- Devising coordination and monitoring mechanisms.

- Creating the implementation process and procedures.

From the process that eventually led to INSPIRE, it was agreed that the key principles to be followed should be that:

- Official spatial data should be collected once only and maintained at the level where this can be done most effectively.

- It must be possible to combine seamlessly spatial data from different sources across the EU and share them among many users and applications.

- It must be possible for spatial data collected at one level of government to be shared between all the different levels of government.

- Spatial data needed for good governance should be available on conditions that do not restrict its extensive use.

- It should be easy to discover which spatial data are available, to evaluate their fitness for a purpose, and to know which conditions apply for their use.

To anyone based in Europe, it is self-evident that some of these conditions and principles are not yet met. Achieving them will necessitate changes to national GI policy in various countries.

The path to acceptance of the INSPIRE Directive involved making impact and risk assessments and establishing an expert group made up of official representatives of all the member states, plus working groups with expertise in the fields of environmental policy and geographic information, to formulate proposals and forge consensus (see Box 19.2). Account had also to be taken of the EU principles of subsidiarity (i.e., the principle that nothing should be done at the community level that can be done more efficiently at the national or lower level). Following three years of intensive consultation among the member states and their experts, a public consultation, and the assessment of the likely impacts of INSPIRE, the European Commission adopted the INSPIRE proposal for a directive in July 2004. An amended proposal was adopted by the Council and European Parliament in March 2007 and was mandated to be part of all national laws by May 2009.

In many respects, the original legitimation for INSPIRE was similar to that of the United States' NSDI (see Section 19.3.1), even if the process for

Milan Konečný, GIS Academic and Entrepreneur

Milan Konečný (Figure 19.10) is professor of cartography and geoinformatics at the Institute of Geography in Masaryk University, Brno, Czech Republic. He was the author of the first GIS textbook published in Eastern Europe, but he is as well known outside his native country as within: he has been engaged in many international projects spanning a variety of cartographic and GIS-related subjects for over 30 years. His partnerships with colleagues in Brazil (where he was honored as a Knight of Brazilian Cartography), China, India, Japan, many countries in Europe, and the United States have often been informal, but he has acted as a conduit for new ideas and good practice across the world. This was particularly evident when he served as the president of the International Cartographic Association, but he has also been vice-president of the International Society on Digital Earth and a member of the board of the International Steering Committee for Global Map as well as being involved in many other organizations. He has represented his country on committees of the European Union's 4th, 5th and 6th Framework (research) Programs and has been involved with the creation of the INSPIRE Directive since its inception (see Section 19.3.3). The Konecny approach to life is to travel to meet people and exchange ideas and experiences, shape organizations, and forge plans for conferences,

Figure 19.10 Milan Konečný (right) with the president of the Czech Republic, Vaclav Klaus (second left), and the rector of Masaryk University (left). (Provided by Milan Konečný)

publications, and research in GIS which will help tackle pressing practical problems such as natural disasters. His irrepressible energy, resilience, connections, and ability to raise funds for meetings make him an academic entrepreneur unique in the GIS world.

---

taking it forward was very different and the outcome applies more widely. All governmental elements of the nation-state are also bound by the Directive. The Directive covers spatial datasets held by public authorities *or their agents* and spatial data services that employ these datasets. It is highly prescriptive in defining the coverage of datasets (see the Directive's annexes in Table 19.1) and setting target dates; for example, it mandates the creation and availability by the end of 2010 of a community geoportal (Section 11.2.2) and the availability of all datasets in the annexes by 2019 in all member states. It defines national and Europe-wide coordination structures and mandates the creation of metadata. Procedures are laid down for establishing harmonized data specifications, so as to facilitate safe added-value operations including unique identifiers for spatial objects across the continent.

Member states of the EU are also mandated to provide certain classes of services. These are discovery services permitting search for spatial datasets and related services, viewing services to display and navigate

data, download services, data transformation services, and the capacity for invoking spatial data services. The first two of these services must be provided free of charge, though public access to the others may be curtailed for a number of reasons (e.g., national security and statistical or commercial confidentiality). One public authority in each country is responsible for liaising with the European Commission on the directive.

It is obvious that, unlike the NSDI where a presidential edict triggered the process—at least across the federal government—the EU process was inherently more democratic in involving huge numbers of multinational and within-country discussions. As a result, it took much longer to formalize. Such a process enables different countries to argue for their own vested interests, and different views across Europe (e.g., on charging for making data available; see Section 18.5.2.2) were evident in the course of passing the legislation. Despite all that, and although it is too early to judge the success of INSPIRE, it stands a good chance of changing the GI situation across a group of 27 countries. This is because adherence to

Table 19.1 The data types to be made available by all EU countries, as defined in the INSPIRE Directive. Datasets in the different annexes are to be made available by different dates.

Annex I
Coordinate reference systems

Geographical grid systems

Geographical names

Administrative units

Addresses

Cadastral parcels

Transport networks

Hydrography

Protected sites

Annex II
Elevation

Land cover

Ortho-imagery

Geology

Annex III
Statistical units

Buildings

Soil

Land use

Human health and safety

Utility and governmental services

Environmental monitoring facilities

Production and industrial facilities

Agricultural and aquaculture facilities

Population distribution—demography

Area management/restriction/
	regulation zones & reporting units

Natural risk zones

Atmospheric conditions

Meteorological geographical features

Oceanographic geographical features

Sea regions

Bio–geographical regions

Habitats and biotopes

Species distribution

Energy Resources

Mineral resources

the Directive will be monitored, and users can address complaints about noncompliance to the European Commission as well as to their national governments.

19.3.3.2 The Permanent Committee on GIS Infrastructure for Asia and the Pacific (PCGIAP)

The PCGIAP is the largest multinational GIS body known at present, at least as judged by the number of countries involved (55). The countries span a wide part of the globe from Iran and Armenia in the west to French Polynesia in the east; from the Russian Federation and Japan in the north, China in the center to New Zealand and Australia in the south. PCGIAP was established by the United Nations Regional Cartographic Conference for Asia and the Pacific (UNRCC-AP) at its triennial meeting in Beijing in May 1994. The Permanent Committee operates under, and reports to, the UNRCC-AP. Membership of PCGIAP comprises directorates of national survey and mapping organizations and equivalent national agencies of the nations from Asia and the Pacific.

The Committee's aims include maximizing the economic, social, and environmental benefits of geographic information in accordance with AGENDA 21 (see Section 19.3.2) by providing a forum for nations from Asia and the Pacific to cooperate in the development of a regional geographic information infrastructure and to share experiences. "GIS infrastructure" is taken by the PCGIAP to include the institutional framework (which defines the policy, legislative, and administrative arrangement for building, maintaining, accessing, and applying standards and fundamental datasets), technical standards, fundamental datasets, and a framework that enables users to access them. For the members of PCGIAP, key applications of GIS are national or regional land administration, land rights and tenure, resource management and conservation, and economic development. An important change was made in 2006 when the cadaster topic was dropped as the theme of an active working group in favor of "spatially enabled government," that is, a less technocratic and wider view of how GIS and GI impacts the whole of government.

Considerable GIS activity is also underway across the individual member nations, including the development of numerous NSDIs. This information is shared through numerous meetings and newsletters.

It is clear that there are huge differences in what is happening between Europe and the Asia-Pacific region. Whereas in Europe a legal approach has been taken, built on EU structures, no such structures exist across the 55 PCGIAP member countries in the Asia-Pacific region. In addition, the Asian organization is founded on national mapping agencies, while the European developments are driven by policy concerns to enhance the availability of GI to users, ensure equity across the continent, and foster innovation to maximize European competitiveness and employment. That said, much valuable work is being done in the Asia-Pacific area, usually with far fewer resources and in an even more complex political framework.

Consistent continentwide NSDIs have proved much more time-consuming to launch in Europe than in the United States because of the power structures spanning 27 European countries, but, once in place, these apply more widely and are much more prescriptive.

19.3.4 SDI Partnerships at the National Level

The original concept of an SDI was focused on the nation-state level. We now review the evolution of three NSDIs, starting with the first formally named one. It is fair to point out that many nations have both a national SDI and many local (e.g., state or county) ones as well.

19.3.4.1 Current Status of the U.S. NSDI

Following the Presidential Executive Order 12906 and early efforts in conceptualizing the NSDI and its implementation, numerous activities in the United States have centered on developing data and metadata standards, raising awareness at all levels, establishing clearinghouses, defining framework data, and creating partnerships to facilitate spatial data availability and access. The results of such efforts include data catalogs and metadata, the U.S. National Map, and the Geospatial One-Stop portal (Figure 19.11). Perhaps most significant (but only partly visible to the public) is the work of the Department of Homeland Security and of the military National Geospatial-Intelligence Agency (NGIA).

In general terms, the NSDI is steered by the Federal Geographic Data Committee, which was originally established in 1990. The FGDC itself is governed by a steering committee drawn from over 30 federal agencies, which set its high-level strategic direction. An Executive Committee of officials from federal government agencies with a major geospatial

Figure 19.11 The U.S. federal government's Geospatial One-Stop portal, which provides access to details of thousands of datasets.

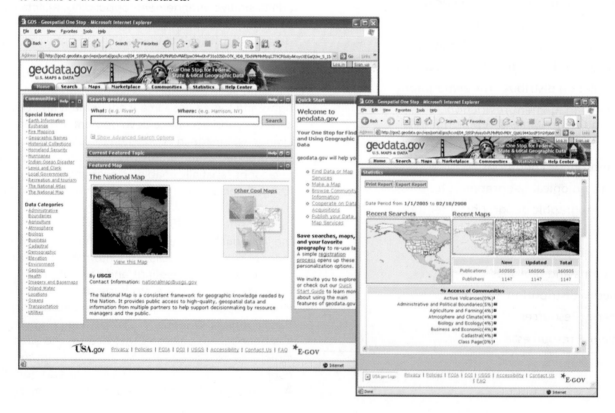

component in their mission provides additional guidance to the steering committee. Advice and recommendations on federal and other national geospatial programs come from the National Geospatial Advisory Committee (NGAC), composed of around 20 individuals drawn from U.S. states and other governments and from the private sector or academia. Important inputs come from a variety of other organizations such as the National States Geographic Information Council, the National Association of Counties, the Open Geospatial Consortium, the University Consortium for Geographic Information Science, the National League of Cities, and the Intertribal GIS Council.

What we can conclude from all of this is that a significant amount has been achieved in the areas identified in the original plans and that the NSDI is in a mature phase with regard to its structures and underpinning bureaucracy. Indeed, the latter achievement has been decried as ossification in some quarters. Therefore, the new challenge—as outlined in the first document produced by the National Geospatial Advisory Committee—is to determine the federal government's best response in policy terms to the dramatic changes in technical capability and user expectations in recent years. Table 19.2 highlights the challenge in contrasting the traditional GIS-powered view of the SDI world with that in a user-driven Web 2.0 world.

Many countries are implementing GI partnerships through varied forms of national spatial data infrastructure.

19.3.4.2 The UK NSDI
Although attempts were made to create a form of UK NSDI in the mid-1990s, little of substance was achieved. Long after the creation of NSDIs in other countries, however, the UK government has agreed to the creation of one under the auspices of its 2008 Location Strategy. Unlike the U.S. case, this was driven not by a grandiose vision but by a pragmatic appreciation of the role of GI and GIS in delivering key government policies. The first illustration in the Location Strategy shows the application of location information to public policy (Figure 19.12). Many elements of the strategy are familiar in NSDI terms, such as the need to know where data are available, the use of common standards, appropriate skills, and appropriate leadership. But it is striking that GIS and NSDI are not explicitly mentioned in it. This reflects a central principle: the strategy had to be approved by a set of ministers, drawn from elected members of the U.K. Parliament, across government. The view was taken that all technical jargon should be excluded and that the language used should be couched in terms that ministers understood and valued—policy imperatives that they were employed to achieve and on whose success they would be measured. As a result, the Location Strategy is the least technical of all the documents worldwide creating an NSDI.

The extent to which an NSDI has to be created by the national or federal government varies by country. In some cases, national associations of local bodies have been set up to create and share knowledge and information at a national level. The 410 local governments in England and Wales, for instance, formed a Geographic Information House to trade in information and have forged a partnership with the Land Registry and central government and other partners to create a National Land Information Service.

19.3.4.3 Singapore's NSDI
Components of Singapore's NSDI have been around for many years, primarily in the security and land-planning authorities. Singapore Public Service's Land Data Hub, for example, has been in place since 1989. Fifteen public agencies share some 125 layers of

---

Table 19.2 The contrast between the traditional GIS-powered view of the SDI world with that in a user-driven Web 2.0 world.

Professional GIS	Web 2.0
Metadata documents	Tagging
Catalog portals	Search engines
Distributed catalogs (distributed searches)	Centralized search engine (caching)
Authoritative GI	Volunteered GI ("wisdom of the crowd"—or "madness of the mob"?)
Geocommunity standards (WMS, GML, etc.)	IT standards (REST, JavaScript, SOAP, etc.)
Producer-centric	User-centric

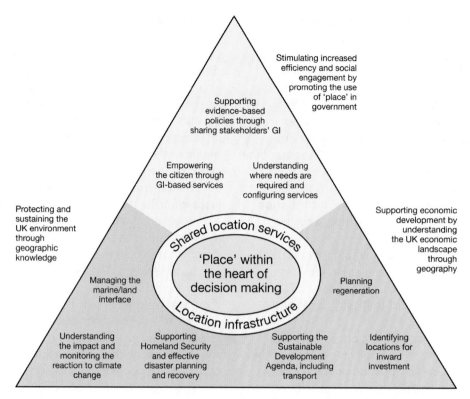

Figure 19.12 How place is seen at the heart of many UK government policies and is the rationale for the national Location Strategy (a type of SDI). (Source: Department of Communities and Local Government. Crown copyright)

geospatial information through the LandNet network. Since April 2008, however, the Singapore Public Service has embarked on a new phase. It has been decided to link up the Land Data Hub and three other data hubs under an NSDI framework. The other data hubs are repositories of data concerning people, businesses, and security matters, owned by various public agencies. Hitherto, all four data hubs have been developed and operated independently.

In Singapore's concept of an NSDI, the country aims to develop an environment in which public agencies can collaborate and share data more easily. The initiative is termed the Singapore Geospatial Collaborative Environment, or SG SPACE. Its implementation is designed to allow geospatial information from many public sector sources to be shared within a consistent reference framework throughout the public sector. However, Singapore's vision for the initiative goes beyond the public sector to encompass the private and public sectors, through licensing and the provision of common goods. Drivers for the initiative are to remove duplication of data collection, encourage reuse of data, enhance policy formulation, and permit more informed decision making and monitoring. Geospatial information is seen as being essential

to support Singapore's handling of emergencies, environmental challenges (e.g., climate change and water issues), disease management and control, and many other challenges.

A multiagency and inclusive committee has been formed to drive and coordinate the various action plans. Industry is being invited to help build a national clearinghouse for geospatial data to link up all four hubs.

19.3.5 SDI Partnerships at the Subnational or Local Level

The size of many organizations at this level and the resources available to them ensure that investment in GIS is often a heavy (or sometimes impossible) burden: The scope for collaboration or partnerships at this level—typically involving operational matters or sharing of knowledge—is therefore considerable.

19.3.5.1 Catalonia
One good example of how a subnational SDI can be made to work effectively is that of the Catalan Region in Spain (Box 19.3).

Costs and Benefits of the Catalonian SDI

A major study was undertaken in 2007 on the socioeconomic impact of the spatial data infrastructure (SDI) of Catalonia, an autonomous community of Spain with a population of 7 million. The Joint Research Center of the European Commission commissioned the study and recommended the methodology. The study was based on a sample of 20 local authorities participating in the Catalan SDI (IDEC) together with three local authorities not participating to act as comparators and 15 end-user organizations, of which 12 were private companies operating in the GI sector and 3 were large institutional users of GI.

Costs: The total direct cost of establishing and operating the IDEC over a five-year period (2002–2006) was approximately $2 million, of which about $500,000 was expended in each of the first two years (2002–2003) to launch the SDI, and around $350,000 per annum was spent to operate and develop the infrastructure in the three subsequent years (2004–2006). Human resources represented 76% of the costs during the launch period (the rest being capital investment), and 91% during operation. These costs do *not* include the creation and updating of topographic data—which is the responsibility of the Cartographic Institute of Catalonia (ICC) and would happen regardless of the development of the SDI. Nor do they include the indirect costs associated with the physical and technological infrastructure (e.g., office space) provided by the ICC. They do include the costs of metadata creation and maintenance, development of geoservices (including the geoportal, catalog, and a Web map service client), preparation of data

for publication, applications, hardware and software, and management.

Benefits: The main benefits of the SDI were found to accrue at the level of local public administration through internal efficiency benefits (time saved in internal queries by technical staff, in attending queries by the public, and in internal processes) and effectiveness benefits (time saved by the public and by companies in dealing with public administration). Extrapolating the detailed findings from the sample of 20 local authorities to the 100 that participate in the IDEC, the study estimated that the internal efficiency benefits account for over 500 hours per month. Based on an hourly rate of approximately $40 for technical staff in local government, these savings totaled over $3.5 million per year.

Effectiveness savings were just as large at another 500 hours per month. Even considering only the efficiency benefits for 2006 (i.e., ignoring those that may have accrued in 2004–2005, as well as the effectiveness benefits), the study showed that the total investment to set up the IDEC and develop it over a four-year period (2002–2005) was recovered in just over 6 months. Wider socioeconomic benefits have also been identified but not quantified. In particular, the study indicates that Web-based spatial services allow smaller local authorities to narrow the digital divide with larger ones in the provision of services to citizens and companies.

Although this study can be criticized for ignoring important elements of the cost base that would be normal in cost/benefit studies, it serves as a valuable real-world example.

19.3.5.2 Australia

Perhaps the oldest and best known subnational SDI partnerships are those of Australia. The states have a strong role in the Australian federation: virtually all large-scale survey and cadastral land registration and many other responsibilities are devolved to the eight states and territories that constitute the regional level. The states pioneered geographic database management systems, some including GIS, from the early 1970s, and an important governance structure (now the Australia New Zealand Land Information Council

or ANZLIC) was set up in 1986 to facilitate the collection and transfer of land-related information between the different levels of government. As an example of the long evolution, the state of Victoria has now published and operated under five successive SDI strategies.

The reality, however, is that, while states lead these SDIs, many partnerships between the state and federal and multiple local governments are inevitably involved. Perhaps unsurprisingly, these partnerships and the power of the states have led to similarities but also some differences in policy and practice. For example, some states have pursued cost recovery strategies (see Section 18.5.2.2) with respect to the data that they hold, while others charge little more than the costs of their duplication. To complicate matters further, the federal government in late 2009 changed policy by making much federally collected GI available for free under creative commons licensing. A study of the situation in three of the largest states in Australia is summarized in Table 19.3. In essence, the study highlighted the need for clear strategic goals and responsive negotiation structures in all the partnerships involved. The study authors also suggest that an important motivator for local government in the early stages is the financial incentives offered. Without such incentives many local governments are unlikely to be able to participate in the critical early stages of SDI partnerships.

19.4 Partnerships of Individual Volunteers

Web 2.0 has transformed the potential for certain types of GI collection, leading to the concept of volunteered GI (VGI: see Chapter 5 and Section 11.2). It may not be appropriate to describe such individualistic collections as partnerships in the strict sense since the people involved are not part of an organization and have not undergone training or work to agreed protocols. But an example where such partnership is definitely involved is in the Audobon Society Christmas Bird Count. In the 107th annual count, no less than 69,354,406 birds of 1894 species were identified across many countries and logged in a GIS and analyzed.

But GISCorps provides a more sustained example of individual volunteering in the GIS domain—albeit

Table 19.3 Main findings in the three Australian case studies of state-led SDIs (Source: McDougall, quoted by Craglia and Compagna)

Collaborative Stage	Victoria Property Information project (PIP)	Queensland Property Location Index (PLI) project	Land Information System Tasmania (LIST)
Establishment and Direction Setting • Goal setting • Negotiation • Agreements	A clear common goal for the project. Well-managed process of negotiation and development of policy and institutional structures.	Business case for the project was limited. Goals unclear, and policy framework worked against data-sharing agreements.	High level strategy and clear overall goals. Policy and negotiations strategy well structured. Agreements very detailed.
Operation and Maintenance • Project management • Maintenance • Resources • Communication	Project management has been good since inception, maintenance infrastructure developed progressively, some resource limitations. Communication with stakeholders and partners has been positive.	Poor institutional arrangements led to poor resourcing and project support. Culture of interjurisdictional sharing only now becoming established. Confused channels of communication due to dispersed organizational structure.	LIST started with strong overall leadership and project support. Project generally well-resourced and technology-focused. Issues of local government communication and data maintenance now starting to emerge.
Governance • Governance structures • Reporting • Performance management	Early project efforts focused on negotiation and data exchange. Performance management now part of the process. Improved governance arrangements now emerging.	There appears to have been little performance management or reporting. No governance structure in place which includes the key stakeholders.	Initial governance and reporting structures were appropriate, but as project matures, new governance models are required.

Shoreh Elhami, Volunteering GIS Expertise Worldwide

Shoreh Elhami's (Figure 19.13) day job is as GIS director in the Delaware County Auditor's Office in Delaware, Ohio, where she manages a large GIS operation in support of the county's activities. In her spare time she co-founded GISCorps (www.giscorps.org), a program of the Urban and Regional Information Systems Association (URISA) that helps bring GIS expertise to the aid of victims of disaster and crises worldwide. GISCorps relies on a pool of volunteers and works through international organizations such as the United Nations to bring suitably qualified and equipped teams into locations where they are most needed in disaster response and recovery as well as in GIS capacity building. Most recently, GISCorps has helped to respond to the Myanmar cyclone and the Missouri floods of 2008, and to events in Darfur, Chad, and Afghanistan. Shoreh is also a part-time adjunct instructor in GIS at the Ohio State University and a former member of the U.S. National Research Council's Mapping Science Committee and the URISA Board of Directors.

Shoreh says:

For the past two decades, I have enjoyed working with the GIS technology to help solve problems and provide alternatives to

Figure 19.13 Shoreh Elhami. (Courtesy Shoreh Elhami)

decision makers; that joy took an amazing turn as GISCorps took form. With hundreds of volunteers from all over the world and tens of implemented missions, we are now witnessing the remarkable impact that our volunteers and the GIS technology are having on communities in need around the globe which is most gratifying.

through a centralized infrastructure. Operating under the auspices of the Urban and Regional Information Systems Association (URISA—see also Section 18.5.4), GISCorps coordinates short-term, volunteer-based GIS services to underprivileged communities. From being launched in 2003, the number of volunteers had risen to over 1,000 by 2009; these were drawn from over 60 countries and have operated in many locations (see Box 19.4 and Figure 19.14).

A particular form of local partnerships that has flourished in some areas is Public Participation in GIS (PPGIS). This form of organization is meant to bring the academic practices of GIS and mapping to the local level in order to promote knowledge production. Most frequently, this is achieved through not-for-profit organizations. Through this activity, PPGIS seeks to provide empowerment and inclusion of marginalized populations, who normally have little voice in the public arena. The potential outcomes of PPGIS uses range from community and neighborhood plan-

ning and development to environmental and natural resource management. Not surprisingly, success in this area requires access to GIS tools and skills, to free or inexpensive data, and to some funding. Recent technological developments such as the semantic Web, AJAX, and simpler interfaces give hope that use of GIS technology can eventually be accessible directly to those without formal training in the area, removing one difficulty in addressing the interests of disadvantaged communities.

19.5 Have SDIs Been a Success?

Not surprisingly, implementation of many NSDIs (and SDIs more generally) has not gone smoothly. Anything so broadly defined and forming such an intangible asset, whose creation involves huge numbers of different organizations and individuals—each with their own views, values, and objectives—and having

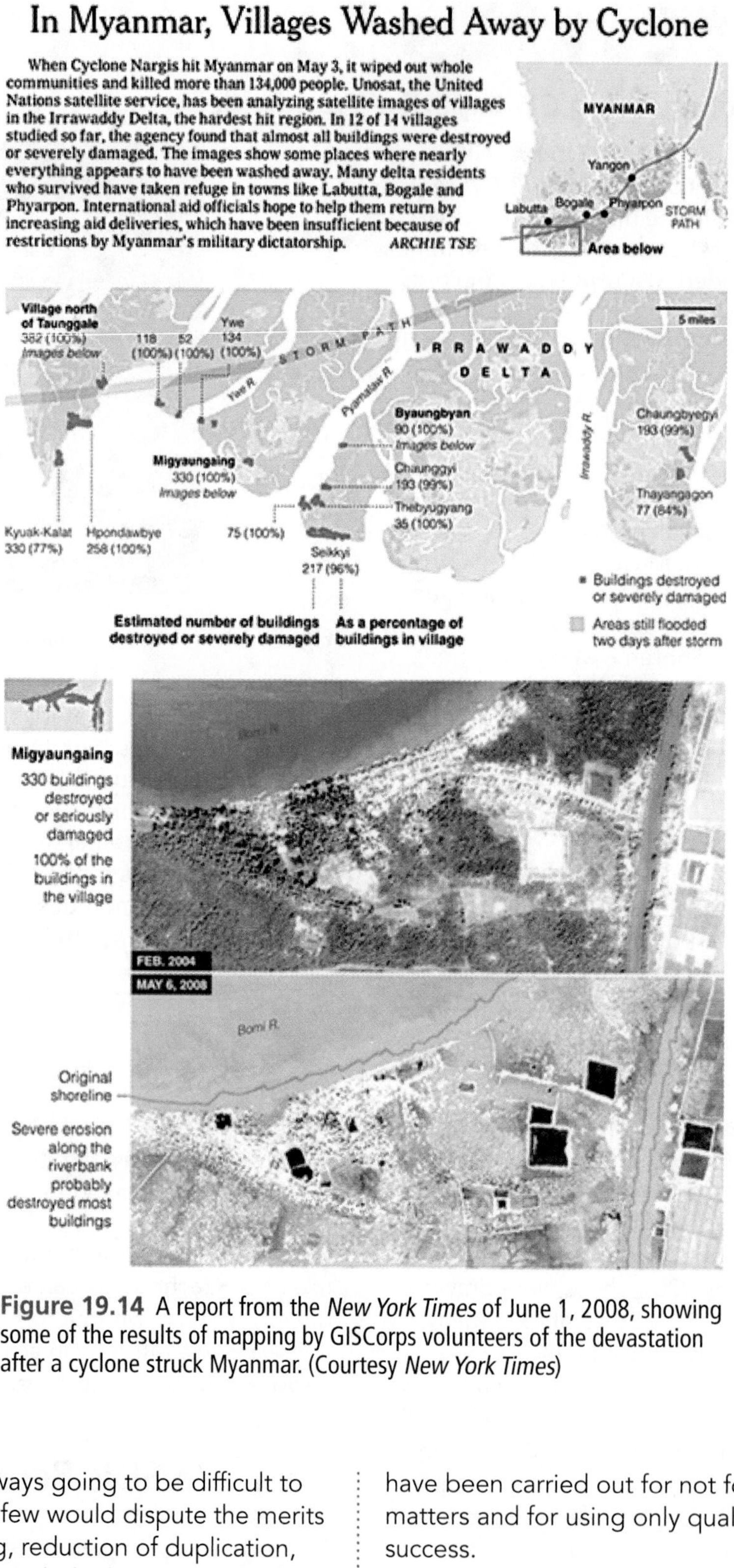
In Myanmar, Villages Washed Away by Cyclone
When Cyclone Nargis hit Myanmar on May 3, it wiped out whole communities and killed more than 134,000 people. Unosat, the United Nations satellite service, has been analyzing satellite images of villages in the Irrawaddy Delta, the hardest hit region. In 12 of 14 villages studied so far, the agency found that almost all buildings were destroyed or severely damaged. The images show some places where nearly everything appears to have been washed away. Many delta residents who survived have taken refuge in towns like Labutta, Bogale and Phyarpon. International aid officials hope to help them return by increasing aid deliveries, which have been insufficient because of restrictions by Myanmar's military dictatorship. ARCHIE TSE
MYANMAR
Yangon
Labutta
Bogale
Phyarpon
STORM PATH
Area below
Village north of Taunggale
Images below
Ywe
118 52 134
(100%) (100%) (100%)
STORM PATH
IRRAWADDY DELTA
5 miles
Byaungbyan
90 (100%)
Images below
Migyaungaing
330 (100%)
Images below
Chaunggyi
193 (99%)
Thebyugyang
35 (100%)
Chaungbyegyi
193 (99%)
Thayangagon
77 (84%)
Kyuak-Kalat
330 (77%)
Hpondawbye
258 (100%)
75 (100%)
Seikkyi
217 (96%)
Estimated number of buildings destroyed or severely damaged
As a percentage of buildings in village
Buildings destroyed or severely damaged
Areas still flooded two days after storm
Migyaungaing
330 buildings destroyed or seriously damaged
100% of the buildings in the village
FEB. 2004
MAY 6, 2008
Bomi R.
Original shoreline
Severe erosion along the riverbank probably destroyed most buildings

Figure 19.14 A report from the *New York Times* of June 1, 2008, showing some of the results of mapping by GISCorps volunteers of the devastation after a cyclone struck Myanmar. (Courtesy *New York Times*)

no one in charge was always going to be difficult to achieve. At a high level, few would dispute the merits of achieving data sharing, reduction of duplication, and risk minimization through the better use of good-quality GI. But how to make it happen for real, how to avoid excessive bureaucracy whilst being accountable to stakeholders and how to measure success are different matters. A number of external-to-GIS researchers have, for instance, criticized the way evaluations of SDIs have been carried out for not focusing on governance matters and for using only qualitative measures of success.

Something as diffuse as NSDI will never be seen as a success by everyone, but it has been a catalyst for many positive developments.

Notwithstanding all these complications, in their short lifetime SDIs have generated high levels of interest in many countries. Some real successes have been achieved, as already indicated. Perhaps the greatest success of SDIs, however, has been as a catalyst, acting as a policy focus, publicizing the importance of geographic information, and focusing attention on the benefits of collaboration through partnerships.

All this is especially important in a country as large and governmentally complex as the United States, which again serves as a good worked example. Since the creation of America's NSDI, many more organizations in the country—from different levels of government and occasionally from the private sector—have formed consortia in their geographic area to build and maintain digital geospatial datasets. Examples include various cities in the United States where regional efforts have developed among major cities and surrounding jurisdictions (e.g., Dallas, Texas), between city and county governments (e.g., San Diego, California), and between state and federal agencies (e.g., in Utah).

The U.S. Geological Survey has been a fertile supporter of new partnerships, especially in relation to improving the National Map, which itself is part of USGS's National Geospatial program. One of the Survey's initiatives was to create the National Map Corps (originally the Earth Science Corps). Individuals applied to join the Corps and were then able to identify areas where the 1:24,000 scale base mapping of the United States needed to be updated (intelligence of change is a major problem for all organizations maintaining a map series). For reasons which are unclear, however, the Corps was suspended in the fall of 2008 after eight years of activity.

The reality is that the U.S. NSDI concept of bottom-up aggregation of data from many, perhaps thousands of sources to form quality national datasets is intrinsically complex, even daunting. In addition, there is a problem with continuity: typically, government initiatives fostered by politicians wane after their authors have moved on. The U.S. NSDI's profile and prestige certainly diminished somewhat around 2000 when it was no longer new and an incoming administration had different priorities. There was also: internal dissension within the federal government about who led on what; some shortcomings in agreements between federal, state, and local government and with the private sector on how to progress; and budget constraints, together with a host of other new initiatives demanding staff and resources. These all contributed to a slowdown of progress of the NSDI in its heartland. A new second phase was triggered, however, by the U.S. e-government initiative launched in 2002,

an initiative designed to improve the value for money of government. Driven by the Office of Management and Budget, it led directly to the Geospatial One-Stop geoportal (www.geodata.gov).

The impetus for greater government efficiency and effectiveness and the advent of the Department of Homeland Security have reignited the cause of NSDI in the U.S.

Confusion about what all the different initiatives were and how they related to each other began to become common in 2003–2004. Various strategy papers were produced by different organizations. Reorganizations of functions in various federal bodies complicated the situation. Reviewing the situation in early 2009, Craglia and Campagna concluded that there was a dearth of sound evaluations of the NSDI benefits, but that:

> [w]hile the efficiencies at the national level are starting to show up in terms of cost savings and increased use of the geoportals, the NSDI strategies and initiatives have not succeeded to reach the local and regional base in a more comprehensive and complete way. This level is the most challenging and requires the strategies to be complemented with clear implementation policies and programs and supported by a continuous stream of funding. . . . The middle level—the states—is addressed through the 50 States Initiative, but is also active in geographic information management regardless of the U.S. NSDI initiatives. . . . In sum, the progress towards the U.S. NSDI vision is substantial, particularly given the country's scale in terms of number of public, private, and non-profit organizations and diversity of institutional setups and cultures. But, there is still plenty to accomplish, especially at the local and regional levels.

Highly partisan views exist on what NSDI has achieved and what should be done next. This is not surprising given its ambitious scope, its nature, and the lack of simple measures of success.

It is difficult to avoid the conclusion that the drive and energy behind the U.S. NSDI stalled somewhat during the Bush administration, while later initiatives (like the INSPIRE Directive and numerous regional SDIs in Europe described by Craglia and Campagna) have demonstrated new concepts and approaches. That said, unexpected new developments in government policy can often give fresh impetus to tired programs; the data.gov initiative of President Obama (see Section 18.5.2.2) may reignite the U.S. NSDI.

The conclusions to be drawn from this discussion are many and varied. Drawing on the international reviews by Craglia and Compagna, Masser, and others, we may set out the main conclusions as follows:

- There are multiple models for an SDI. In particular, the regional level of SDIs is often not simply an intermediate level from global to local, subservient to the higher administrative authority. SDIs at this level are often leading the field, predating national developments, or setting the example and framework, including technical specifications, for the national levels. In Italy, Spain, Belgium, and Germany they are the key building blocks of the national SDIs, with the national level providing a thin layer on the regional infrastructures.

- Adequate funding is crucial. In many SDIs this support has been lacking, in part because so many organizations are involved and all see themselves as contributors rather than taking lead responsibility. It helps if there is a business plan with clear benefits projected to the public-sector organizations intimately involved. Where the benefits accrue to organizations other than those that contribute resources, this slows progress.

- Surprisingly few professional quality assessments of the social and economic impacts of SDIs have been carried out. We have much anecdotal and some financial indicators of success. but these need to be supplemented.

- Inter- and intraorganizational conflict is inevitable in implementing an SDI. Thus successful SDIs are above all networks of people and organizations, in which technology only plays a supporting role. Building the technological foundation can be relatively easy, but building and maintaining the social back-end is much harder, takes longer, and is more resource intensive.

- Since no single organization can build an SDI, the success of all SDIs is totally dependent on the quality and effectiveness of the partnerships on which they are founded.

- If SDI has been the big idea of the last 15 years, the landscape is now more complicated: the Open Source movement has produced a different, bottom-up model for creating content on a collaborative basis which is very different to the largely top-down, public sector-driven SDI model. Alongside this is the rapidly increasing role of the private sector in providing content. At the time of writing it is not at all clear which—if any one—of these will triumph or whether they will co-exist.

19.6 Nationalism, Globalization, Politics, and GIS

In this chapter, we have shown how many nation-states, led initially by the United States, and many other governmental structures have attempted to enhance the quality, utility, accessibility, and awareness and exploitation of geographic information so as to provide benefits to citizens, governments, and businesses. This initiative has been carried out through the creation of institutional structures and umbrella bodies, holding together disparate partners. But much GI has hitherto been created to suit national needs (at best). Its wider use is constrained by its historical legacy and the need for continuity through time for comparative purposes. Other than that collected via satellite remote sensing and (arguably) road transport networks, truly global coverage GI is currently still mostly the sum (and sometimes less) of the national parts and is not always readily available. Yet the need for globally or regionally available, easily accessible, and consistent GI—especially the geographic framework variety (see Chapter 1)—is increasingly evident.

Even though few people other than those in UN agencies currently have management responsibilities that are global, the present shortcomings in global GI have severe consequences for many actual and potential users who operate beyond national frameworks. In the next chapter we show how GIS and GI can make contributions to resolving the many challenges that face the billions of people on this planet. Little of this can be achieved without GIS partnerships.

Other than for imagery and some roads data, global GI is currently little more than the sum of the highly varied national parts, and detailed consistent information is rarely readily available.

Questions for Further Study

1. Why might you wish to act in GIS partnership with other organizations? What are the likely benefits and disbenefits?

2. Describe what is meant by spatial data infrastructures.

3. Is there an SDI in your country, state, or locality? If so, what has each achieved?

4. Does it seem likely to you that the most effective SDIs are those for parts of a country (e.g., states) rather than whole countries? If so, why?

Further Reading

Craglia, M. and Campagna, M. 2009. *Advanced Regional Spatial Data Infrastructures in Europe.* Joint Research Center of the European Union; available at sdi.jrc.ec.europa.eu/ws/Advanced_ Regional_SDIs/arsdi_report.pdf.

Masser, I. 2005. *GIS Worlds: Creating Spatial Data Infrastructures.* Redlands, CA: ESRI Press.

Masser, I., Rajabifard, A., and Williamson, I. 2007. Spatially enabling governments through SDI implementation. *International Journal of Geographical Information Science*, 21: 1–16.

Much SDI material is updated frequently and is out of date in textbooks. Useful Web sites include:

Global Spatial Data Infrastructure www.gsdi.org.

International Journal of Spatial Data Infrastructures Research http://ijsdir.jrc.ec.europa.eu/.

The International Society for Digital Earth www.digitalearth-isde.org.

The International Steering Committee for Global Map www.iscgm.org/cgi-bin/fswiki/wiki.cgi.

The European Commission INSPIRE Directive Web site www.ec-gis.org/inspire.

The (U.S.) Federal Geographic Data Committee www.fgdc.gov.

The ESRI portal to SDI and related sites www.esri.com/software/arcgis/geoportal/live_user_ sites.html.

20 Epilogue: GIS&S in the Service of Humanity

The world is a very differentiated place: 1% of the world's population owns 50% of the world's ownable assets. Whilst many people live in comfort, many millions of others live in great poverty, at risk of natural or human-made disasters and succumb to avoidable diseases that kill or maim; children often suffer the most.

Simply being experts in GIS and the underlying science is not enough. We should be able to use our skills, tools, experience, and the knowledge set out in this book to make a positive difference to the world. This chapter describes how this might be done, starting with a definition of the Grand Challenges to which Geographic Information Systems and Science (GIS&S) can contribute and using both example and theory. Of all the steps we can take, fostering and spreading education is perhaps the single most productive one, and we stress the importance and value of inculcating spatial thinking in all strands of education.

◯ LEARNING OBJECTIVES

After studying this chapter, you will understand:

- Just how differentiated world geography is, notably in the incidence of poverty and disease.

- The biggest challenges facing humanity.

- The interdependency between many factors causing these problems.

- How GI systems and science (GIS&S) can help us tackle these challenges.

- A number of issues that we need to resolve to progress.

20.1 Introduction

In earlier chapters we have identified the science underlying the use of GIS, the technologies that are transforming what we can do, and some of the constraints that need to be circumvented. We have given details of individual projects where GIS has made a difference, and we have celebrated the work of people who have advanced the cause. In this final chapter, however, we ask what GIS can, should, or will be able to do for humanity faced with profound global challenges—a world that is divided, uneven, and troubled. To set the scene and ensure that everyone starts from a common base, we provide a thumbnail sketch of what we (and others) see as the major challenges. Then we draw on past experience, foresight, and intuition to show how GIS practitioners can help improve matters. Put simply, this active citizen approach is envisioned as being built on four big contributions that geographical information systems and science (GIS&S) can make:

- The ability to help us discover and share new understandings in the physical, environmental, and social sciences.

- From these, the means to help us devise new products and services that improve the quality of life, especially for the disadvantaged of the world.

- Use of these new products or services to enhance the efficiency of public and private tasks so as to release resources for other valued things.

- Achievement of all this in as sustainable a way as possible.

20.2 The Differentiated World

As suggested earlier, the world is a very differentiated, heterogeneous, and hence unequal place (Section 4.2). From topography through natural hazards to political systems, economic well-being, and adequacy of food, our life experiences and opportunities are shaped—and in some cases determined—by geography. For example, our expectation of *how long* we will live differs greatly depending on *where* we live (Table 20.1). Wealth and consumption are very unequally distributed: globally, 10% of the world accounts for 59% of total private consumption, with the poorest 10% consuming just 0.5% (see Figure 20.1). Rich people are also geographically concentrated within their

Table 20.1 The geography of life expectancy—life expectancy at birth in selected locations. As discussed in Section 6.4.3, large areas (such as the whole of India) are likely to be more heterogeneous than small areas (e.g., Glasgow city in the UK and neighborhoods) and to be more prone to ecological fallacy. But the general point is correct: life experience, wealth, and cultural values in different areas strongly influence this outcome. (Source: World Health Organization (WHO) Report, *Closing the Gap in a Generation*, August 2008.)

UK, Glasgow (Calton)	54
India	62
United States, Washington, DC (black)	63
Philippines	64
Lithuania	65
Poland	71
Mexico	72
United States	75
Cuba	75
United Kingdom	77
Japan	79
Iceland	79
United States, Montgomery County (white)	80
UK, Glasgow (Lenzie N.)	82

own countries: Figure 20.2 shows the highly differentiated local geodemography (Section 2.3.3) of Glasgow and the 7.8 mile (12.6 km) route that links two communities with a 28-year difference in life expectancy (Table 20.1). Food shortages are highly

Figure 20.1 Share of the world's private consumption in 2005. (Source: World Bank Development Indicators 2008: www.globalissues.org)

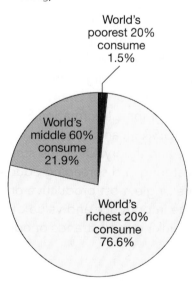

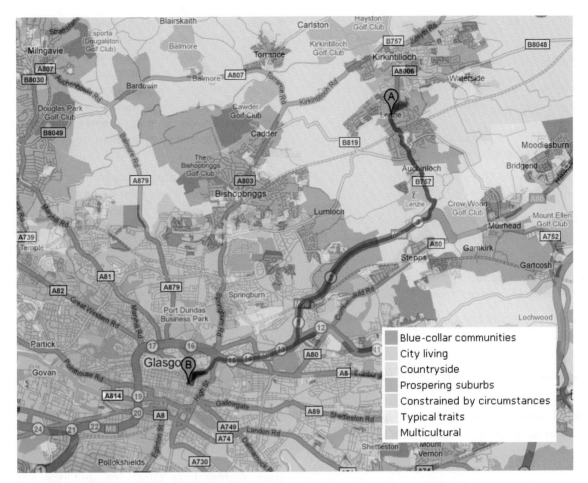

Figure 20.2 The local geodemography of Glasgow, showing the 7.8-mile route (dark blue line) that links communities (A: Lenzie and B: Calton) with wildly different life expectancies (see Table 20.1). (Background data: Google Maps)

differentiated across the world. Differentiation has become greater in recent years both within and between countries: without strenuous efforts by governments and nongovernment organizations (NGOs), the situation is going to get more differentiated still.

As authors, we believe strongly in the role of science in generating evidence to support decision making in public policy and in the merits of transparency. Accordingly, this book has set out some of the principal ways in which Geographic Information Systems (GIS&S) is fundamental to the conception, representation, and analysis of geographic variations in ways that are robust, transparent, and potentially open to scrutiny by every stakeholder. But we also believe in the positive potential contribution of science to ameliorating big problems as well as little, local ones—the Grand Challenges.

The world is very differentiated on almost every criterion; if it were not, geography would not exist!

20.3 Grand Challenges

We now identify a selection of these major problems—the Grand Challenges—that in some way or other face governments (at all levels), businesses, democratic institutions, and individual citizens. From this identification, including examples of past successful use of GIS, we draw out generic ways in which we can contribute and the challenges for our community in so doing.

20.3.1 The Global View of Governments

The Grand Challenges facing humanity are extensively discussed by national governments, regional bodies, and at global fora. An important example is the signature by 189 governments to the Millennium Development Goals. These goals, agreed to in 2000 and scheduled to be achieved by 2015, are as follows:

- Goal 1: Eradicate extreme poverty and hunger.

- Goal 2: Achieve universal primary education.

- Goal 3: Promote gender equality and empower women.

- Goal 4: Reduce child mortality.

- Goal 5: Improve maternal health.

- Goal 6: Combat HIV/AIDS, malaria, and other diseases.

- Goal 7: Ensure environmental sustainability.

- Goal 8: Develop a Global Partnership for Development.

Each goal is expanded into (sometimes) quantitative targets. Goal 7, for instance, comprises four targets:

- Integrate the principles of sustainable development into country policies and programs; reverse loss of environmental resources.

- Reduce biodiversity loss, achieving, by 2010, a significant reduction in the rate of loss.

- By 2015, halve the proportion of people without sustainable access to safe drinking water and basic sanitation.

- By 2020, achieve a significant improvement in the lives of at least 100 million slum-dwellers.

Although these goals were launched with many good intentions, progress in achieving them has been much slower than planned. This is depressing but simply emphasizes the difficulty of effecting change in a multilateral world where there are always other calls on resources. The trade-offs and the multiplicity of inter-connected issues involved in such circumstances were strongly evident in the 2009 Copenhagen Climate Change conference designed to agree on ways of reducing carbon dioxide emissions and minimizing global warming, attended by 115 national leaders. Such difficulties do not deter us from what we advocate, but in what follows we emphasize the practical rather than the exotic and overambitious.

World governments agreed to ambitious millennium goals in 2000, but progress toward them has been slow.

20.3.2 Challenges Amenable to Use of GIS&S

There are many different views on the most important challenges facing humanity: some might point to the growth of urbanization (while only 13% of the world's population lived in cities in 1900, over 50% now do so)

as creating particular hazards (and opportunities). Our own selection, described in the following, overlaps with but is not identical to governmental views. Our aim is to describe the challenges in such a way as to highlight interdependencies and to identify where, with our skills, knowledge, and technologies and determination, we GIS practitioners might make a contribution to ameliorating them.

20.3.2.1 Poverty and Hunger

Poverty is a difficult concept to measure in a meaningful way, but the UN and other bodies have devised widely used thresholds based on measures of average personal income adjusted for purchasing power parity. Figure 20.3 shows how the numbers in poverty vary depending on the income threshold used. Almost half the world—over 3 billion people—live on less than $2.50 a day, while at least 80% of humanity lives on incomes of less than $10 per day. Figure 20.3 shows the geographical distribution of those living on income of less than $1 per day. Moreover, over 80% of the world's population lives in countries where income differentials are widening. Numerical comparisons are valuable: Box 20.1 presents some selected, grim, figures concerning the state of global poverty and inequality. Mapping using GISystems adds greater richness and diversity to statistics such as these. GIScience can further enrich the representation of geographic variation. For example, Figure 20.4 is a cartogram (see Section 13.3.2) in which territory size is varied to represent the proportion of all people living on US$1 or less a day; this gives a

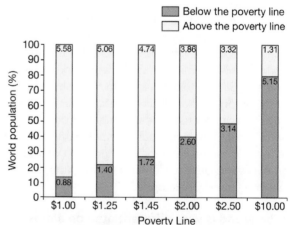

Figure 20.3 The proportions of people in poverty at different income thresholds. (Source: www.globalissues.org)

Poverty and Inequality

- The GDP (gross domestic product) of the 41 Heavily Indebted Poor Countries (567 million people) is less than the combined wealth of the world's 7 richest people.

- Nearly a billion people entered the Twenty-First Century unable to read a book or sign their names.

- Less than 1% of what the world spends every year on weapons is needed to put every child into school.

- One billion children live in absolute poverty (1 in 2 children in the world). 640 million live without adequate shelter, 400 million have no access to safe water, 270 million have no access to health services. 10.6 million children died in 2003 before they reached the age of 5 (or roughly 29,000 children per day).

Source: www.globalissues.org/issue/2/causes-of-poverty.

much more accurate perception of the magnitude of poverty than does conventional choropleth mapping based on the actual areal extent of a nation.

Is poverty inevitable, is it a consequence of poor management of national economies, or is it simply a result of our failure to conceive, represent, analyze, understand and communicate what is happening? All the evidence is that much poverty arises from a cocktail of poor governance, corruption, tough environmental conditions, poor levels of education, powerlessness in the global financial system, and an apparent lack of alternative options. Whatever the causes, the consequences can be highly disruptive, even dangerous—for example, a response of many living in poverty has been to migrate to developed areas, such as from North Africa to Europe; many

deaths have resulted when ships of illegal migrants have sunk and many tensions have arisen where large numbers of migrants have added to indigenous populations.

The combined wealth of the world's seven richest people exceeds the gross domestic product of 567 million people in 41 indebted countries.

Poverty also generally means poor levels of nutrition or, in rural areas at least, dependence on one crop. Failure of that crop can mean starvation and death or enforced migration. Food security is therefore a key risk faced by many people worldwide: Such security varies greatly at a national level, though it is obvious

Figure 20.4 The geography of human poverty. Territory size is used to represent the proportion of all people living on less than or equal to US$1 in purchasing power parity a day. (Source: www.worldmapper.org. © Copyright 2006 SASI Group, University of Sheffield, and Mark Newman, University of Michigan)

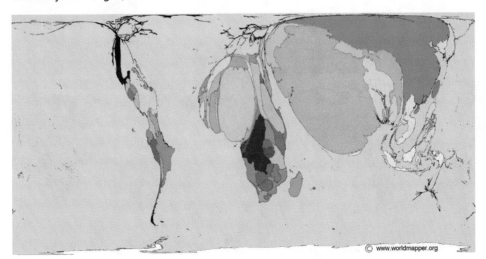

that variations in food insecurity are even more significant at subnational levels. Indeed, recent evidence shows that, contrary to the situation over the past 200 years, within-country inequality has been growing even faster than that between countries.

To make matters worse, the Intergovernmental Panel on Climate Change (see www.ipcc.ch) has used its projections of future average temperatures and other climatic variables to analyze how the production of cereal crops would change around the world. Figure 20.5 shows the projected results consequent on different levels at which carbon dioxide is stabilized in the atmosphere; the consequences of this happening are particularly severe for many poor countries.

GIS contributes to our understanding of poverty and inequality because it improves our ability to integrate, analyze, and portray multiple datasets and to support logistics through the use of our tools. But, more importantly, and as set out in Section 16.5, the science underpinning GIS enables us to move from what-is descriptions to what-if analysis at a full range of interactive scales from the global to the local. Our abilities to do this in practice remain constrained by data that are of highly variable quality and inadequate modeling tools. Nevertheless the evidence and accumulated experience suggest that we can formulate better policies through evidence-based approaches, though implementing them would inevitably require us to engage closely with the political process.

20.3.2.2 Population Growth
Eighteenth-Century views that population growth would outrun food supply have been averted or delayed by technological improvements and other factors. But the problem of feeding the world's population remains. The size of global, national, and local populations and economic or political migrations between them manifestly drives consumption of food and many natural resources, some of which are finite, though in theory may be substitutable. Furthermore, if any of the inequalities described in this section can be rectified, consumption will increase still further, placing even greater stresses on the environment and its sustainability.

Reaching the first billion population took almost all of human history; the second billion took just over a century up until 1930; while the third and fourth billions took 30 years and 15 years, respectively. The global population as of May 2010 was estimated by the U.S.

Figure 20.5 Projected changes in crop yield consequent upon different levels of CO_2 stabilization achieved. (Source: ec.europa.eu/environment/climat/adaptation)

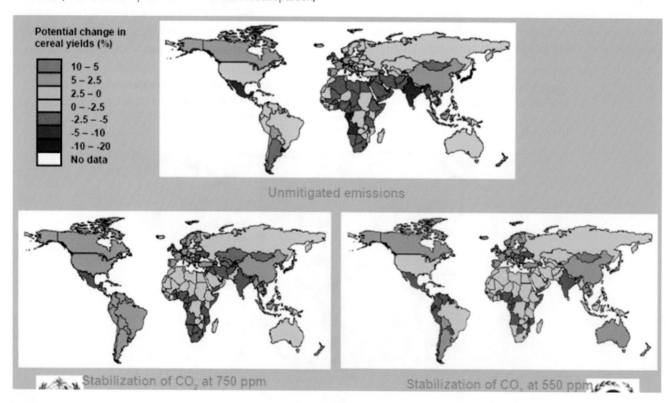

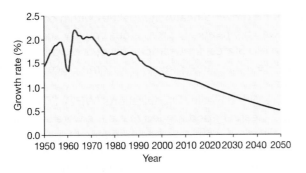

Figure 20.6 World population annual growth rates according to the U.S. Bureau of Census. (www.census.gov/ipc/www/idb/worldgrgraph.html)

Bureau of Census to be 6.821 billion. The good news is that the *rate of increase* of global population has been falling since the 1960s (see Figure 20.6). Many demographers now expect the total to level off to somewhere around 10 billion after the middle of the Twenty-First Century, that is, with about 50% more people on the planet than now. Much of the net increase will occur in the poorest areas of Africa and Asia. The factors accounting for this deceleration of growth are highly relevant for this epilogue: they include the consequences of better sanitation and health care, increased education and employment opportunities for women, and the availability of family planning and contraception.

The geography of population change, as well as population density, is far from uniform: in many developed countries population growth is less than one percent or even negative (Figure 20.7). This has serious implications for sustaining the tax base to provide public services, providing enough labor for commerce, or supporting the increasing numbers of aged people. In a narrow sense, migration provides a solution, and it more than doubled between 1970 and 2000, with the largest proportion of migrants moving to countries in the developed world. In Europe, for example, almost all population growth is due to in-migration and the fecundity of the new arrivals. However, political tensions frequently arise from such migrations and from the changes they make in local culture and values. It is also the case that migrants are disproportionately the most able and skilled in the country of origin, particularly given that developed countries place entry restrictions on in-migration of low skilled workers. This means that migration ensures that origin countries themselves usually become less well equipped to develop, thereby further accelerating the differentiation between rich and poor countries.

In principle, GIS provides an excellent instrument for measuring the ebbs and flows of population movements. However, measuring people numbers is becoming more difficult as response rates to censuses and surveys decline in most countries and proxy methods (e.g., remote sensing) remain unreliable. What works in some societies (e.g., the population registers of Scandinavian countries) will not be acceptable or effective in others. Thus there is a major challenge for scientists and GIS and other computer specialists to produce better methodologies—almost certainly based on linkage of many different kinds

Figure 20.7 Annual percentage growth in the world's population by country. (Source: commons. wikimedia.org/wiki/File:Population_growth_rate_world.PNG based on estimated figures in the 2006 CIA Factbook)

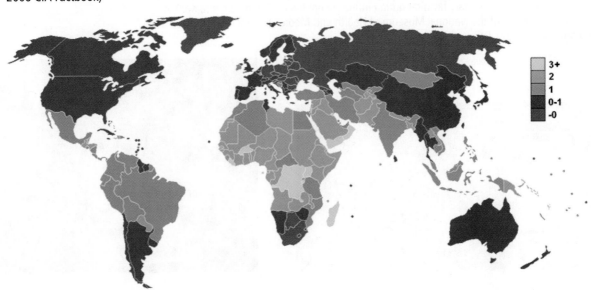

of data together—for measuring populations and changes to them. The time dimension is also important in estimating the cumulative stock of past migrants as well as the flows of more recent ones. GIS projects such as www.publicprofiler.org/worldnames (see Box 1.2) demonstrate that it is possible to estimate the cumulative impact of migration flows over human history.

Once again, however, the use of GIS to measure and represent what is need not be the end of the story. Understanding *why* people move, what entices and constrains them, and how land-use patterns change as a consequence of population change is a major challenge for GI scientists. Together, GIS&S leverages the benefits of better data and modeling tools that are necessary to predict the impacts of changed laws, barriers, or incentives. Without this, understanding of migration in particular and population growth in general is impossible, and policies are likely to be ill-founded.

20.3.2.3 Disease

The spread of disease, access to care, and the treatment and prevention of illness are unevenly distributed across the globe, partly related to poverty. The substantial progress in overall global health improvement over recent decades has however been deeply unequal, with convergence toward improved health in a large part of the world but with a considerable number of other countries falling further behind. Every year there are 350–500 million cases of malaria in the world, with 1 million fatalities: Africa accounts for 90% of malarial deaths, and African children account for over 80% of malaria victims worldwide. Furthermore, there is now ample documentation, which was not available 30 years ago, of considerable

and often growing health inequalities within countries. Even within a relatively prosperous country such as the United States, for example, cancers of various types are much more prevalent in some places than in others.

The World Health Organization's (WHO) 2008 Annual Health Report argued that there were several main causes of ineffective health treatment: poor people have the least access to health care and leave treatment to the last moment because they cannot easily pay for it. Many health services for the poor are highly fragmented and thus provide poor service, often where hospital-acquired infections flourish. The WHO is calling for the redirection of available resources toward the poor and toward primary rather than expensive tertiary services, that is, toward the local everywhere. This presupposes that we can identify the highest-priority areas for such support, and create systems that are efficient and effective. Use of GIS to describe the geographies of need and of quality of treatment is a key factor in deciding and acting on priorities.

> **Many of the Grand Challenges—poverty, disease, hunger—are closely connected.**

The science underlying GIS is once again important for what-if analysis of processes and scenarios. Aside from the ongoing disease avoidance and treatment measures, physicians and the WHO need to plan to cope with potentially disastrous global pandemics. Between 1918 and 1920, between 20 and 100 million individuals worldwide were killed by the so-called Spanish Flu—many of them healthy young adults. Nearly a third of the world's population was affected (see Figure 20.8). The

---

Figure 20.8 (A) Chart showing mortality from the 1918 influenza pandemic in the United States and Europe. (B) Emergency military hospital during influenza epidemic, Camp Funston, Kansas, United States. (Courtesy of the National Museum of Health and Medicine, Armed Forces Institute of Pathology, Washington, DC)

(B)

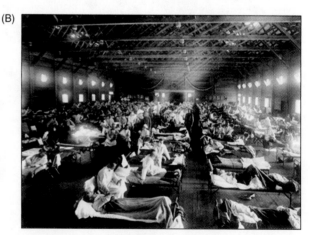

(A)

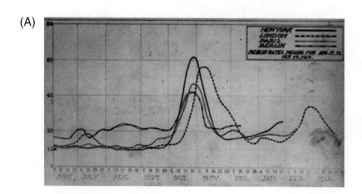

threat of new epidemics and pandemics such as SARS and swine flu suggests that potentially catastrophic events can occur at any time. Rapid identification of the location of outbreaks of infectious diseases is crucial to containment or remedial action. This response requires good information networks and the ability to aggregate information geographically and report different aggregations routinely and speedily to different audiences.

As a consequence, health is one of the most urgent and compelling areas in which GIS&S can contribute to society. Whether the focus is malaria, HIV-AIDS, cholera, avian influenza, children's obesity, access to health care, cancer, or the impact of climate change on health, developing a more thorough understanding and response to critical health issues cannot be achieved without the approaches and concepts of different branches of science. GIS&S provides the organizing and analytical framework that accommodates and focuses our different approaches to science in ways amenable to practical problem solving. Examples of past successes include efforts to identify disease clusters; computation of correlations between disease outcomes and their possible causes; the formulation and testing of spatially explicit models of disease spread that incorporate concepts such as distance and connectivity directly into the model (often in the form of cellular automata or as agent-based models: see Section 16.2.3 and Box 16.3); and research on the significance of proximity in determining successful treatments (e.g., the distance between the home of a stroke victim and the nearest emergency center).

GIS&S has played a significant role in almost all modern work of this kind.

20.3.2.4 Access to Potable Water

Water problems affect half of humanity: some 1.1 billion people in developing countries have inadequate access to water, and 2.6 billion lack basic sanitation. About 1.8 million child deaths occur each year as a result of diarrhea. Close to half of all people in developing countries are suffering at any given time from a health problem caused by water or poor sanitation. Millions of women find themselves spending several hours a day collecting water.

Water shortages are already increasing in many parts of the world as a result of competing uses, growing populations, and international disputes over riverine water flowing across national boundaries. Moreover, if the Intergovernmental Panel on Climate Change (the IPCC: see Section 20.3.2.6) is correct, the situation can only get worse: the Panel's 2007 report projected with high confidence that dry regions will get drier, wet regions will get wetter, drought-affected areas will become larger, heavy precipitation events are very likely to become more common and will increase flood risk, and water supplies stored in glaciers and snow cover (Figure 20.9) will be reduced. In these circumstances, disputes over who has the rights to water resources can only grow still further. One example of attempts to minimize such disputes is the agreement between China and India announced in August 2009 to monitor jointly the state of glaciers in the Himalayas. Seven of the world's greatest rivers, including the Ganges and the Yangtse, are fed by

Figure 20.9 (A) A photograph of Muir Glacier taken on August 13, 1941, by glaciologist William O. Field; (B) is a photograph taken from the same vantage point on August 31, 2004, by geologist Bruce F. Molnia of the United States Geological Survey (USGS). According to Molnia, between 1941 and 2004 the glacier front retreated more than 12 km (7 mi) and thinned by more than 800 m (875 yd). Ocean water has filled the valley, replacing the ice of Muir Glacier; the end of the glacier has retreated out of the field of view. (Courtesy U.S. National Snow and Ice Data Center)

(A)

(B)

glaciers. They supply water to about 40% of the world's population. In 1962 the two countries went to war over disputed territory in this sensitive area.

Aside from being involved in modeling of the IPCC projections using the best available data and models, GIS practitioners have made significant contributions to the definition of many boundaries. For example, the International Boundaries Research Unit (IBRU) at Durham University has built up internationally recognized expertise over several decades in supporting boundary demarcations, training those to be involved in such demarcations anywhere in the world and maintaining—increasingly in GIS form—a geographical record of boundaries.

The International River Boundaries Database (IRBD) is particularly interesting from an historical standpoint, since rivers have been a popular choice for boundary makers with concerns for potential defensive capability and their clarity on the ground. However, as dynamic natural features, the movement of rivers has generated frequent disputes over the position of boundaries that are often understood to require rigidity. In addition, as vital natural resources that are both divided and shared, river boundaries are vulnerable to dispute over aspects of water management as well as boundary definition. Compiled by the IBRU to provide the most accurate information available related to river boundary sections, including their length and definition, the IRBD includes all recognized international boundaries as well as known de facto boundaries (e.g., North Korea–South Korea, Israel–Palestine, Cyprus–Northern Cyprus: see Box 13.1). The lengths of many river boundary sections (recognizing the scale-dependency of these measures: see Box 4.6), within the IRBD have been estimated using paths created within Google Earth and are available as KML files. GIS&S again provides not only the technology for measuring the trajectory and scale-dependent lengths of features such as rivers, but also an environment within which general hydrologic models can be applied to unique landscapes. We have already discussed the differences between uniform and functional regions (see Section 6.2.1) and the extent to which they can be used to support what-is and what-if analysis. GIS&S permits functional regions from the human (e.g., journey to work areas) and environmental (e.g., watersheds) domains to be analyzed alongside one another.

Such boundary work can have major political significance. In August 2007 Russian scientists sent a submarine to the Arctic Ocean seabed at 90° North to gather data in support of Russia's claim that the North Pole is part of the Russian continental shelf. The expedition provoked a hostile reaction from other Arctic littoral states and prompted media speculation that Russia's action might trigger a new Cold War over the resources of the Arctic. Following the publication of numerous poorly informed articles about jurisdiction in the Arctic, in August 2008 IBRU prepared a detailed map and briefing notes offering an objective overview of the current state of play in the region. The map, shown as Figure 20.10, was constructed using specialist GIS tools. It identifies agreed maritime boundaries, known jurisdictional claims and disputes, and potential areas of seabed that may be claimed in the future. The map received global media attention and in the three days following its publication was downloaded more than 40,000 times.

20.3.2.5 Natural Disasters

Television brings us frequent updates on natural disasters from many parts of the world. The most extreme in recent years was probably the December 2004 tsunami originating from a submarine earthquake whose epicenter was off Banda Aceh in Indonesia; approximately 250,000 people were killed (see Figure 20.11). But many other natural disasters have occurred, notably but not solely along plate margins in the Earth's crust. Figure 20.12 shows how the statistics of reported disasters have changed over a 20-year period.

Most natural disasters are caused by hydrological and/or meteorological factors. Until recently, the most respected international database was held in tabular form. However, the Belgian coordinating center for details of natural disasters recognized in 2008 that "GIS technology has imposed itself as an essential tool for all actors involved in the different sectors of the disaster and conflict management cycle." As Figure 20.13 shows, these disasters have had huge economic, as well as human, impact. The global reinsurer Munich Re described 2008 as the third-worst year on record, with total losses of about $200 billion and 130,000 people killed by Cyclone Nargis in the Irrawaddy Delta area of Burma plus 70,000 killed and millions made homeless by an earthquake in Sichuan Province of China. But even such dire figures disguise the real extent of natural disasters because many of the losses are uninsured or otherwise unquantifiable in areas with poor populations.

Mapping the geography of such losses is useful to insurers in order to understand their risk profiles. Yet here again the main scientific focus of research efforts is on forecasting the occurrence of natural disasters. While real-time monitoring of hurricane tracks has proved relatively valuable, both for enhancing evacuations and predicting insurance losses

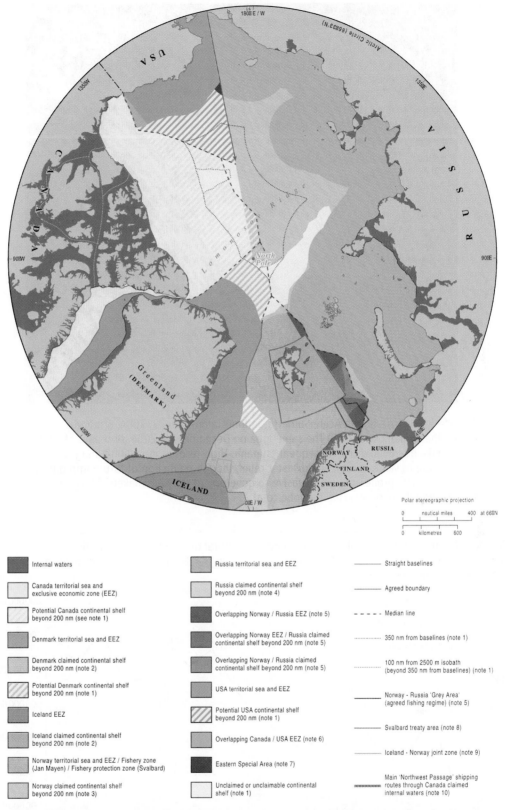

Polar stereographic projection

0	nautical miles	400 at 66°N
0	kilometres	600

Legend		
Internal waters	Russia territorial sea and EEZ	———— Straight baselines
Canada territorial sea and exclusive economic zone (EEZ)	Russia claimed continental shelf beyond 200 nm (note 4)	———— Agreed boundary
Potential Canada continental shelf beyond 200 nm (see note 1)	Overlapping Norway / Russia EEZ (note 5)	- - - - Median line
Denmark territorial sea and EEZ	Overlapping Norway EEZ / Russia claimed continental shelf beyond 200 nm (note 5)	············ 350 nm from baselines (note 1)
Denmark claimed continental shelf beyond 200 nm (note 2)	Overlapping Norway / Russia claimed continental shelf beyond 200 nm (note 5)	·········· 100 nm from 2500 m isobath (beyond 350 nm from baselines) (note 1)
Potential Denmark continental shelf beyond 200 nm (note 1)	USA territorial sea and EEZ	———— Norway - Russia 'Grey Area' (agreed fishing regime) (note 5)
Iceland EEZ	Potential USA continental shelf beyond 200 nm (note 1)	———— Svalbard treaty area (note 8)
Iceland claimed continental shelf beyond 200 nm (note 2)	Overlapping Canada / USA EEZ (note 6)	———— Iceland - Norway joint zone (note 9)
Norway territorial sea and EEZ / Fishery zone (Jan Mayen) / Fishery protection zone (Svalbard)	Eastern Special Area (note 7)	════════ Main 'Northwest Passage' shipping routes through Canada claimed internal waters (note 10)
Norway claimed continental shelf beyond 200 nm (note 3)	Unclaimed or unclaimable continental shelf (note 1)	

Figure 20.10 Maritime jurisdiction and boundaries in the Arctic area. The map identifies known claims and agreed boundaries, plus potential areas that might be claimed in the future by the countries bordering the Arctic, that is, Canada, Denmark, Iceland, Norway, Russia, and the United States. (Courtesy: International Boundaries Research Unit, Durham University)

Figure 20.11 (A) Banda Aceh, Sumatra, Indonesia on June 23, 2004, 6 months before the tsunami struck. (B) The same area on December 28, 2004, two days after the earthquake nearby and the consequent tsunami. High-resolution satellite images such as these are of value in directing disaster relief, though prediction of future infrequent events are error prone and unpersuasive to government decision makers. (Image Courtesy of DigitalGlobe, www.digitalglobe.com)

Figure 20.12 Trends in the number of occurrences of natural disasters and their victims. (Source: The Center for Research on the Epidemiology of Disasters (CRED) at the Université catholique de Louvain in Belgium. www.emdat.be)

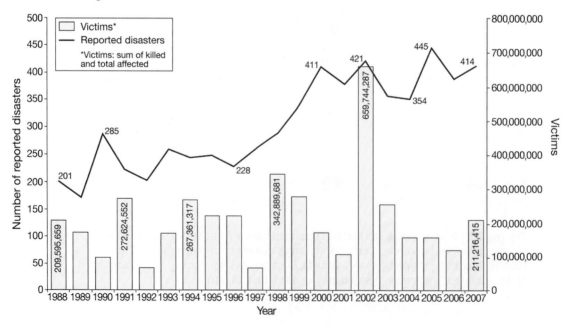

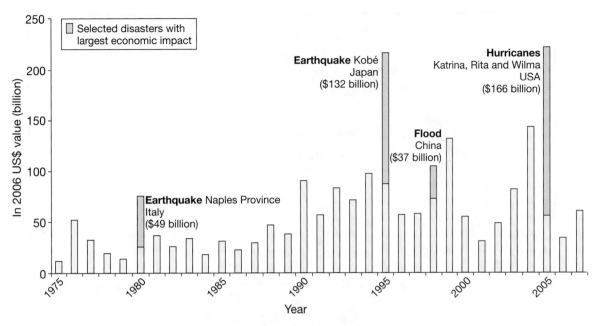

Figure 20.13 The annual reported damages from natural disasters between 1975 and 2007.
(Source: www.emdat.be/Activities/press_conference.html)

(Figure 20.14), predicting other events remains mostly unreliable. Even where probabilistic statements of the likelihood of a disaster have been made, these have been inadequately persuasive for governments to take action (Figures 20.15A and 20.15B). Thus there is much for GIS&S experts, in partnership with Earth and social scientists, to do.

Where GIS&S experts have made major contributions in recent years is through improving the effectiveness of post-disaster recovery work. For example, immediately after the December 2004 tsunami disaster (see Figure 20.11), the Asian Development Bank initiated the Earthquake and Tsunami Emergency Support Project. This Australian-led group gradually

Figure 20.14 Projected insurance losses to members of Lloyds of London in one year attributable to hurricanes of different categories in the southeastern United States. (Image NASA, Image © 2008 TerraMatrices, Image © 2008 Digital Globe)

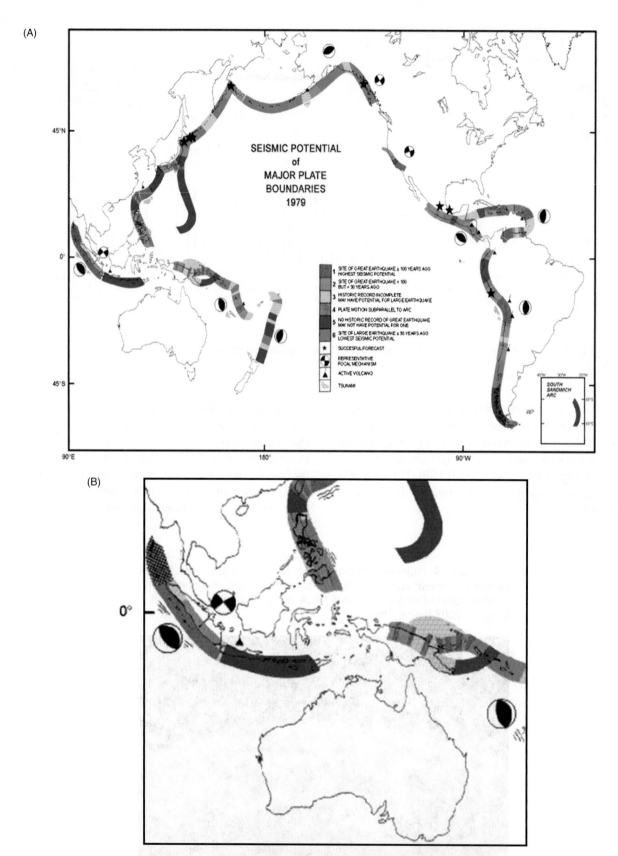

Figure 20.15 (A) 1979 map showing global distribution of seismic potential at plate boundaries. (B) Enlargement showing likelihood of tsunami identified near Banda Ache—which occurred 25 years after publication of this map. (Source: W. R. McCann, S. P. Nishenko, L.R. Sykes and J. Krause, Seismic gaps and plate tectonics: Seismic potential for major boundaries, *Pure and Applied Geophysics* 117, 6, November 1979)

converted its focus from supporting recovery activities in the short term into continuous support to development planning activities over the longer term. Maps and plans were devised speedily from every available source to aid with the recovery by international and national teams. As time went on, increasingly sophisticated multifactor maps showing areas suited for redevelopment were produced for government agencies.

GIS experts have made major contributions to improving the effectiveness of post-disaster recovery work.

The UK charity MapAction is a standing organization active in the same role that makes much use of highly trained volunteers. It argues that aid that ends up in the wrong place is of no use in relieving human suffering and that, in a humanitarian crisis, relief agencies need rapid answers to questions such as "Where are the greatest needs?" and "Where are the gaps that need to be filled?" To provide such information, the volunteer teams skilled in use of GIS and GPS produce frequently updated situation maps for all other agencies active in the area and publish them on the Web (see Figure 20.16). The teams have operated in countries such as Angola, Bolivia, China, Indonesia, Jamaica, Kenya, Myanmar, Pakistan, Sri Lanka, Suriname, and Tajikistan. That this has major benefits is shown by an independent report produced after a deployment in Haiti following the passage of a hurricane. This report concluded that MapAction's activities:

● Helped relief agency staff arriving in the country to orient themselves and to start productive work more quickly.

Figure 20.16 Situation map of Chinese earthquake at Sichuan, produced and distributed widely to summarize known situation on May 12, 2008. (Courtesy: MapAction)

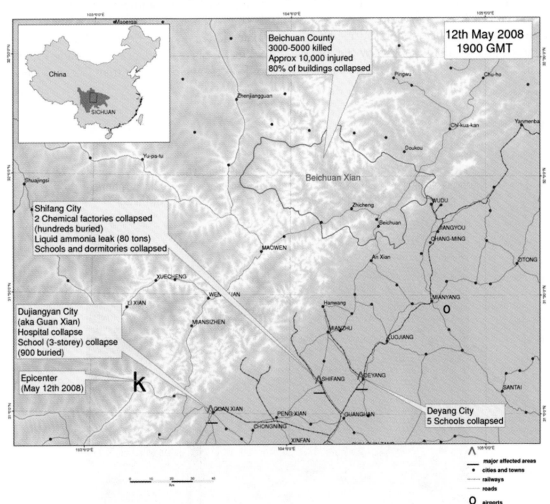

- Supported donors in making informed decisions so that funding and resource-allocation decisions were made more rapidly.

- Made complex situation information easier to understand, therefore reducing time spent in meetings discussing data and thereby enabling a focus on actual programming and delivery of aid.

- Improved coordination through a visual who?—what?—where? that highlighted gaps and overlaps in response.

- Demonstrated priority needs to donors and brought in more support for the reconstruction phase of the emergency.

- Influenced the choice of resources supplied by relief agencies.

20.3.2.6 Environmental Sustainability

The book *Silent Spring*, published by Rachel Carson in 1962, is widely credited with helping launch the environmental movement. Within a few years of the publication, satellite remote sensing and GIS began to play a significant role in monitoring the state of the environment globally and more locally. But the perceived seriousness of environmental sustainability and, in particular, the threats caused by climatic change—which seem to have been at least exacer-bated by human action—are now widely perceived as major challenges. A further complicating factor is that some of the worst-affected areas are likely to be those in which some of the poorest people on Earth live: the most vulnerable regions are Africa, the Asian mega-deltas, small islands, and the Arctic. The most vulnerable sectors are held to be the availability of water (especially in the dry tropics); agriculture (especially in low latitudes); human health in countries with low adaptive capacity; some ecosystems, notably coral, sea-ice biomes, coastal mangrove and salt marshes, and those in tundra/boreal/mountain areas. The debate and controversies are now wide-ranging and encompass coping strategies as well as avoidance ones. As GIScientists, however, we have no doubt that climate change is underway and that some ameliorative and adaptive action is essential, if difficult. One example of possible disastrous consequences of failure to act to counter predicted rising sea levels is shown in Figure 20.17.

Even aside from climate change, human-induced environmental stress is now much more widespread and severe as a result of human action, whether it is manifested in the deforestation of the Amazon Basin or the progressive disappearance of wild areas and the extinction of species. Between May 2000 and August 2006, Brazil is said to have lost nearly 150,000 sq km of forest—an area larger than Greece—and, since

Figure 20.17 Projected effects of rising sea level on central London, assuming breaching of the Thames Barrier and a 6-m higher sea level than is normal at present (an extreme prediction). (Courtesy GMJ Ltd © 2009 GMJ Ltd.)

1970, over 600,000 sq km of Amazon rainforest have been destroyed (these figures are based on analysis of satellite imagery time series, linked to on-ground observations). It seems self-evident to us that ways of reducing this stress and at least adapting to climate change (if amelioration proves impossible to enact) are essential.

Here again, mapping and measurement of environmental change is valuable, yet still greater value can be added through explanatory and predictive modeling. One area where GIS&S has made and will continue to make a contribution is in regard to preserving biological diversity and protecting endangered ecosystems. Nearly 850 known species have become extinct or disappeared in the wild over the past 500 years, apparently because of human activity. About two-thirds of endangered mammal and bird species owe their threatened status to habitat destruction and fragmentation. This matters because ecosystem functioning and the services provided by ecosystems are critical to human welfare. These functions include sequestration of carbon, production of oxygen, primary production of chemical energy from sunlight, soil formation, nutrient cycling, food production, wood and fiber production, fuel production, and regulation of water flow and transfer to the atmosphere.

Scientists working in GIS&S are playing a significant role in documenting and explaining the current rate and magnitude of biodiversity and ecosystem loss and in developing possible coping strategies. Astonishingly, we still know surprisingly little about such matters. Successful preservation of biodiversity and ecosystem functioning also requires anticipation of how and where land-use and climate change will produce increased extinction rates. Built on that, it also requires the development of models that can estimate potential biodiversity and the selection of conservation areas and conservation management strategies to mitigate extinctions and declining ecosystem functioning. Typically, the models have to be at different scales and levels of generalization—global, national, and local—and linked. The same geographical scientists have already contributed significantly in recent years to advances in methods and applications linking land cover changes not only with biophysical processes and consequences, but also with human circumstances.

The U.S. National Research Council has argued that we need reliable and relatively fine-scale global atlases of the world's threatened species and habitats/ecosystems in order to develop a series of geographically explicit, biophysically sound conservation strategies for the most endangered areas. Even that is not enough: we also need socially acceptable strategies that take account of locally differing cultural approaches to nature valuation, conservation practices, and compliance. In short, GIS&S provides a framework for integration of knowledge and information about the Earth or parts of it and has a crucial role to play in so doing. We again stress that in this area, as in all others, the GIS&S specialist cannot resolve all the problems on his or her own. Typically, close partnerships (Chapter 19) are needed with physical, biological, and social scientists to describe and understand what is going on and to seek solutions.

To succeed, GIS&S specialists must work with other natural, environmental, and social scientists, and with decision makers.

20.3.2.7 Terrorism and Crime

Attempts to overthrow governments or ways of life by violence are commonplace in many parts of the world. Arguably the best solutions to this violence lie in diplomacy, changing perceptions of grievances, and statesmanship, as in South Africa and Northern Ireland. But since the first duty of any state is to protect its citizens, monitoring of threats and taking appropriate actions to thwart their realization are the tasks of national organizations equivalent to the U.S. Department of Homeland Security. Most of these organizations are now assiduous users of GIS technology and related sciences.

The result has been increasing use of surveillance technologies and linkage of all the sensors used together through GIS to portray the movements of suspects or monitor suspicious activities around sensitive sites. The sensor arrays range from those on satellites and unmanned airborne vehicles to CCTVs, telephone interceptions, and much else. Even at the less serious level of crime against individuals or their properties, monitoring of activity in high-risk areas is important, as is post *hoc* analysis of the characteristics of the crime scene. Geodemographic profiling (Section 2.3.3) of the characteristics of areas to quantify risk is commonplace, and matching of DNA samples from suspects or databases with that found at crime scenes is permitted in some jurisdictions. Figure 18.1A shows what can now readily be achieved in mapping crime on a detailed basis. All of this raises trade-offs between potentially enhanced security and loss of privacy (see Section 18.5.3). Even though some of it is classified, we know that spatial analysis and GIS&S specialists are central to much of the work of agencies seeking to prevent terrorist or criminal activities.

Surveillance of individuals can reduce risk of harm but endangers privacy.

20.4 Seeking the Root Causes

It is normal in science to look for the root causes rather than the symptoms of problems and then tackle them. The problems we have identified, however, are bewildering in the range of causal factors they encompass, the uncertainty involved in characterizing their geographic nature is considerable, and their interdependencies and feedback loops are only partly understood. Some of the factors are not easily quantifiable. For example, changing global power relationships impact on how the world economy and politics operate. Perhaps the first obvious sign of such changes was the demise of the Soviet Union in 1991, but, especially since 2000, major shifts of economic power have occurred to the rapidly developing BRIC (Brazil, Russia, India, and China) countries and to petro-states such as those in the Middle East. In essence, the West is surrendering centuries of economic and political hegemony.

The partly connected globalization of business has also had a major effect, some beneficial, on the major problems. Much employment in the developed world has moved out of manufacturing into service provision; low-margin manufacturing and many commoditized services have become the preserve of low-wage economies. Global firms or partnerships have been able to outsource or to offshore work through the advent of the Web and search engines, the creation of work-flow software that enabled collaborative working across the globe, integrated supply chains between producers and retailers, plus the rise of community involvement in creating information ranging from Wikipedia to street maps (see Box 18.8). The ongoing consequences are not hard to predict: low-wage economies will seek to develop higher value-added activities, global competition will increase still further (assuming no resurrection of tariff barriers), and the most highly skilled and innovative players will become sought after globally while the rewards for unskilled workers will not rise.

Part of the taken-for-granted world that undergirded all this was the global trade and financial system. Over a 50-year period, trade between nations has multiplied hundreds of times. Seasonal availability was all but eradicated as we imported exotic flowers (and much else) by air from countries on the other side of the world. Measures of the world's wealth grew dramatically between 1945 and 2006. Underlying all of that was an increasingly globalized financial system in which stocks, commodities, foreign exchange, and much else were traded constantly, especially through the three major financial centers in New York, London,

and Tokyo (their success owing much to their locations equally distributed around the world and permitting 24-hour trading). The entire system provided credit (bank loans and other sources of funding) to enable new businesses to start, to smooth cash flows, and to enable individuals to better themselves through purchase of housing and the like. Without such credit, societies cannot change. Technology played a major role: the ability to move funds near-instantaneously across the world in response to opportunities large and small enabled speculators to profit from minute differentials in pricing needs of financial instruments and currency.

All of this led to massive instability. The financial boom, followed by a crisis from 2007 on, affected every citizen in every country. The price of oil rose from $70 per barrel in August 2007 to $147 a year later, leading to near-bankruptcy for two of the three largest U.S. automobile manufacturers as consumers fled to less fuel-hungry transport. Commodity prices headed for the stratosphere, and central banks worried about inflation. Russia, Venezuela, and sovereign wealth funds in the Middle and Far East were the destinations of substantial inflows of money. A few months later, oil and other commodity prices crashed to less than one-third of their peaks as demand fell across the world and every economy became affected by a major global downturn in demand. The outcome was a global recession, with unemployment and bankruptcies soaring and demand for goods and services slumping. Governments have committed trillions of dollars to stem the crisis. Consequently, the debt levels of many governments have ballooned and their sovereign debt credit ratings have been reduced—raising the cost of borrowing hugely in Greece and elsewhere. At least one country—Iceland—had inadequate reserves to repay debts to foreign creditors and became effectively bankrupt. Truly the world's economies, as well as the state of the environment and much else, are now closely coupled.

If nothing else, all this demonstrates that global interdependency on most fronts—economic, environmental, and much else—is a fact of life, that virtually everyone on Earth is affected by it, and that its consistent stability is an illusory concept. But the impacts are not the same everywhere: geography matters. We need to grasp the implications of this—the need for constant monitoring using a variety of sources of information, assessing feedback loops, modeling and forecasting outcomes, testing what-if scenarios in planning, and communicating the results effectively to key decision makers—and the closely related implications for the deployment of GIS&S.

Today's world is tightly interlinked: a disaster in one geographic area can trigger global impacts.

20.5 Meeting the Challenges

The most casual reading of this brief summary of the global challenges should tell us various things. The first is that, whatever their local manifestations, these global problems face us all: we can no more isolate ourselves from the impact of global pandemics or climate change than from the spread of terrorism. The second is that, while some challenges might seem consistently threatening and long-lasting, sudden changes of a catastrophic nature will occur. In an ideal world, therefore, precautions should be planned beforehand (while accepting that it is difficult to anticipate the next crisis, business continuity planning is important for every organization).The third is that many of the challenges—and solutions—are interconnected, notably poverty, food supply, disease, and population numbers. These challenges are truly formidable and complicated. Analyzing and anticipating threats involves use of both quantitative and qualitative information varying widely in reliability. No one set of skills—of geographers, Earth or computer scientists, economists, or others—can resolve these problems on their own. Indeed, the collaborations involved must be multinational, multidisciplinary, and geographically extensive in scale. That said, there are many niche areas where the contributions we can make as individuals or small groups can be profound and far-reaching.

As a result of this interconnectivity and complexity, there is already a vast and partly connected set of organizational silos charged with responsibility for tackling these problems. That responsibility is discharged mainly through national and local governments, multinational bodies like the United Nations and its agencies, commercial enterprises, and non-governmental organizations (NGOs).

A fourth conclusion follows from everything that has been said: good communications of all sorts is a necessary (but not sufficient) condition for success, especially in emergencies. The impact of the Internet and Web has already been astonishing in providing new tools to describe, communicate, and tackle problems. Interestingly, geography has influenced how the plumbing of the Internet has developed (see Box 20.2).

Finally, based on the earlier part of this chapter and indeed on the examples throughout the rest of the book, we contend that GIS&S in general and we

---

Applications Box 20.2

The Geography of the Cloud

Central to much of the vision of this book is the role of GIS as a ubiquitous provision, with location-based services available everywhere, and with large amounts of information available on-tap to facilitate decision making and fast response rates. In short, GIS is seen as a crucial part of an information utility.

This presently demands large farms of servers distributed around the world, with high-speed connections to clients—the *computing cloud* concept. It results in formidable power consumption to drive and cool the systems of thousands of servers. Microsoft's data center in Chicago, for instance, needs 200 megawatts of power—comparable to aluminum smelters. At the time of writing, the Internet and its consequences are estimated to consume over 1% of U.S. electricity, with the United States accounting for around 40% of all global consumption, and a global growth in power use of about 17% per annum. The need for cheap electricity is therefore a major decision factor: many U.S. server farms are located in remote areas to access hydroelectricity from the Columbia River. And countries as far afield as Iceland and Siberia are making attractive offers based on hydroelectricity, cool environments, and so on, to the owners of cloud facilities. Only where speed is of the absolute essence—typically where highly time-critical transactions are made in major financial centers—does the advantage of proximity overwhelm the geography of cost and political stability. The legislative setting is also important: the bank-transfer consortium SWIFT, for instance, chose neutral Switzerland for its data center so that personal data collected in Europe will not be stored in a U.S. facility where it can be subpoenaed by the U.S. government. Like any technology-induced geography, the geography of the cloud is constantly changing, and for a multiplicity of reasons.

as practitioners can manifestly each make a small but important contribution to meeting these challenges. Because the contributions of individual people are vital, we have included in this chapter boxes describing two individuals, Foote and Palin, (see Boxes 20.3 and 20.4) from outside of the traditional GIS domain who serve as valuable role models. These show that our skills, technology, knowledge of what works and who else has useful knowledge, plus our commitment *can* make a difference. But it is worth stepping back from all the examples and summarizing in generic terms why this is so.

Multinational problems require multinational collaborations between scientists and governments.

20.5.1 Why GIS&S *Should* Enable Us to Make a Difference

We now take utterly for granted many GIS functions embedded in services, notably those in Web-based mapping and car guidance systems. All users of these systems are therefore users of GIS functionality, even if they do not appreciate it. In total, such users already may well number over a billion people worldwide.

Beyond this ubiquity are the professional users of GI systems. Our guesstimate is that there are over a million of us around the world. What follows is aimed at these active GIS professionals. The essence of working in GIS&S is that we are committed to the use of evidence to underpin decisions. Evidence must be assembled from the best available sources; tested to ascertain its veracity; used in defined ways with tools whose internal mechanics are well understood; and shared so our findings can be replicated by others. In short, we are scientists. But the nature of the Grand Challenges outlined earlier enables those active in GIS&S to make particular contributions for the following reasons:

- Many problems are manifested initially through geographical variations.

- Studying the geographical manifestation can help us to propose and test causal factors and hence identify possible solutions to the problems.

- The mechanics by which we have to tackle problems are normally geographically structured (e.g., administrations that control access or provide resources or that need persuading).

- There has been a significant change, at least in Western democracies, toward the requirement for quantifiable and published evidence to support and justify policy-making.

Fortunately, GIS&S have developed greatly in the last 30 years. Given our skills, knowledge, and technologies, the demonstrable advantages we have are as follows:

- The integrative capacity of GIS, enabling us to link multiple datasets, analyze the results spatially, and generate added value—then redo the whole operation at short notice whenever the situation changes.

- Use of the same GIS tools at local, regional, national, and global levels simplifies linking models operating at different levels of abstraction.

- GIS's modeling capability, which has grown enormously in recent years. This is part of a wider development: a recent study of science and technology in the services sector showed that the common tools valued in many subsectors from retailing through financial services to health care and transport planning and design were data mining, mathematical modeling, simulation, and visualization.

- Superb visualization and user interaction capabilities.

- Our ability and willingness to share data, ideas, and concepts, increasingly on a global basis.

- A growing awareness and acceptance of the value of GIS among policy-makers/government leaders, public servants, and the business and education communities.

We can best structure our potential contributions to the Grand Challenges under the simple model of different stages shown in Figure 20.18. Rarely is such a process so linear: this ignores the many feedback loops and the soft factors involved in innovative resolution of problems. But it will serve for present purposes.

20.5.1.1 Stage 1: Defining and Describing the Issue

This seems basic, even prosaic, but assembling evidence to monitor evolving situations or understand problems is fundamental to success. Here we include not merely collecting any possibly relevant data from whatever sources—internal to an organization, from national or international statistical bodies, from commercial sources, or from the Internet—but assembling and integrating them in as consistent, documented, and professional a way as is possible in the circumstances. Beyond that, making them as

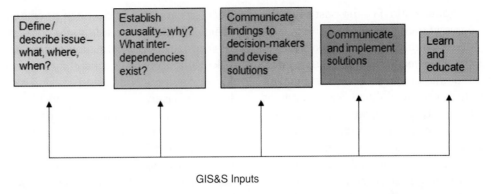

Figure 20.18 Simplistic illustration of how GIS&S inputs can contribute to tackling the Grand Challenges. Note that this ignores many feedback loops and many other complications; GIS has a more limited area of contributions.

widely available as possible to others with an interest is potentially invaluable because it provides for independent scrutiny and critique. Where data are widely used they become of higher quality as shortcomings become identified: better investigations ensue. We recognize that the confidentiality of work in the commercial sector and the need to be first to market sometimes precludes such widespread sharing. Yet in today's service-driven economy, the speed of innovation is such that it is often churlish, time-consuming, and ultimately pointless to become consumed by issues of intellectual property rights. This is acknowledged, for example, in commercially funded work such as the Willis Research Network, funded by a commercial insurance organization, which has successfully embraced an open publication approach involving more than 20 universities worldwide.

20.5.1.2 Stage 2: Analyzing and Modeling Interrelationships

Many examples of spatial analysis have been presented in this book. Section 20.4 has already discussed in a high-level way the need to seek the root causes of problems. But the reality is that detailed analysis of correlations and apparent relationships, leads and lags between changes in different variables, and changes in these relationships through time and over space are required if we are to go beyond superficial judgments. Some of the team members at least must be mathematically and statistically literate. Modeling drought frequency on a global basis or even demonstrating the relationship between the incidence of diarrhea and the availability of piped water in Africa is not a trivial endeavor and requires high caliber data-handling and analytical skills.

20.5.1.3 Stage 3: Conveying Results and Possible Solutions to Those with Responsibility for Action

This stage includes two of the boxes in Figure 20.18: the communication of the results of analyses and of possible solutions. Quite often these are separate stages, with important discussions between the analysts and the decision makers in between. Although this is generic to dealing with all the challenges we have described, nowhere is its importance greater than in dealing with emergencies. The role of small, technologically accomplished organizations (such as MapAction) in natural disasters has proven hugely important. Fast in action and less encumbered by bureaucracy than many state organizations, they typically exploit whatever information is available but succeed only insofar as they do not alienate the local authorities who determine access to the area. This means that they have to be able to communicate the nature of the problem and proposed actions to those authorities in a way that minimizes misunderstandings, facilitates discussion and enables acceptance of recommendations.

20.5.1.4 Stage 4: Learn and Educate

In our view, the world of education needs to evolve substantially if we are to tackle the challenges we have outlined with a decent chance of success. In the first instance there are not enough skilled GIS&S protagonists and practitioners in every country to contribute meaningfully to meeting all of the challenges. For this reason, we must learn from real-world experiences, distil the lessons from them and then seek to educate and train others in the scientific concepts underlying GIS and in the lessons learned so they too can help meet the challenges.

Most fundamentally of all, however, we need to respond to a very different world full of *digital natives—*

people born into a digital world. For the past several decades, it has been standard practice to think of GIS education as a process of training professionals to acquire substantial skills in manipulating the user interfaces of GIS and in understanding the science that lies behind them. But today many of these GIS capabilities are familiar even to young children, who have become accustomed to the user interfaces of Google Maps and similar services. What skills, if any, are needed by a general public that is now increasingly able both to produce and to consume geographic information? The answer to this involves a different educational paradigm to that which has been in place till now.

More widespread education is a necessary condition to reduce the ills of the world

Remarkably, one of the fundamental forms of human reasoning—spatial thinking—is given virtually no attention in the curriculum, except when it is encountered by undergraduates in courses in GIS. Yet there is abundant research to show that early attention to these concepts can lead to improved performance in a range of subjects. One benefit of nurturing a spatial thinking approach is that it offers the potential of a redesigned user interface to GIS that is much easier to learn and to navigate. We believe spatial thinking should be embedded in many courses.

But what is spatial thinking? It is embedded implicitly in many geography courses, but we need to make it explicit if it is to be recognized and adopted elsewhere. Such a public reexamination may well lead to parochial questioning of what is traditionally included in geography curricula. For instance, is there any point in teaching navigation when our hand-held GPS receivers do it all for us (see Section 11.3)? Do we need to carry out field work if or when augmented reality can deliver an adequate understanding of parts of the real world with minimum cost and reduced risk? What forms of spatial thinking do we really need for a population increasingly taking a multiplicity of location-based services (Section 11.3.2) for granted?

For the authors, the quality, nature, availability, geographical extent, excitement, and take-up of education in GIS&S will strongly influence our ultimate success in tackling the Grand Challenges. To achieve that success, we do not just need people who know about GIS concepts, practices, and past pathfinder successes. We need people who will innovate and use our tools to both describe current geographies and to formulate and communicate possible new, alternative, and better geographies that may occur naturally or can be created.

Some of its delivery of this ambitious curriculum may be best suited to the private sector. However, the longer term can only be safeguarded through education in schools. Some useful steps forward have been made in certain countries. In the UK, for instance, where a national curriculum is applied in all state funded schools, a revision in 2007 led to GIS being recognized as a fundamental basis for understanding the contemporary world. The relevant government document included the statement: "Geography inspires pupils to become global citizens by exploring their own place in the world, their values and responsibilities to other people, to the environment and to the sustainability of the planet. . . . Pupils should have opportunities to learn about GIS."

We should not be under any illusions, however: where the need is greatest—for example, among the poorest people in the world—education is immensely challenging. What provides the greatest chance of success is to have inspired and inspiring teachers (see Box 20.3). For us, those who see GIS&S as an

---

Biographical Box (20.3)

Jeff Foote, Inspired Teacher

Jeff Foote has taught science to Grades 6 through 8 at Kermit McKenzie Middle School (Figure 20.19) in Guadalupe, California, for 17 years. Guadalupe is located west of Santa Maria in a predominantly agricultural area, and most of his students are the children of recently immigrated Hispanic families. He writes:

As a very new science teacher in the early 90s I knew it was a good thing to get kids outside for science learning . . . a rich and inexpensive laboratory. We'd take trips to the upper Santa Ynez River, test water quality, map plants near the river, and ask questions about the observable geology. A good teacher friend, Walt Bunning from Righetti High, took me up in his little airplane, and I took video of the whole river course, to help students put into context what they learned in one small place.

▶

Figure 20.19 Jeff Foote and his class. (Courtesy Mike Goodchild)

As part of a University of California, Santa Barbara (UCSB) project, our multischool team all had students recording multiple weather data points every day and sharing via a pre-Web mechanism.

A very cool element was when students matched their local ground cloud observations with the satellite capture the high school saved from above us. This kind of layering of information seemed like magic. So, when I saw a demo of GIS software at the UCSB geography lab, I immediately "got it" . . . the ability to use layers and relationships of information—associated with a place—is exactly the tool that young thinkers need to build contextual understanding.

However, matching this value with the technology interface, and mental building blocks my 11–13 year old students need, has been a challenge. Our new partnership with US Fish and Wildlife and UCSB's Spatial Center is helping us work through that from the ground up . . . exploring what spatial thinking looks like for kids this age, both in the classroom and in the field. . . . and using some friendlier high-tech tools like handheld GPS and Google Earth mapping to answer our questions. As a hands-on science teacher, I'm intent on figuring out how math and language learning can also come alive with this approach to building context and an understanding of how we connect to our environment.

important contribution to that education will help citizens contribute to resolving the challenges we have identified.

Finally, we need all those who are involved in GIS&S to become ambassadors for what we can contribute. This is easiest where we converse with those commissioning work. Beyond those, however, we have a need to reach the general public. To do so requires recognizing that our lexicon is a foreign language to others (see Section 19.3.4.2) and using their language to inform or persuade them. This is not easy and requires top-class communication skills, as well as experience of and access to the media. But we already have some role models (see Box 20.4 for one example) and need others.

20.6 Conclusions

This epilogue could have set out to do many alternative things: most obviously, it could have anticipated the future developments of technology in our field. We have chosen to ignore a technological approach in this final chapter and to focus on the Grand Challenges. In so doing, we recognize that we are only part of the solution—but we think those active in GIS&S are important contributors.

Not all of us can be major players in regulating financial institutions or leading the World Health Organization to play key roles in the diminution of risk to humanity. But two circumstances enable us to make our contribution. The first is that some global problems (e.g., a flu pandemic) start very locally: only monitoring at the local level can isolate such problems early enough to be effective. The second is that we possess the tools—data mining, mathematical modeling, simulation, and visualization—widely held to be necessary to help analyze problems and find possible solutions.

The challenges are formidable and require the GIS community to address particular questions:

- How do we support decision making where both quantitative and qualitative information needs to be combined in the analysis?

Michael Palin, President of the Royal Geographical Society, Comedian, Actor, Writer, and Television Presenter

In 2009 the Royal Geographical Society (RGS) celebrated the 150th anniversary of receiving its charter for "the advancement of geographical science." The Society also welcomed a new president, Michael Palin (Figure 20.20), who is well known for his work as a comedian (*Monty Python, Ripping Yarns, A Fish Called Wanda*, etc.) and numerous travel documentaries (*Around the World in 80 Days, Pole to Pole, New Europe*, etc.). He has been traveling the world for more than three decades and says that his love of geography stems from school field trips as a boy living in Sheffield, UK. In a press release issued by RGS on his appointment as president he said: "I am immensely excited about taking this leading role in which I can use my passion for learning about the world to promote what I regard as the most important subject for the future—geography."

Michael took over the presidency at a time of great debate about the very purpose of the Society. One group of Fellows, the "explorers," felt that the Society should once again invest in large multidisciplinary expeditions focused on single locations. Another, the "scientists," argued that the present policy of funding a wide range of research projects and scientific expeditions studying important contemporary issues in many different places across the world was more appropriate.

That debate takes us back to the opening Section 1.3.1 of this book where we said that geography was concerned about both the uniqueness of places (idiographic) and general processes (nomothetic). Each has its place in geography, though the interests of protagonists of either approach can lead to conflict, even in a mature and sensible organization. The RGS debate was measured and often thoughtful; the scientists won a vote of the

Figure 20.20 Michael Palin. (© AbbieTrayler-Smith/eyevine/ZumaPress)

13,000 fellows, but the explorer's position was well-supported. This served neatly to highlight the importance of the science of problem solving alongside the significance of the fascinating places that Michael Palin has communicated to so many people. We are clear that GIS brings together the best from the idiographic and nomothetic traditions in the interests of practical problem solving.

- How do we help to break down the silo thinking between different scientists and others who are necessarily working on meeting these challenges?

- How do we change the educational paradigm to include spatial thinking in all disciplinary studies?

- How do we encourage all GIS professionals to become ambassadors to their clients and to the public at large for the contributions we can make?

We are convinced that such an activist approach is the right one. We also hope that, like us, you use your abilities to improve the lot of our children and grandchildren and those of others across the world. We wish you well in that noble endeavor.

Questions for Further Study

1. Why is the world so differentiated?

2. What do *you* think are the biggest challenges to humanity where GIS can make a contribution?

3. You are invited on to a local radio or TV station to talk about what GIS can do to solve local or global problems. What would you say?

4. What is spatial thinking? Can you give some examples?

Further Reading

Friedman, T.L. 2007. *The World Is Flat: A Brief History of the Twenty-First Century* (3rd ed.). London: Picador.

MapAction. 2009. Field guide to humanitarian mapping. Available at www.mapaction.org/more-news/183-new-gis-field-guide.html.

NRC. 2009. *Understanding and Adapting to a Changing Planet: Strategic directions for the Geographical Sciences in the Next Decade*. Washington, DC: National Research Council, National Academy of Sciences.

Sachs, J.D. 2008. *Common Wealth: Economics for a Crowded Planet*. London: Penguin.

INDEX

Page numbers in *italics* refer to figures; page numbers in **bold** refer to tables; ***bold, italicized*** page numbers refer to boxes.